Ryszard Jabłoński, Mateusz Turkowski, Roman Szewczyk (Eds.)

Recent Advances in Mechatronics

Ryszard Jabłoński, Mateusz Turkowski, Roman Szewczyk
(Eds.)

Recent Advances in Mechatronics

With 487 Figures and 40 Tables

Springer

Ryszard Jabłoński
Mateusz Turkowski
Roman Szewczyk

Warsaw University of Technology
Faculty of Mechatronics
Św. Andrzeja Boboli 8 street
room 343
02-525 Warsaw
Poland
Email: yabu@mchtr.pw.edu.pl
m.turkowski@mchtr.pw.edu.pl
szewczyk@mchtr.pw.edu.pl

Library of Congress Control Number: 2007932802

ISBN 978-3-540-73955-5 Springer Berlin Heidelberg New York

Springer is a part of Springer Science+Business Media

springer.com

Typesetting: Digital data supplied by the Editors
Production: LE-TEX Jelonek, Schmidt & Vöckler GbR, Leipzig
Cover: Erich Kirchner, Heidelberg/WMXDesign, Heidelberg

SPIN 12034321 89/3180/YL - 5 4 3 2 1 0 Printed on acid-free paper

Preface

The International Conference MECHATRONICS has progressed considerably over the 15 years of its existence. The seventh in the series is hosted this year at the Faculty of Mechatronics, Warsaw University of Technology, Poland. The subjects covered in the conference are wide-ranging and detailed. Mechatronics is in fact the combination of enabling technologies brought together to reduce complexity through the adaptation of interdisciplinary techniques in production.
The chosen topics for conference include: Nanotechnology, Automatic Control & Robotics, Biomedical Engineering, Design Manufacturing and Testing of MEMS, Metrology, Photonics, Mechatronic Products. The goal of the conference is to bring together experts from different areas to give an overview of the state of the art and to present new research results and prospects of the future development in this interdisciplinary field of mechatronic systems.
The selection of papers for inclusion in this book was based on the recommendations from the preliminary review of abstracts and from the final review of full lengths papers, with both reviews concentrating on originality and quality. Finally, out of 182 papers contributed from over 15 countries, 136 papers are included in this book.
We believe that the book will present the newest applicable information for active researches and engineers and form a basis for further research in the field of mechatronics

We would like to thank all authors for their contribution for this book.

Ryszard Jablonski
Conference Chairman
Warsaw University of Technology

Contents

Automatic Control and Robotics

Biomedical Engineering

Mechatronic Products – Design and Manufacturing

Nanotechnology, Design Manufacturing and Testing of MEMS

Metrology

Photonics

Dynamical behaviors of the C axis multibody mass system with the worm gear

J. Křepela (a)*, V. Singule(b)

(a) Brno University of Technology, Faculty of Mechanical Engineering, Technická 2, Brno 616 69, Czech Republic

(b) Brno University of Technology, Faculty of Mechanical Engineering, Technická 2, Brno 616 69, Czech Republic

Abstrakt

This paper describes mathematic model of multibody mass system of the C-axis over mentioned machine. C-axis is controlled with position feedback and its mathematic model is determined for observation of dynamic characteristic in the loadings working cycles before of machine prototype realisation. This multifunction turning centre is determinate for heavy duty roughing cutting of forged peaces, where is problem with dynamic stability of cutting process. Dynamic stability influeces the eigen frequences the complete torsion system. Positive effect of this conception is for dynamic stability damping on the worm gear.

1. Introduction

Main Spindel of the machine, on the witch is implemented the C axis, is for turning operations driven by asynchronous motor with power 71kW. For high torque moment necessity is gearing reduced through two steps planetary gearbox and constantly belt gear. For milling and drilling operation is this main motor uncoupled through neutral position in this gear box and spindle is hydraulic coupled with worm gears, where are geared with two synchronous servomotors controlled in mode Master – Slave (fig.1.). This mode assures to change the parameters of electrical preloading between both servomotors from the machine control. Preloading of servomotors holds positions of coupling eliminated the production backlash of worm gearing. This is arranged with leaned teeth flanks against both worms opposite teeth of worm gears react in opposite direction of torque moments both servomotors. By this rotating movement is changed step by step the direction of torque moments actuating from contradirection to the same direction, but always continues the constant difference of this moments, which produces preloading even by rotating.

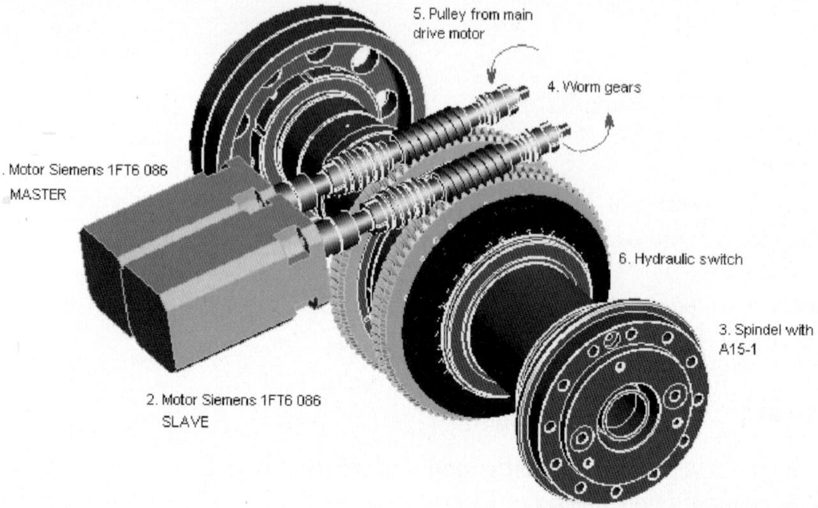

Fig. 1. Desigen of the C axis

Cutting procedure realised many times with more cutting edges tool causes the oscillation of necessary torque moment and it is absolutely necessary to technologically predefine the servomotor preloading value in advance from reason of elimination of contact damaging on the worm gear teeth flank caused from variable loading [3].

2. Multibody system

The simplification is possible just under following condition: The servomotors are connected with worm gears by shafts and they turn them in opposite directions. The turning causes taking up clearances and preload of the worm gear assembly. The construction from the servomotor to the worm gear is turned over an angle $\varphi_{M.}$ The switching-on of the C axis causes the preload and the slewing into a defined angular position. The preload remains constant during machining with the moving C axis and the increase of moments is used just for start up of the servomotors. A torque deviation would lead to a creation of a gap in a tooth space. The spindel remains at the position set up by a CNC control. Please see attached the block diagram of the preloaded mechanical system of the C axis on the figure 2. The interface for multibody mass model is created at the boundry of the parts. The typ of the worm gear is ZA with a gear ratio 40,5. The worm gear was designed as self-locking.

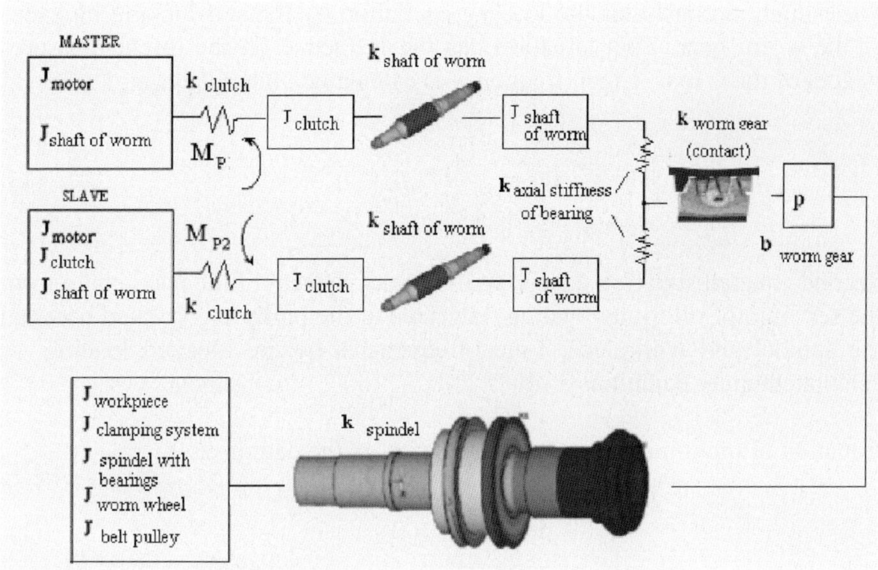

Fig. 2.: Diagram of the mechanical system by the preloading Master-Slave

A worm gear holder is taken for simplification as perfectly torsion rigid because its torsion rigidity is multiple higher than the rigidity of the components chained on the worm shaft. It is necessary to calculate a torsional rigidity and a moment of inertia of individual components as well as approximately calculate damping on the worm gear for the mathematical model of the C axis. The torsional rigidity of the spindle and the worm shaft are calculated with the help of the FEM (Finite Element Method). The calculation of contact stiffness on tooth of the worm gear is highly simplified to a 2D model. The contact rigidity is solved in the plane perpendicular to the pitch of tooth of the worm wheel. The model includes the coefficient of the tooth in the grip 2,94 with help of three tooth in the gip. The force creating the deformation at this plane is calculated as the force between two built-in solids of the worm and the worm wheel. The worm wheel is siplification by the fixation of a meshing segment at the position of the interface between the bronze metal and stell holder. The overall stiffness of the chain of the components between the servomotor and the worm gear is necessary to calculate for the definition of the preload torque. A moment of inertia of the component parts is directly detected in the 3D model of the design of the C axis. The coefficient of damping is necessary for the description this mechanical system.

For influence evaluation of eigen mechanical frequencies of the C axis control system is necesery this problem separated to the two situations.

First situation consist of the loading oscilation by the self-locking blokade of the worm gear. This situation has the influence on the direkt measure system of the C axis. Eigen frequency is calculated under equation 1.

$$f_{rez} = \sqrt{\frac{k_{s+e}}{J_w}} \tag{1}$$

Second equated situation consist of multibody system of the parts chain from the servomotor till to the loading. Interface of the blokade is created between the spindel and workpiece. Eigen frequencies by the blocked loading is calculated under equations 2 till 7.

Equation of motion in the matrix form without the damping:

$$M\ddot{q} + Kq = 0 \tag{2}$$

Transfer to the complex plane:

$$q = v.e^{j\omega t} \tag{3}$$

$$M^{-1}.K.v = \Omega^2.v \tag{4}$$

Determinant of left site the equation must be to equal 0 for the solution of the eigen frequencies:

$$\left| M^{-1}.K - \Omega^2.E \right| = 0 \tag{5}$$

Matrix of mass:

$$M = \begin{bmatrix} J_{sw} & 0 & 0 \\ 0 & J_c & 0 \\ 0 & 0 & J_m \end{bmatrix} \tag{6}$$

Matrix of stiffness:

$$K = \begin{bmatrix} k_{s+w} + k_{sw} & -k_{sw} & 0 \\ -k_{sw} & k_{sw} + k_{c+m} & -k_{c+m} \\ 0 & -k_{c+m} & k_{c+m} \end{bmatrix} \tag{7}$$

3. Manuscript submission

First eigen frequency of the C axis determines the size of the proportional amplification K_v. Further eigen frequencies influnce the the proces of the respances on the torque steps or dynamic of the run-up. In the table 1. are written values for eigen frequencies.

	Situation 1	Situation 2
Eigen frequency of loading [Hz]	57	
1. eigen frequency by the blocated loading [Hz]		31,4
2. eigen frequency by the blocated loading [Hz]		32,3
3. eigen frequency by the blocated loading [Hz]		40

Tab 1. Eigen frequency

On the stability by the step changes has advantageous influnce the dumping of the worm gear. Big ratio of the worm gear reduces the influence of the moment of inertia of the workpiece on the eigen frequency. The knowledge of the eigen frequencies for this mechanical system enables accurater regulators optimization both motors.

References

[1] F. Procházka, C. Kratochvíl, Úvod do matematického modelování pohonových soustav. Cerm Brno, 2002, ISBN 80-7204-256-4.
[2] P. Souček, Servomechanismy ve výrobních strojích, ČVUT Praha, 2004, ISBN 80-01-0292-6.
[3] J. Křepela, V. Singule, Mathematic model of C-axis drive for identification of dynamic behaviour horizontal multifunction turning center, Engineering mechanics 2007, Svratka, Institute of Thermomechanics Academy of Sciences of the Czech Republic,2007, ISBN978-80-87012-03-6-2.
[4] Siemens: Speed/Torque Coupling, Master-Slave (TE3). Function Manual, Siemens, 03/2006 Edition, 2006, 6FC5397-2BP10-1

Control unit architecture for biped robot

D. Vlachý, P. Zezula, R. Grepl

Institute of Solid Mechanics, Mechatronics and Biomechanics,
Faculty of Mechanical Engineering, Brno University of Technology,
Czech Republic

Abstract

This paper deals with the design of a control unit for biped robot „Golem 2" with 12 DOF. It contains information about the topology of electronic system, including description of communication between MC units, sensing elements and PC. The kinematic models used for drive robot and the information about their importance for static walking are mentioned. Some information about actuators (Hitec servos) and specificity their control are also described.

1. Introduction

Design and implementation of autonomous locomotion robots belongs to the important areas of academic as well as commercial research and development. Actually, we are focused on the design of the control unit for biped robot named „Golem 2", with following features: implemented kinematic models, ready to acquisition sensor data (from Acc, camera, FSR etc.), wireless control of robot and possibility for integration of high level AI algorithms (Image processing and understanding, neural network approximator, agent oriented software etc.).

2. Static walking and importance of kinematic models

Without a kinematic models, we have a few simple methods, how to obtain a elemental static walking, for example setting each servomotor separately to get a one stable position of robot and consequently make a step as a sequence of that positions. This „uninformed methods" spent much time and results may not be acceptable, because of non system admittance. In

case of advanced non elemental walking, there aren't simple methods usable and we crave help of kinematic models.
The robot „Golem 2" [2] , developed at Laboratory of mechatronics is comprehended as an open tree manipulator and therefore standard algorithms for forward and inverse kinematics was used to get the appropriate kinematics models. Forward kinematic model (FKM) and Inverse kinematic model (IKM) are described in [1] in details.
By the help of FKM, we can get the position of legs against the body of robot (and reversely) from information about actual servo states.
By the help of IKM, we can get the relevant servo states, relativly to desired position of legs. So we can easy define a vector, which means the changes in position of body from the last taken position:

$$\overline{\Delta} = \begin{bmatrix} \Delta x \\ \Delta y \\ \Delta z \\ \Delta \varphi_R \\ \Delta \theta_P \\ \Delta \psi_Y \end{bmatrix} \quad (1)$$

Where $[x, y, z]$ is cartesian position and $[\varphi_R, \theta_P, \psi_Y]$ is spatial orientation of foot in roll–pitch–yaw notation (Euler angles XYZ).

We can now define a sequential moving of robot as a vector:

$$moving \approx \left[\begin{pmatrix} \overline{\Delta}_1 \\ \iota_1 \end{pmatrix}, \begin{pmatrix} \overline{\Delta}_2 \\ \iota_2 \end{pmatrix}, \cdots, \begin{pmatrix} \overline{\Delta}_n \\ \iota_n \end{pmatrix} \right] \quad (2)$$

Where ι is needle, determining target of using Δ:

$$\iota \in \langle Left, \quad Right, \quad Both \rangle$$

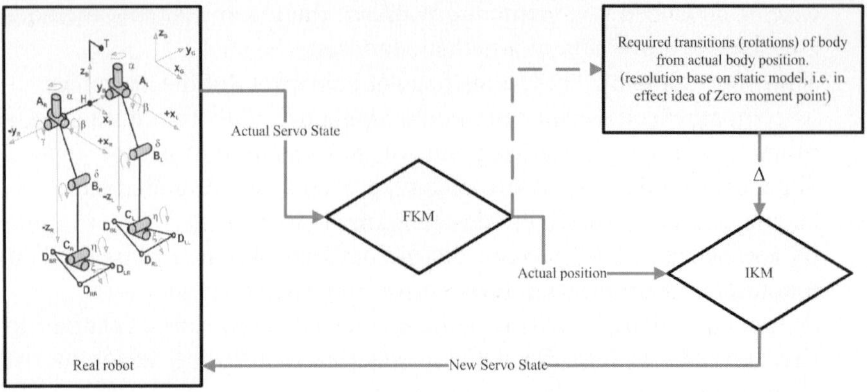

Fig. 1. Schema of using kinematic models

This formulation of moving is much better to develop any locomotion, e.g. static walking. This idea is shown in Fig.1.

3. Control unit

Complete control unit consists of several cooperating units – PC, AT Mega 128 „main unit“, AT Mega 8 „servo control unit“, sensor modules. The backbone network is serial line with our original protocol (variable packet size, master-slave architecture), interconnection between main unit and sensor modules run over SPI interface. The topology is shown in Fig.2.

PC – The main brain. In PC are implemented both kinematic models (because high hw requirements) and GUI for manipulating robot. PC is connected via bluetooth adapter. Implementation in Delphi, some parts uses outputs from Matlab.

AT Mega 128 „main unit“ - Keep wireless connection between robot and PC. Interpolate continuous positions for servos in relation to desired speed. Share data form all connected peripherals together. In progress: Data acquisition from Battery, Accelerometers and FSR sensing modul. Communicate with EyeBot controller.

AT Mega 8 „servo control unit“- Individually control 12 servos, by signals based on the length of pulse. Hardware peripherals (1x16bit Timer/Counter + 1x8bit Timer/Counter, USART) are exploited and software is fully event-driven by the help of interrupts, it's important to precise servos control. There are 4 types of servos made by Hitec and controlled by the length of pulse, generated every 20ms (50Hz) .To get the best resuts, we have to precise servos control, i.e. find the right mutual characteristic between desired position and final pulse length.

We observed, that the each type of servo used, need itself mutual characteristic that is defined by linear function, and is necessary to find out the right constants for each type of servo. Our control now proceed with precise to 1°, except the 2nd ankle, there is a double precision needed, so the extra mutual characteristic is defined for relevant servos.

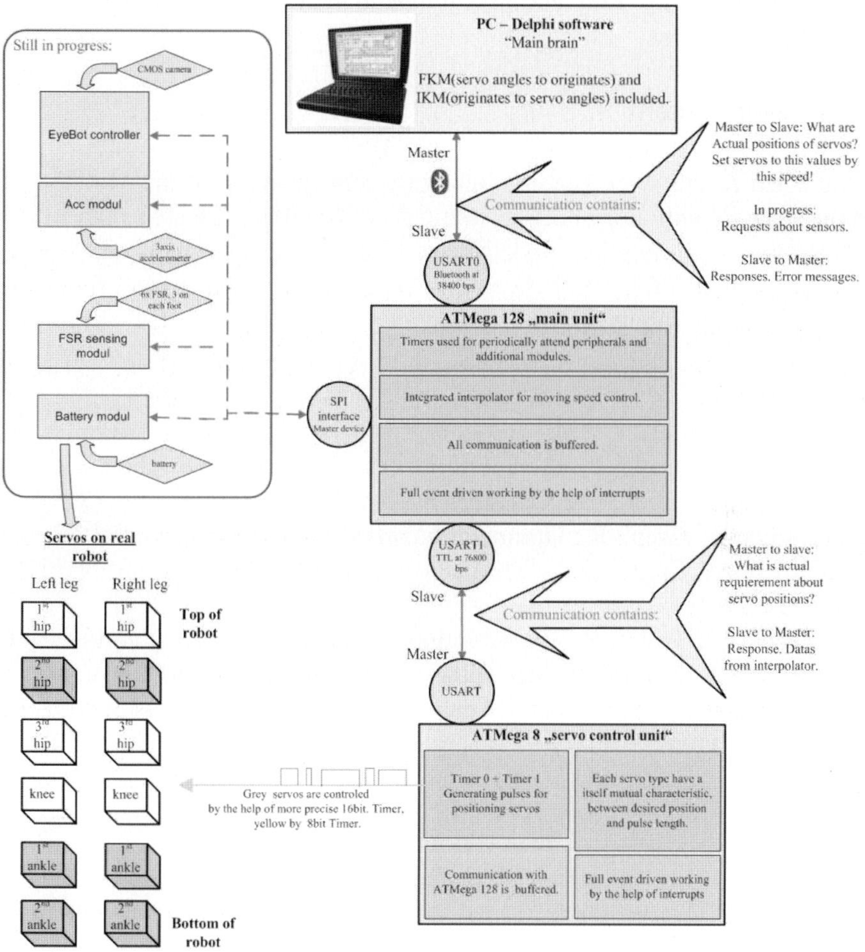

Fig. 2. Topology of control unit

Sensor modules – These modules administer sensors. Each sensor modul have own microcontroller, comunicate with main unit via SPI and provide appropriate data on demand. Eyebot controller have other usable properties, which we plan to exploit.

5. Conclusion

The control unit was designed, successfully tested and a first goal, static walking, was accomplished. The future work
will be headed to fully integration of sensors and enhanced static walking (uneven surface, barriers in trajectory etc.).
The dynamic walk using complex dynamic model will be the next goal in robot development.

Acknowledgment

Published results were acquired using the subsidization of the Ministry of Education, Youth and Sports of the Czech Republic, research plan MSM 0021630518 "Simulation modelling of mechatronic systems" and the project GACR 101/06/P108 "Research of simulation and experimental modelling of walking robots dynamics".

References

[1] Grepl, R., Zezula, P.: Modelling of kinematics of biped robot, Dynamics of machines 2006, ISBN 80-85918-97-8, 2006
[2] Zezula, P., Grepl, R.: Optimization and design of experimental bipedal robot, Journal Engeneering Mechanics, Volume 12, Nr. 4 , ISSN 1210-2717, 2005
[3] Zezula, P., Grepl, R.: Construction design and stability control of humanoid robot, Dynamics of machines 2006, ISBN 80-85918-97-8, 2006
[4] Y. Tagawa, T. Yamashita, "Analysis of human abnormal walking using zero moment joint", 2000

Quantifying the amount of spatial and temporal information in video test sequences

A. Ostaszewska, R. Kłoda

Warsaw University of Technology, Faculty of Mechatronics,
Sw. Andrzeja Boboli 8 Str., Warsaw, 02-525, Poland

Abstract

In case of compressed video quality assessment, the selection of test scenes is an important issue. So far there was only one conception for evaluation the level of scene complication. It was given in International Telecommunication Union recommendations and was broadly used. Authors investigated features of recommended parameters. The paper presents the incompatibility of those parameters with human perception that was discovered and gives a proposal of modification in algorithm, which improves accordance of parameters with observers' opinion.

1. Introduction

The rapid growth of digital television, DVD editions and video transmission over the Internet has increased the demand for effective image compression techniques and the methods of coding/decoding systems evaluation. There are two alternative ways of compressed video quality evaluation: perceptual (sometimes called subjective) and computational (also referred to as objective). No matter what the method is, the crucial role in results of a coder evaluation is played by the scene selection. The algorithm (or the whole system) performance is strictly dependant on the amount of perceptual information that the picture contains. In case of a video, the perceptual information can be divided into spatial and temporal. Test sequences must span the full range of spatial and temporal information of interest to users of the system under test. Considering test sequence selection, the need to quantify the amount of this information seems to be obvious.

2. SI and TI according to ITU Recomendations

The spatial and temporal information measures proposed by International Telecommunication Union [1] are represented by single values for the whole test sequence.

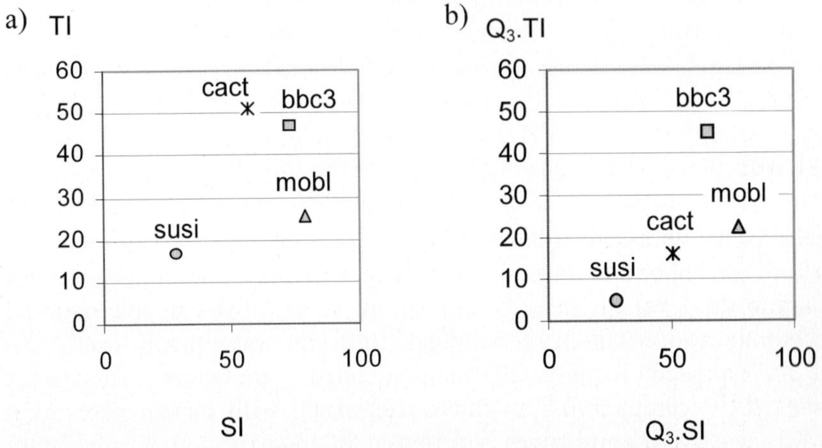

Fig.1. The comparison of SI(TI) plot (a) and Q_3.SI(Q_3.TI) (b)

The SI (Spatial perceptual Information) takes into consideration the luminance plane only and is computed on the base of Sobel filter. Each video frame at time n (F_n) is transformed with the Sobel filter [Sobel(F_n)]. Then the standard deviation over the pixels (std_{space}) in each Sobel-filtered frame is computed. This operation is repeated for each frame in the video sequence and afterwards the maximum value in the time series (max_{time}) is chosen:

$$SI = \max_{time} \{ std_{space} [Sobel(F_n)] \} \tag{1}$$

The TI (Temporal perceptual Information) is also based on a luminance plane and calculates the motion. The motion is considered to be the difference between the pixel values at the same location in space but at successive frames: $M_n(i, j)$. $M_n(i, j)$ is therefore a function of time (n) and it is defined as:

$$M_n(i,j) = F_n(i,j) - F_{n-1}(i,j) \tag{2}$$

where $F_n(i, j)$ is the pixel value at the ith row and j^{th} column of nth frame in time.

The measure of TI is calculated as the maximum over time (max_{time}) of the standard deviation over space (std_{space}) of $M_n(i, j)$ over all i and j.

$$TI = \max_{time}\{std_{space}[M_n(i,j)]\} \tag{3}$$

SI and TI are usually computed for the whole sequence, so each scene is described by two parameters. Higher values of SI and TI represent sequences which are more difficult to decode and are more likely to suffer from impairments. In order to choose scenes which will span as wide range of information to decode as possible, usually SI and TI are put in the TI(SI) plot and the scenes with the uttermost values are selected.

3. SI and TI new approach

Authors conducted Single Stimulus Continuous Quality Evaluation method [2, 3, 4], using 4 sequences (each 15 seconds long) coded with 13 GOP, all three possible GOP structures (with 1, 2 or without B frames) and with 5 levels of bitrate in a range of 2 Mbps to 5 Mbps. Hence, the test material was 30 minutes long and contained 15 variants of coding each of 4 test sequences. 45 subjects participated in the research. The voting signal was sampled at 2 Hz frequency.

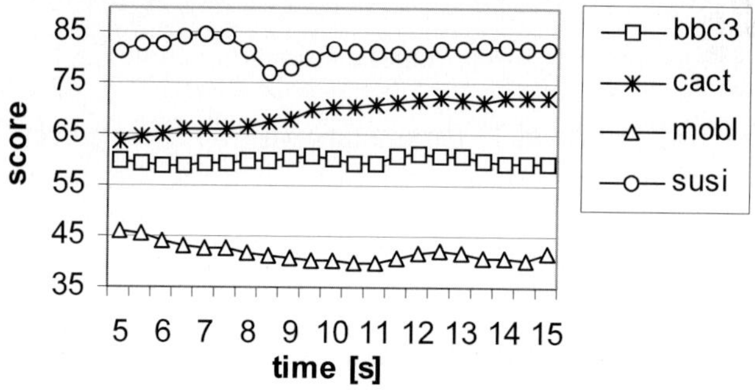

Fig.2. The average score given in time
to 4 sequences across the whole bitrate range

The interesting observation was that the lowest grade was always given to the sequence 'mobl' or 'bbc3', while 'cact' scene used to get scores close to the easiest to decode – 'susi' (fig. 2). According to SI and TI parameters, 'cact' was the sequence with the highest TI value and should contain clearly visible impairments (fig. 1a), which were supposed to affect the mean score given by observers.

This phenomenon impelled authors to investigate the variability of SI and TI in time. For this purpose both parameters were calculated on fame by frame basis. The intriguing discovery was that the high level of TI for 'cact' sequence was caused by one extraordinary peak, which falls on the frames with scene cut (fig. 3). Although it may cause some problems with coding, observers seem not to react to this incident at all (fig. 2).

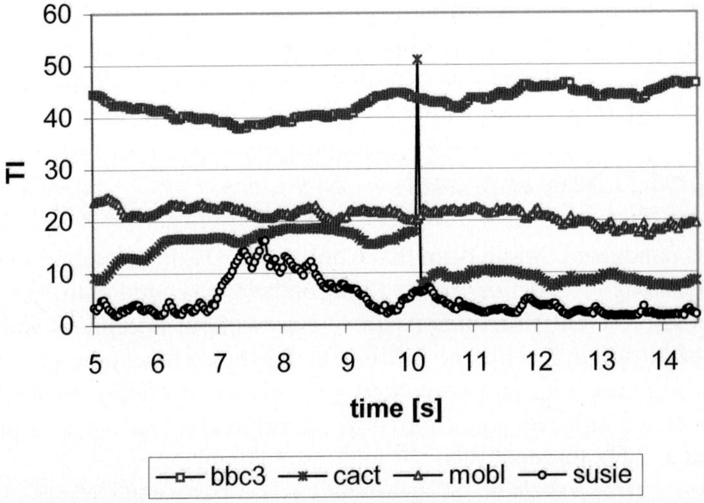

Fig.3. The TI (computed for each frame) distribution in time for 4 test sequences

As the initial role of SI and TI parameters was to reflect perceptual amount of information, authors propose a slight modification in the way those parameters are computed, so that the values were in accordance with the level of scene complication perceived by the observer:

$$Q_3.SI = \text{Upper quartile}_{time}\{std_{space}[Sobel(F_n)]\} \quad (4)$$

$$Q_3.TI = \text{Upper quartile}_{time}\{std_{space}[M_n(i,j)]\} \quad (5)$$

Fig. 1 shows that Q_3.SI and Q_3.TI placed the "cact" sequence in a new position on the plot – now it can be identified as the scene just slightly more difficult to decode than "susi", while "mobl" and "bbc3" kept their position of critical for the coding system. Hence Q_3.SI and Q_3.TI reflect perceptual amount of information in a better way in comparison with traditional SI and TI.

4. Conclusions

As the end-user of the video coding system is the observer himself, on each step of investigation it is important to mind the features of human visual perception, which is disposed to average the stimuli rather than to react to short time values. Hence, the idea of computation the upper quartile values of information in a scene seems to reflect human perception in a more adequate way than the maximum value. Still the conception of evaluating the amount of perceptual information seems to be imperfect and should be under investigation in the future. Before it's done, authors advise to take into consideration the upper quartile value or to study the plots of spatial-temporal information on a frame-by-frame basis. The use of information distributions over a test sequence also permits better assessment of scenes in case of continuous assessment, which is a new mainstream in the area of subjective quality evaluation of compressed video.

References

[1] ITU-Telecommunications Standardization Sector: Two criteria of video test scene selection, Geneva, 2-5 December 1994.

[2] ITU-T Recommendation P.911 (1996), Subjective audiovisual quality assessment methods for multimedia applications.

[3] Ostaszewska A., Żebrowska-Łucyk S., Kłoda R.: Metrology tools in subjective quality evaluation of compressed video, Mechatronics Robotics and Biomechanics Trest, Czechy 2005.

[4] Ostaszewska A., Żebrowska-Łucyk S., Kłoda R.:Metrology properties of human observer in compressed video quality evaluation, XVIII IMEKO WORLD CONGRESS, Metrology for a Sustainable Development Rio de Janeiro, Brazil 2006.

Genetic Identification of Parameters the Piezoelectric Ceramic Transducers for Cleaning System

Paweł Fabijański, Ryszard Łagoda

Institute of Control and Industrial Electronics, Warsaw University of Technology, Koszykowa 75, 00-662 Warszawa, Polska

Abstract

Source of ultrasounds in technical cleaning system are Sandwich type piezoelectric ceramic transducers. They have the ability to radiate in an ultrasonic medium with maximum acoustic power when the vibration is activated by a voltage generator with frequency equals the mechanical resonance of the transducer. In resonant the transducer are the oscillating element, for which equivalent electrical circuit consist of connection in parallel: Co end RLC. To optimization of structure and parameters piezoelectric ceramic transducer can been used the Genetic Algorithm. GA have been shown to be effective in the resolution of difficult problems. The resonant frequency of the real circuit varies during the operation in function of many parameters, among others, the most important are temperature, time, and the surface of the cleaned elements.

1. Introduction

The Sandwich-type piezoelectric ceramic transducers are the most frequently applied sources of ultrasound. They have the ability to radiate in an ultrasonic medium with maximum acoustic power when the vibration is activated by a current whose frequency equals the mechanical resonance of the transducer. This transducer is made of piezoelectric ceramics PZT.

3. Power Circuit Configuration

Block diagram and the main circuit of the converter with piezoelectric ceramic transducer (Fig. 1) consists of: converter AC/DC, full-bridge inverter FBI, isolating transformer T, where $z_2/z_1=n_1$, $z_1/z_3=n_2$, special filter F, transducer PT, sensor of vibrations S, control and identification system.

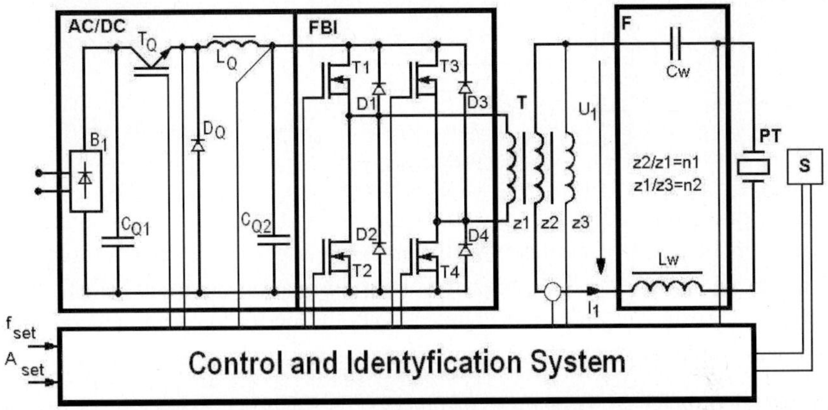

Fig. 1. The block diagram of the ultrasonic generator

The control unit consists of two independently parts: in FBI inverter, is the frequency feed-back control loop and in AC/DC converter, is the amplitude feed-back control loop. Signal f_{set} make possible to set up manually frequency switching inverter FBI and signal A_{set} establish amplitude ultrasonic oscillation.

3. Digital Model of Inverter-Special Filter-Transducer Group System Circuit

Genetic Algorithm is used to identification of structure and parameters of Piezoelectric Ceramic Transducer. Simple genetic algorithm was applied and shown on Fig. 2.

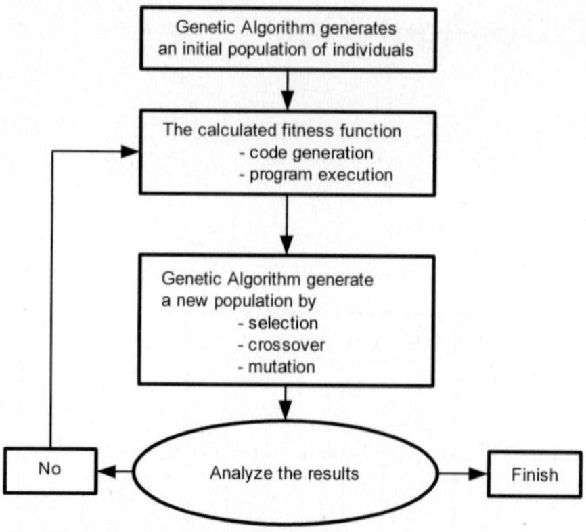

Fig.2. Simply Genetic Algorithm GA

In resonant the transducer are the oscillating element, for which equivalent electrical circuit consist of connection in parallel: Co end RLC, this simple non-linear circuit is numerically oppressive. To modification the basic Euler's interpolation algorithm is proposed classical genetic algorithm. Our genetic algorithm was implemented in DSP simulation programing language.

```
//Program piezoelectric ceramic transducer
input { Circuit parameters [ R,L,C,u(t)];
        Simulation parameters( t_p , t_k ,Δ,v(0));
        Evolution parameters(Num,Size, NumM,Initial,NumG,stop);)
output{ t, u; t_p ; u_o =u(o);};
for (t, t_k , t++){ New function form;
                    start initial population;}
for(integer i=1,Num,i++){New population,
                    Select the best}
```

where:
Genotype dimension : Num,
Size of the generation cerated by mutation : Size,
Number of the best selected genotypes for mutation: NumM,
Number of the best selected genotypes for crossing: NumC,
initial parameter mutation range : initial,

The equivalent circuit of the piezoelectric ceramic transducers with frequency close to resonant frequency is shown in Figure 3, where:Co - static capacity of the transducer, C - equivalent mechanical capacity, L - equivalent mechanical inductance, R - equivalent resistance, Rp = Rm + Ra, where: Rm - equivalent mechanical loss resistance, Ra - equivalent acoustic resistance.

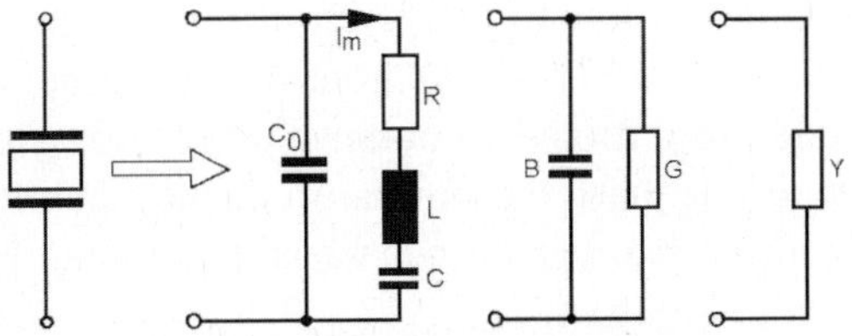

Fig. 3. The equivalent circuit of piezoelectric transducer.
Exemplary values Co = 4 nF, L = 246 mH, C = 182 pF, R = 392 Ω

The frequency fm of mechanical vibration may be calculated by equation:

$$fm = \frac{1}{2\pi}\sqrt{LC} \tag{1}$$

The digital model of the inverter-special filter- transducer group system circuit, work in PSpice language. Exemplary results are presented in Fig. 4.

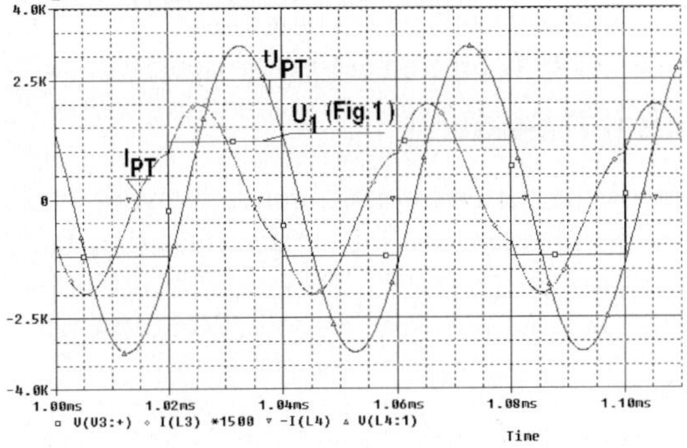

Fig. 4. Current and voltage waveform for f ≈ fm

4 Frequency Feed-Back Control Loop

To obtain the maximum value of converter efficiency is necessary to assure the optimal frequencies of inverter. Changing the cleaning medium results in the variations of output power of inverter and in output resonant frequency we have four case (Fig. 5).

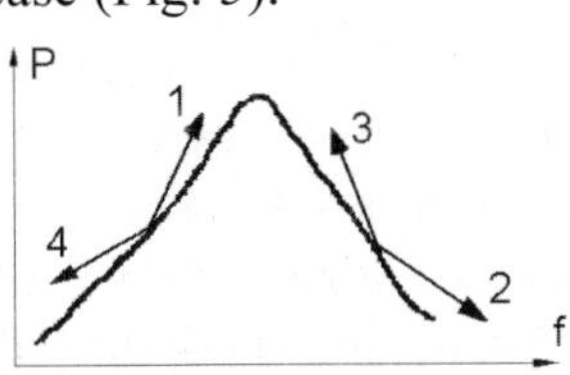

Fig. 5. Four case variations of change output power of inverter and resonant frequency

To obtain the maximum value of converter efficiency, its important role of fuzzy logic control system. The logic control system define derived mark of signal amplitude proportional to output power and output resonant frequency and change adequate the output frequency of inverter. The relationship between the input and output variable contain Tab. 1 and fuzzy controller is shown in Fig. 6.

Table 1. Control rules

	derived of power	derived of frequency	rules of logic control
1	dP/dt > 0	df/dt > 0	frequency increase
2	dP/dt < 0	df/dt > 0	frequency decrease
3	dP/dt > 0	df/dt < 0	frequency decrease
4	dP/dt < 0	df/dt < 0	frequency increase

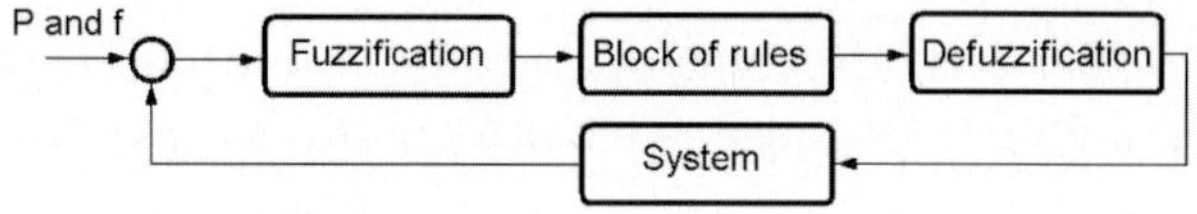

Fig. 6. Structure of Fuzzy Controller

5. Experimental Research

The main circuit of the series-resonant converter with piezoelectric ceramic transducer and special filter system (Fig. 1) was modeling, build and tested.

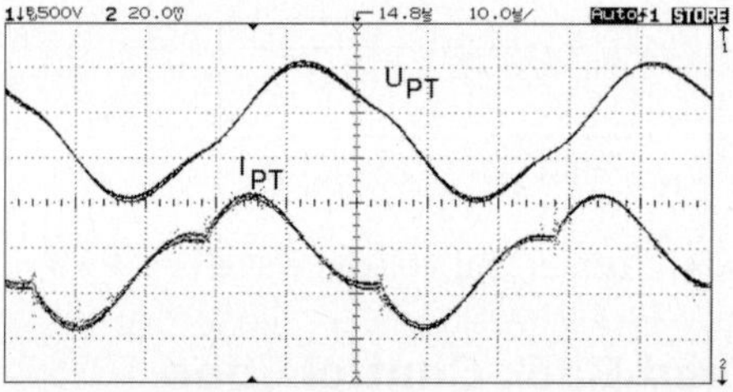

Fig. 7. Current and voltage waveform in the case when f = fm

The experimental result was obtain in the case of tuning and untuning of output frequency of inverter and the mechanical frequency of the transducer.
Exemplary experimental current and voltage waveform of transducer in the case when f = fm are shown in the Fig. 7.

6. Conclusions

The presented control method make possible to supply the piezoelectric ceramic transducers group with the quasi-sinusoidal current and voltage waveform with self-tuning frequency to mechanical resonance. To the identification of structure and parameters of piezoelectric ceramic transducer was used Genetic Algorithm. The frequency feed-back control system was to tested in the case of tuning the generator frequency to the mechanical resonance frequency of transducer.

In the case of untuning the output frequency of inverter according to piezoelectric ceramic transducer mechanical frequency the current and voltage waveforms of piezoelectric transducer are non-sinusoidal. Especially current waveform is deformation.

The results of analysis of piezoelectric transducer and of the system of the resonance converter with control loop of frequency have been compared with experimental results in real piezoelectric transducer system and satisfactory results has been obtained.

References

[1] P. Fabijański, R. Łagoda: Control and Application of Series Resonant Converter in Technical Cleaning System. Proceedings of the IASTED, International Conference Control and Application, Cancun, Mexico, (2002)

[2] P. Wnuk: Genetic optimisation of structure and parameters of TSK fuzzy models. Elektronika 8-9/2004

Simulation Modeling and Control of a Mobile Robot with Omnidirectional Wheels

T. Kubela (a) *, A. Pochylý (b)

(a), (b) Institute of Production Machines, Systems and Robotics, Faculty of Mechanical Engineering, Brno University of Technology, Technicka 2, Brno, 616 69, Czech Republic

Abstract

This contribution summarizes the results of work carried out during the proposal and simulation modelling of the mobile robot undercarriage equipped with omnidirectional wheels.
In order to make an accurate design and appropriate dimensioning of driving units, there was carried out a simulation of dynamic properties of the undercarriage with omnidirectional wheels in Mathworks MATLAB 7.
There is described a kinematical and a dynamical model of the whole robot platform. By using this functionality we can relatively overcome the problem of miss-alignments of the wheels during the assembly. Results of the simulation were summarized in form of a state space model of the whole robot platform.

1. Introduction

The main advantage of mobile robot undercarriages with omnidirectional wheels deals mainly with very good maneuverability. Using omnidirectional wheels was the basic and initial assumption for the project. Particularly for the reason of very good maneuverability, it can be considered as an ideal tool for verification of various types of algorithms determined to local navigation, path planning, mapping and further development with respect to university-indoor environment and robotics classes.
At the Institute of Production Machines, Systems and Robotics (IPMSaR) has been approached to a decision to design a mobile robot with this type of undercarriage. However, each design process should be preceded by optimally chosen parameters that can influence the resulting behavior of

the whole platform; mechatronic and systemic approach. The simplest form how these parameters can be achieved deals with a description of a complex simulation model which results in an assessment of the final platform behavior with these parameters. The purpose of this contribution is to describe extended kinematical and dynamical model of the mobile robot with a possibility to implement skew wheel angles overcoming the problem of miss-alignments of the wheels having impact on the real robot's trajectory.

Further there is shortly described the design of the robot, selected wheels, driving units, gear boxes and incremental encoders and a simulation of power demands with respect to specified conditions and limitations.

2. Kinematical model of the robot

Provided that the robot is moving only within 2D environment, its absolute position in a global coordinate system is defined by the vector [ξ, ψ, θ] as shown in Fig. 1. The inverse kinematics equations are given by

$$(\omega_1 \quad \omega_2 \quad \omega_3)^T = \frac{1}{r} \cdot A \cdot R(\theta) \cdot (\dot{x} \quad \dot{y} \quad \dot{\theta})^T \tag{1}$$

where ω, i = 1, 2, 3, is the angular velocity of each wheel of the robot, r is the wheel radius,

$$A = \begin{pmatrix} 1 & 0 & L_1 \\ -1/2 & \sqrt{3}/2 & L_2 \\ -1/2 & -\sqrt{3}/2 & L_3 \end{pmatrix}, \quad R(\theta) = \begin{pmatrix} \cos\theta & \sin\theta & 0 \\ -\sin\theta & \cos\theta & 0 \\ 0 & 0 & 1 \end{pmatrix} \tag{2),(3}$$

where L_i is the radius of the robot platform. By means of this equation (1) we can compute angular velocities of all wheels that are defined by desired trajectory of motion.

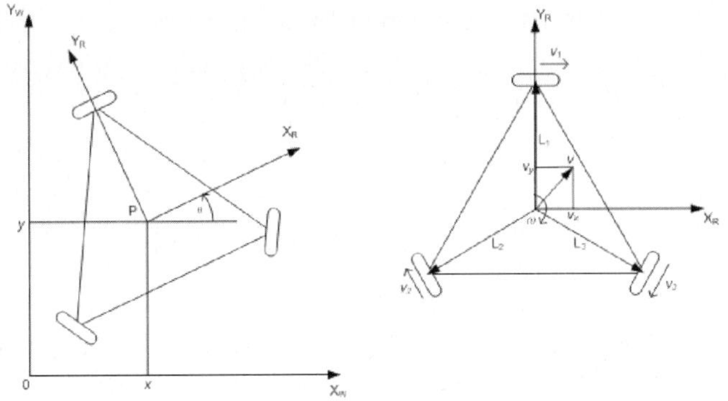

Fig. 1. Kinematical scheme of the omnidirectional mobile robot

3. Robot dynamics modeling

Forces that are caused by the motion of the whole platform can be described as follows:

$$(m\ddot{x} \quad m\ddot{y} \quad J_R\ddot{\theta})^T = A^T \cdot (f_{R1} \quad f_{R2} \quad f_{R3})^T$$
$$(f_{R1} \quad f_{R2} \quad f_{R3})^T = (A^{-1})^T (m\ddot{x} \quad m\ddot{y} \quad J_R\ddot{\theta})^T \qquad (4),(5)$$

where m is the robot mass, J_R is the robot moment inertia, f_{Ri} is the traction force of each wheel.
Dynamic model for each wheel:

$$J_w\dot{\omega}_i + c\omega = nM - rf_{wi} \qquad (6)$$

where J_w is the inertial moment of the wheel, c is the viscous friction factor of the omniwheel, M is the driving input torque, n is the gear ratio and f_{wi} is the driving force due to each wheel.
The dynamics of each DC motor can be described using the following equations:

$$L\frac{di}{dt} + Ri + k_1\omega_m = u$$
$$J_m\dot{\omega}_m + b\omega_m + M_{ext} = k_2 i \qquad (7),(8)$$

where u is the applied voltage, i is the current, L is the inductance, R is the resistance, k_1 is the emf constant, k_2 is the motor torque constant, J_m is the inertial moment of the motor, b is the vicious friction coefficient, M_{ext} is the moment of an external load and ω_m is the angular speed of the motor.
By merging equations (2)-(8) we obtain a mathematical model of each essential dynamic properties of mobile robot undercarriage.

4. Robot model

The trajectory of motion is described by a list of points, each with four important parameters *[x, y, v, ω]*. From these points obtained needed vector of velocities $[v_x, v_y, \omega]$ by the Trajectory controller module. Inverse kinematics is used to translate this vector into individual theoretically required velocities of wheels. Dynamics module is then used to compute inertial forces and actual velocities of wheels. By means of Direct Kinematics module, these velocities are re-translated into the final vector of velocities of the whole platform. Simulator module obtains the actual performed path of the robot by their integration.
The whole model [1] was designed as a basis for modeling of mobile robot motions in order to analyze the impact of each constructional parameter on its behavior. For this reason, there is not used any feedback to control the

motion on a desired trajectory. This approach allows choosing key parameters more precisely for better constructional design.

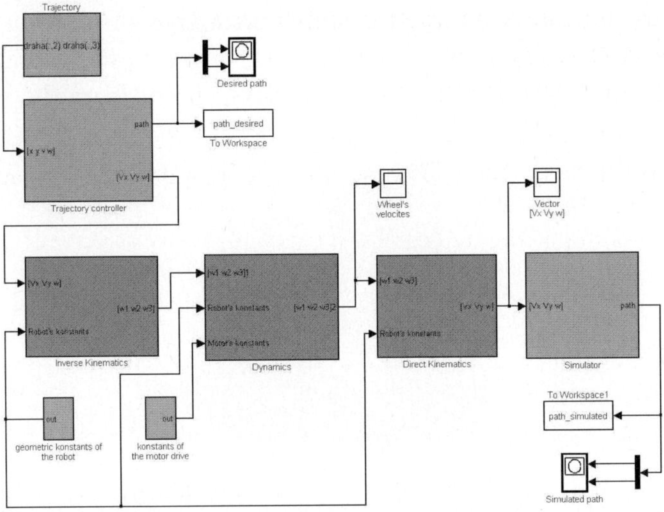

Fig. 2. Robot model in Matlab Simulink

In order to create this simulation model there was used software MATLAB Simulink 7.0.1. The emphasis was laid particularly on its schematic clearness and good encapsulation of each individual module. It has an important impact on an extensibility of the model in the future in order to create and analyzed other function blocks.

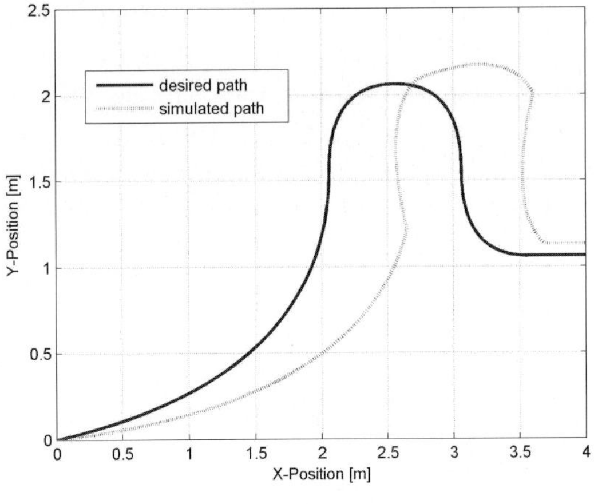

Fig. 3. Example of the impact of dynamic properties on a performed path of the robot

5. Design of the mobile robot

There was chosen a symmetric undercarriage with omnidirectional wheels in order to simulate the behaviors, properties and characteristics of the mobile robot. It was the starting point from the previous year.

Fig. 4. OMR IV design – view on the whole system

6. Conclusion

This contribution summarizes a kinematical and dynamical model of the mobile robot without a feedback (open-loop) simulated in Matlab Simulink environment. The whole model was designed as a basis for motions modelling of a mobile robot undercarriage in order to analyze its key factors that influence the final motion and allow an optimal choice of these parameters which are required for a constructional proposal.

Published results were acquired using the subsidization of the Ministry of Education Youth and Sports of the Czech Republic research plan MSM 0021630518 „Simulation modelling of mechatronic systems".

7. References

[1] Kubela, T., Pochylý A., Knoflíček R. (2006) Dynamic properties modeling of mobile robot undercarriage with omnidirectional wheels: *Proceedings of International Conference PhD2006, Pilsen, pp 45-46.*

[2] Rodrigues, J., Brandao, S., Lobo, J., Rocha, R., Dias, J. (2005) RAC Robotic Soccer small-size team: Omnidirectional Drive Modelling and Robot construction, *Robótica 2005 – Actas do Encontro Científico, pp 130-135.*

Environment detection and recognition system of a mobile robot for inspecting ventilation ducts[1]

A.Timofiejczuk, M. Adamczyk, A. Bzymek, P. Przystałka,

Department of Fundamentals of Machinery Design,
Silesian University of Technology, Konarskiego 18a
Gliwice, 44-100, Poland

Abstract

The system of environment detection and recognition is a part of a mobile robot. The task of the robot is to inspect ventilation ducts. The paper deals with elaborated methods of data and image transmission, image processing, analysis and recognition. While developing the system numerous approaches to different methods of lighting and image recognition were tested. In the paper there are presented results of their applications.

1. Introduction

Tasks of a system of environment detection and recognition (vision system VS) of an autonomous inspecting robot, whose prototype was designed and manufactured in Department of Fundamentals of Machinery Design at Silesian University of Technology is to inspect ventilation ducts [2]. VS acquires, processs and analyzes information regarding environment of the robot. Results, in the form of simple messages, are transmitted to a control system (CS) [3]. Main task of the vision system are to identify obstacles and their shapes. The robot is equipped with a single digital color mini camera, which is placed in front of the robot (Fig.1.) Two groups of procedures were elaborated:

- procedures installed on a board computer (registration, decompression, selection, recognition and transmission);

---

[1] Scientific work financed by the Ministry of Science and Higher Education and carried out within the Multi-Year Programme “Development of innovativeness systems of manufacturing and maintenance 2004-2008”

- procedures installed on the operator's computer (visualization, recording film documentation).

Fig.1. Single camera is placed in front of the robot

All procedures of VS were elaborated within Matlab and C++, and operate under Linux. Three modes of VS operation are possible [2] [3]:

- manual –images (films) are visualized on the operator's monitor and film documentation is recorded. The robot is controlled by an operator. Image processing and recognition is inactive.
- autonomous – recorded images are sent to the operator's computer as single images (they are selected after decompression realized on the board computer). Image processing and recognition is active.
- „with a teacher" – films are sent to the operator's computer. Robot is controlled by the operator. Realization of operator's commands is synchronized with activation of image and recognition procedures. A goal of this mode is to gather information on robot control. The information is used in a process of self-learning that is in-the middle of elaborating.

2. Image registration, transmission and visualization

Since duct interiors are dark, closed and in most cases duct walls are shiny a way of lightening has a significant meaning [4]. Image recording was preceded by examination of different lightening. Most important demands were small size, low energy consumption. Several sources were tested (Fig.2) (a few kinds of diodes, and white bulbs). As the result a light source consisting of 12 diodes was selected (it is used in car headlights).
Image recording was performed by a set consisting of a digital camera and (520 TV lines, 0.3 Lux, objective 2,8 mm, vision angle 96 degrees) and frame grabber CTR-1472 (PC-104 standard). Film resolution is 720x576

pixels in MPEG-4 compressed format [1]. Transmission is active in the manual mode. In autonomous mode decompression and selection are performed. An image is processed only in case a change in the robot neighborhood occurred.

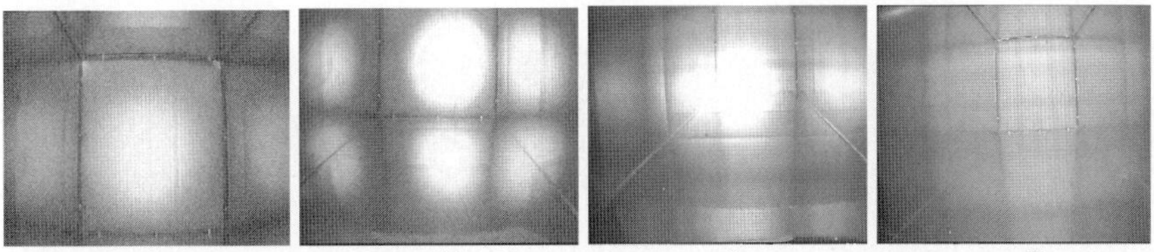

a) b) c) d)

Fig.2. Selected examples of light sources, a) white LED, b) blue LED, c) LED headlamp, d) car headlight.

Only images transmitted to the operator's monitor can be visualized. Depending on a mode the operator observes transmitted films (manual and „with teacher" modes) or single decompressed images (autonomous mode). Transmitted images are stored as a film documentation. Collecting such documents is one of the main tasks of the robot [2].

3. Image processing and analysis

Two main approaches to image analysis were elaborated. A goal of the first one was to extract some image features. Monochrome transformation, reflection removal, binarization, histogram equalization and filtration were applied. Results of these procedures were shown in Fig. 3.

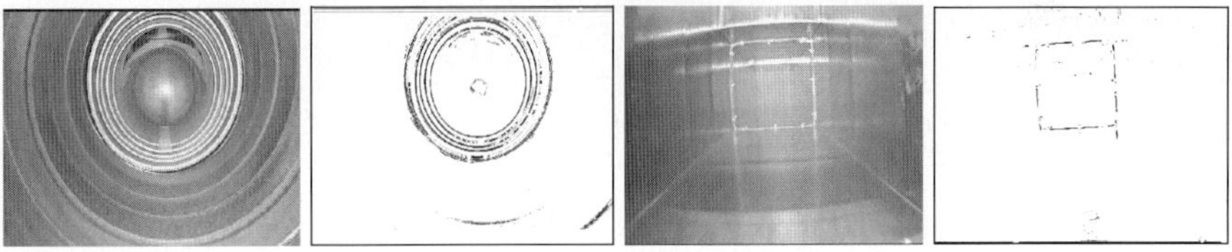

Fig.3. Examples of images and results of their processing

Analysis procedures were applied to the images shown in Fig. 3. Results of analysis are 5 features (shape factors and moments) calculated for identified objects. These features have different values for different duct shapes and obstacles what makes it possible to identify them. However, performed research shown that image recognition on the basis of these values in case of composed shapes (curves and dimension changes) does not give expected results. Moreover it requires that advanced and time consuming methods of image processing have to be applied. As the result of that another approach was elaborated. Images were resampled in order to obtain possible low resolution that was enough to distinguish single objects visible in the image (Fig. 4). These images were inputs to neural networks applied at recognition stage.

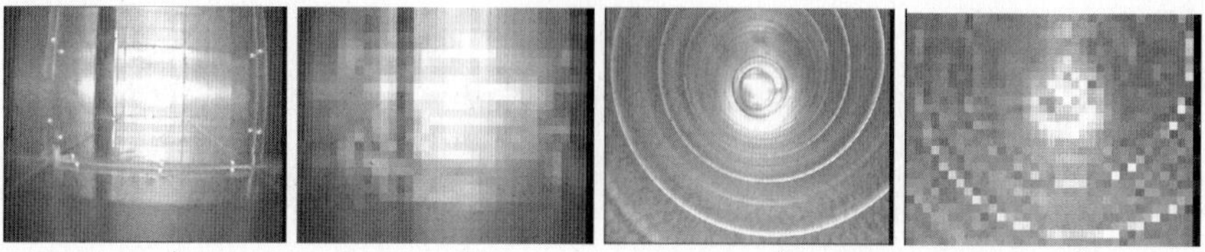

a) b) c) d)

Fig.4. Resolution resampling a) and c) 720x576, b) and d) 30x30

4. Image recognition

Image recognition was based on the application of neural networks that were trained and tested with the use of images recorded in ducts of different configuration (Fig.5).

[PP1]

Fig.5. Ventilation ducts used as test stage

A library of images and patterns was elaborated. As a result of a few tests of different neural networks a structures consisting of several three layer perceptrons was applied. Each single network corresponds to a distinguished obstacle (shape or curve). Depending on the image resolution a single network has different numbers of inputs. The lowest number (shapes are distinguishable) was 30x30 pixels. All networks have the same structure. It was established by trial and error, and was as follows: the input layer has 10 neurons (tangensoidal activation function), the hidden layer has 3 neurons (tangensoidal activation function) and the output layer has 2 neurons (logistic activation function). As training method Scaled Conjugate Gradient Algorithm was used. For each network examples were selected in the following way: the first half of examples – the shape to be recognized and the second half of examples – randomly selected images representing other shapes. This approach is a result of numerous tests and gave the best effect.

It must be stressed that results of neural network testing strongly depend on lightening and camera objective, as well as a number of examples and first of all image resolution. Results of present tests made it possible to obtain classification efficiency about 88%. Such low image resolution and numbers of neurons in a single network required that 54000 examples had to be used during network training. In order to increase the number of testing images a few different noises were introduced to images.

5. Summary

The most important factor that influences recognition correctness is too low resolution. However, its increase leads to non-linear decrease of a number of examples necessary to network training. At present stage of the research the application of cellular network is tested. One expects that outputs of these networks can be applied as inputs to the three layer perceptron. The most important is that these outputs seem to describe shapes more precisely than shape factors and moments and simultaneously their number is lower then the number of pixels of images with increased resolution.

References

[1] M. Adamczyk: "Mechanical carrier of a mobile robot for inspecting ventilation ducts" In the current proceedings of the 7th International Conference "MECHATRONICS 2007".

[2] W. Moczulski, M. Adamczyk, P. Przystałka, A. Timofiejczuk: „Mobile robot for inspecting ventilation ducts" In the current proceedings of the 7th International Conference "MECHATRONICS 2007".

[3] P. Przystałka, M. Adamczyk: "EmAmigo framework for developing behavior-based control systems of inspection robots." In the current proceedings of the 7th International Conference "MECHATRONICS 2007".

[4] A. Bzymek:"Experimental analysis of images taken with the use of different types of illumination" Proceedings of OPTIMESS 2007 Workshop, Leuven, Belgium.

Calculation of robot model using feed-forward neural nets

C. Wildner, J. E. Kurek

Institute of Robotics and Control Warsaw University of Technology
ul. Św. Andrzeja Boboli 8, 02-525, Warszawa, Poland

Abstract

Neural nets for calculation of parameters of robot model in the form of the Lagrange-Euler equations are presented. Neural nets were used for calculation of the robot model parameters. The proposed method was used for calculation of robot PUMA 560 model parameters.

1. Introduction

Mathematical model of industrial robot can be easily calculated using for instance Lagrange-Euler equation [1]. However, it is very difficult to calculate the real robots inertia momentums, masses, etc. Mathematical model of industrial robot is highly nonlinear. In this paper for assignment of model coefficients we will use neural nets. Neural nets can approximate nonlinear functions. Neural model of robot has been built using only information from inputs and outputs of robot (i.e. control signals and joint positions) and knowledge of model structure.

2. Lagrange-Euler model of robot

Mathematical model of robot dynamic with n degree of freedom can be presented in the form of Lagrange-Euler equation as follows:

$$M(\theta)\ddot{\theta} + V(\theta,\dot{\theta}) + G(\theta) = \tau \quad (1)$$

where $\theta \in R^n$ is a vector of generalized robot variable, $\tau \in R^n$ is a vector of generalized input signal, $M(\theta) \in R^{n \times n}$ is inertia matrix of robot, $V(\theta,\dot{\theta}) \in R^n$ is a vector of centrifugal and Coriolis forces and $G(\theta) \in R^n$ is a homogenous gravity vector with respect to the base coordinate frame. Calculating the derivatives in the following way [1]:

$$\dot{\theta}(t) \cong \frac{\theta(t) - \theta(t - T_p)}{T_p}, \ \ddot{\theta}(t) \cong \frac{\dot{\theta}(t + T_p) - \dot{\theta}(t)}{T_p} \tag{2}$$

where T_p denotes the sampling period, one can obtain the discrete-time model of the robot based on (1) as follows:

$$\theta(k+1) = 2\theta(k) - \theta(k-1) + A[\theta(k), \theta(k-1)] + B[\theta(k)] + C[\theta(k)]\tau(k) \tag{3}$$

where k is a discrete time, $t=kT_p$, and

$$\begin{aligned} A[\theta(k), \theta(k-1)] &= -T_p^2 M^{-1}[\theta(k)]V[\theta(k), \theta(k-1)] \\ B[\theta(k)] &= -T_p^2 M^{-1}[\theta(k)]G[\theta(k)] \\ C[\theta(k)] &= T_p^2 M^{-1}[\theta(k)] \end{aligned} \tag{4}$$

3. Neural model for robot

The set of three, three-layer feed-forward neural nets was used for calculation of unknown parameters A, B, C of model (3), fig. 1.

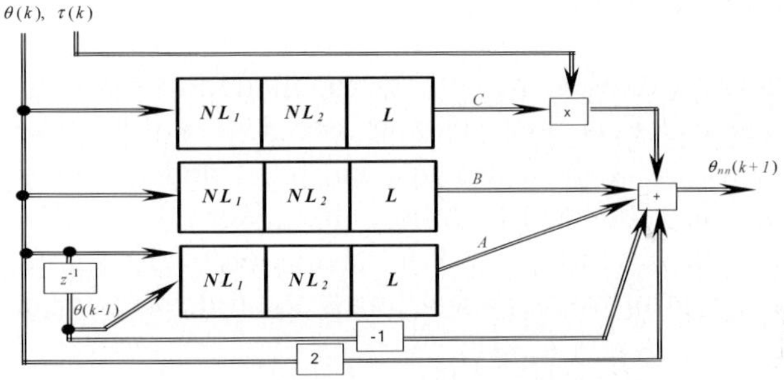

Fig. 1. The strukture of neural nets

Each net consists of 3 neuron layers: the first and the second layer is nonlinear (NL_1, NL_2), and output layer is linear (L). A general neuron equation is as follows

$$y(k) = f[v(k)] \tag{5}$$

where v is a generalized neuron input

$$v(k) = \sum_{i=1}^{h} w_i u_i(k) + w_0 \tag{6}$$

and u_i is input to the neuron, w_i is input's weight, w_0 is bias, and $f(*)$ is a transition function. Nonlinear neuron layer consists of neurons with sigmoidal hyperbolic tangent transition function

$$y = f_{NL}(v) = \text{tansig}(v) = \frac{2}{1+e^{-2v}} - 1 \quad y \in [-1,1] \tag{7}$$

and neurons in linear neuron layer have the following transition function:

$$y = f_L(v) = bv \quad b = const \tag{8}$$

We have assumed that the input signals to every layer are connected with all neurons in this layer. Input signal to the nets is generalized robot variable $\theta(k)$, $\theta(k-1)$. The following performance index was used for network learning

$$J(d) = \frac{1}{2}\sum_{k=1}^{N}\sum_{j=1}^{n}[\theta_j(k) - \theta_{nnj}(k,d)]^2 \tag{9}$$

where $\theta_{nn} \in R^n$ is an output vector of the network, d is the learning iteration number, N is the length of learning data sequence. The backpropagation method [2] were used for network learning.

4. Calculation of neural model

We have used the proposed method for calculation of the model parameters of Puma 560 robot [1]. The robot has n=6 degree of freedom: six revolute joints. Sequence of the learning data, input and output signals, had length N=600. The robot had to follow the given reference trajectory in time t=6 [sec] with T_p=0.01 [sec]. For calculation of the learning data the trajectory of every joint was set according to the following formula

$$\theta_{r,i}(t) = \frac{1}{3}\sin(t), \quad i=1,...,6 \tag{10}$$

Then, for network testing we have used another data. The reference testing trajectory was as follows

$$\theta_{t,i}(t) = \frac{1}{2}\sin(3t) \quad , i=1,...,6 \tag{11}$$

Both reference trajectories for learning and testing of the network are presented in fig. 2, note that the trajectories values are given in degrees.

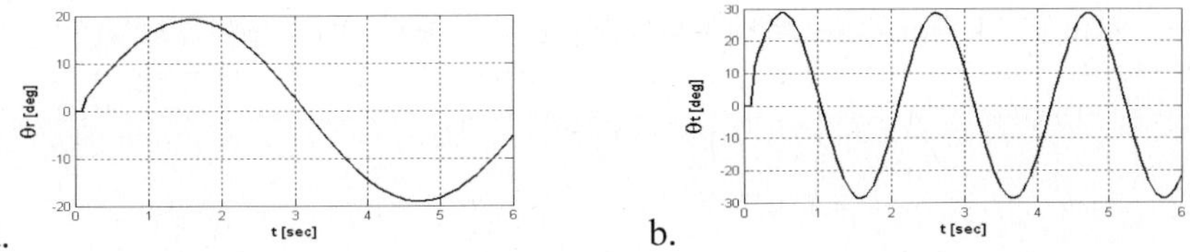

Fig. 2. The reference trajectories: a) learning data output, b) testing data.

Neural nets for calculation of model (3) unknown parameters A, B, C, had respectively 5, 4, 8 neurons in layers NL_1 and NL_2. The number of neurons in output layer L was equal to number of elements in matrices A, B, C, respectively 6, 6 and 36. The results obtained after 2000 learning iterations are presented in fig. 3 and in tab. 1. In fig. 3 the difference between the reference trajectory θ_r and output of the neural network θ_{nn} is given in degree. In tab. 1 maximal errors between θ_r and θ_{nn} are given.

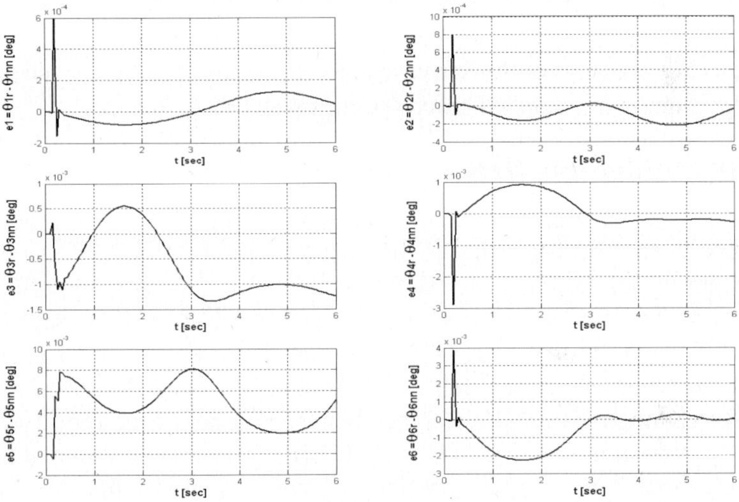

Fig. 3. Difference between the reference trajectory θ_r and output θ_{nn} obtained from neural model for the learning data after the learning process.

Tab. 1. Maximal errors after 2000 learning iterations $\left|\theta_r - \theta_{nn}\right|_{\max}$, $\theta_r(t)=\sin(t)/3$

Joint	1	2	3	4	5	6
	0.0006 [°]	0.0008 [°]	0.0013 [°]	0.0029 [°]	0.0081 [°]	0.0039 [°]

The results of testing of the network are presented in fig. 5 and tab. 2. Difference between θ_t and θ_{nn} is shown in fig. 5. Maximal errors between θ_t and θ_{nn} are given in tab. 2.

Tab. 2. Maximal errors for testing data $\left|\theta_r - \theta_{nn}\right|_{\max}$, $\{\theta_i(t) = \sin(3t)/2\}$

Joint	1	2	3	4	5	6
	0.1687 [°]	0.2048 [°]	0.2732 [°]	0.6810 [°]	0.8188 [°]	0.7869 [°]

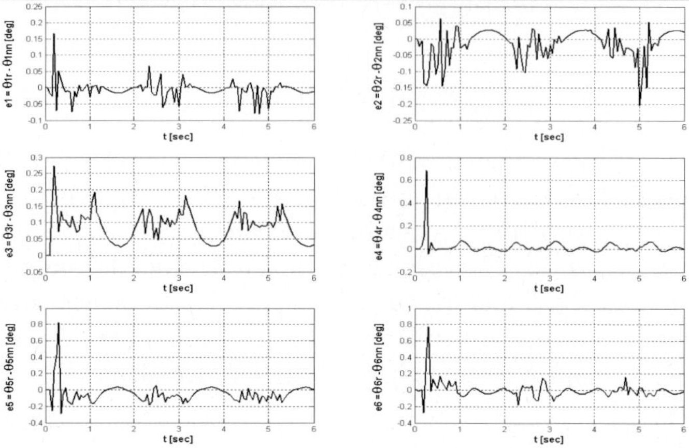

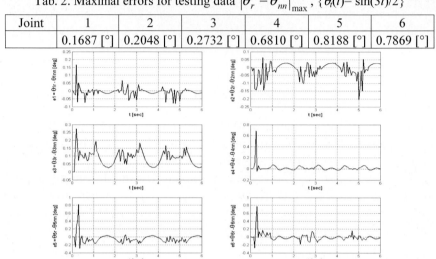

Fig. 5. Difference between the reference trajectory θ_r and output θ_{nn} obtained from neural model for the testing data.

5. Concluding remarks

Neural nets were used for calculation of the Puma 560 robot model parameters. Neural nets learning is difficult to execute. From the results obtained during learning process with the reference trajectory described by equation (10) it follows that an average difference between output of the robot and neural model after 2000 learning iterations was approximately 0.003 [deg]. For testing data with the reference trajectory described by (11) we have obtained maximal difference between output of the robot and output of the neural model approximately 0.5 [deg]. From the obtained results it follows that it is possible to obtain a neural model of the robot based only on robot outputs and inputs. We plan to use other techniques calculation of the neural nets.

References

[1] Fu K. S., Gonzalez R. C., Lee C. S. G., "Robotics, control, sensing, vision, and intelligence", McGraw-Hill Book Company, 1987.

[2] Osowski S., „Sieci neuronowe w ujęciu algorytmicznym", WNT, Warszawa, 1996.

[3] Wildner C., Kurek J. E., „Identyfikacja modelu robota za pomocą rekurencyjnych sieci neuronowych – synteza układu sterowania robota", XV Kraj. Konf. Automatyki KKA, Warszawa, Poland 2005.

EmAmigo framework for developing behavior-based control systems of inspection robots[1]

P. Przystałka, M. Adamczyk

Silesian University of Technology, Department of Fundamentals of Machinery Design, 18A Konarskiego Str., 44-100 Gliwice, Poland

Abstract

The paper deals with the implementation of the behavior-based control and learning framework for autonomous inspection robots. The presented control architecture is designed to be used in the mobile robot (Amigo) for the visual inspection of ventilation ducts. In this paper, various problems are discussed including brief description of hardware components (PC-104 modules), low-level hard real-time operating system (RTAI), high-level manual and autonomous mode control interface (Linux, KDE). The main aim of this paper is to present the framework for rapid control prototyping in the case of the inspection robot.

1. Introduction

Modern mobile robots are usually composed of heterogeneous [WM1]hardware and software components. In a large amount of cases, during developing stage when a construction of a robot is still in an early development phase, hardware and software components should allow rapid prototyping of a control system. The proposed control architecture may be used either in early development phase or in a final robot prototype. This work focuses only on such key aspects as the hardware layout, low-level real-time operation system and some software components implemented on the robot.

[1] Scientific work financed by the Ministry of Science and Higher Education and carried out within the Multi-Year Programme "Development of innovativeness systems of manufacturing and maintenance 2004-2008"

The paper is organized as follows. Section 2 describes PC/104 embedded modules and a real-time operating system as the core of the discussed framework. There is also given remote control software enabling typical capability of the robot. Finally, the paper is concluded in Section 3.

2. Control system architecture

This section describes the hardware layout and software components that are necessary to make the general-purpose framework for developing various control strategies of an inspection robot.

2.1 Hardware layout

As demonstrated in the related paper [1, 2, 3], the construction of the robot has been modified from a four-legged walking robot (12 servomechanisms) to a wheeled robot (at first with 4 and next with 6 stepper motors). This was the main reason that PC/104 technology was chosen to be used as the hardware platform. The main module of the hardware layout is PC/104 motherboard with power supply DC/DC 5V. It is equipped with 300 MHz Celeron processor (Ethernet interface, USB 1.1, RS-232, PCI bus), 512 MB RAM, 2 GB Flash.

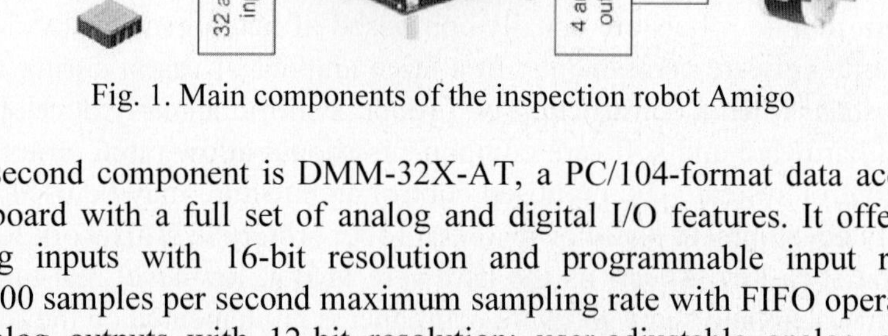

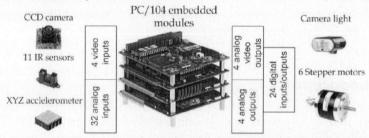

Fig. 1. Main components of the inspection robot Amigo

The second component is DMM-32X-AT, a PC/104-format data acquisition board with a full set of analog and digital I/O features. It offers 32 analog inputs with 16-bit resolution and programmable input range; 250,000 samples per second maximum sampling rate with FIFO operation; 4 analog outputs with 12-bit resolution; user-adjustable analog output ranges; 24 lines of digital I/O; one 32-bit counter/timer for A/D conversion and interrupt timing; and one 16-bit counter/timer for general purpose use. The last component is a high performance four channel MPEG-4 video compressor that supports real-time video encoding. These modules are used to handle all the I/O data for the whole system and to calculate the control parameters.

2.2 Hard real-time operating system

Real-time operating system (RTOS for short) has to provide a required level of service in a bounded response time (e.g. a real-time process control system for a robot may sample sensor data a few times per second whereas stepper motors must be serviced every few microseconds). A so-called hard real-time system is one that misses no timing deadlines. The authors considered such RTOS as: RTLinux [5], KURT [10], RTAI [4], Windows CE [6], VxWorks [9]. Finally, RTAI was selected as extension of standard Linux Kernel for the simple reasons that it offers extended LXRT (hard-hard real time in user space) and is free of charge.

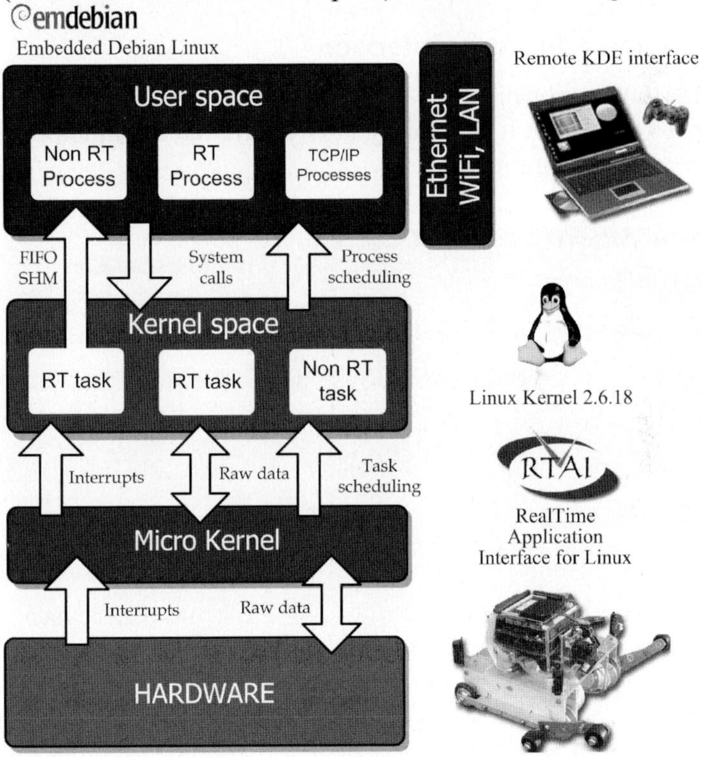

Fig. 2. Micro Kernel real-time architecture for the inspection robot Amigo

Figure 1 shows a diagram of the discussed operating system implemented on the robot. Inter-process communication is provided by one of the following mechanisms: UNIX PIPEs, shared memory, mailboxes and net_rpc. In the first stage of the project only named pipes are applied. It can be used to communicate real-time process between them and also with normal Linux user processes.

Device	RTAI FIFO (O_NONBLOCK)
Sensors	/dev/dmm32x/sensors/sensor...
Actuators	/dev/dmm32x/actuators/motor...
Controller	/dev/dmm32x/control/behaviour...
	/dev/dmm32x/control/coordinator

In this way the access to every device connected to the robot may be simply available by putting/reading data into/from RTAI FIFOs. Moreover, Embedded Debian Linux which is installed on the robot gives all the advantages of standard operating systems (TCP/IP, Wireless LANs, SSH, NFS, SQL databases, C/C++, Python, Java application, etc.).

2.3 EmAmigo 3.14 – user interface

EmAmigo 3.14 application is a task-oriented interface for end users interacting with the robot. It is implemented on the remote Linux-based host using C++ and Qt libraries [11].

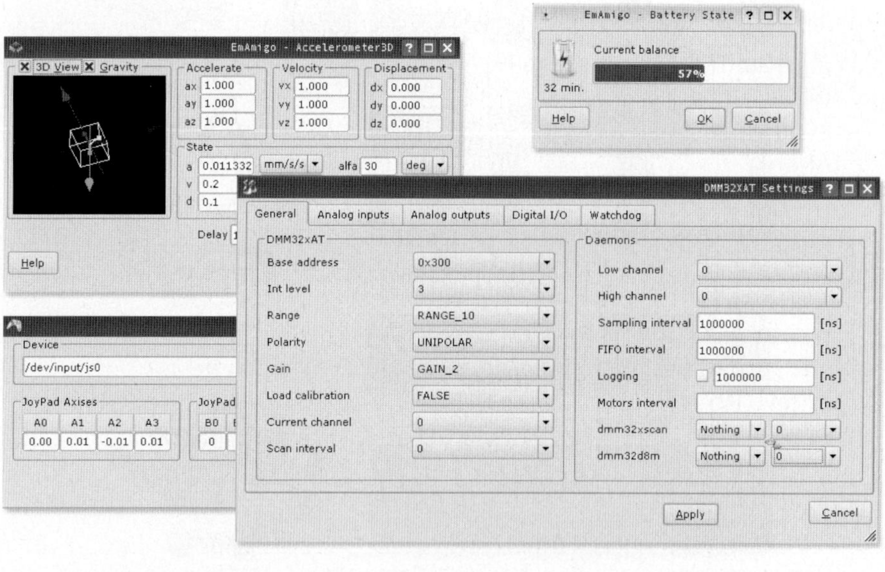

Fig. 2. EmAmigo 3.14 - KDE user interface of the Amigo robot

This application enables some typical capability of the robot: remote control by human operator (using keyboard or joypad controller), monitoring different robot parameters (robot states, real-time video footage), and gathering data needed for learning behavior-based controller.

3. Conclusions

In this paper, the authors described the hardware and software framework (free of charge for non-commercial purposes) that can be used for rapid control prototyping of such mechatronic systems as mobile robots. Putting together PC/104 technology and RTAI hard real-time operating system one obtains a tool which is easy to adapt and easy to develop.

References

[1] Adamczyk M.: "Mechanical carrier of a mobile robot for inspecting ventilation ducts" In the current proceedings of the 7th International Conference "MECHATRONICS 2007".
[2] Adamczyk M., Bzymek A., Przystałka P, Timofiejczuk A.: "Environment detection and recognition system of a mobile robot for inspecting ventilation ducts." In the current proceedings of the 7th International Conference "MECHATRONICS 2007".
[3] Moczulski W., Adamczyk M., Przystałka P., Timofiejczuk A.: „Mobile robot for inspecting ventilation ducts" In the current proceedings of the 7th International Conference "MECHATRONICS 2007".
[4] The homepage of RTAI - the RealTime Application Interface for Linux, 2006. https://www.rtai.org/.
http://hegel.ittc.ukans.edu/projects/proteus/docs/RTAS-2000.pdf.
[5] FSM Labs Inc. First real-time linux, 2002.
http://www.fsmlabs.com/community.
[6] Microsoft. Windows CE .NET Home Page, 2002.
http://www.microsoft.com/windows/embedded/ce.net/default.asp.
[7] Frederick M Proctor and William P Shackleford. Embedded real-time Linux for cable robot control. In Proceedings of DETC'02 ASME 2002 2002. http://www.isd.mel.nist.gov/documents/publist.htm.
[8] Steve Rosenbluth, Michael Babcock, and David Barrington Holt. Controlling creatures with Linux. Linux Journal, November 2002. http://www.linuxdevices.com/articles/AT6435065918.html.
[9] Wind River Systems. VxWorks 5.x, 2002.
http://www.windriver.com/products/vxworks5/index.html.
[10] Kansas University. KURT: Real Time Linux, 1997.
http://www.ittc.ku.edu/kurt.
[11] Trolltech: Leader in cross-platform C++ GUI software development and embedded Linux solutions. http://www.trolltech.com/

Simulation of Stirling engine working cycle

M. Sikora, R. Vlach

Institute of Solid Mechanics, Mechatronics and Biomechanics,
Faculty of Mechanical Engineering, Brno University of Technology,
Czech Republic, ysikor0l@stud.fme.vutbr.cz

Institute of Solid Mechanics, Mechatronics and Biomechanics,
Faculty of Mechanical Engineering, Brno University of Technology,
Czech Republic, vach.rl@fme.vutbr.cz

Abstract

This paper describes model of Stirling machines. The model will be used for optimalization of power station. The power station consists of Stirling engine and electric generator. The genetic algorithms can be used for identification parameters of engine.

1. Introduction

The engines with external heat inlet were never widespread in the past [3] except steam engine. Nowadays, it is necessary to solve some global problems and to look for new alternative sources of energy.
The aim is a design of small combined heat and power unit which is driven by Stirling engine. Achieving of good thermodynamically Stirling engine efficiency represents relatively difficult optimizing task. The design of accurate thermal model is very important. We cannot neglect many heat losses, so the theory of ideal cycles is not usable. The computational models dividing working gas volume into two or three sub-areas are not too accurate. The engine dividing into many volume elements (final volumes method) makes better results. There are few simplifications of gas properties in the model. The above mentioned method of calculation does not achieve the CFD accuracy. However, this model is faster and more suitable for future optimalization of engine parameters.

2. Thermal model characteristic

The properties of the developed Stirling engine model (γ-modification) are as follows:

- Numerical model is used. It is able to simulate non-stationary and transient processes.
- The ideal gas, for which the state equations are applied [2], are considered as working medium.
- The friction and inertial forces are not considered in working gas.
- The leakages of the gas from engine working part are not considered.
- The pressure losses due to displacement of the working gas are omitted, the same holds for the gas warming due to friction.
- The model so far does not include the re-generator of the working gas temperature.
- The engine is divided into volume elements for thermal processes modelling (final volumes method), see fig. 1

3. Numerical calculation system of this model

The Stirling engine model is represented by system of several non-linear differential equations and by another auxiliary relation. The regular dividing of the engine makes possible using of the matrix system of many variables. The main calculation is solved by numerical integration with fixed time step [4]. For calculation of some variables we must use results from previous step because actual value has not yet been calculated. Errors due to this fact are negligible at sufficiently small time step.
Three related problems are solved in this iteration calculation:

1) The behaviour of working gas temperatures and pressure. The pressure value is used for output power determination.
2) The determination of working gas flow among gas elements. It is solved by using other sub-iteration cycle.
3) We must still consider the thermal processes in solid parts of engine in each solution step. The interaction of solid parts elements is solved by thermal network method.

We consider also addition of electrical generator model. The whole model is implemented in MATLAB software.

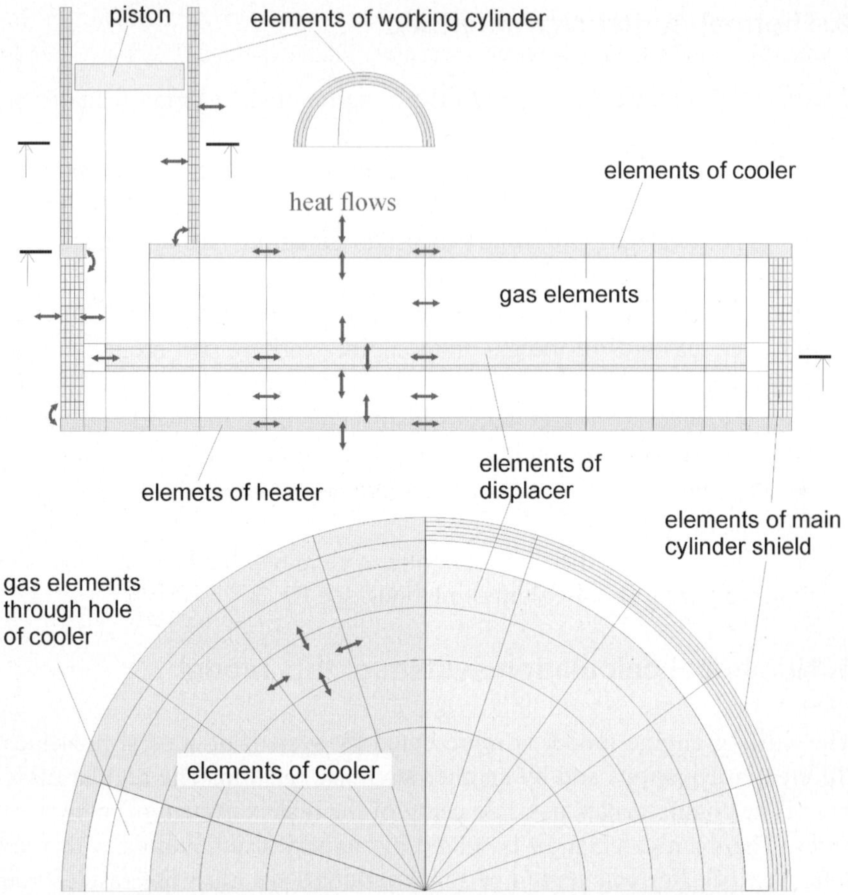

Fig. 1. Small sample Stirling engine division to elements

4. Simulations results

The input parameters correspond with dimensions and properties of Stirling engine small real model. The values of some parameters have been only estimated (e. g. heat transfer coefficients) and these parameters should be verified by experiments.

Figure 2 show the time (and crank tilting angle) relations of torque, pressure and temperatures of working gas in two crank movements. These temperatures belong to elements: under the displacer, above displacer and in working cylinder.

Figure 3 shows distribution of temperature and weight flow for tilting crank angle 198° from its upper dead point. (the displacer goes up, the piston goes down).

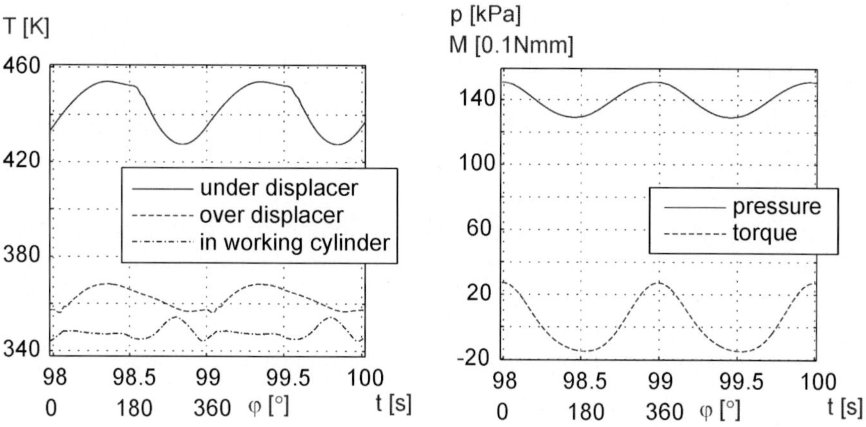

Fig. 2 The temperatures of working gas (left). The pressure and engine torque (right), all in relation of time and of crank tilting angle of displacer.

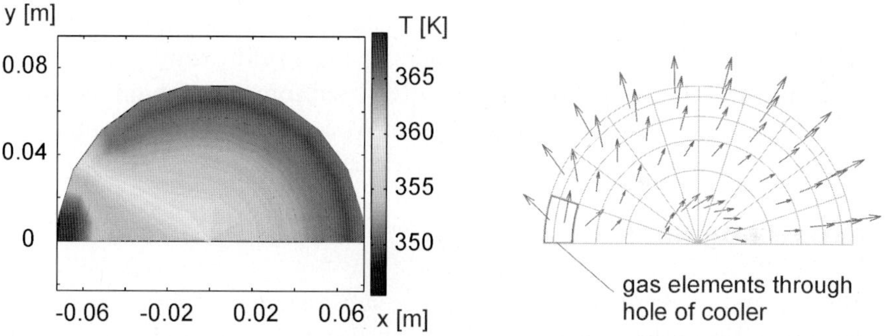

Fig. 3 The gas temperatures (left) and the vector field of weight flow above the displacer (right).

5. Optimalization by using genetic algorithm

The maximal power output and efficiency of the combined heat and power unit are our aims. We must find the parameters of engine which give the best compromise of both requirements. An extensive zone of good efficiency is not necessary, because one operating point is supposed. The design of engine is naturally limited by weight and dimensional requirement.

The resulting virtual model will work with many variables and parameters which determine engine behavior. We can sort these parameters into two classes:
-*the parameters given in advance* – e. g.: the maximal and minimal temperatures, the mean value of pressure, the properties of working gas and materials, minimal required power output, the main dimensions, characteristic of generator.
-*the variable parameters* – e. g.: some dimensions, the angle between cranks, the area and volume of regenerator, etc.

Using of genetic algorithm is planed for finding of variable parameters optimal values. The success of design will be determined by simple function which will evaluate (by numerical weights) results of simulations. The model with designed parameters will be used. We will have to evaluate especially the efficiency for set power, the output power for maximal efficiency, the important dimensions, etc.

6. Conclusion

The biggest disadvantage of built numerical model is determination of heat transfer coefficients between solid parts and gas. The existing empirical formulas for it's determination are usable in the specific conditions. They are not mostly realized in Stirling engine. Therefore, we should estimate the coefficients by using experiments or FEM software. Then, the numerical model will be considered like adequately accurate.

Acknowledgment

Published results were acquired using the subsidization of the Ministry of Education, Youth and Sports of the Czech Republic, research plan MSM 0021630518 "Simulation modelling of mechatronic systems".

References

[1] M. Jícha "Přenos tepla a látky" Akademické nakladatelství CERM, Brno, 2001
[2] M. Pavelek "Termomechanika" Akademické nakladatelství CERM, Brno, 2003
[3] G. Walker "Stirling-cycle machines" Clarendon press, Oxford, 1973
[4] B. Maroš, M. Marošová "Základy numerické matematiky" PC-DIR Real, Brno, 1999

Mobile robot for inspecting ventilation ducts[1]

W. Moczulski, M. Adamczyk, P. Przystałka, A. Timofiejczuk

Silesian University of Technology, 18a Konarskiego str.,
Gliwice, 44-100, Poland

Abstract

The paper deals with a concept and design of a mobile robot capable of inspecting ventilation ducts made of steel sheet. The robot can operate in several modes including: autonomous, manual, and training one. Some subsystems are briefly outlined. Mobility of the robot is achieved by four wheels that include permanent magnets. There are 2 DOFs associated to each wheel. The detection system assesses the internal state of the robot and its subsystems, and perceives the surrounding environment, providing the control system with vital data that allows completing inspection tasks. The robot is also equipped with environment recognition system that collects data whose meaning is twofold. Primarily, it allows assessing actual condition of the ducts being inspected. Further on, the data makes possible creating plans of long-term movements that are required in order to complete the mission. The work of all these systems is coordinated by the control system. It is based on behaviors that are selected or combined by a neural controller. The controller is able to learn better behaviors from examples. These issues are discussed in details in other papers presented at this event.

1. Introduction

Recent EU and also domestic law requires that each ventilation duct located in majority of public buildings such as hospitals, restaurants, or even supermarkets undergo periodic inspection that should allow assessment of

[1] This research has been financed by the Ministry of Science and Higher Education and carried out within the Multi-Year Programme PW-004 "Development of innovativeness systems of manufacturing and maintenance 2004-2008" – grant No. PW-004/02/2005/2/UW-2005.

the overall condition of the ventilation system with special attention paid to its cleanness and integrity. To this end, manual inspection is carried out by human personnel. To allow exhaustive control of the complete ventilation system including its parts of small cross sections, it becomes ever more and more popular to apply remote-controlled mobile robots. Since elements of the ventilation system are made of steel, inspection robots are controlled by using wired remote control, which restricts operation range of the robot to the maximal length of the uncoiled wire cable that connects the robot with the control box operated by a member of human personnel, which is usually equal to maximum a dozen or so meters.

These drawbacks could be overcome by applying a truly autonomous robot that could be able to travel along separate paths of the ventilation system, and acquire relevant data, which after completing the route could be sent to the control unit to allow its careful examination by the experienced human operator.

The paper deals with an original design of the autonomous robot capable of inspecting ventilation ducts. Since the complete range of problems to be solved is enormously broad, the most general issues are only addressed. In Section 2 we formulate requirements. Then in Section 3 the prototype of the mobile robot is presented. The paper ends with conclusions and future work. The issues mentioned in this paper are subject of more detailed description in additional papers included in the proceedings.

2. Identified requirements

The goal of the project is to built non-commercial prototype version of a mobile robot capable of inspecting ventilation ducts diversified with respect to their cross-sections, location, direction or number of junctions and elbows. The robot should be equipped with an intelligent control system and a system for detecting and recognizing the robot's environment.

It has been assumed that the robot's environment are ventilation ducts built from galvanized steel tubes of circular and/or square cross-sections whose dimensions are contained within Ø 250 and 600 mm, and Ø 300 and 700 mm, respectively. The minimal radius of elbows is 300 mm. Different sections of ventilation ducts can be connected by means of suitable adapters. Both horizontal and vertical sections of the ducts are possible.

The robot itself should be able to carry load at least 0.6 kg and drive through horizontal ventilation ducts with minimal speed of 0.02 m/s. Its design should allow energy-saving operation and assure reliability of operation by simultaneous preserving its simplicity.

3. Prototype of the autonomous mobile robot

The reported project included the complete conceptual design stage. Many conceptions were formulated and then carefully evaluated using respective systems of criteria. In this Section the final design is briefly described.

3.1. Mechanical chassis

Basing on exhaustive analyses, magnetic wheels containing permanent magnets, and an additional system of tearing off the wheels of the walls of the ducts have been applied. Fig. 1 shows the final solution.

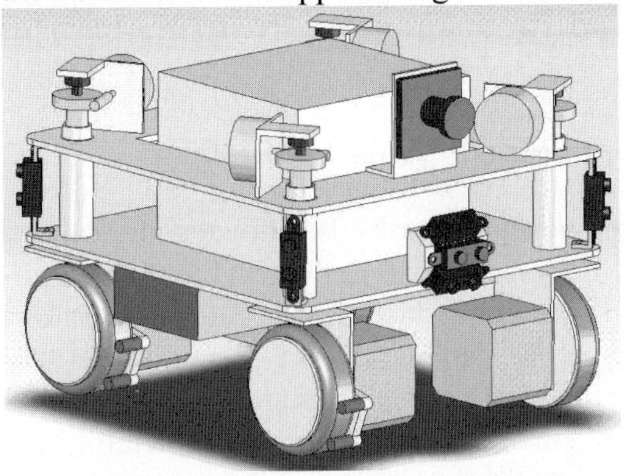

Fig. 1 The overall view of the mobile robot

There are 8 DOFs, 4 of them associated to wheels, and the other to axes that allow turning the wheels. Each DOF is powered by its own, individually controlled drive composed of a step DC motor and possibly a worm gear. To allow the robot to clear joints of walls, wheels are equipped with special mechanisms that tear off the magnetic wheels of the steel walls. Two parallel plates constitute right basis for electronic equipment of the robot, which performs control functions and allows detecting and recognizing the environment. Details of design solutions are described in [1].

3.2. Control system architecture

The control system consists of a few major parts (hardware and software components): a main control computer, data acquisition boards, actuators, sensors, video CCD camera, a remote control host, WiFi/LAN based communication protocols, low/high-level controller, real-time operating system, EmAmigo - Qt based user interface. The major function of the main control computer is to supervise sub-systems during the navigation and perform behavior-based control and position estimation. The authors adapted a PC/104 technology in order to do so. The mobile robot is equipped with internal sensors (four encoders, a battery sensor, two load sensors), external sensors (a camera, eleven infrared sensors, 3-axial accelerometer). Eight stepper motors are controlled by a real-time process in order to drive four wheels. The CCD camera together with sensors is used for recognizing and detecting obstacles around the mobile robot. It is also employed for capturing video footage. EmAmigo interface is implemented on the remote linux-based host for end users interacting with the robot.

The authors based on information in the literature decided to use a behavior-based control schema (so-called co-operative/competitive architecture [3, 4]). The behavior-based layer consists of a set of behaviors and a coordinator. As the coordinator, competitive and feed-forward neural networks are ideal candidates for use in the behavior selection because of their ability to learn state-action mapping and their capability of fast adaptation to the new rules required by the environment of the robot. More details may be found in the related papers [5, 6] presented at this event.

3.3. System for environment detecting and recognizing

This system carries out several different tasks. It estimates internal and external state of the robot and delivers data to the control system. Additionally, it detects shape of the surrounding duct, further course of the duct (such as changes of the cross-sections and elbows, T-junctions and others), and obstacles in the duct. Finally, the system captures videos and photos, extracts single frames from video streams, compresses pictures, and transmits video information to the external computer of the operator.

Several issues are worth mentioning due to their originality. One of them is the video capturing system. This system allows linguistic summarization of acquired pictures. Furthermore, since the pictures are taken quite rare due to slow speed of the robot and energy saving, a movie is generated for the operator basing on acquired individual pictures. The environment recognition is based upon neural networks. To this end, cellular automata, Kohonen's self-organizing maps and simple perceptrons were applied. To assure safe operation of the robot, a subsystem of nearest environment recognition has been developed. This system uses a system of infrared detectors (cf. Fig. 1) and a 3-axis accelerometer which allows the robot to estimate its position in the absolute coordination system. More information about the system can be found in [2].

All the data is send to the system database which in its part serves as a blackboard for communication purposes.

The data collected on-line may be presented to the operator by using the operator's desktop.

4. Conclusion

In the paper a prototype mobile robot for inspecting ventilation ducts is shortly described. The robot takes advantage of magnetic force for driving along steel pipes of a ventilation system. The control system is able to control 8 DOFs of the robot. The system for environment detection and recognition identifies both the internal state of the robot and its nearest environment that allows the control system to select proper actions in order to complete the task. On the other hand, this system collects huge amount of data – videos and pictures of the interior of ventilation ducts to be inspected, and transmits this data to the external computer.

In the future we are going to develop this non-commercial prototype in order to get ready a new design that would be suitable for manufacturing. Further on, software of the mobile robot and the operator's desktop is to be developed in order to facilitate manual analysis of data collected by the mobile robot while inspecting ventilation systems.

The authors would like to thank to all the members of the research team who developed the conception of the mobile robot, and then implemented individual system's components.

References

[1] Adamczyk M.: "Mechanical carrier of a mobile robot for inspecting ventilation ducts". Proc. of the 7th Int. Conference "Mechatronics 2007".
[2] Adamczyk M., Bzymek A., Przystałka P, Timofiejczuk A.: "Environment detection and recognition system of a mobile robot for inspecting ventilation ducts." Proc. of the 7th Int. Conference "Mechatronics 2007".
[3] Arkin, R. C. 1989 Neuroscience in motion: the application of schema theory to mobile robotics. [In:] Visuomotor coordination: amphibians, comparisons, models and robots (ed. J.-P. Ewert & M. A. Arbib), pp. 649-671. New York: Plenum.
[4] Brooks, R. A., "A Robust Layered Control System for a Mobile Robot." IEEE Journal of Robotics and Automation, Vol. RA-2, No. 1(1986), 14-23.
[5] Panfil W., Przystałka P.: "Behavior-based control system of a mobile robot for the visual inspection of ventilation ducts." Proc. of the 7th Int. Conference "Mechatronics 2007".
[6] Przystałka P., Adamczyk M.: " EmAmigo framework for developing behavior-based control systems of inspection robots." Proc. of the 7th Int. Conference "Mechatronics 2007".

Applications of augmented reality in machinery design, maintenance and diagnostics

W. Moczulski, W. Panfil, M. Januszka, G. Mikulski

Silesian University of Technology, Department of Fundamentals of Machinery Design, 18A Konarskiego Str., 44-100 Gliwice, Poland

Abstract

The paper deals with technical applications of Augmented Reality (AR) technology. The discussion starts with the conception of AR. Further on, three applications connected with different stages of existence of a technical means (machinery or equipment) are presented. They address design process, maintenance and diagnostics of different objects. Finally, further research is outlined.

1. Introduction

This paper presents possible applications of oncoming technology called augmented reality (AR). It rises from the well-known Virtual Reality (VR). It can be said that the first AR systems were those applying Head-Up Displays (HUD) in the fighter planes. The main idea of AR is to facilitate performing some task by the user in the real world. This goal is achieved by providing additional, hands-on pieces of information carried through several information channels, by different forms of messages. Selection of the specific information channel and, additionally, the form of message delivered to the user, can be an additional task of optimization.
The research on applications of AR in mechanical engineering has been carried out in the Department since 2004. Before that, AR applications were implemented for training the personnel, and for supporting human installers by assembling mechanical systems. The authors report concisely their own research works. The first application of AR system concerns the possibilities of aiding the user in the design process. The second one is the implementation of the portable AR system in maintenance/diagnostics of a system of mechatronic devices. Finally, the AR system for reasoning in

machinery diagnostics is presented. The paper ends with conclusions and future work.

2. AR interface in intelligent systems

Nowadays expert systems and other intelligent, knowledge-based applications are widely used to aid human personnel in carrying out complex tasks. The user communicates with the running program by means of a *user interface.* AR technology allows creating highly user-friendly interface for providing the user with hands-on information about the problem to be solved, accessible in the very comfortable and easy-to-adopt form. Pieces of information can be delivered to the user in an automated way, since the AR-based system is able to catch or even recognize which problem the user is going to solve and which piece of information is the most relevant to this problem.

AR systems base mainly on graphical (virtual) information partially covering the view of the real world. The role of the virtual object is to present the user some information to facilitate his/her tasks. But AR interface can deliver not only visual information seen by the user. AR employs also many other *user input/output information channels* [3]. Besides visual information, the user can be influenced by other output interfaces such as haptic devices, acoustic or motion systems. As the input interfaces are considered tracking systems, image acquisition systems and input devices (data gloves, 3D controllers – mice, trackballs, joysticks). Input interfaces are sources of data for intelligent systems. Tracking systems provide information about position/orientation of some objects (e.g. human limbs). Image acquisition systems are used for registering view of the real world seen by the user. Thanks to the input devices the communication between the system and the user becomes more interactive.

3. Examples of implementation

In the following three applications developed in the Department are presented. All the three projects were carried out in the framework of MSc Theses supervised by W. Moczulski.

3.1. Application of AR in machinery design

AR mode for changing views of the model and completely understanding the model content is more efficient, intuitive and clear than the traditional one. Therefore, the AR technology, as a kind of new user interface, should introduce completely new perspective for the computer aided design systems [1]. Our research is focused on AR system, which among other things enables the user to easily view the model from any perspective.

The most important and difficult part of AR system proposed by us is software. The AR viewing software allows the user to see virtual 3D models superimposed on the real world. We base on public-domain AR tracking library called ARToolKit [4] with LibVRML97 parser for reading and viewing VRML files. ARToolKit is software library that uses computer vision techniques to precisely overlay VRML models onto the real world. For that purpose software uses markers. Each marker includes different digitally encoded pattern, so that unique identification of each marker is possible. In our conception the markers are printed on the cards of the catalog. We can compute the user's head location as soon as the given marker is tracked by the optical tracking system. The result is a view of the real world with 3D VRML models overlaying exactly on the card position and orientation.

To see 3D VRML models on the cards the user needs a video or optical head mounted display (HMD) connected to the computer by a cable or wirelessly and integrated with a camera. The user wears HMD with the video camera attached, so when she/he looks at the tracking card through the HMD a virtual object is seen on the card. A designer can pick up the catalog and manually manipulate the model for an inspection.

In our conception of AR system the user with the HMD on head sits in a front of a computer. Moreover, a catalog with cards to be tracked is necessary. The user looks over the catalog with standard parts (for example rolling bearings, servo-motors etc.) and by changing pages can preview all the parts. When the user chooses the best fitting part, she/he can export this part to the modeling software (CATIA V5R16). Having selected the "EXPORT" button, the designer can see the imported part in the workplane of CATIA. After this the user can go back to modeling in CATIA or repeat the procedure for another part. When the design process is accomplished the user can export the finished 3D model back to the AR software and preview results of her/his work.

3.2. Application of AR in the maintenance of an equipment

The exemplary AR system for aiding personnel during maintenance of equipment is a PDA-based system, developed to support maintenance of Electronic Gaming Devices (EGD). The primary objective was to create a knowledge base and system that are capable of using mobile devices like PDA. The role of this system is to fulfill two distinctive tasks. One being to enable the selection of the troubled device and the other to empower individuals with the capability of identifying the specific malfunction. The first function is designed to enable the selection of a faulty device, via a touch-screen tool, while also displaying information concerning the proper function of the respective system. Information will be transmitted as sound through headphone(s), and in the same time displayed on the screen of the PDA.

The second function of the AR system, as identified previously in the text, is the ability of the tool to locate the failure within the given device. To this end, the screen will display detailed inquiries with possible Y/N answers. Based on the answers, the user is informed about potential solutions to the problem/s at hand. A structure of questions and answers plays the role of knowledge base represented by a tree structure. This solution allows further development.

Information contained in the knowledge base of the system have been generated through two distinct sources: employee research from previous encounters with that particular model or a similar model of a given piece of equipment, and the manufacturer's instructions and other publications. Pieces of knowledge are stored in sound and text files.

3.3. Application of AR in machinery diagnostics

The last example is the application of AR in machinery diagnostics [2]. The main role of this AR system is to aid the user in the process of measuring noise level around the machine and in the diagnostic reasoning. There are 21 measurement points placed on the half-sphere surrounding the machine. The elaborated system consists of a five main parts: USB camera, monitor, PC, printed marker, and the tracking library ARToolKit for Matlab [5]. The operation of the system is divided into a few steps. Firstly the camera registers a view of the real world and sends it to computational software. Further on, basing on the marker size and shape in the image, the program estimates the relative pose between the camera and the marker. Dimensions and shape of the marker must be known. Then the image of the real world is overlaid by virtual objects. Finally, a combined result is sent to the computer display.

The elaborated system fulfills two main tasks. Before all, the aim of the system operation is to help the user to place the microphone in the right place in space. The system indicates the measurement points around the machine and theirs projection points on the ground. The second task of the system is to aid the user in the process of reasoning about the machine state basing on the previously performed measurements. The system presents the measurement results using circles whose filling color corresponds to these results.

4. Recapitulation and conclusions

In the paper a very modern technology of Augmented Reality was briefly presented. Moreover, three technically important implementations of AR have been introduced. All these applications have been developed in the Department of Fundamentals of Machinery Design. AR seems to be brilliant interface to many intelligent systems that are developed in the area of computer-aided design, manufacturing, maintenance, training, and many others. The authors expect that in the very next time every important computer application will be equipped with components such as user interfaces which will take advantage of AR technology.
AR as such is quite young domain of research. Therefore, many issues still remain unsolved or even unidentified. One of the most important ones concerns methods of knowledge engineering specific to developing AR-based intelligent applications. The authors are going to carry out an exhaustive research on this methodology in the very next time.

Bibliography

[1] Dunston P. S., Wang X., Billinghurst M., Hampson B., "Mixed Reality benefits for design perception", 19 th International Symposium on Automation and Robotics Construction (ISARC 2002), Gaithersburg, 2002
[2] Panfil W., "System wspomagania wnioskowania diagnostycznego z zastosowaniem rozszerzonej rzeczywistości", MSc Thesis (in Polish), Silesian University of Technology at Gliwice, 2005
[3] Youngblut C., Johnson R. E., Nash S. H., Wienclaw R. A., Will C. A., "Review of Virtual Environment Interface Technology" , Institute for Defense Analyses – IDA, Paper P-3186, 1996,
[4] The Human Interface Technology Laboratory at the University of Washington; URL: http://www.hitl.washington.edu/research/shared_space/
[5] URL: http://mixedreality.nus.edu.sg/software.htm (July 2005)

Approach to Early Boiler Tube Leak Detection with Artificial Neural Networks

A. Jankowska

Institute of Autom. and Robotics, The Warsaw University of Technology
ul. św. A. Boboli 8 pok.253, 02-525 Warsaw, Poland

Abstract

The early boiler tube leak detection is highly desirable in power plant for prevention of following utility destruction. In the paper the results of artificial neural network (ANN) models of flue gas humidity for steam leak detection are presented and discussed on example of fluid boiler data.

1. Introduction

The boiler tube failures are major cause of utility forced outages and induce great economical costs. The early detection of faults can help avoid power plant shut-down, breakdown and even catastrophes involving human fatalities and material damage [1,4,5]. Steam leaks can take values between 1 000 – 50 000 kg/h[1]. Reconditioning cost are much lower, when steam leak is early detected. Tube failures in steam generators are typically caused by one the following general categories [1]: metallurgical damage caused by hydrogen absorption, erosion caused by impacts from solid ash particles, corrosion-fatigue, overheating, etc.

1.1 Industrial methods of tube leak detection

The methods of steam leak detection can be enumerated as[1]:
-1)acoustic monitoring devices-drawbacks: little to medium leaks (<10 000 kg/h) aren't detected; sensors- expensive and require benchmarking;
-2) steam/water balance testing – drawbacks: time consuming, insensitive to small leaks, frequency of tests to low for preventing serious damage;

-3) monitoring of flue gas humidity – it's any information about real humidity source, i.e. measured humidity can be caused by water added to combustion chamber, soot blowing, changing fuel hydrogen, etc. or steam leaks. However, model of humidity built on data composed for many nominal states of plant job, i.e. including many disturbances and variation of fuel contents, can be tested to detect boiler faults moments. In the paper will be studied ANN models, which attribute is generalization propriety.

1.2 Use of artificial intelligence models to fault detection

Nowadays the faults detection and isolation (diagnosis, i.e. FDI) systems are applied in many industrial plants[3]. Very often their main idea is based on continues comparison of measured, in on-line mode, signals and models output of observed variable. Inconsistencies between these two variables, named *residuum*, give information of faulty work of observed, monitored plant. The models of complex, real, nonlinear, multi inputs industrial systems are often approached using artificial intelligence (AI) methods[1,3,5,6]. The advantages of using AI methods (ANN, fuzzy logic, neuro-fuzzy, genetic algorithms) approach to steam leak detection can be named as[1]: new devices or signals (besides DCS) aren't necessary, expected earlier leak detection (vs. steam/water balance method) because of using many measured signals and no apparent interdependencies, expected solutions portability between like plants.

2. Humidity of flue gas model

The previous research [1] involving sensitivity analysis of process variables vs. steam leak specified basic signals for leak detection while leak was modeled. The possibilities of reduction of inputs variables set [4] were studied for humidity ANN model using PCA method and cross correlation functions without expected results. Only experts opinion and sensitivity analysis of trained model were successful in this matter.
The ANN models of flue gas humidity, H_2O[%], were build in 3 structures: linear nets (LIN, without hidden layer), *Radial Basis Function* (RBF) and feedforward *Multilayer Perceptron* (MLP). RBF nets are less useful for humidity modeling [4]. Only LIN and MLP nets were considered later.

2.1 Data series compounding

For industrial application of ANN models (to work with long time horizon) we should train models with data compounded from long period of time and next decimated[2]. Because of many missing and faults records in process databases[4], the training set was compounding with 9 periods from different months of year 2005, each of them involved about 4 thousands of correct data records for nominal state of plant job. The learning, testing and validation subsets were distinguished. Reconstruction, validation of missing and fault values of measured data is necessary stage in off-line and special in on-line mode of models application [2].
There was 34000 samples summary in training set. Due to non equal delays between variables in each month the averaged values of delays were applied i.e. columns were shifted vs. each other in data sets[2,6]. Model dynamic is obtain by external buffers. The investigated ANN was static[4].

2.2 Model MLP in a structure 15:15-16-1:1

Two models are built and tested [4]: linear LIN_08 and MLP_19. They were trained with backpropagation method and then learned once more with quasi-Newton algorithm (6000 epochs). Results are given in Tab.1.

Model	Inputs	Hidden	Err. learn.	Err. val.	Err. tst	Qual. learn.	Qual. val.	Qual. tst
LIN_04	17	-	0,263	0,265	0,267	0,188	0,190	0,189
LIN_08	17	-	0,380	0,376	0,374	0,247	0,243	0,245
MLP_18	19	16	0,184	0,188	0,191	0,132	0,135	0,136
MLP_19	19	16	0,167	0,169	0,167	0,109	0,109	0,110

Tab.1. Results of the best ANN built at 1 month (LIN_04, MLP_18) and 1 year data (LIN_08, MLP_19) for ~ 34000 records-cases.

Linear net quality decreased vs. LIN_04, the best one trained at 1 month data, about 28%. It's due to over complex dependencies between inputs and output in modeled humidity and more information (more variation) on 1 year data. Quality of new MLP_19 model is better vs. previous MLP_18 more then 19%. It's resulted from more training cases, which represented nominal job conditions involving greater variation of data.
The task of detection steam leak faults was tested later with ANN model in MLP structure only. Sensitivity analysis was made for the best MLP model [4,5] and four least ranged variables were omitted. The net was trained again with quasi-Newton method by 2000 epoch [4]. Quality of this model (ratio of RMS error to standard deviation of pattern) was kept. Next the

number of neurons in hidden layer was chosen from number under Kolmogorow rule equal 39 changing to 10 neurons. The best quality (ca.0,110) was achieved at 16 neurons in hidden layer-model MLP_20 (15:15-16-1:1). Averaged quality values for unknown (not used during training) data from 12 months of 2005 year hesitate between 0,11 (April) to 0,23 (December). The quality of resulted models is sufficient good for leak detection task. The example of 1 day of January 2006 (quality 0,17) is presented at Fig.1.

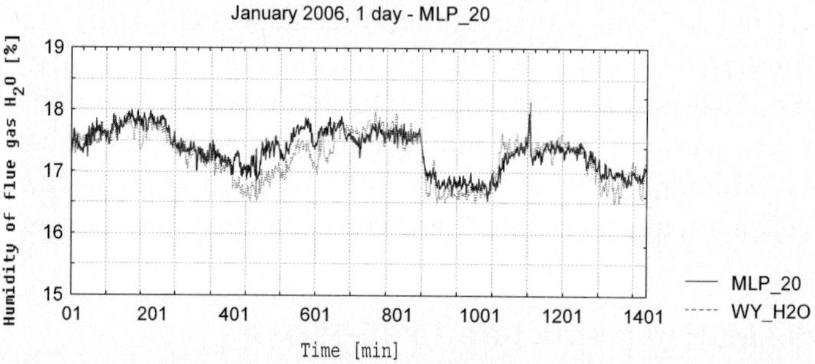

Fig 1. Measured and modeled MLP_20 trajectory of flue gas humidity H_2O[%].

The dependence of model output vs. pattern is shown at Fig. 2. Only few points are visible as distinct from straight line with coefficient close to 1.

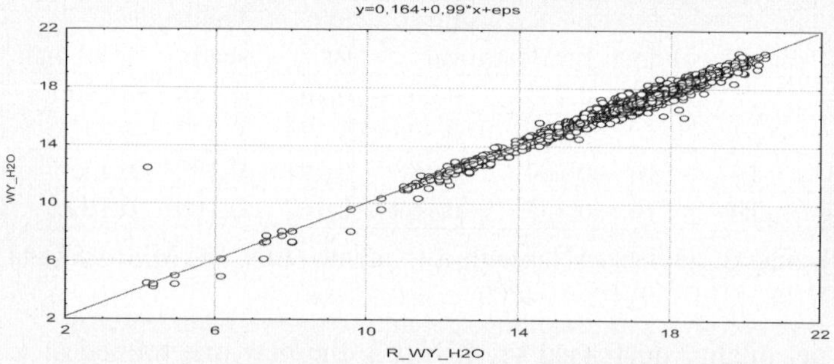

Fig. 2. Output of MLP_20 model vs. measured humidity R_WY_H_2O.

The 3 first variable in sensitivity ranking are: flow of raw steam, O_2 concentration in flue gas, power of boiler.

3. Residuum of humidity for 3 registered faults cases.

The ANN model MLP_20 was run for input data to be 3 days ahead of steam leak moments. An example is presented at Fig.3. The moment of residuum appearance is indicated by dotted line. The humidity residuum (H_2O concentration >2%, i.e. |residuum|/avg. H_2O concentration >0,1) were detected in 3 studied cases with 120 to 220 minutes prediction vs. shut-down moments.

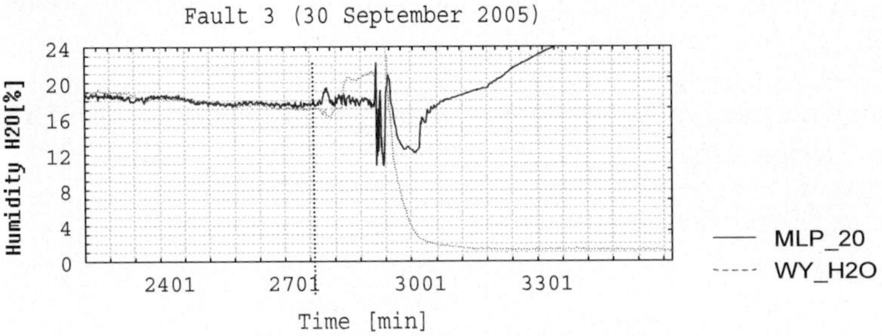

Fig 3. Trajectory of modeled (MLP_20) and measured (WY_H2O) humidity -H_2O [%] for steam leak fault at 30 Sept. 2005. Indices at time axis every 60 minutes.

4. Final remarks

Due to averaging and generalization properties of ANN external process disturbances, like variation of fuel contents (hydrogen and water) etc., were sufficient well represented in model. The tested ANN model gave promising results in early detection of tube boiler faults, but very limited number of faults cases was in disposal. Unfortunately, only erosion faults cases were recorded and available for testing.

References

[1] A. T. Alouani, P. Shih -Yung Chang "Artificial Neural Network and Fuzzy Logic Based Boiler Tube Leak Detection Systems" USA Patent No: 6,192,352 B1, Feb 20, 2001.

[2] S. Kornacki, „Neuronowe modele procesów o zmiennych właściwościach". VII KK. Diag. Proc. Przem.,Rajgród 12-14.09.2005. PAK 9/2005.

[3] J.M. Kościelny, „Diagnostyka zautomatyzowanych procesów przemysłowych" Akadem. Oficyna Wyd. EXIT, Warszawa, 2001.

[4] K. Olwert, „Opracowanie i analiza modelu wilgotności spalin w zadaniu wczesnej detekcji nieszczelności parowej kotła bloku energetycznego" praca dyplomowa PW D-IAR -306, 2006, praca niepublikowana.

[5] R. J Patton, C.J Lopez-Toribio, F.J Uppa "Artificial Intelligence Approaches to Fault Diagnosis" Int. Jour. of Applied Mathematics and Computer Science. Vol.9No 3. 471-518 (1999).

Behavior-based control system of a mobile robot for the visual inspection of ventilation ducts[1]

W. Panfil, P. Przystałka, M. Adamczyk

Silesian University of Technology, Department of Fundamentals of Machinery Design, 18A Konarskiego Str., 44-100 Gliwice, Poland

Abstract

The paper deals with the implementation of a behavior-based control and learning controller for autonomous inspection robots. The presented control architecture is designed to be used in the mobile robot (Amigo) for the visual inspection of ventilation systems. The main aim of the authors' study is to propose a behavior-based controller with neural network-based coordination methods. Preliminary results are promising for further development of the proposed solution. The method has several advantages when compared with other competitive and/or co-operative approaches due to its robustness and modularity.

1. Introduction

In this paper, the authors are particularly interested in ventilation ducts inspection using a mobile robot to assist in the detection of faults (mainly dust pollutions). The visual inspection of ventilation ducts is currently performed manually by human operators watching real-time video footage for hours and finally, it is a very boring, tiring and repeatable duty. Human operators would benefit enormously from the support of an inspection robot able to advise just in the unusual condition. On the other hand, a robot might act autonomously in (un)known environments gathering essential data. There are many problems to deal with this type of fault inspection task which are referred in corresponding papers [1, 2, 7]. This work focuses on such key aspects as the high-level behavior-based controller of

[1] The research presented in the paper has been supported by the Ministry of Science and Higher Education and carried out within the Multi-Year Programme "Development of innovativeness systems of manufacturing and maintenance 2004-2008"

the inspection robot (so-called co-operative/competitive, or "Brooksian" architectures [4, 5]) and neural network-based coordination methods. Behavior-based controllers give advantageous features such as reliable and robust operation in (un)known environments. The main problem in such control systems is the behavior selection problem. A comprehensive survey of the state of the art in behavior selection strategies may be found in [9]. The authors categorized different proposals for behavior selection mechanisms in a systematic way. They discuss various properties of cooperative or competitive, implicit or explicit and adaptive or non-adaptive approaches. In this project, the authors basing on information in the literature decided to use a behavior-based control schema which is very similar to these presented in [6, 8] but some further modifications are introduced.

2. First steps into development of behavior-based control system

This section describes the main idea of a behavior-based schema, simulation framework and kinematical rules applied for determining the movement of the robot.

2.1 Behavior-based schema

The main idea of the considered control architecture is as follows. There are three layers: the low-level layer, the behavior-based layer and the deliberative layer. The behavior-based layer consists of a set of behaviors and a coordinator. There are two methods given based on competitive and feed-forward neural networks. The first one is used for selecting behaviors (a), whereas the second one is used to learn behavior state-action mapping (b). In this way two independent modes are available: competitive and co-operative mode.

2.2. Simulation framework

In the second step of the research the authors propose a simulation framework for developing the behavior-based controller of the inspection robot. It allows obtaining simulation results being the base for further development. The proposed software consists of the MATLAB/SIMULINK with Stateflow toolbox and MSC.visualNastran 4D simulation environment.

The SIMULINK part of the framework consists of Stateflow charts, each concerning one of the following behaviors: *turn left, turn right, obstacle avoidance, right wall following* and so on. Every chart (behavior) obtains the following types of information. The first is a distance between the robot and obstacles coming from eleven infrared sensors. The second one is the information from three axes of the accelerometer about the orientation of the robot. There is also available data (a few features of the image computed by the recognizing subsystem) from a camera mounted on the front of the robot. Furthermore, the charts are provided with information about the current intensity of the motors, level of the battery charge and, the most important, the planned tasks.
Basing on the mentioned information each chart generates information about the linear velocity of point S and the turning radius R for its behavior. This information serves as input for a kinematical model of the robot. It enables to compute a turning angle and angular velocity of every wheel of the robot.

2.3. Kinematics of the robot

Below presented is a conception which allows determining the kinematical rules for the movement of the robot. Basing on the geometry of the robot there was assumed that the robot consists of the main body of 200x200 [mm] (*LxB*) dimensions and four driving units. Each driving unit is an assembly of a motor and driving wheel of radius R_w=30 [mm]. Of course, it is a quite big simplification – main body and driving units consist of many other parts.
Each driving unit has two degrees of freedom. The motor drives the driving wheel. Furthermore, every driving unit can rotate round axes (passing through points A,B,C,D) perpendicular to axes of the wheels. The wheel base is l=160 [mm] and wheel track is b=160 [mm].
The base for the next considerations is an assumption that the robot is controlled by two parameters: the linear velocity of the point *S* and the turning radius *R*.
It was proposed that the robot moves in the following manner. When it goes straight every motor rotates with the same speed but with inverse direction with respect to the side it is mounted on. Rotary planes of the wheels have to be parallel to each other. Turning radius goes to infinity. When the robot turns on the radius *R* the rotation axis of every wheel is coincident in O point which is the instantaneous turning point of the moving robot. Rotating speed of every point of the robot is ω equal to

$$\omega = \frac{V_s}{R} = \frac{V_{out}}{R_{out}} = \frac{V_{in}}{R_{in}}.$$

The angle between the rotary plane of the outer (inner) wheel and instantaneous moving direction equals to

$$\beta_{out(in)} = arctg\left(\frac{l/2}{R \pm b/2}\right).$$

The instantaneous radius of the circle covered by points A and D (B and C) is

$$R_{out(in)} = \sqrt{\left(\frac{l}{2}\right)^2 + \left(R \pm \frac{b}{2}\right)^2}.$$

When the robot turns on the radius R with the linear speed V_s of the point S (centre of robot area) then the linear speed V_{in} of points B and C and linear speed V_{out} of points A and D equal to

$$V_{in} = V_s \frac{R_{in}}{R} \quad \text{and} \quad V_{out} = V_s \frac{R_{out}}{R}.$$

Taking into account considerations presented above speed of rotation of the inner wheel ω_{iw} and the outer wheel ω_{ow} can be expressed as follows:

$$\omega_{iw} = \frac{V_{in}}{R_w} \quad \text{and} \quad \omega_{ow} = \frac{V_{out}}{R_w}.$$

4. Conclusions and future work

For this time the framework allows manually controlling (using joystick, game pad) the virtual robot. Thanks to the MSC.visualNastran 4D software, there can be obtained the information about the kinematics (also dynamics) of the robot – positions, orientations, linear/angular velocities/accelerations of any part of the robot or any point placed on it. The second advantage of this approach is that the behavior of the robot can be assessed visually.

The main disadvantage of the proposed solution is a time-consuming operation. It results from the complexity of computing the contact joints between the wheels of the robot and the ducts.

Since Stateflow enables to model and simulate event-driven systems and also to generate C code implementation in the future the authors are going to further develop the behavior-based control system of the mobile robot. This research will start with the simulation of simple behaviors, e.g. *turning left* where the robot moves to the corner, stops, turns the driving units, moves on the assumed radius to the assumed point, turns back the driving units and goes straight. The others simple behaviors will be trained and then joined together. When the simulation results are promising the code will be implemented into the control system of the real robot.

References

[1] Adamczyk M.: "Mechanical carrier of a mobile robot for inspecting ventilation ducts" In the current proceedings of the 7th International Conference "MECHATRONICS 2007".
[2] Adamczyk M., Bzymek A., Przystał ka P, Timofiejczuk A.: "Environment detection and recognition system of a mobile robot for inspecting ventilation ducts." In the current proceedings of the 7th International Conference "MECHATRONICS 2007".
[3] A. D'Amico, Ippoliti G., Longhi S.: "A Multiple Models Approach for Adaptation and Learning in Mobile Robots Control" Journal of Intelligent and Robotic Systems, Vol. 47, pp. 3 – 31, (September 2006).
[4] Arkin, R. C. 1989 Neuroscience in motion: the application of schema theory to mobile robotics. In Visuomotor coordination: amphibians, comparisons, models and robots (ed. J.-P. Ewert & M. A. Arbib), pp. 649-671. New York: Plenum.
[5] Brooks, R. A., "A Robust Layered Control System for a Mobile Robot." IEEE Journal of Robotics and Automation, Vol. RA-2, No. 1, 1986, pp. 14-23.
[6] Carreras M., Yuh J., Batlle J., Pere Ridao: "A behavior-based scheme using reinforcement learning for autonomous underwater vehicles." Oceanic Engineering, IEEE Journal, April 2005, Vol. 30, pp. 416- 427.
[7] Moczulski W., Adamczyk M., Przystałka P., Timofiejczuk A.: „Mobile robot for inspecting ventilation ducts" In the current proceedings of the 7th International Conference "MECHATRONICS 2007".
[8] Rusu P., Petriu E.M., Whalen T.E., Cornell A.: Spoelder, H.J.W.: "Behavior-based neuro-fuzzy controller for mobile robot navigation" Instrumentation and Measurement, IEEE Transactions on Vol. 52, Aug. 2003 pp.1335--1340.
[9] Scheutz M., Andronache V.: "Architectural mechanisms for dynamic changes of behavior selection strategies in behavior-based systems" Systems, Man and Cybernetics, Part B, Dec. 2004, Vol. 34, pp. 2377- 2395.

Simulation and Realization of Combined Snake Robot

V. Racek, J. Sitar, D. Maga

(a) Alexander Dubcek University in Trencin, Studentska 2,
Trencin, 911 50, Slovakia

Abstract

The paper is deals with verification of mechanical construction design by simulation of combined snake robot. This robot can be used for various applications. Universality of the solution is assigned by special construction of snake robot. This construction is consisting of independent segments design. Each of designed segments can realize not only linear movement but curving movements too. Verification of designed structure is realized in program Matlab/Simulink. Obtained results are presented in video and picture format. Designed and simulated model can be realized from lightweight materials mainly from duralumin, bronze and from nylon.

1. Modeling and simulation of snaking system

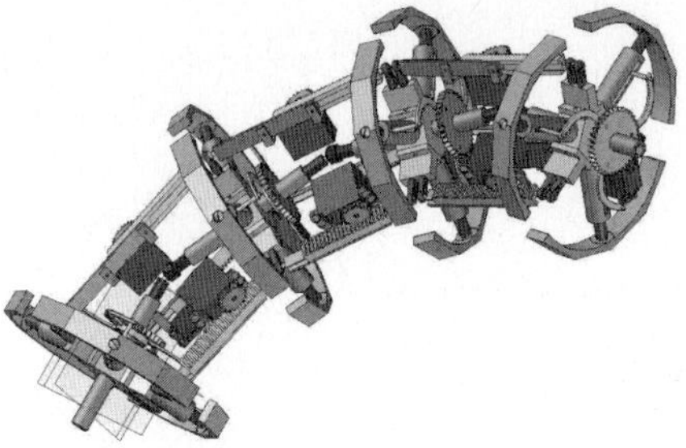

Fig. 1: Model of combined snake robot construction. Model is consisting of four independent segments.

Mathematical model of snake robot is realized in Matlab/Simulink program and is based on designed construction (Fig. 1). Complete model is consisting of subsystems. These subsystems are described all prismatic and rotary bonds, movement definitions for different environments types and different controls system. As is mentioned before all robot movements are based on prismatic and rotary bonds. These bonds are arranged in lines as is shown in Fig. 2. All this lines models are connected to the central hexagonal part. In the final solution only two basic type of arm mechanisms are used (Fig. 2). First one type is substitution of cogged dovetail guide way. Each end is finished with rotary joints. Rotation angle of these joints is 25°. Model is consisting of two rotating and one prismatic bond as is shown in Fig. 2a). Movement and angle displacement is defined by drive control (joint sensor and joint actuator). Joint sensor is used for measurement of actual bond position and the joint actuator is used for bonds movement control. Rotary bonds are substituted by universal bonds. With universal bonds is possible create revolution in three axis of Cartesian coordination system. Second mechanism type is substitution for central connection part. This part is used to stabilization of mutual position between two independent segments. This model part is without drive unites (joint actuators) and is consisting of two simple prismatic bonds which are connected by rotary bond (universal rotary bond). In Fig. 2 b) the internal structure of central connection part with kinematics block diagram is presented.

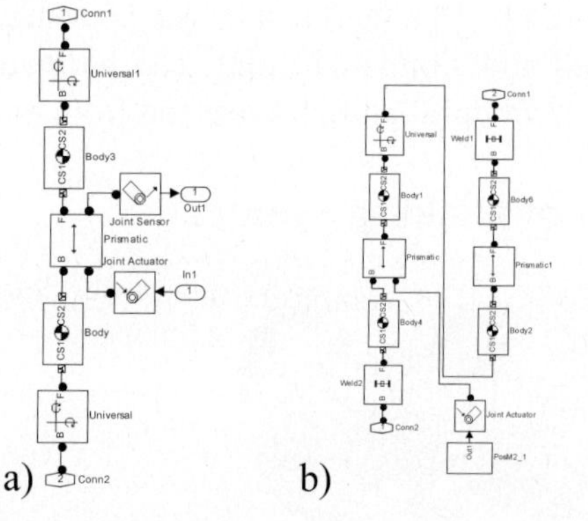

Fig. 2: Internal structure of individual mechanisms and arms of snake robot system (prismatic and rotary bonds): a) cogged dovetail guide way structure b) structure of central connection part.

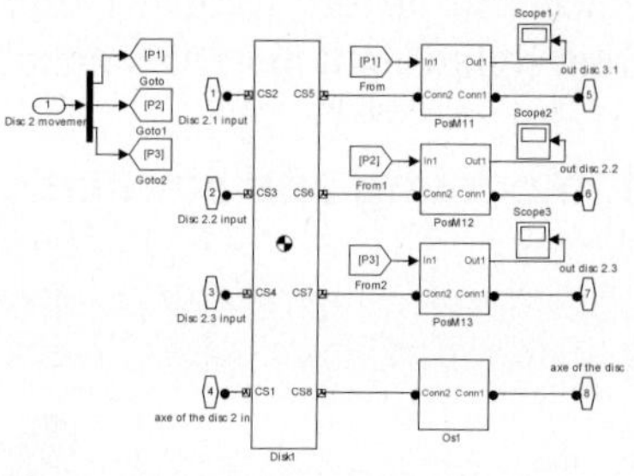

Fig. 3: Model of independent snake robot segment with position control system together with his kinematics block diagram.

From both of these interconnections are created simple subsystems PosM and Os. In subsystem PosM are described all connections and bonds in cogged dovetail guide way. In Os subsystem is defined structure of central connection part. After connection of three PosM subsystems and one Os subsystem to the one central item described in Body block with name Disk1 the model of one independent snake robot segment can be created. The output block diagram is presented in Fig. 3 together with control system and alternative kinematics block diagram. Final mathematic model of combined snake robot is realized by four independent robot segments. Segments are connected together as is shown in Fig. 3. Internal structure connection of snake robot mechanism is shown in Fig. 4. Complete movement is realized in block machine environment and is set into the kinematics calculation. Snake robot movement is possible thanks to the prismatic and weld bonds which are connected with machine environment. With assistance of these bonds is possible realized rotational and translational movement in Cartesian coordinate system.

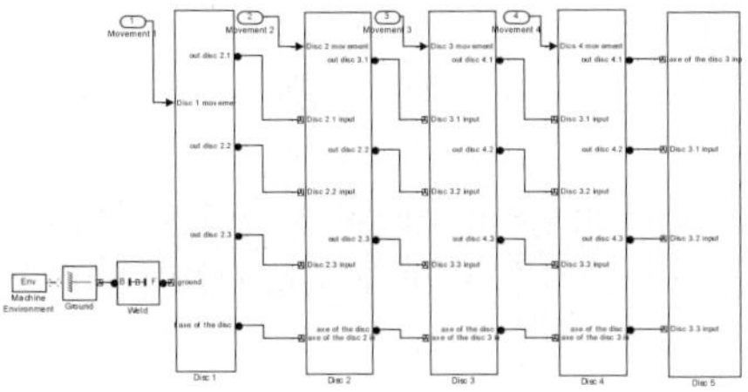

Fig. 4: Model of internal structure of combined snake robot assembly.

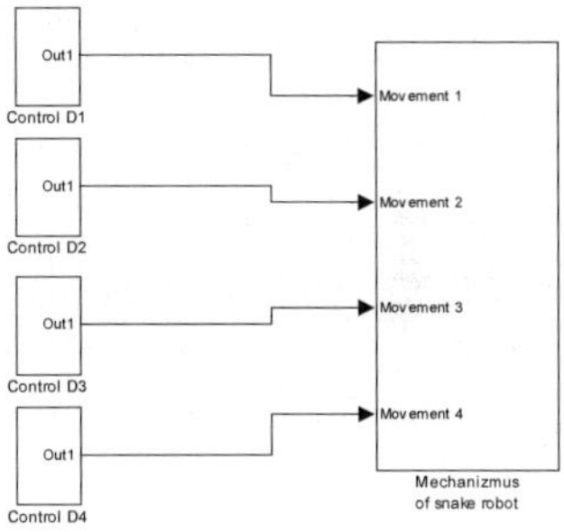

Fig. 5: Complete model of combined snake robot with control system.

In Fig. 5 the final model of snake robot mechanism is shown and is consist of presented subsystems. Control system for complete set is realized by generating of input control signals.

2. Simulation results

These signals are generated in blocks Control D1-D3. In control block Control D1 the caterpillar movement is generated. Side waving movement is generated by control block Control D2 and worm's movement and harmonic movement is generated by block Control D3. Simulation results are presented in Fig. 6 and Fig. 7. Fig. 6 is representing the movement from initial position with minimal length of snake robot. During this time all prismatic bonds are bring together on minimum. Contrary to this in Fig. 6 is presented maximal length of snake robot. In this case all prismatic bonds are protuberant to the maximum possible expanse state.

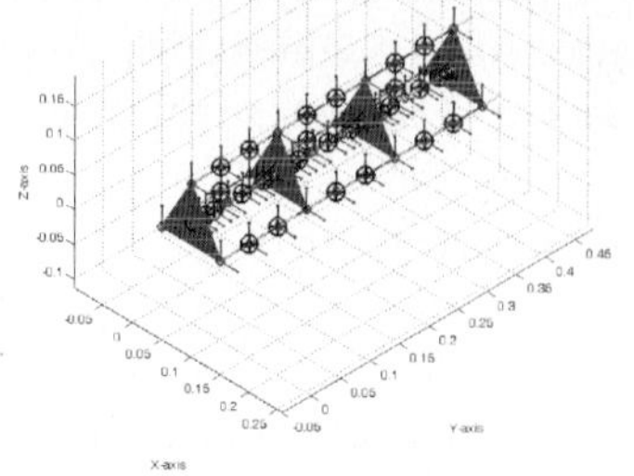

Fig. 6: Simulation of snake robot activity (caterpillar movement), primary position – minimal length of snake robot is turned into the final position – maximal length of snake robot.

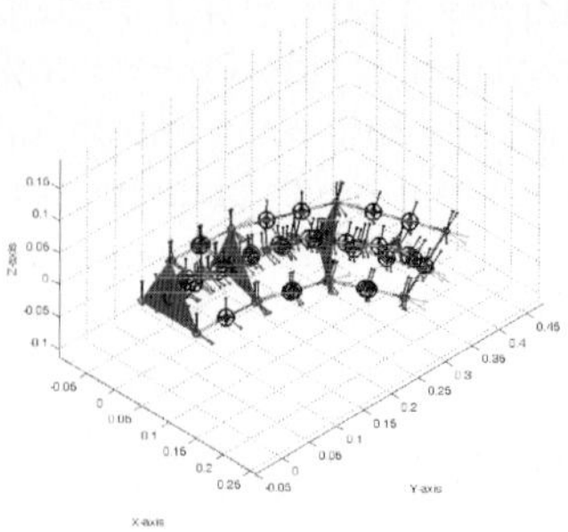

Fig. 7: Simulation of four segment snake robot (orientation angle between two segments is maximally 30°).

Changes in angular position between snake robot segments can be seen in Fig. 7 and is realized by motion control of individual prismatic bonds. In reality these prismatic bounds are created from cogged dovetail guide ways and servomotors with cogwheel. Gear drive in servomotor is equilibrating actual prismatic position.

Realization of snake robot

For verification process two independent segments are created (Fig. 8). Connection between segments is realized by guideways with servomotor (distance changes) and ball joints (rotation in all directions).

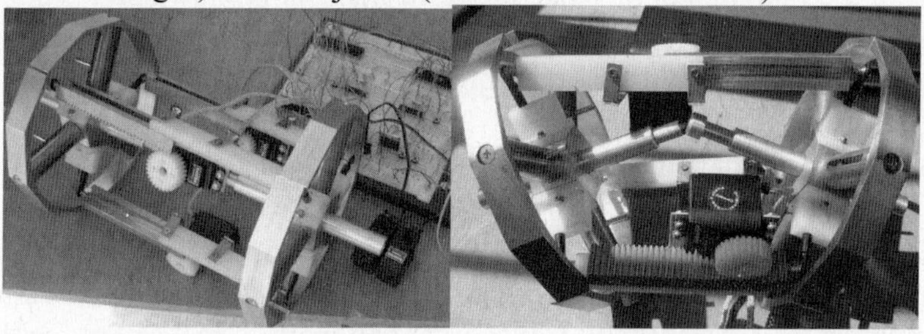

Fig. 8: The experimental set of snake robot (set is consist of two segments)

The maximal possible angle between these two segments is 35° and is limited with central connection joint. Distance between segments is from 12cm to 20cm. Verified model have instabilities in ball joints (rotation). For this reason the ball joints are displaced by cardan universal joints.

Fig. 9: Design of cardan universal joint without axis rotation.

Conclusion

The paper is focused on construction design verification, basic motion type simulation of combined snake robot. Model is simulated by Matlab/Simulink program for several types of movement (caterpillar movement, side waving, worm's movement and harmonic movement). These movements' types pertain to the different robot activities. These combined snake robots can be use for many applications as inspection and service activities of unavailable equipment, for pipes inspection, for explore of underground and thin passages.

Acknowledgement

Combined snake robot is the result from support of Research Grant Agency VEGA, project number: 1/3144/06: Research of Intelligent Mechatronics Motion Systems Properties with Personal Focus on Mobile Robotic Systems Including Walking Robots.

References

[1] Matlab, Simulink - Simulink Modeling Tutorial - Train System

[2] K. Williams, „Amphibionics Build Your Own Biologically Inspired Reptilian Robot“

[3] Copyright © 2003 by the McGraw-Hill Companies, Inc. 0-07-142921-2

[3] L. Karnik, R. Knoflicek, J. Novak Marcincin, „Mobilni roboty“, Marfy Slezsko 2000

[4] http://www.fzi.de/divisions/ipt/WMC/walking_machines_katalog/walking_machines_katalog.html

Design of Combined Snake Robot

V. Racek, J. Sitar, D. Maga

(a) Alexander Dubcek University in Trencin, Studentska 2,
Trencin, 911 50, Slovakia

Abstract

The paper is deals with mechanical construction design and simulation of designed structure of combined snake robot. This robot can be used for various applications. Universality of the solution is assigned by special construction of snake robot. This construction is consisting of independent segments design. Each of designed segments can realize not only linear movement but curving movements too. Obtained results are presented in video and picture format. Designed and simulated model can be realized from lightweight materials mainly from duralumin, bronze and from nylon.

1. Introduction

Basis inspiration for construction of snake robots are life forms – snakes, who populating in large territory on Earth. To move used variously methods of movement that are depended from medium in which are (sand, water, rigid surface et al.). They can move in slick surface or slippery surface, climb on barrier and so negotiate it. Snake robots architecture in conjunction with large numbers degree of freedom makes is possible to three-dimensional motion. Snake robots are defined slender elongated structures that consist of in the same types of segment that are together coupled. The mode of moves flowing from two basic motion models of the animals – snake and earthworm. The bodies of these animals are possible think it an open kinematics chain with a large number of segments that are coupled by joints. It is making possible between this segments actual rotation around two at each other vertical axis. The advantage of this design is high ability at copied broken terrain. The snake robots are used in compliance with choices construction and movement in terrain with large surfaces bumps, different types of surfaces etc. The main disadvantage is low speed and energy title that directly relate with type and number of used engines. The snake robots with large number of segments are used for inspection and service activity within hardly accessible conveniences, pipes in underground and narrow spaces. At the present time is began implement also in fire department and automobile industry. Additional zone usable snake robots are by motion in a very broken terrain that is unsuitable for wheeled or walking robots. In this case is construction of the snake robots it features small number of segments with rigid structure at which is able to outmatch barriers that are superior to half-length.

TABLE I: Advantages of the snake robots

Mobility in terrain	Makes it possible to movement through rough, soft or viscous terrain, climb to barrier
Tractive force	Reptiles can used all the long of body
Dimension	Low diameter hull
Multiplicity	The snake robots consist of number of similar parts. Defection some of mechanism part can be compensated all the others.

TABLE II: Disadvantages of the snake robots

Actual load	Complicated transportation of materials
Degrees of freedom	A large number of driving mechanisms is needed. Problem with movement control.
Thermal control	Complicated measurement of the heat in internal parts of robots.
Speed	The snake robots are much slower than natural rivals and wheeled robots.

2. Basic movement possibilities of snake

As previous was say the snake robots may move in multiple environs. Then at design is strong to analyses environs and method move of the snake robots. In our design we try to combine multiple types of movement and so achieved more universality and taking advantage of the snake robots. Types of elementary motion are:

- Serpentine motion
- Concertina motion
- Side winding motion
- Slide shifting motion
- Caterpillar rather motion

- Worm motion

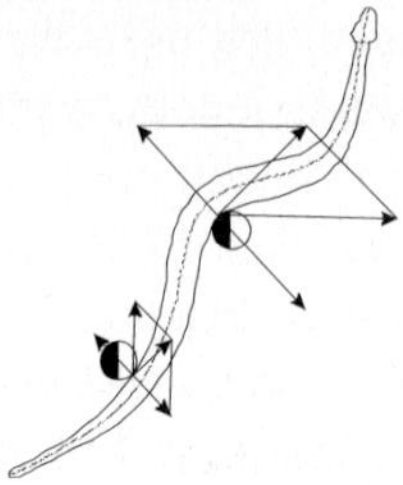

Fig. 1: Serpentine motion.

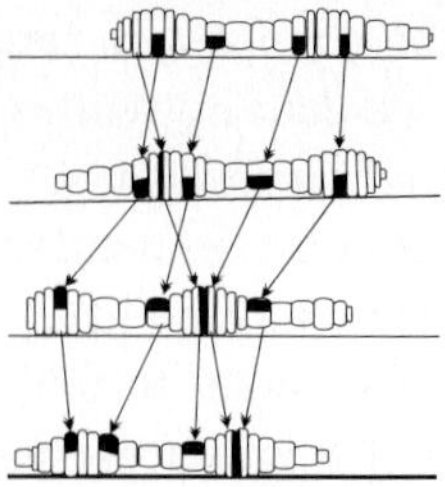

Fig. 2: Worm motion

In an analysis we are focus on creating combined type the snake robot that was demonstrated no fewer than three types of motion (worm motion, caterpillar and serpentine). In snake robots are not realized on basis of wheels but as by shrinking and tension of individual parts, rolling, wave motion etc. These forms of motion are request used special type's drives. In most of examples are use dc electric motors or servomotors with low energy severity. The exiguousness drive units allow also reduction general robot dimensions. In the design is important make provision for also friction between surface and external robot cover. In some types are requires that the using special structure of surface which emulates function real snakes skin (e.g. bigger friction in reverse motion and less in direct motion). The external cover has to conformation motion robot body and is flexible or not allowed to leak water yet.

3. Design structure of the snake robot

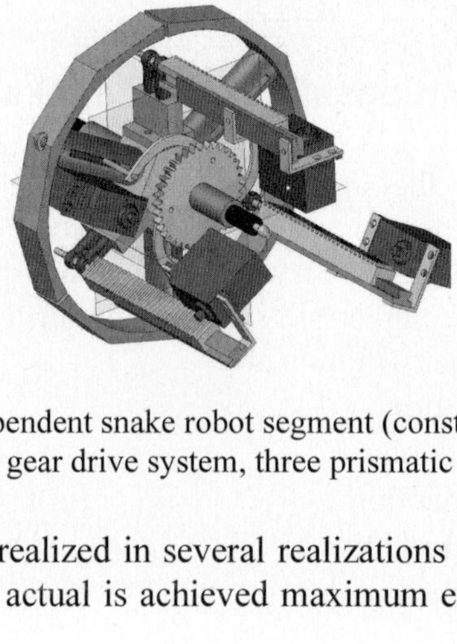

Fig. 3: Model of independent snake robot segment (construction with four servomotors, gear drive system, three prismatic systems).

The snake robots are realized in several realizations so that is minimizing number of drives and actual is achieved maximum effectiveness, moment and force. The important parameter in design structure is its locomotion. At the present time is beginning to use special modified joins so called gearless design, angular bevel, double angular bevel and orientation preserving bevel.

In our design try combine several types of snake motion (caterpillar, serpentine, worm motion and concertina), that is requested great requirement on degree freedom. It is demonstrative mainly on number of drive units in final realization of the snake robot. In Fig. 3 is presented model for one element of the snake robot. Consist of the master hexagons on which are attachments all the moving parts (movable and rotary joins). Cross connection between parts is created by toothed dovetail groove. On of both backends are rotary joints modified in the master hexagon.

The motion is realized by four DC servomotors with performance 35-45 Ncm. Three servomotors are used for drive toothed dovetail grooves. In this manner achieve disengagement and swiveling part of the snake robot. Total length of the dovetail groove is 110 mm and its maximum extension is 190 mm. That means single part is possible extension about 80 mm that present 88 % lengths of one parts. As was firstly mentioned is possible realized not only protraction but twirling of the individual segments towards themselves. Movement realization is based on optimal control algorithms selection. This algorithm is dependent mainly on surrounding environment, obstructions and on selected movement type. Maximal rotation angle of one segment is from 25° to 35°.

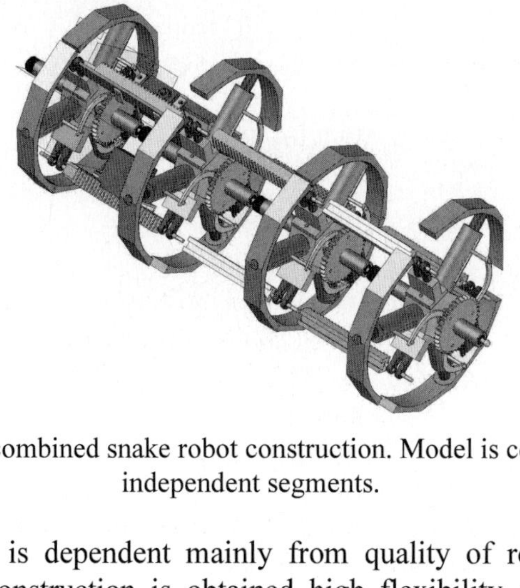

Fig. 4: Model of combined snake robot construction. Model is consisting of four independent segments.

Rotational angle is dependent mainly from quality of rotational joints. With designed construction is obtained high flexibility of snake robot. Fourth servomotor is used to increasing and decreasing of segment dimension. With gear drive assistance the moment from servomotor is delegated on prismatic bonds. These prismatic bonds are located in three arms connected to the central hexagon. Servomotor is located in position where don't hobble the next three servomotors in their work. Diameter of one segment is 160mm and can be resized to 220mm. Construction is created from lightweight materials as nylon, duralumin, bronze and aluminum. With this materials can be reached the lover weight of snake robot. This is proving on power and dimensions of used servomotors.

Conclusion

The paper is focused on construction design and basic motion of combined snake robot. Basic construction of snake robot is designed for various environments which are presented by various robot movements. Model is realized in construction program for several types of movement (caterpillar movement, side waving, worm's movement and harmonic movement). These movements' types pertain to the different robot activities. These combined snake robots can be use for many applications as inspection and service activities of unavailable equipment, for pipes inspection, for explore of underground and thin passages.

Acknowledgement

Combined snake robot is the result from support of Research Grant Agency VEGA, project number: 1/3144/06: Research of Intelligent Mechatronics Motion Systems Properties with Personal Focus on Mobile Robotic Systems Including Walking Robots.

References

[1] Matlab, Simulink - Simulink Modeling Tutorial - Train System

[2] K. Williams, „Amphibionics Build Your Own Biologically Inspired Reptilian Robot" Copyright © 2003 by the McGraw-Hill Companies, Inc. 0-07-142921-2

[3] L. Karnik, R. Knoflicek, J. Novak Marcincin, „Mobilni roboty", Marfy Slezsko 2000

[4] http://www.fzi.de/divisions/ipt/WMC/walking_machines_katalog/walking_machines_katalog.html

Design of small-outline robot - simulator of gait of an amphibian

M. Bodnicki (a) *, M. Sęklewski (b)

(a) Institute of Micromechanics and Photonics, Warsaw University of Technology, 8 Św. A. Boboli Str. 02-525 Warsaw, Poland

(a) Graduate of Institute of Micromechanics and Photonics, Warsaw University of Technology, 8 Św. A. Boboli Str. 02-525 Warsaw, Poland

Abstract

A subject of presented work ware design and construction of a prototype of robot, which can move, like an amphibian, for example salamander. A lot of issues were considered in this work connected to quadrupeds, particularly amphibians. Robot, which has been built, generates both types of gait: walking gait and swimming gait. Modular structure is one of its features, and constructed modules are fully interchangeable. This feature makes that construction easy to reconstruction and makes it multitasking. Considered device is a great base for a further development of animals-like robots and is a very valuable tool for didactic purposes. It is a typical example of mechatronic device structure, which combines mechanical and electronic parts with software.

1. Introduction – four legs microrobots inspired by biology

A walk or a swim are typical form of the number of animals. Some of them connect both forms of the movement. A movement of them is studying by biologists, biomechanics and now – specialists in robotics. There is popular tendency in microrobotics to design objects inspired by mechanical solutions of the Nature. There are analyzed a walk structures [1] as well as maintained amphibious ones [2,3,4]. The directly inspiration for authors were works of Ijspeert at team [3,4], especially presented algorithms of the motion and analysis of the walk/swim phases.

2. General characteristics of the robot build in IMiF

The works realized in Institute of Micromechanics and Photonics, Warsaw University of Technology had following stages: design of electromechanical components, adaptation of control algorithms and implementation of them on PC and test of work of prototype. The modular structure of the robot was assumed (its block diagram is presented on Fig 1). There are used two kinds of modules:

- A – type – basic element of the body, with characteristic details: symmetric design, coupling element, rotary servodrive (with gear) – for realisation rotating movement between module and the following one.
- B – type – the leg module built on A – type and equipped with additional two leg units; each leg unit is driven by next two servodrives with gears.

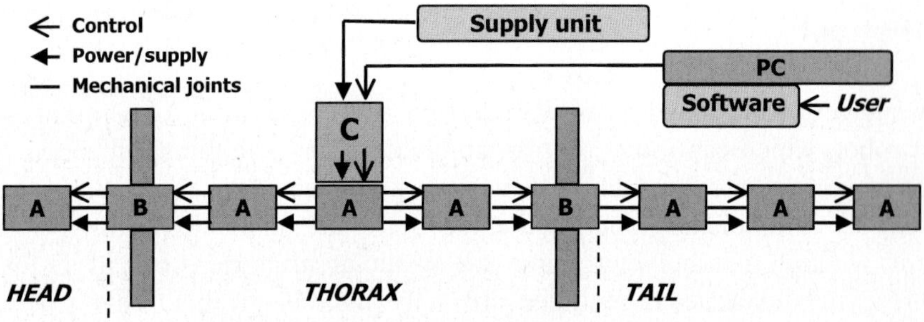

Fig. 1. The scheme of the lizard-robot
A – basic module of a body (head, thorax, tail), B – legs module; C – controller

The fundamental stage of the design process was analysis of the kinematics of the legs and – in effect – assumption of the structure of the B module first, and then – a control algorithm.
As the actuators systems of 18 servodrives (hobby-type) with SK18 controller are used. Transmission from PC "master unit" is realised via RS232 in typical transmission protocol. Servodrives are control by PWM signals. The structure of the modules possible the battery supply, but prototype is supplied from outside source (6V).

3. Analysis of a joint structure of legs.

The possibility and quality (realism) of the walk depends on the degrees of freedom in joints of the legs. The structure apply in the robot consist from two joints and two-segment legs (a tight and a shinbone, without a feet). A scheme of the leg is shown on Fig. 2 and the general view on robot – on Fig. 3.

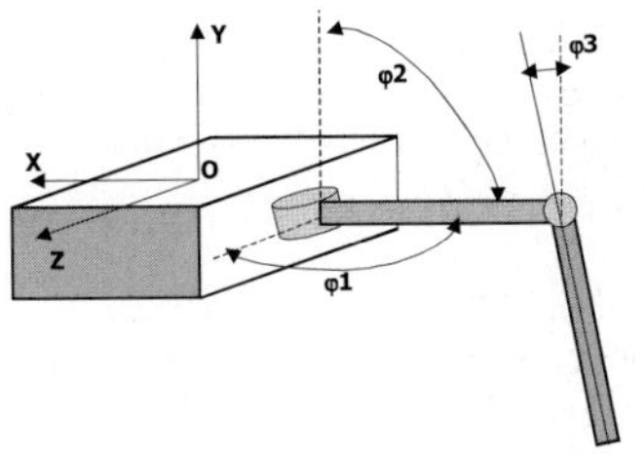

Fig. 2. Scheme of the leg structure (description in the text of chapter)

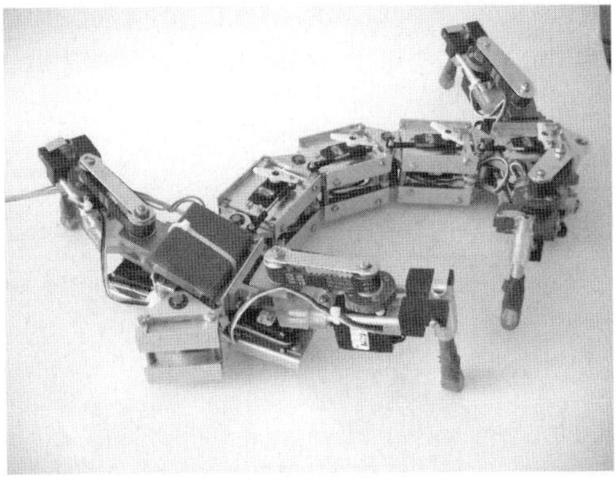

Fig. 3. General view on lizard-microrobot built in IMiF PW

A **φ1** angle (in shoulder joint) - this degree of freedom establishes length of the step of quadruped. From point of view of realization of a translation changes of this angle is the most important. The range of the φ1 angle is usually about π, but could measure to 2π. In robots is possible to generate the movement using only this degree of freedom – with legs sliding on a surface (but there is necessary blocking mechanism for support phase), e.g. by change of a friction coefficient according to move direction like "seal skin" for skiing.

A **φ2** angle (in the shoulder joint) - this degree of freedom establishes rise of the leg in carriage phase, and is very important during the movement. The φ2 angle gives the change of leg position in carriage phase – which enables avoiding of obstacles as well as to distinguish legs in support phase. The range of the φ2 angle is usually about 1/4π, but value to 2/3π is better for bigger obstacles.

A **φ3** angle (in the elbow joint) - during the walk of the quadruped this angle makes possible change of the trajectory – movement by the line, curve or sideways (than the length depends on φ3 and φ1 is an equivalent of φ3). The end of the leg can be located in the selected point in the space (according to kinematic of the mechanism). The range of the φ2 angle is usually about 1/4π, and for folding of the legs this angle has to measure to π. It means that the realization both swim /walk phases and realistic turn depends on this angle.

4. Control algorithms and software

The control software is an integrated part of the robot. In presented phase of the works operator of PC microcomputer realizes all control options. The block diagrams of the main algorithms is presented on Fig. 4.

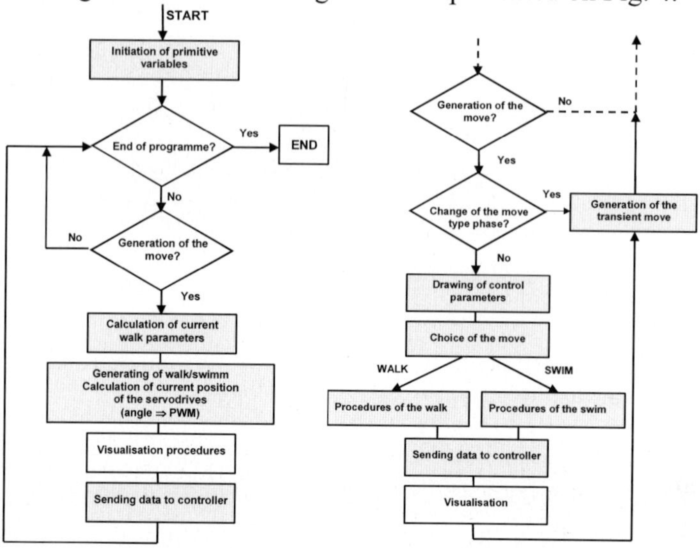

Fig. 4. The scheme of control algorithms
left – main program, right – generation of the move

The main window of the program implemented on PC gives the operator possibility to control all function of the robot. There are nine basic fields on this window (see Fig. 5). The field no 1 has three overlaps for choice of form of the move – "walk", "advanced walk" and "swim". The 2 field makes possible to set fit a length of move cycle and amplitude of a bow of the body (in effect to assume a speed of the move – both walk and swim). The role of the field no 3 is visualization of the robot. There are plan view from above and position of the legs from behind presented. This field is usually also for selftests of the software. The running time of the cycle of the move is shown in window no 4. Option of the control with use of arrows button from the keyboard is switch on/off by field no 5. The control by main window is realized via buttons (fields) no 6. The 7 button initializes the move. For a change of the variant walk/swim. the button 8 is used. Initialization of the button 8 starts the special procedure, which begins change of the legs position after full cycle of the move. The amplitude of the bow of the body is automatically and fluently minimized to zero and than returns to nominal value.

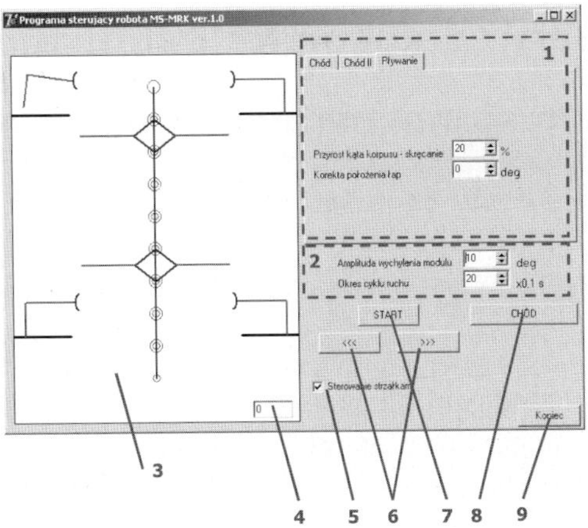

Fig. 5. The main window of the control software (description in the text)

5. Summary

The building and activating of the microrobot was successful. Tests confirm correctness of algorithms and their implementation. Now the next stages are realized with following goals. In mechanical part: reduction of the weight and design of waterproof casing (for tests in water), as well as design of full section-mechanisms for the legs are necessary. There is also plan to build the head module with microcameras. In electronic part an independent control (instead of central control unit) is planned. The improvements in control are going to implementation of the algorithms of dynamic movement, with use of signals from sensors. There is plan to build-in miniature accelerometers into modules of the body and force (pressure on ground) sensors in each leg.

References

[1] T. Zielińska "Maszyny kroczące – Podstawy, projektowanie, sterowanie i wzorce biologiczne" PWN, Warszawa, 2003 (in Polish)

[2] A.J. Ijspeert, A. Crespi, J. Cabelguen, Neuroinformatics, Vol.3 no 3, (2005)

[3] Ijspeert A.J.: "A 3-D Biomechanical Model of the Salamander". Brain Simulation Laboratory & Computational Learning and Motor Control Laboratory, University of Southern California

[4] M. A. Ashley-Ross Miriam, Journal Exp. Biol. v. 193. (1994). pp. 255-283

The necessary condition for information usefulness in signal parameter estimation

Grzegorz Smołalski

Department of Biomedical Engineering and Instrumentation,
Faculty of Fundamental Problems of Technology, Wrocław University of Technology, Wybrzeże Wyspiańskiego 27, 50-370 Wrocław, Poland

Abstract

The entire knowledge available of the investigated signal has been represented as a set of specific constraints imposed in the signal space. The notion of the subsets' cluster was introduced and used for formulating the necessary condition for both direct and indirect usefulness of the given information item in estimating the needed parameter of the signal. Since the checking procedure for the presented necessary condition is quite simple, it seems to be a practical tool for elimination of useless information items.

1. Introduction

A one-dimensional signal is a typical object of measurement and the value of certain parameter of such a signal is a typical measurement purpose. Here, the parameter of interest E is referred to as the estimated parameter and the maximum, acceptable value of this parameter uncertainty, which is usually given or tacitly assumed, will be denoted as ΔE .

The procedure of the parameter estimation is always performed in circumstances of a preliminary knowledge of the investigated signal (see, e.g., [1-6]). This knowledge is usually composed of the set of individual information items. A signal investigation consists in the measurement of the value of certain parameter M or the whole set of them. The estimation of the value of the required parameter E is finally performed thanks to all the acquired knowledge of the signal. A crucial point in the procedure is then the verification of an information item usefulness in the reduction of the estimated parameter uncertainty.

2. Information item as a restriction in the signal space

If an adequate mathematical model of the investigated signal *u(t)* is necessary for the time interval of the finite length only, the generalized Fourier series may be used:

$$u(t) = \sum_{n=1}^{\infty} c_n b_n(t), \tag{1}$$

where $b_n(t)$ denotes the complete set of orthogonal functions, and the coefficients c_n are obtained as inner products of the investigated signal *u(t)* and the consecutive base functions $b_n(t)$. Since in practice any investigation of the signal may be carried out with a limited accuracy only, the above signal model does not have to be exact either, which allows a series (1) truncation to the first *N* terms. This way, the finite-dimensional numerical representation $\{c_n\}_{n=1}^{N}$ is obtained for the signal. Any signal segment investigated in practice may then be mapped into the *N*-dimensional vector space which will be referred to as the signal space.

The entire available knowledge of the investigated signal is usually composed of a number of individual information items:

$$II_1, II_2, ..., II_J \quad . \tag{2}$$

These items may refer not only to various signal properties but also to various signal components originated from various physical phenomena [3,7]. All individual information items are, in turn, connected by the appropriate logical functions which determine the logical structure of the available knowledge. The most typical relation, which is often tacitly assumed, is the logical conjunction.

Each information item II_i imposes a specific constraint in the signal space, of the form:

$$\psi_i(c_1, c_2, ...c_N) \;\; >=< \;\; 0, \qquad i = 1,2,...,J \quad . \tag{3}$$

In some cases, a certain information item may constrain only a single dimension in the signal space, e.g., $\underline{c_n} < c_n < \overline{c_n}$, where $\underline{c_n}$ and $\overline{c_n}$ denote the appropriate bounds known for c_n. Typically, however, the ψ_i function binds coefficients from the specific subset of signal space dimensions:

$$\kappa_i = \{c_k : \;\; c_k \text{ is an argument of the appropriate } \psi_i \text{ relationship}\} \tag{4}$$

The subset κ_i of variables bounded together by the function ψ_i, describing the given information item II_i, is an important attribute of the latter.

3. Usefulness of an information item in signal parameter estimation

The purpose of measurement is a sufficiently accurate estimation of the parameter E under the conditions of the availability of the given set (2) of information items. The additional, new information item, say II_{J+1}, will be considered as useful for this purpose if appending it to the set (2) reduces the resulting uncertainty of the estimated parameter, i.e. if:

$$\Delta E_{II_1, II_2, \ldots, II_J, II_{J+1}} < \Delta E_{II_1, II_2, \ldots, II_J} \quad , \qquad (5)$$

where symbols $\Delta E_{II_1, II_2, \ldots, II_J}$ denote the estimated parameter uncertainty when the set of information items specified in the index is available. Since checking the sufficient condition (5) of usefulness for some new information II_{J+1} may be analytically and computationally laborious (compare, e.g., [8,9]) finding out the necessary condition for such a usefulness seems to be profitable. Namely, it may be used in initial elimination of those information items which undoubtedly are useless for the estimation of E.

We say that the information item II_{J+1} may be directly useful in the estimation of the parameter E if the restriction (3) generated by this information binds at least some coefficients from the subset κ_E, containing c_k on which the parameter E depends. In other words, it is necessary for direct usefulness that: $\kappa_{J+1} \cap \kappa_E \neq \varnothing$. Nevertheless, the fact that the given information item II_{J+1} restricts only these coefficients c_k which are not present in the set κ_E, of course does not imply the uselessness of this information for the estimation of E. It happens so because the influence of some parameter c_k on the value of E may also be indirect. Since such an indirect effect may take many forms, the question arises when the influence of some information item on the value of the estimated parameter is not possible at all.

4. The necessary condition for information usefulness

As has been stated, the information item II_{J+1} can not be directly useful in the estimation of the parameter E if the subsets of parameters κ_E and κ_{J+1} have an empty common part (see, e.g., the subsets $\kappa_2, \ldots, \kappa_6$ in Fig. 1). This remark may be intuitively extended on the case of indirect usefulness. This, however, needs an introduction of the concept of **a cluster of sets generated by the given set** κ_E.

The set of coefficients' subsets (4) describing all the information items already available is considered:

$$\{\kappa_i\}_{i=1}^{J} \quad . \tag{6}$$

From (6) such subsets are chosen for which the condition: $\kappa_i \cap \kappa_E \neq \varnothing$ is fulfilled, and the following sum is created: $C_{\kappa_E}^{(1)} = \kappa_E \cup \bigcup_{i:\,\kappa_i \cap \kappa_E \neq \varnothing} \kappa_i$, which will be called the cluster of the first stage. Next, the clusters of the consecutive stages $C_{\kappa_E}^{(m+1)}$ are formed, according to the following algorithm: from the rest of subsets (6) all those are chosen for which the intersections: $\kappa_i \cap C_{\kappa_E}^{(m)}$ are not empty and they are appended to the cluster of the previous stage:

$$C_{\kappa_E}^{(m+1)} = C_{\kappa_E}^{(m)} \cup \bigcup_{i:\,\kappa_i \cap C_{\kappa_E}^{(m)} \neq \varnothing} \kappa_i \quad . \tag{7}$$

Because the number of subsets in (6) is limited, the sequence $C_{\kappa_E}^{(m)}$, m=1,2… will settle or the set (6) will be exhausted. The sum (7) of the highest stage will be called the cluster of the subsets generated by the set κ_E and will be denoted by C_{κ_E} . In Fig. 1, e.g., the cluster generated by κ_E consists of subsets: κ_E , κ_1, κ_2 and may be constructed in two stages. The necessary, but not sufficient, condition for the usefulness of some new information item II_{J+1} in the estimation of E is that the corresponding set of coefficients κ_{J+1} must belong to the cluster of subsets generated by κ_E in $\{\kappa_i\}_{i=1}^{J+1}$: $\kappa_{J+1} \in C_{\kappa_E}$. In Fig. 1, e.g., the indirect usefulness of II_2 in the estimation of E cannot be excluded, whereas the information items $II_3,\ldots,II_6$ without any additional knowledge, are useless for this purpose.

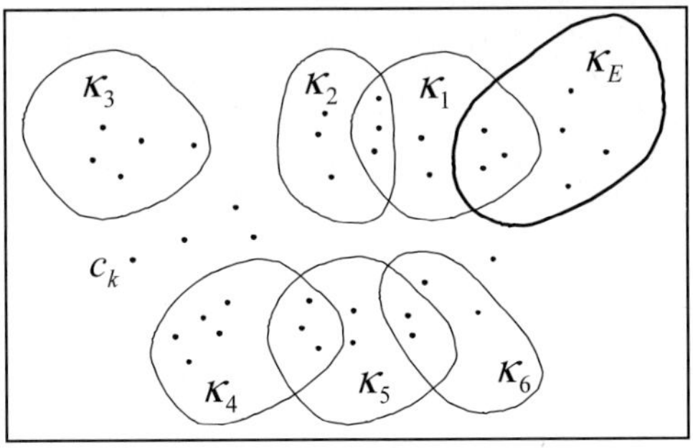

Fig. 1. Signal expansion coefficients' subsets corresponding to individual information items and their clusters.

It can be easily seen from the above definition that the relation of belonging to a cluster is reflexive, symmetrical, and transitive. It thus reveals all formal properties of the equivalence relation [10] and divides the set of coefficients' subsets (and – in the same way - the set of information items) into equivalence classes. The subsets κ_i corresponding to mutually useless information items belong to different classes. In the example presented in Fig. 1, there are three clusters representing three classes of information items which cannot be mutually useful.

It also follows from the above consideration that the usefulness of any information item is related to the entire knowledge already available. E.g. the usefulness of II_2: it cannot be excluded in the presence of II_1. However, when II_1 is not available, II_2 becomes useless, since in that case $C_{\kappa_E} = \kappa_E$ in the example presented in Fig. 1.

5. Conclusions

The entire knowledge available of the investigated signal has been divided into the set of individual information items. Each information item has been modeled as a specific constraint imposed in the signal space. The subset of the dimensions in the signal space which are related by the given information item, has been found to be an important attribute of the information item model. It has been shown that the information item usefulness may be direct or indirect, and the necessary condition for both kinds of usefulness has been proposed. The notion of the subsets' cluster was introduced for this purpose. Because the checking procedure of the necessary condition is quite simple, it seems to be a practical tool for preliminary elimination of such information items which are useless for the estimation of a given parameter.

References

[1] L. Finkelstein, Measurement 14 (1994) 23-29.
[2] J. Sztipanovits, Measurement 7 (1989) 98-108.
[3] T.L.J. Ferris, Measurement 21 (1997) 137-146
[4] P.H. Sydenham, M.M. Vaughan, Measurement 8 (1990) 180-187.
[5] E. McDermid, J. Vyduna, J. Gorin, Hewlett-Packard Journal Feb. 1977 11-19.
[6] A. Zayezdny, I. Druckmann, Signal Processing 22 (1991) 153-178.
[7] T. Ishioka, M. Takegaki, Measurement 12 (1994) 227-235.
[8] J. Beyerer, Measurement 25 (1999) 1-7 and Measurement 18 (1996) 225-235
[9] M. Parvis, Measurement 12 (1994) 237-249.
[10] J. Pfanzagl, "Theory of measurement" Physica-Verlag, Würzburg, 1968.

Grammar Based Automatic Speech Recognition System for the Polish Language

Danijel Koržinek , Łukasz Brocki

Polish-Japanese Institute of Information Technology,
ul. Koszykowa 86, 02-008, Warsaw

Abstract

Automatic Speech Recognition (ASR) is gaining significance in the fields of automation and user-to-machine interaction. In this paper, the authors present a working framework for a grammar based ASR system. The paper discuses the state-of-the-art speech recognition technology used in the system. Three applications of this technology are discussed: in robotics, forms filling and telephony.

1. Introduction

Speech is the essential method of interaction for humans. That is why ASR has been the forefront of computer science for decades. This paper describes the state-of-the-art speech recognition technology in chapter 2. It shows the use of Artificial Neural Networks (ANNs) in chapter 3. Finally, it presents three working implementations of this technology in chapter 4.

2. Speech recognition basics

Speech recognition begins with splitting raw signal into equal sized frames. A frame contains several hundred samples that are converted using well known signal processing techniques into several real-valued features. The features used in our system are called Mel-Frequency Cepstrum Coefficients (MFCC) [4]. We use 12 MFCCs combined with an energy feature with first and second order derivatives of these values. This gives 39 features overall.

Having parametrized the signal one can use an ANN [1, 2] to get posterior probabilities of phones which are present in the given utterance. This network essentially performs the mapping of one sequence (speech features windows) into another (a sequence of posterior probabilities of phones). One could correlate the sequence of posterior phone probabilities with the phone labels of each individual frame.

Most automatic speech recognizers implement the above mentioned procedure using Hidden Markov Models (HMM) [3]. The algorithm uses the posterior probabilities of sub-word units to recognize words and in the following step it uses a language model to constraint the search space of possible word sequences.

```
$start = $person ( buys | sells ) ( $item [and] ) <1-> ;
$person = Mary Johnson | Drew Black | John Stewart | professor ;
$item = a house | a car | a pen | the field | the company | stock ;
```

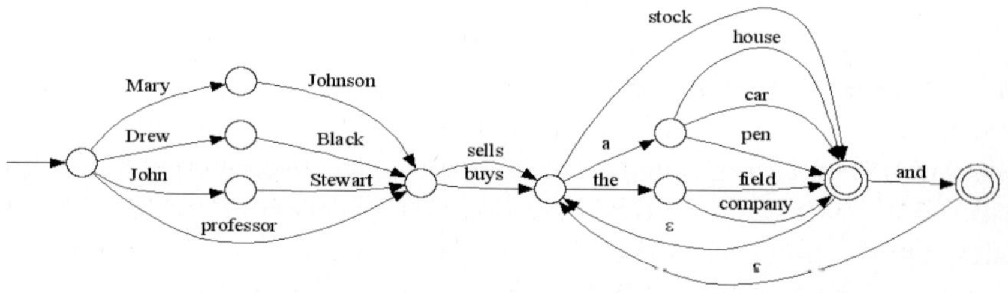

Drew Black sells a car a pen and the company
Mary Johnson buys a pen and a pen a car and stock ...
...

Fig 1. An example of a simple grammar.

In domain-constrained speech recognition systems, grammar-based language models are used. A grammar is an automaton that accepts or rejects word sequences. Using regular-expression syntax, it allows the user to define the exact utterances the system is to recognize. In figure 1, the three grammar rules on top are converted to a Finite-State Machine (FSM) in the middle. Using this simple automaton one can recognize many utterances, like the ones on the bottom of the figure.

3. ANN as a phoneme probability estimator

Context makes the speech recognition task challenging. When people speak fluently a blurring of acoustic features occurs. It is known as coarticulation effect. To make things worse acoustic realization of phonemes depends not only on the context, but also on speech rate, accent, dialect, background noise and equipment that is used.

ANNs are known for their good handling of distorted and noisy data. They can cope with most of the above mentioned problems. However several adjacent speech frames need to be given as input to the Multi-layer perceptrons (MLPs) [1] to amend the coarticulation effect. The acoustic context is handled much better when the dimensionality of the observation vector is increased, but this introduces an increase in the size of the network and lowers its performance. A Recurrent Neural Network (RNN) [2] was therefore used.

4. Grammar based automatic speech recognition applications

Application 1: a mobile robot controlled by voice

Our system was used in the Robotics Lab in the Polish-Japanese Institute of Information Technology for a mobile, radio controlled robot built using a tank model. The robot has two independent electric motors that control the tank's tracks, 8 sonars and a CCD camera. The robot has a two way wireless communication with a PC through an RS-232 port. A program that runs on a computer can control the tank's movement as well as receive information from various onboard sensors (sonars and light detecting diodes). We used the tank platform to check if it would be possible to build a system that allows a person to control the tank's movement using voice. Several simple sentences were designed, eg: “move forward two meters”, “turn right 65 degrees”, “move backward 60 centimeters”. There is a slight delay between the utterance and the execution of each command. The system is speaker independent and it allows the user to control the tank hands free. Of course, this application is not very convenient – it is only a proof of concept. However, given a more elaborate set of commands, the system might prove useful in cases where a hands free interface is needed for controlling devices in environments with moderate noise.

Application 2: forms filling by voice

We have built a system that runs on an ordinary PC and allows a person to fill out various forms by voice. Our application simulates an investment trust manager. Speaker can purchase and sell shares and bonds, make bank transfers, change currencies, move people between different investment risk groups, dictate telephone numbers, and press almost all possible keys on the keyboard, all using just pure speech. The application is speaker independent and it fully depends on grammars. Every command must be spoken exactly according to the grammar. For example one could say: "Teresa Mazur buys 3452 shares of Techmex company". If speaker says a sentence that is out of grammar, the application can either fill out the form badly, or use the so called "garbage model" to ignore the sentence completely.

Application 3: telephone voice portal

Our recent and most advanced application of grammar based speech recognition is the Primespeech telephony server. The server runs on an ordinary PC with a Linux operating system. It uses special telephony hardware to connect the computer to a telephone line. This allows the user to call in from any telephone and use the ASR features of the server. Our preliminary tests have shown no substantial difference in the performance of the system when the user calls from different stationary and mobile phones. It also works relatively well when the speakerphone is used. Moreover, one can use VoiceOverIP to communicate with the server.

The server contains a web-based interface. This makes it possible to monitor the recognition process over the Internet. Our first telephony application implements a simple garbage model and can recognize 20 names from continuous speech without pauses.

In the near future, we plan to implement speech synthesis in our server. This would allow us to make simple dialogs with the callers. Also, we want to integrate the server with a database, to be able to synthesize and recognize items from the database. The simplest example for the application of such a server would be in a large cinema complex. It would allow people to call and book tickets automatically using speech. Such systems already exist, however they rely on touch tone technology. Having speech recognition and synthesis makes the whole process much more convenient.

5. Conclusion

ASR is an emerging technology that is already used in many large call centers, and it is still gaining significance. New exciting applications of this technology are currently implemented. Unfortunately, people that use less common languages, like Polish, still cannot take full advantage from this technology, as only basic applications are available.

References

[1] Z. Michalewicz, D.B. Fogel, "How to Solve It: Modern Heuristics", Springer Verlag, 1999
[2] A. J. Robinson, (1994). An application of recurrent nets to phone probability estimation. IEEE Transactions on Neural Networks, 5(2):298–305.
[3] L.R. Rabiner, "A tutorial on hidden Markov models and selected applications in speech recognition". Readings in speech recognition, pages 267296, 1990.
[4] S. Young,. The HTK Book, Cambridge University Press, 1995

State Controller of Active Magnetic Bearing

M. Turek (a) *, T. Březina (b)

(a) Faculty of Mechanical Engineering, Brno University of Technology, Technická 2, Brno, 616 69, Czech Republic

(b) Faculty of Mechanical Engineering, Brno University of Technology, Technická 2, Brno, 616 69, Czech Republic

Abstract

A state controller of rotating shaft levitated by an active magnetic bearing is described in this contribution. It is shown that the state controller is able to slow down and stabilize the response of the controlled system. Furthermore an error compensation represented by an integrative controller connected on input to the state controller allows compensation of external forces. It is shown that such controller is able to control the shaft even at high rotational speed. Study of dependence of controller parameters on rotational speed of the controlled shaft is done.

1. Introduction

An active magnetic bearing (AMB) inhibits the contact between the rotor and stator and so it eliminates the limitations of classic bearing. Therefore it is possible to use AMB in specific and extreme circumstances where classic bearing is inapplicable. Electromagnets located in stator of the bearing create a magnetic field. The force caused by magnetic field keeps the rotor levitating in desired position in the middle of air clearance. So the control of magnetic field is necessary.

Although AMB is highly nonlinear, linear model can be developed. Afterwards standard methods to design state controller can be used. Two different state controllers are designed. The first one is used to place poles of AMB to desired positions, i.e. stabilize AMB. The second one then minimizes control error. Additionally an integrative gain is added parallely to the second controller to compensate constant control error.

2. Active magnetic bearing model

Model used for control design is composed of two parts – model of levitated rotor and model of magnetic force. Behavior of rotor can be described by linear second order differential equation

$$M\frac{d^2q}{dt^2} + \omega G\frac{dq}{dt} = Bf + f_g + \omega^2 d_u(\varphi), \tag{1}$$

Magnetic force is composed from forces caused by opposite electromagnets. As can be seen from equations 2, magnetic force depends on feeding currents and position of the rotor and is highly nonlinear.

$$\begin{aligned} F_{m,x} &= \frac{Ai_{x,2}^2}{\left(\frac{d}{2} - x + a\right)^2} - \frac{Ai_{x,1}^2}{\left(\frac{d}{2} + x + a\right)^2} \\ F_{m,y} &= \frac{Ai_{y,2}^2}{\left(\frac{d}{2} - y + a\right)^2} - \frac{Ai_{y,1}^2}{\left(\frac{d}{2} + y + a\right)^2} \end{aligned} \tag{2}$$

For detailed model description and parameters of used AMB see [3].

3. Controller

Linear model of AMB is needed to design state controller, but behavior of magnetic force is highly nonlinear. One of methods to acquire linear model of AMB is linearize its behavior by auxiliary nonlinear controller connected on input of AMB given by equations 3. F_p is input of linear model of AMB and u is feeding voltage of electromagnets of AMB.

$$u_1 = \begin{cases} \sqrt{-\frac{F_p}{A}}\left(\frac{d}{2}+x+a\right) & \text{if } F_p < 0 \\ 0 & \text{otherwise} \end{cases}$$
$$u_2 = \begin{cases} \sqrt{\frac{F_p}{A}}\left(\frac{d}{2}-x+a\right) & \text{if } F_p > 0 \\ 0 & \text{otherwise} \end{cases} \quad (3)$$

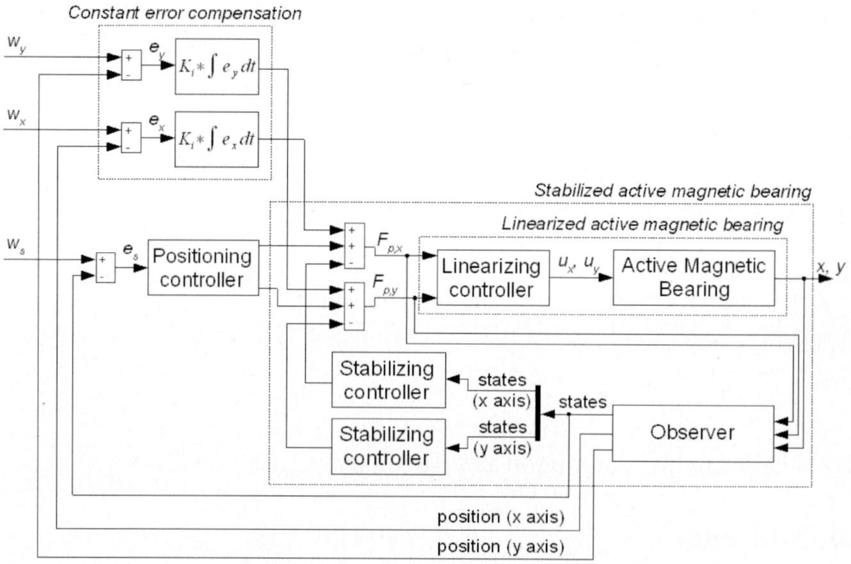

Fig. 1. Controller interconnections

Afterwards, behavior of magnetic force can be described by linear second order differential equation. Its parameters depend on parameters of a controlled AMB.

When linear description of AMB exists it can be stabilized by state controller. The stabilizing controller is designed by pole placement method (see [1]). The stabilizing controllers have to be designed independently for each axis of AMB otherwise it would affect the movement of rotor in opposite axis.

Stabilizing controller does not allow easily define significance of separate controlled states. So, additional positioning controller is designed to minimize control error. The positioning controller is designed by LQ design [2]. Behavior of rotor of AMB depends on its speed of rotation as can be seen from its model given by equation 1. It means that optimal parameters of positioning controller also depend on its speed of rotation. So positioning controller should consist of set of controllers for each possible speed of rotation. Fortunately the dependence is linear or can be considered zero so the whole set of controllers can be described by set of simple linear equations.

On figure 2 are graphs with dependence of two from sixteen parameters (two outputs times eighth states) of positioning controller on speed of rotation. The first graph shows dependence of gain from deviation in vertical (y) axis to feeding voltage to electromagnets in horizontal (x) axis. The second one shows dependence of gain from deviation in horizontal axis to feeding voltage to electromagnets in horizontal axis. As can be seen one is linear and the second is less than one percent, i.e. can be considered zero.

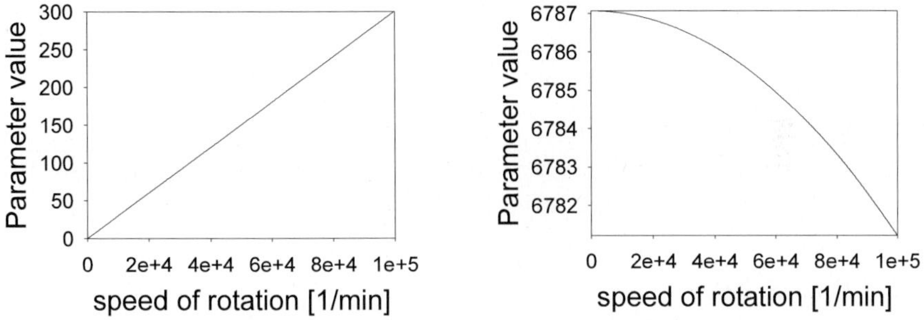

Fig. 2. Dependence of two of controller parameters on speed of rotation

Finally an integrator is connected parallely to positioning controller. The integrator allows compensate constant control error. Its gain is designed by trial-error method.

Final interconnections of controller are given by figure 1.

4. Results

Performance of controller is verified by simulation. A small unbalance of rotor and influence of gravitation are simulated. Initially the rotor has maximal deviation in vertical axis (thanks to gravitation). A driving motor causing rotation of rotor is switched on when the rotor is in center of air gap (three seconds after start of control). The simulations were done for different speeds of rotation of rotor.

The influence of rotor unbalance is minimal for speeds of rotation from zero to 1000 min^{-1} (which corresponds to sampling frequency of controller, i.e. 1 kHz), for higher speeds of rotation is the influence of unbalance significant. The uncontrollable high frequency forces caused by unbalance are damped by inertia of rotor, but the driving motor start causes force impulse which is not damped and has to be controlled. The controller stabilizes the rotor position with sufficient performance (see fig. 3).

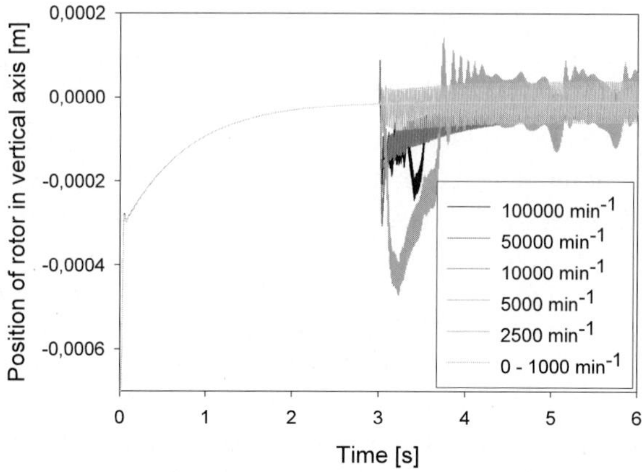

Fig. 3. Response to control (vertical axis)

5. Conclusion

The designed controller is capable of control of AMB with high performance. Although it is seemingly complicated it in reality consists of observer, twenty linear gains and simple nonlinear linearizing controller so it can be easily implemented.

6. Acknowledgement

Published results were acquired using the subsidization of the Ministry of Education, Youth and Sports of the Czech Republic, research plan MSM 0021630518 "Simulation modeling of mechatronic systems".

References

[1] W. L. Brogan "Modern Control Theory, 3rd Ed." Prentice Hall, 1991.
[2] P. Dorato, C. Abdallah, V. Cerone "LinearQuadratic Control: An Introduction" Englewood Clis, New Jersey: Prentice-Hall Inc., 1996.
[3] M. Turek, T. Březina "Control Design of Active Magnetic Bearing by Genetic Algorithms" Engineering Mechanics, 2007, CDROM, in print.

Fuzzy set approach to signal detection

M. Šeda

Brno University of Technology, Faculty of Mechanical Engineering, Institute of Automation and Computer Science, Technická 2, Brno 616 69, Czech Republic

Abstract

Automated supervision and fault diagnosis are important features in design of efficient and reliable systems. Detection algorithms are generally optimised with respect to a particular set of cost functions chosen for the specific application. In the last few years in the field of detection systems there have been an increasing number of applications based on algorithms using methodologies, which belong to a subclass of Artificial Intelligence called Soft Computing.

In this paper, we propose a fuzzy method for the detection of dangerous states based on matching a predefined database of these states with periodically measured or estimated parameter values.

1. Introduction

The operation of technical processes requires increasingly advanced supervision and fault diagnosis to improve reliability, safety, and economy. When testing a complex technical equipment, we try, besides measuring its parameters, to determine if the equipment behaves in a "normal" way, or if its characteristics signalise "abnormal" behaviour that can result, in specific situations, even in its destruction.

As the description of technical parameters may include imprecise expressions containing, e.g., linguistic modifiers, formal management of uncertainty and imprecision is needed [1], [5]. For example the application of fuzzy logic to fault diagnosis for nonlinear systems is described in [6] and [7]. In [4], a fuzzy logic-based algorithm for a predictive model of an evolving signal in nuclear systems is introduced.

2. Data processing with uncertainties

Parameter values can be crisp or imprecise when they cannot be measured directly. In the second case, the values can also be modified by one or more linguistic modifiers, e.g. *very*, *highly*, *more-or-less*, *roughly* and *rather*. These modifiers are usually defined by fuzzy operations *dilation* (DIL), *concentration* (CON) and *intensification* (INT).
If A is a fuzzy set in universe X and μ_A is its membership function, then these operations can be defined as follows [1], [2]:

$$\mathrm{DIL}_1(A) = A^{0.5},\ \forall x \in X\colon \mu_{\mathrm{DIL}(A)}(x) = [\mu_A(x)]^{0.5} \tag{1}$$

or more precisely [5]

$$\mathrm{DIL}_2(A) = 2A - A^2,\ \forall x \in X\colon \mu_{\mathrm{DIL}(A)}(x) = 2\mu_A(x) - [\mu_A(x)]^2 \tag{1'}$$

$$\mathrm{CON}(A) = A^2,\ \forall x \in X\colon \mu_{\mathrm{CON}(A)}(x) = [\mu_A(x)]^2 \tag{2}$$

$$\mathrm{INT(A)},\ \forall x \in X\colon\ \mu_{\mathrm{INT}(A)}(x) = \begin{cases} 2[\mu_A(x)]^2, & \text{for} \quad \mu_A(x) < 0.5 \\ 1 - 2[1 - \mu_A(x)]^2, & \text{otherwise} \end{cases} \tag{3}$$

Zadeh proposed the main linguistic modifiers in this form:

$$very(A) = \mathrm{CON}(A) \tag{4}$$

$$highly(A) = A^3 \tag{5}$$

$$more\_or\_less(A) = \mathrm{DIL}_1(A) \tag{6}$$

$$roughly(A) = \mathrm{DIL}_2(\mathrm{DIL}_2(A)) \tag{7}$$

$$rather(A) = \mathrm{INT}(\mathrm{CON}(A)) \tag{8}$$

These proposals are not accepted in general and more sophisticated definitions are introduced in the literature. For instance, if A is a fuzzy set and m is a linguistic modifier, then $\mu_m(A)$ can be defined as follows [5]:

$$\mu_m(A) = \mu_m \circ \mu_A \circ q_m \tag{9}$$

where $\mu_m : [0,1] \to [0,1]$, $q_m :\ X \to X$ is a translation and $\circ$ denotes the composition of functions.
Besides linguistic modifiers, Boolean operators conjunction, disjunction, negation denoted by symbols $\wedge$, $\vee$, $\neg$, or by operators AND, OR, NOT may also occur in queries. If A, B are fuzzy sets, then the operation A AND B is usually interpreted as the intersection of such sets, A OR B as the union operation and NOT A as the complement. More formally, it can be expressed by the following formulas:

$$\forall x \in X\colon \mu_{A \cap B}(x) = \min\ \{\mu_A(x), \mu_B(x)\} \tag{10}$$

$$\forall x \in X\colon \mu_{A \cup B}(x) = \max\ \{\mu_A(x), \mu_B(x)\} \tag{11}$$

$$\forall x \in X: \mu_{\bar{A}}(x) = 1 - \mu_A(x) \tag{12}$$

3. Detection of dangerous states

Let us assume that a set P of parameters of a technical equipment is given and their values are periodically measured or estimated. If a set of dangerous states D is specified, then the natural aim is to detect each occurence of such a state and generate a corresponding action to prevent undesirable effects. As a *dangerous state* can be understood a value exceeding some parameter threshold, occurence of a certain set of parameters with values close to their threshold or such combinations of these values as are not acceptable for the application in question.

For simplicity, we assume that the database of dangerous states is represented by tuples whose number of elements is equal to number of parameters. Some of the elements can have null values, which means that the value of such parameters is not substantial for safety operation. For example the tuple (100, , 80, {high, very high}, , 50, low) has seven elements: three of them are crisp (100, 80, 50), the second and the fifth value is null and the remaining two are imprecise. The value {high, very high} shows that we admit that it can be expressed by a subset of a domain instead of only by a singleton. Similar uncertainty can be included in parameter values. This extension provides more flexibility for matching the threshold values of dangerous states with parameters values. In order to measure the results of these comparisons we define the following way of calculating the membership values for each tuple of dangerous state factors.

If d_{ij} is the factor of the i-th dangerous state referring to the j-th parameter p_j, then their *similarity measure* is defined as follows:

$$SM_j(d_{ij}, p_j) = \begin{cases} 1, & \text{if } d_{ij} \le p_j \\ \dfrac{p_j}{d_{ij}}, & \text{if } d_{ij} > p_j \\ 0, & \text{if } d_{ij} \text{ is null} \\ \max\{S_j(d,p) \mid d \in d_{ij}, p \in p_j\}, & \text{otherwise} \end{cases} \tag{13}$$

In real situations a dangerous state can be represented not only by exceeding a given threshold, but also by decreasing under an acceptable level, e. g. in combustion processes the air and gas pressure must satisfy requirements of this type. In this case (15) can be easily modified in a complementary way.

The similarity relation S_j used for matching imprecise values is a binary relation $S_j : D_j \times D_j \to [0,1]$ which is

(i) reflexive: $S_j(a, a) = 1$,

(ii) symmetric: $S_j(a, b) = S_j(b, a)$, and

(iii) transitive: $S_j(a, c) \geq \max\{\min(S_j(a, b), S_j(b, c)) \mid b \in D_j\}$.

The total similarity of measured or estimated parameter values with respect to the tuples in the database of dangerous states then will be determined by evaluating the similarity measures of the corresponding factors. In these evaluations, linguistic modifiers precede Boolean operations. If, for a tuple in the database, a threshold depending on the application area is exceeded, then an action preventing serious consequences or even destruction must be generated.

This approach is summarised in the following pseudopascal code. By the return value we can conclude whether a dangerous state was detected or not. The aggregation$_i$ depends on the i-th dangerous state specification. It may be represented by a conjunction of its factors but it also can include a disjunction of factors and its negations.

It is obvious that various technological parameters need different time periods for their measurements. For simplicity, we can suppose that all measurements are "synchronized" by the parameter that needs the most frequent measurements. In time intervals between two measurements we suppose that parameter values are constant and they are given by the values from the last measurement.

```
danger := false ;
repeat  make measurements or estimations of all parameters from P
        i := 1;
        while (not danger) and ( i ≤ |D| ) do
            begin j := 1;
                while (not danger) and ( j ≤ |P| ) do
                    begin determine SM_j(d_ij, p_j) ;
                          apply linguistic modifiers on SM_j(d_ij, p_j);
                          danger := (SM_j(d_ij, p_j) ≥ threshold_j) ;
                          j := j+1;
                    end ;
                if not danger
                  then SM_i := aggregation_i {SM_j(d_ij, p_j), j = 1, … , |P|}
                danger := (SM_i ≥ threshold_i) ;
                i := i+1;
            end ;
        Delay(time period)
until danger or stop ;
if danger then return(i–1) else return(0)
```

4. Conclusions

Classical techniques for determining the key properties of methods based on an analysis of behaviour can be classified by monitoring their dynamic characteristics. In this paper, we focused on a specific problem of fault detection in one or more parameters that can cause danger states resulting e.g. in damage to expensive testing tools or technological equipment. We have proposed a method based on matching the typical dangerous states stored in a database with parameters measured or estimated on a periodic basis. As all these data can contain imprecise information, the algorithm processing such data is based on fuzzy logic. However, the proposed similarity measure mechanism can be used also for the case of crisp data.

Acknowledgments

The results presented have been achieved using a subsidy of the Ministry of Education, Youth and Sports of the Czech Republic, research plan MSM 0021630518 "Simulation modelling of mechatronic systems".

References

[1] G. J. Klir, B. Yuan "Fuzzy Sets and Fuzzy Logic. Theory and Applications", Prentice Hall, New Jersey, 1995.

[2] L. D. Lascio, A. Gisolfi, V. Loia, "A New Model for Linguistic Modifiers", International Journal of Approximate Reasoning 15 (1996) 25-47.

[3] S. W. Leung, J. W. Minett "The Use of Fuzzy Spaces in Signal Detection", Fuzzy Sets and Systems 114 (2000) 175-184.

[4] M. Marseguerra, E. Zio, P. Baraldi, A. Oldrini "Fuzzy Logic for Signal Prediction in Nuclear Systems", Progress in Nuclear Energy 43 (2003) 373-380.

[5] V. Novák "Fundamentals of Fuzzy Modelling (in Czech)", BEN, Praha, 2000.

[6] S. Oblak, I. Škrjanc, S. Blažič "Fault Detection for Nonlinear Systems with Uncertain Parameters Based on the Interval Fuzzy Model", Engineering Applications of Artificial Intelligence 20 (2007) 503-510.

[7] G. I. S. Palmero, J. J. Santamaria, E. J. S. de la Torre, J. R. P. González "Fault Detection and Fuzzy Rule Extraction in AC Motors by a Neuro-Fuzzy ART-Based System", Engineering Applications of Artificial Intelligence 18 (2005) 867-874.

The robot for practical verifying of artificial intelligence methods: Micro-mouse task

T. Marada (a)

(a) Faculty of Mechanical Engineering, Brno University of Technology, Technická 2896/2, Brno, 616 69, Czech Republic

Abstract

Robot localization and path planning belong to actual problems in robotics. The paper is focused on design of small autonomous robot for practical verifying artificial intelligence methods concretely on Micro-mouse task. The physical model was designed with respect to its simple construction, unpretentious production and relatively little cost but sufficient capability for performing different experiments.

1. Introduction

By reason of practical verification of artificial intelligence methods, localization and navigation has been built autonomous mobile robot on the institute of automation and informatics. We can see this robot in the figure 1. This robot can be used at micro-mouse task. The main part of this paper is focused on design of small autonomous robot for practical verifying artificial intelligence methods concretely on Micro-mouse task.

Figure 1: Micro-mouse robot

2. Micro-mouse robot

Robot construction has been built with a view to low cost price. The conception is based on two stepper motors drive. These supporting move forward, move backward and navigation on the differential control principle. Direction of motion change is realized by different wheels speed of rotation. Robot is provided by two balls supports so can't be turnover. The main technical parameters of robot are summarized in the table 1.

Parameter	Value
Construction	Two wheels truck
Drive	2x stepper motor TEAC KP39HM2-025
Sensors	3x sensor SHARP - GP2D120
Power supply	4x accumulator Li-Ion CGR18650/4.2V
Communications	2x RS232 / TTL / 38400Bd
Length	130mm
Width	105mm
High	80mm
Wheels spacing	95mm
Weight	990g

Table 1: Robot technical parameters

Robot is divided on four parts: Main CPU board, sensors board, motors control board and operating board with LCD display. Micro-mouse block diagram is in the figure 2. Power supply of all boards is providing by four storage cells Li-Ion CGR18650 / 4.2V connected in series. Total voltage is 16.8V. High voltage is using for supply motors by reason of reach higher gyroscopic moment.

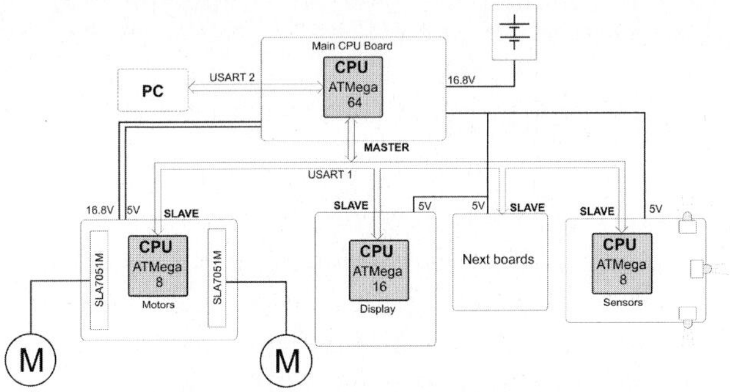

Figure 2: Micro-mouse block diagram

3. Maze storing

One of the more useful properties of the maze is its size. For a full sized maze, we would have 16 rows by 16 columns = 256 cell values. Therefore we would need 256 bytes to store the distance values for a complete maze. A single byte can be used to indicate the presence or absence of a wall in the maze. The first 4 bits can represent the walls. A typical cell byte can look like this:

Bit No.	7	6	5	4	3	2	1	0
Wall					W	S	E	N

When we are using the binary bit value for each wall position, we have North = 1, East = 2, South = 4 and West = 8. Now any combination of walls in a cell can be represented by a number in the range 0 to 15. For example if some cell have wall on the West and on the East then this cell can be represent by value 10 which is 0x0A in hex or 00001010 in binary.

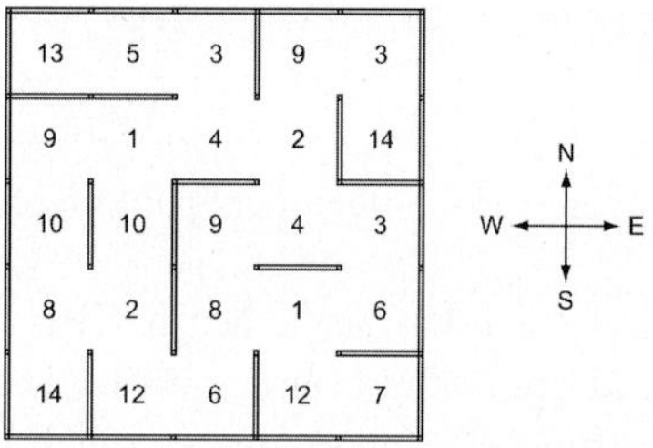

Figure 3: Example of maze storing

Every interior wall is shared by two cells so when we update the wall value for one cell then we have to update the wall value for its neighbor too.

4. The Flood-fill algorithm

The idea of flood-fill algorithm is to start at the goal (centre of the maze) and fill the maze with values which represent the distance from each cell to the goal. When the flooding reaches the starting cell then we can stop and follow the values downhill to the goal. In the figure 4 we can see the sequence of the maze being flooded. This maze is completely mapped and we know where all walls are. We can clearly see how dead ends are handled and what happens when there is more than one way through maze.

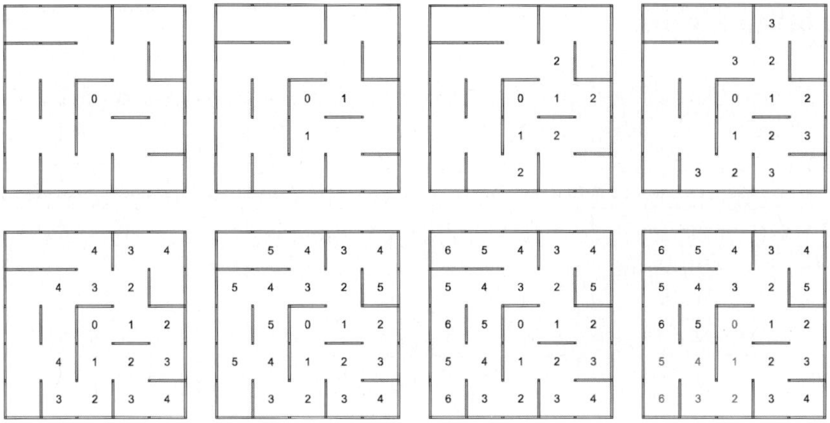

Figure 4: Sequence of the maze being flooded

For a full sized maze, we would have 16 rows by 16 columns = 256 cell values. Therefore we would need 256 bytes to store the distance values for a complete maze. Because the micro-mouse can't move diagonally, the values for a 5x5 maze without walls would look like this:

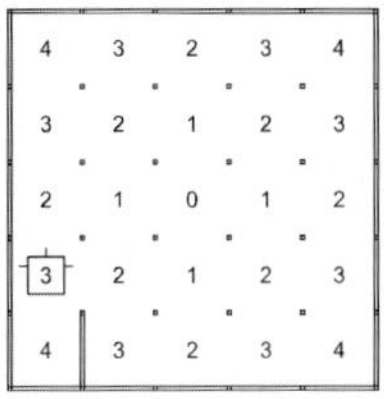

Figure 5: Flood-Fill example without walls

When it comes time to make a move, the robot must examine all adjacent cells which are not separated by walls and choose the one with the lowest distance value. In our example in the figure 5, the robot would ignore any cell to the West because there is a wall, and he would look at the distance values of the cells to the North, to the East and to the South since those are not separated by walls. The cell to the North has a value 2, the cell to the East has a value 2 and the cell to the South has a value 4. That means that the robot can go to the North or to the East and traverse the same number of cells on its way to the destination cell. Because turning would take time, the robot will choose to go forward to the North cell. When the new walls are found, the distance values of the cells are affected and we have to update them. Look at the example in the figure 6.

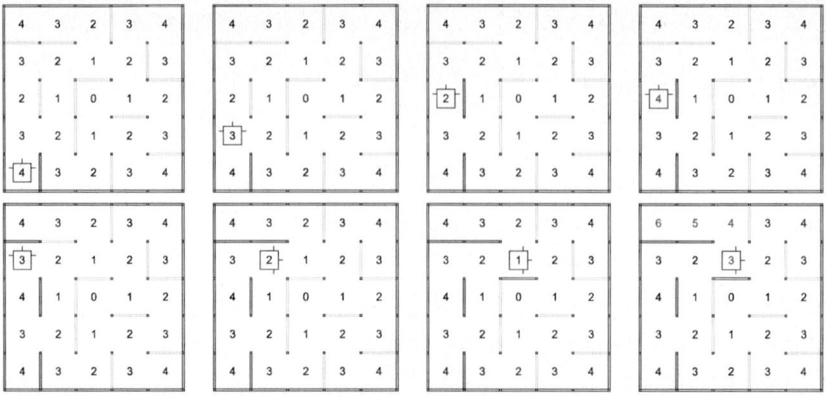

Figure 6: Sequence of the regular flood-fill algorithm

In the third step the robot has found a wall. We can't go to West and we cannot go to the East, we can only travel to the North or to the South. But going to the North or to the South means going up in distance values which we do not want to do. So we need to update the cell values as a result of finding this new wall. To do this we "flood" the maze with new values (step fourth). The same case we can see in the seventh step and eighth step where robot find new wall and distance values had to change.

5. Conclusion

We have implemented regular flood-fill algorithm in to the robot. This algorithm is very well applicable when the maze includes the single islands. The flood-fill algorithm is a good way of finding the path from the start cell to the destination cells, but he is very slow.

Acknowledgements

This work was support by project MSM 0021630518 "Simulation modeling of Mechatronics systems".

References

[1] Marada T., Houška P., Paseka T.: Small autonomous robot for practical verifying of artificial intelligence methods, Engineering mechanics 2006, Svratka 2006.

The enhancement of PCSM method by motion history analysis.

S. Věchet, J. Krejsa, P. Houška

Brno University of Technology, Faculty of Mechanical Engineering,
Technická 2, 616 69, Brno, Czech Republic

Abstract

This paper deals with the identification of wheel robot position and orientation when dealing with the global localization problem. We used a method called PCSM (Pre-Computed Scan Matching) for solving this problem for autonomous robot in known environment. This method was developed for small robots. The identification of the position and orientation of the robot is based on the fusion of pre-computed match data and the analysis of the history of robot motion. The paper provides information about this fast yet simple method.

1. Introduction

Navigation of mobile robots is an actual problem in robotics. Many successful applications of mobile robots contribute to the further expansion of robots to the ordinary life. Rescue, survey or delivery robots in dangerous environment are standard applications.

Identification of robots position relative to the environment is basic task in navigation. This problem is called localization. Mobile robots localization is divided into three main parts: the first and simplest is the position tracking, the second is local localization (the initial position of the robot is known) and the last and the most complicated is the global localization (the initial position of the robot is unknown). Pre-Computed Scan Matching method (PCSM) is presented in this paper. The method was introduced in [1]. PCSM method belongs to the group of global localization methods. Presented method solves among other a robot kidnapped problem, when the robot is taken (kidnapped) from correctly localized position to another position without any information about the position change.

PCSM method is designed mainly to solve a robot kidnapping problem with no respect to previous localization results. The method itself is fast and highly efficient, it failed only in couple of cases. When the localization fails, the position of the robot is found in totally different position and history analysis of robot motion can be incorporated to correct the true position of the robot in a fast and simple way.

2. Localization method

PCSM (Pre-computed Scan Matching) algorithm was first described in [1]. The algorithm is based on pre-computed world scans and matching of the scans with actual neighborhood scan. The key idea is to define a value function used to describe the difference between two scans over the state space. This is typically called the "Match" and is denoted as

$$M(x) = r(x, a, S) \tag{1}$$

where x is the state (robot pose), a denotes the actual perceptual data reading by robot in given state (such as infrared sensor measurements), S represents a set of m samples distributed uniformly in state space, r is a reward function which returns the „match" for given inputs x,a,S.
The match is computed for each sample as follows:
Let's assume the robot's pose is x, and let o denote the individual sensor beam with skew α relative to the robot then the distance d read for this beam is given according to

$$d_j = g(x, o_j) \tag{2}$$

$g(x, o_j)$ denotes the measurement of an ideal sensor

$$d = \{d_j\}_{j=1,\ldots,n} \tag{3}$$

then the set S of m samples is

$$S = \{x^{(i)}, d^{(i)}\}_{i=1,\ldots,m} \tag{4}$$

and the reward function r is

$$r(x^{(i)}, a, S) = \sum_{j=0}^{n} \left(d_j^{(i)} - a_i\right)^2 \tag{5}$$

3. History analysis

Presented method was tested in static environment with known map. The aim of the method is to successfully identify robot correct position in known map from neighborhood scans and odometry.
During the beginning of the localization process the robots position in the map is unknown. When the robot gets the first neighborhood scan, the localization method identifies a number of possible locations (see figure 1).

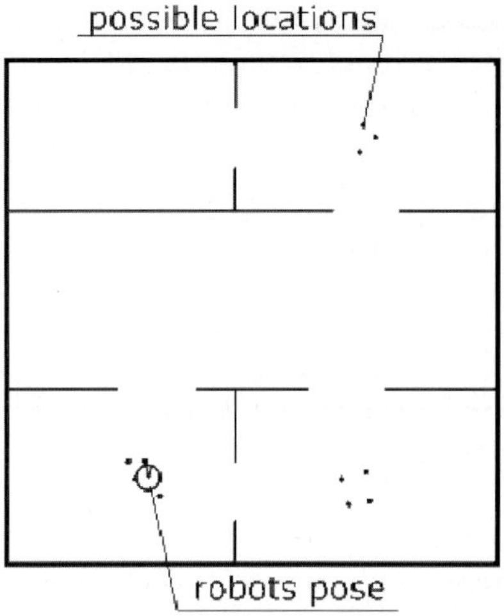

Fig. 1: possible location for the robot

Each location is defined as position and orientation $P=\begin{bmatrix} x & y & \varphi \end{bmatrix}^T$ and the localization method produces a set of probable locations $S_0=\{P_{01},P_{02},..,P_{0n}\}$. When the robot performs a single movement in given direction the neighborhood scan is changed and the localization method produces another set $S_1=\{P_{11},P_{12},..,P_{1n}\}$ of possible locations for the robot. After the movement the robot has also the information about the traveled distance from odometry. The comparison of probable locations from both steps S_0 and S_1 with traveled distance result in a restricted set of possible robots locations (see figure 2). The algorithm works as follows:

1. Initialization of pre-computed scans from know map
2. Get the first range scan of robots neighborhood
   $S_i=\{P_{i1},P_{i2},..,P_{in}\}$
3. Single movement, read the odometry information

4. Get the range scan $S_{i+1} = \{P_{i+1,1}, P_{i+1,2}, .., P_{i+1,n}\}$
5. Perform a history analysis
6. Continue with step 3

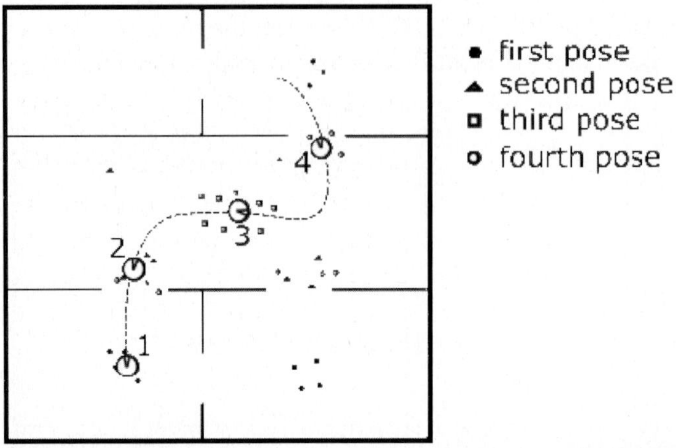

Fig. 2: localization process with history analysis

3. Conclusions

We present a localization method PCSM for mobile robots. PCSM method is used for localization in known static environment and was successfully used in simulation experiments. The method itself failed to localize the robot in several cases therefore it was improved by motion history analysis. The capability to successful identify robots position is enhanced and outliers in robot position are eliminated.

This work was supported by Czech Ministry of Education by project MSM 0021630518 "Simulation modelling of mechatronic systems".

References

[1] Věchet S., Krejsa J. (2005) Real-time localization for mobile robot, Mechatronics, robotics and biomechanics 2005, pp 3-13.

Mathematical Model for the Multi-attribute Control of the air-conditioning in green houses

Wojciech Tarnowski, Prof. Dr. Habilit. (a), Bui Bach Lam, MSc (b)

(a) (b) Control Engng Dept
Technical University, Koszalin, 75-620 Poland

Abstract

In the paper an extended model is presented, which includes all substantial phenomena in the green-house: conversion of mass and energy, and necessary boundary conditions, and a transportation of water and heat between the air, soil and plants. The mathematical model of partial differential equations is proposed.

1. Introduction

The effective growth of plants in green-houses requires that many conditions and constraints are to be met, these are: humidity both of the soil and of the air, as well as the illumination and the temperature and, what more these requirements are related each to other (see Fig. 1, for example). Besides, values of the requirements are changing in time depending on the phase of the development of the plant, and are different for various plants and even for various species [3], [4].

To design the control system and its algorithm, a mathematical model of the green-house is necessary to determine current data for the control algorithm, to define adequate instrumentation and to complete verification experiments. For the optimal real time control an efficient numerical model is compulsory, too.

2. The object

A modern green-house is a complex of many building segments, joint to create a broad common inner space, usually of hundreds square meters of the size. On account of the extensiveness of the greenhouse, usually in the same time in different zones there are planted various plants with different climate requirements. To achieve these variety, in each section there are separate heaters, ventilators, sprinklers, humidifiers, and/or folding windows Therefore it is rational technologically and economically to implement a dispersed control system with a few valves or heaters and with few independently controlled devices (what is the MIMO system). Also, the controlled object must be modelled as the space-continuous unit (i.e. with distributed parameters) in the 3D space.

3. Requirements for the model

Mathematical model is necessary to design the control installation and then to control the air conditioning process in a real time. So it must be fast computable for the predictive control, for example. Next, it should be valid within the operations limits of disturbances and control variables (temperature 0 – 40 °C., humidity 50 – 100%, wind velocity 0 – 30 m/s etc) and the model must deliver an explicit functions of design variables and control variables: temperature, humidity and velocity.

The model must be valid only for dry air, otherwise - if the condensation occurs, an emergency control program (with another model) is to be switched on, because it is very harmful for plants.

Besides, the model must offer an adequate accuracy, for example 1 °C for the temperature, 0,1 m/s for the inner air velocity, and 5 % for the air humidity.

For the research purposes the model should be of an analytical, not of an experimental character [2], [5].

4. Nominal (physical) model

Processes to be described are: mass and heat conversion within the green house (in the air and with plants) and through the walls, and between the walls and the ambient air.

Physical phenomena that are to be considered are:

1. heat conversion,
2. water/steam conversion (evaporation and condensation), mass and heat diffusion and thermo-diffusion flow of the air inside and outside of the object, and via folding windows and ventilators,
3. sunshine radiation on the soil and on plants,
4. evaporation of plants and the soil.

Critical **assumptions** for the model design are:

1. 3D model is necessary for modern greenhouses due to their extension in all dimensions;
2. Air humidity and temperature is off the dew-point (saturation point);

3. Small drops of the pressure, thus small the air velocity (Mach < 0,3);
4. No internal sources of mass or energy, except heaters, and/or water sprinklers;
5. The green-house is leak-proof and air-tight.

Simplifications

On the basis of the above assumptions, the following simplifications may be adopted.

1. Mass and heat diffusion in the inner air is neglected;
2. Air is a viscose, one-phase fluid;
3. Laminar flow of the air;
4. Mass and energy interchange with the outside atmosphere only by folding windows and ventilators;
5. No heat conduction along the walls;
6. Constant wind outside the green-house.

5. Mathematical model

Symbols

a -heat diffusion coefficient ($m^2 s^{-1}$); $Srośr$ - planting area (m^2); $Sgrz$ - heating surface (m^2); C - specific heat of the air ($Jkg^{-1}K^{-1}$); Cpr - specific heat of vaporization (Jkg^{-1}); d - steam diffusion coefficient in the air($m^2 s^{-1}$); Is - sunshine radiation intensity (Wm^{-2}); Vx, Vy, Vz, Vp - air velocity components & heating water (ms^{-1}); M - absolute air humidity inside the green-house ($kg(H_2O)kg^{-1}$); Mro - steam transpiration efficiency of plants ($kg(H_2O)s^{-1}m^{-2}$); $Nro, Ngrz$ - binary signal of the presence of vegetables/heaters; q_m - steam evaporation stream ($kg(H_2O)s^{-1}m^{-2}$); R - gas constant; T - air temperature (K); $Tro, Tgrz$ -plants, heater temperature (K); $t, \Delta t$ - time, step of time (s); $x, y, z, \Delta x, \Delta y, \Delta z$ - Space coordinates, step values (m) i, j, k - indexes of nods for coordinates ox, oy, oz; $i_{\max}, j_{\max}, k_{\max}$ - end indexes in coordinates ox, oy, oz; n -last time step index; $\alpha_{pk}, \alpha_{ro}, \alpha_p, \alpha_w$ -convection heat coefficients ($Wm^{-2}K^{-1}$); ρ - specific material density (kgm^{-3}); μ - dynamic viscosity ($m^2 s^{-1}$); ε - coefficient of the sunshine radiation absorption; ϖ - tilt angle of roof; φ -angle of the sun light.

The heat and mass conservation equations are [1]:

$$\frac{\partial Vx}{\partial t} = \frac{\mu}{\rho} \cdot \nabla^2 Vx - Vx \cdot \frac{\partial Vx}{\partial x} - Vy \cdot \frac{\partial Vx}{\partial y} - Vz \cdot \frac{\partial Vx}{\partial z} - R \cdot \frac{\partial T}{\partial x} \tag{1}$$

$$\frac{\partial Vy}{\partial t} = \frac{\mu}{\rho} \cdot \nabla^2 Vy - Vx \cdot \frac{\partial Vy}{\partial x} - Vy \cdot \frac{\partial Vy}{\partial y} - Vz \cdot \frac{\partial Vy}{\partial z} - R \cdot \frac{\partial T}{\partial y} \tag{2}$$

$$\frac{\partial Vz}{\partial t} = \frac{\mu}{\rho} \cdot \nabla^2 Vz - Vx \cdot \frac{\partial Vz}{\partial x} - Vy \cdot \frac{\partial Vz}{\partial y} - Vz \cdot \frac{\partial Vz}{\partial z} - R \cdot \frac{\partial T}{\partial z} \tag{3}$$

$$\frac{\partial T}{\partial t} = -Vx \cdot \frac{\partial T}{\partial x} - Vy \cdot \frac{\partial T}{\partial y} - Vz \cdot \frac{\partial T}{\partial z} + \frac{1}{\rho \cdot C} q_\upsilon \tag{4}$$

$$\frac{\partial M}{\partial t} = -Vx \cdot \frac{\partial M}{\partial x} - Vy \cdot \frac{\partial M}{\partial y} - Vz \cdot \frac{\partial M}{\partial z} + q_m \tag{5}$$

$$\frac{\partial Tp}{\partial t} = -Vp \cdot \frac{\partial T}{\partial x_m} + \alpha_p \cdot (Tp - T_{grz}) \tag{6}$$

$$\frac{\partial T_{grz}}{\partial t} = \alpha_w \cdot (Tp - T_{grz}) - \alpha_z \cdot (T_{grz} - T) \tag{7}$$

$$q_m = \frac{Nzw \cdot Mzw + Nro \cdot Mro}{\Delta x \cdot \Delta y \cdot \Delta z} \tag{8}$$

$$q_\upsilon = \frac{Ngrz \cdot \alpha_{grz} \cdot S_{grz} \cdot (T_{grz} - T) + Ngrz \cdot S_{grz} \cdot \rho \cdot Is \cdot \cos\varphi + Nro \cdot \alpha_{ro} \cdot S_{ro} \cdot (Tro - T)}{\Delta x \cdot \Delta y \cdot z} - \frac{C_{pr}}{C} \cdot \frac{\partial T}{\partial t} \tag{9}$$

Boundary Conditions

Boundary Conditions (*BCs*) and Initial Conditions must be defined for velocity, temperature and humidity. For example BCs for the roof in the partly incremental form are:

$$Vx = Vy = Vz = 0 \tag{10}$$

$$\frac{\partial T}{\partial t} = a \frac{\partial^2 T}{\partial z^2} + 2a \left(\frac{T^{i+1,jkn} - T^{ijkn}}{(\Delta x)^2} + \frac{T^{i,j-1,kn}}{(\Delta y)^2} \right) - \frac{\alpha_{pk}}{\rho \cdot C} \frac{(T^{ijkn} - T_z)}{\Delta x} + \frac{2 \cdot \varepsilon \cdot \cos\varphi}{\rho \cdot C \cdot \cos\varpi} \cdot \frac{Is}{\Delta x} \tag{11}$$

$$\frac{\partial M}{\partial t} = d\frac{\partial^2 M}{\partial z^2} + 2d\left(\frac{M^{i+1,jkn} - M^{ijkn}}{(\Delta x)^2} + \frac{M^{i,j-1,kn} - M^{ijkn}}{(\Delta y)^2}\right) \tag{12}$$

6. Digital model

To solve the mathematical model a computer technique is necessary. Generally, there are two possibilities:

1) to arbitrary mesh the object, elaborate a set of time continuous ordinary differential equations as incremental equations for discrete space and/or time variables and to devise a specific user-individual code for the mesh of final elements;

2) to apply a commercial MES package.

For some practical reasons the first approach was chosen. The model was converted to a fully incremental form and coded in Visual Basic language [6]. Graphical user interfaces are devised, also. User may observe results of computations: the temperature, the humidity and three components of the velocity in a specific point of the green-house as a function of time.

References

[1] W. Tarnowski, "Modelowanie systemów" (Modeling of systems in engineering) Wydawnictwo Uczelniane Politechniki Koszalińskiej (2004)

[2] H. Latała, "Wpływ zewnętrznych warunków klimatycznych na dynamikę zmian temperatury i wilgotności powietrza w szklarni" (On the influence of ambient conditions on the micro-climate in green-houses), Praca doktorska, (PhD Thesis), Akademia Rolnicza, Kraków (1997)

[3] T. Pudelski "Uprawa warzyw pod osłonami" Praca zbiorowa. Państwowe Wydawnictwo Rolnicze i Leśne (1998)

[4] C. Stanghellini, W.Th.M.van Meurs, "Environmental Control of Greenhouse Crop Transpiration" J. Agric. Eng Res (1992) 51, 297-311

[5] K. Popowski, "Greenhouse climate factors" Faculty of Technical Science, Bitola University, Bitola, Macedonia (2004)

[6]. W. Tarnowski, Bui Bach Lam "Computer simulation model of green houses for the multi-attribute control of the air-conditioning", ISSAT International Conference on Modeling of Complex Systems and Environments July 16-18, 2007 Ho Chi Minh City, Vietnam, 2007.

Kohonen Self-Organizing Map for the Traveling Salesperson Problem

Łukasz Brocki, Danijel Koržinek

Polish-Japanese Institute of Information Technology,
ul. Koszykowa 86, 02-008, Warsaw

Abstract

This work shows how a modified Kohonen Self-Organizing Map with one dimensional neighborhood is used to solve the symmetrical Traveling Salesperson Problem. Solution generated by the Kohonen network is improved using the 2opt algorithm. The paper describes briefly self-organization in neural networks, 2opt algorithm and modifications applied to Self-Organizing Map. Finally, the algorithm is compared with the Evolutionary Algorithm with Enhanced Edge Recombination operator and self-adapting mutation rate.

1. Introduction

The aim of the Traveling Salesperson Problem (TSP) is thus: given a set of *n* cities and costs of traveling between all pairs of cities, what is the cheapest route that visits each city exactly once and returns to the starting city. This problem is the leading example of NP-hard problems. Its search space is exceptionally huge (*n!*) and given that some engineering problems, like VLSI design, need as many as 1.2 million cities [5], a fast and effective heuristic method is desired.

In this paper, we present a neural based algorithm and compare it to an effective heuristic method: Evolutionary Algorithm with the Enhanced Edge Recombination operator.

2. Kohonen Self-Organizing Map for the TSP

In 1975, Teuvo Kohonen introduced a new type of neural network that uses competitive, unsupervised learning [1]. The principle of his algo-

rithm is to adapt a special network to a set of unorganized and unlabeled data. After the training phase, this network can be used for clustering and simple classification tasks.
Interesting results of self-organization can be achieved with networks that have a 2-dimensional input vector and a 1-dimensional neighborhood. In this case the input to the network can be regarded as coordinates in a 2-dimensional space: x and y. Using this technique, one can map a line over an arbitrary binary image. Furthermore, if one provides the algorithm with the same number of neurons as the number of cities it will output an efficient tour of the cities, as depicted in Figure 1.

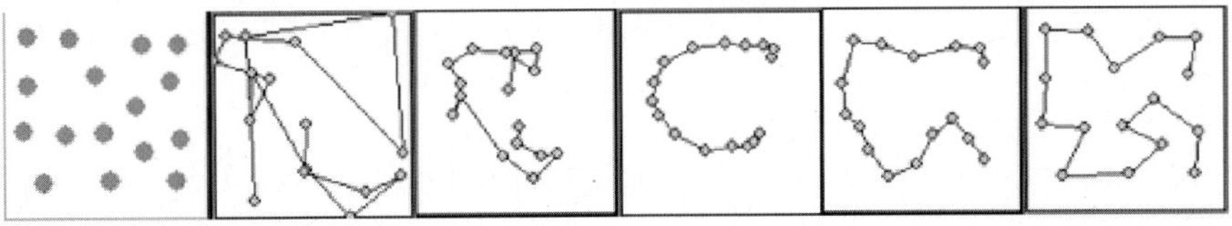

Figure 1. Solving a simple TSP problem. The example consists of six squares. The first one shows an object that is to be learned. The second square illustrates the network just after the randomization of all neural weights. Following squares illustrate the learning process. Please note that each neuron (a circle) represents a point whose coordinates are equal to the neuron's weights.

3. Modifications

Given a solution, like the one above, two things can be done to further enhance the result. First, because the algorithm works by altering the real-valued weights of the neurons it may never achieve the exact values that match the coordinates of the cities. A simple procedure was therefore created to restore the 1-1 mapping between the cities and the individual neurons.
Another improvement can be achieved by applying the well-known and fast 2-opt algorithm. This algorithm works by rearranging pairs of paths connecting the cities in a way that yields a cheaper overall tour. 2-opt provides locally optimal solutions and when starting from a random arrangement of cities, doesn't yield a perfect result. However, thanks to its simplicity it is often used in optimizing already good solutions.

4. The Experiment

Two types of tests were administered: using city sets taken from the TSPLIB [6] and using randomly chosen cities. TSPLIB city sets are quite

difficult. The reason for this is that in many cases cities are not chosen in random. Often larger city sets consist of smaller patterns. The optimal tour is therefore identical in each of the smaller patterns. SOM, on the other hand, tries to figure out a unique tour in each smaller pattern.
Testing using randomly chosen cities is more objective. It is based on the Held-Karp Traveling Salesman bound [4], which is an empirical relation between expected tour length, number of cities and the area of square box on which cities are placed. Three random city sets were used in this experiment (100, 500, 1000 cities). Square box edge length was 500.
All statistics for SOM were generated after 50 runs on each city set. Average tour lengths for city sets up to 2000 cities are around 5 to 6 percent worse than the optimum. SOM approach can generate solutions that are almost always less that 10% worse from the optimal tour. However, in most cases the difference is just a few percent.
SOM has been compared to EA coupled with the Enhanced Edge Recombination (EER) operator [2, 3], Steady-State survivor selection (where always the worst solution is replaced), Tournament parent selection with tournament size depending on number of cities and population size. Scramble mutation was used. Optimal mutation rate depends on amount of cities and state of evolution. Therefore, self-adapting mutation rate has been used. Every genotype has its own mutation rate, which is modified in a similar way as in Evolution Strategies. This strategy adapts mutation rate to number of cities and evolution state automatically, so it is not needed to manually check which parameters are optimal for each city set. Evolution stops when the population converges. Population size was set to 1000 (as in [3]). When EA stopped its best solution was optimized by the 2-opt algorithm. Results for both SOM and EA are shown in Table 1.
All statistics for SOM were generated after 50 runs on each city set. For EA there were 10 runs of the algorithm for sets: EIL51, EIL101 and RAND100. For other sets EA was run only once. Optimum solutions for instances taken from TSPLIB were already given and optimum solutions for random instances were calculated from the empirical relation described above. All computations were performed on an AMD Athlon 64-bit 3500+ processor.

		Self-Organizing Map			Evolutionary Algorithm		
Instances	**Optimum**	**Ave. Result**	**Best Result**	**Ave Time**	**Ave Result**	**Best Result**	**Ave Time**
EIL51	426	444	431	0.068	428.2	426	10
EIL101	629	662	646	0.127	653.3	639	75
TSP225	3916	4192	4106	0.302	----	4044	871

		Self-Organizing Map			Evolutionary Algorithm		
Instances	Optimum	Ave. Result	Best Result	Ave Time	Ave Result	Best Result	Ave Time
PCB442	50778	56634	55138	0.703	----	55657	10395
PR1002	259045	278481	274036	2.425	----	286908	25639
PR2392	378037	418739	411442	12.965	----	----	----
RAND100	3851.81	4051	3882	0.131	3931.4	3822	69.6
RAND500	8203.73	8888	8697	0.824	----	9261	11145
RAND 1000	11475.66	12483	12343	2.311	----	12858	56456

Table 1. The results of the experiments. Time is given in seconds. First six experiments are from the TSPLIB and the last three are random.

5. Conclusions

It seems that SOM-2opt hybrid is not a very powerful algorithm for the TSP. It has been outperformed by EA. On the other hand it is much faster. There are many algorithms that solve permutation problems. Evolutionary Algorithms have many different operators that work with permutations. EER is one of the best operators for the TSP [3]. However, it was proved that other permutation operators, which are worse for the TSP than EER, are actually better for other permutation problems (like warehouse/shipping scheduling) [3]. Therefore, it might be possible that SOM-2opt hybrid might work better for other permutation problems than for the TSP.

We are grateful to Prof. Zbigniew Michalewicz for influencing and helping us to write this paper.

References

[1] Kohonen T. (2001), Self-Organizing Maps, Springer, Berlin.
[2] Michalewicz Z. (1996), Genetic Algorithms + DataStructures = Evolution Programs, Springer – Verlag.
[3] Starkweather T., McDaniel S., Whitley C., Mathias K., Whitley D., (1991), A Comparison of Genetic Sequencing Operators.
[4] Johnson, D.S., McGeoch, L.A., and Rothberg, E.E., Asymptotic experimental analysis for the Held-Karp traveling salesman bound.
[5] Korte B., (1988), Applications of Combinatorial Optimization.
[6] Reinelt G., (1995), TSPLIB 95 documentation, University of Heidelberg.

Simulation modeling, optimalization and stabilisation of biped robot

P. Zezula, D. Vlachý, R. Grepl

Institute of Solid Mechanics, Mechatronics and Biomechanics,
Faculty of Mechanical Engineering, Brno University of Technology, Brno, Czech Republic
Institute of Thermomechanics, Academy of Sciences of the CR, branch Brno, Czech Republic

Abstract

This paper deals with proposal of humanoid robot construction. The construction of two legs (six DOF each) is described. Computational modelling was used, particularly forward and inverse kinematic model. By help of these models was produce several functions for the control of moving robot´s body. Coordination of robot move was simulated in environment VRML.

Key words: biped robot, computer modeling,

1. Introduction

The scientific field of mobile robotics represents an interesting branch for the research and development in mechatronics. This topic becomes more and more actual for the population, because by using of the different robotic manipulators and walking machines the people can make easier a lot of their work. Also, the development of manipulators bring new knowledge in mechanics, electronics, neural network and others.
In this paper, there is briefly described the design of mechanical construction biped robot Golem 2, that was built at FME Brno, University of Technology. There is issue about the stability of robot. The solution of stability allows the gait on irregular terrain.

we have solved all problem by using computer simulation in MATLAB/Simulink/SimMechanics and visualization of results in VRML.

.

2. Mechanical construction

During the design of mechanical construction we have used a several tools. First of them is forward kinematic model (FKM). This model can be symbolically described by equation (1).

$$[\alpha, \beta, \gamma, \delta, \eta, \xi] = [x, y, z, \varphi_x, \varphi_y, \varphi_z] \qquad (1)$$

Where
$\alpha, \beta, \gamma, \delta, \eta, \zeta$ are angles of joints
x, y, z are cartesian position of foot
$\varphi_x, \varphi_y, \varphi_z$ is spatial orientation of foot

Kinematic structure is shown in Fig. 1. The construction has total 12 DOF. Their disposition is following. In each hip joint, there are placed three DOF, each knee has one DOF and in area of each ankle are situated two DOF.

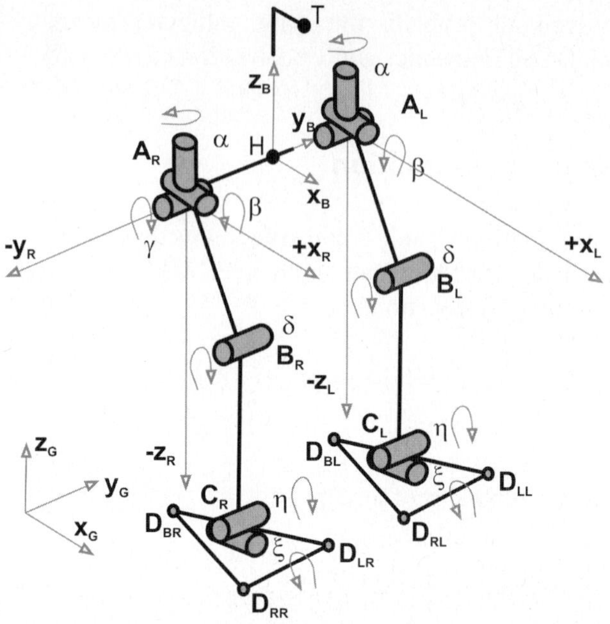

Fig. 1. Geometry of robot

If we want to obtain new servos position and we know Cartesian position of foot and spatial orientation of foot we use inverse kinematic model (IKM), that we can describe by equation (2).

$$[x, y, z, \varphi_x, \varphi_y, \varphi_z] = [\alpha, \beta, \gamma, \delta, \eta, \xi] \tag{2}$$

Further information about FKM and IKM is possible to find in [1].

3. Modelling of robot and visualization

The construction has been made true to scale in program system Solid-Works 2001 after optimalization geometric and mass parameters. So we have obtained visual image and we were able to observe crash states too. The final model of construction is shown in Fig. 2.

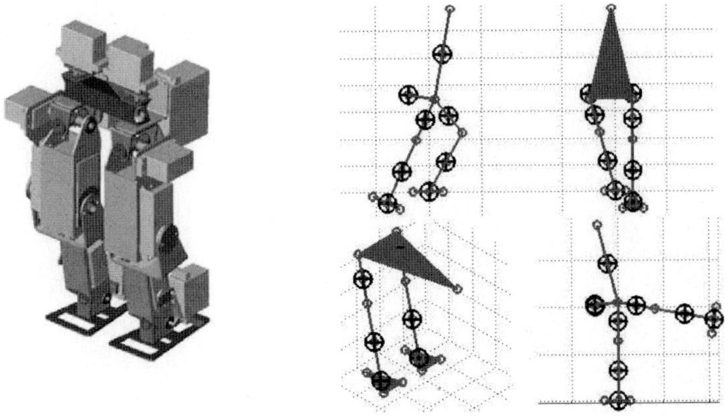

Fig. 2. Model of robot (left in SolidWorks, right in Matalb)

On base of this model were made some functions in MATLAB. We are able to coordinate move of robot by help these functions. In following text is briefly described, how inputs and outputs the functions have.
The inputs are actual positions of servos and new coordinate of body with respect to global coordinate system O-XYZ. Outputs of functions are new position of servos. By help of new servo positions we obtain require move of robot body. Those functions are based on FKM and IKM [1]. Now we are able to interpolate step between actual and new servos position and obtained data we used for visualization in VRML.

4. Stability of robot - Criterion ZMP

During the suggestion of the mechanical design, it is necessary to solve the stability problem of the robot. There is no unique definition of the problem for biped balance control. There are many position, which can become unstable and the robot can fall down. We can solve this problem by the using of workspace and we search through this workspace for stable states of robot. The criterion zero moment point (ZMP) is usually used. This point has to lie in supporting area of foot, when the robot stands on one leg or in supporting area of feet and between the feet, when robot stands on two leg. It means that we have to know position of YMP with respect 0-XYZ in every moment during walk.

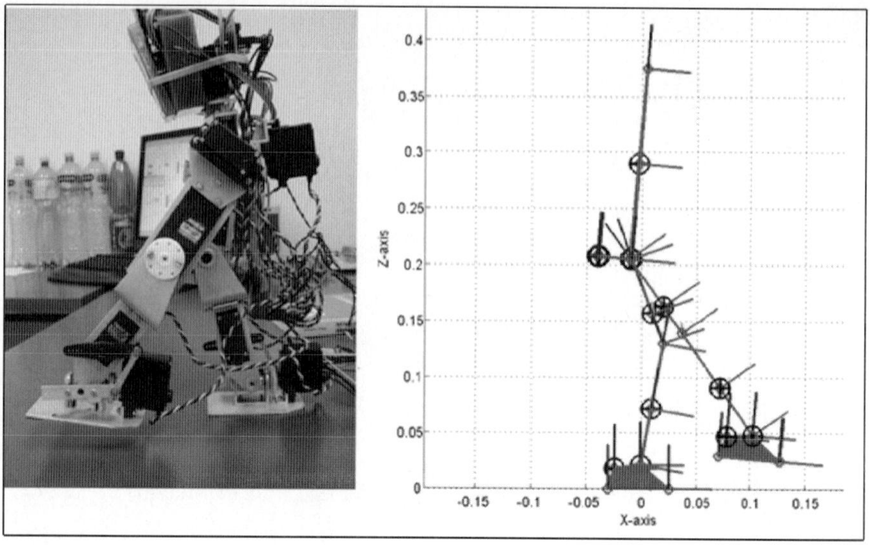

Fig. 3. Experiment

We have solved this problem by using the MATLAB environment, where we created the algorithm of computing ZMP. With help of statical model (Fig.2) we obtain contact forces between foot and ground. This contact forces have to remain of compression character and it means, that the position is stable. We measure those forces in area of tee, little toe and heel. Next we compute the actual position of ZMP from obtaining forces. This actual coordinate of ZMP is compared with required coordinated of ZMP. Required coordinates of ZMP are referred to local 0-XYZ, which is connected with ankle. We used the actual of ZMP as input to function, which calculate new position of servos. This function contains analytical inverse kinematic model of robot.

In program MATLAB/Simulink/SimMechanics we define a position, which is stable. Then we demonstrated this position on real construction of robot and we were able to verify the numerical results. The described experiment is shown in Fig. 3.

4. Conclusion

The most important tools for coordination move of robot FKM and IKM have been created in program MAPLE. For verification of FKM and IKM we used MATLAB/Simulink/SimMechanics and sequentially VRML.

Acknowledgment

The published results have been acquired be support of project AV0Z20760514 and GA□R 101/06/0063.

References

[1] R. Grepl, P. Zezula, "Modelling of kinematics of biped robot", Colloquium Dynamics of Machines 2006, Prague, February 7-8, Czech Republic, ISBN 80-85918-97-8, 2006
[2] Ch.Zhou, Q. Meng, "dynamic balance of a biped robot using fuzzy reinforcement learning agents", 2005
[3] Y. Tagawa, T. Yamashita, "Analysis of human abnormal walking using zero moment joint", 2000

Extended kinematics for control of quadruped robot

R.Grepl

Institute of Solid Mechanics, Mechatronics and Biomechanics,
Faculty of Mechanical Engineering, Brno University of Technology,
Czech Republic, grepl@fme.vutbr.cz

Abstract

This paper deals with the extended kinematical model for quadruped mobile walking robot. The model has been built based on homogenous coordinates approach. The robot is comprehended as an open tree manipulator and therefore standard algorithms for forward and inverse kinematics can be employed. However, the inverse model named 12-18 works with the redundant manipulator structure and pseudoiverse of Jacobian matrix should be used. The inverse model 3-3 uses regular manipulator and allows separate positioning of each leg. After processing and simplification of the equations in Maple, the algorithms have been implemented in Matlab environment. The model automatically built in SimMechanics has been used for verification. Finally, the results have been tested using VRML visualization.

1. Introduction

This paper describes briefly the extended kinematical model for quadruped walking robot based on homogenous coordinates. The work is aimed to particular project of robot Jaromír (Fig. 1) built at FME Brno University of Technology. Robot has in total 12 controlled servo drives, further details has been published in [2].
The control algorithms for irregular terrain as well as advanced gait and movement controllers inevitably require the 6D control of the robot body. Simple models designed relatively to body c.s. can be hardly used for such a task.

In following text, the forward and two inverse kinematical models are introduced as well as a few notes about visualization in VRML.

Fig. 1. Four legged robot Jaromír during its regular walk

2. Forward kinematical model (FKM)

The FKM positions and orients the body and all four legs of the robot relatively to global coordinate system. So, the robot is understood as an open tree manipulator with four end effectors *P*. Forward kinematics of all legs can be apparently separated to individual ones, and described by parameters $x_T, y_T, z_T, \varphi, \vartheta, \psi, \alpha, \beta, \gamma$:

$$\begin{aligned} \mathbf{r}_1^\mathbf{P} &= [x_1^P, y_1^P, z_1^P] = \mathbf{T}_{81}\mathbf{r}_8^\mathbf{P} \\ \mathbf{r}_8^\mathbf{P} &= [L_3, 0, 0, 1]^T \qquad (1) \\ \mathbf{T}_{81}&(x_T, y_T, z_T, \varphi, \vartheta, \psi, \alpha, \beta, \gamma) = \mathbf{T}_{21}\mathbf{T}_{32}\cdots\mathbf{T}_{87} \end{aligned}$$

where $\mathbf{r}_1^\mathbf{P}$ is the position of end of robotic leg in global c.s. 1, $\mathbf{r}_8^\mathbf{P}$ in leg c.s. 8. Matrix $\mathbf{T}_{81}$ incorporates the homogenous transformation between those two systems and depends on position x_T, y_T, z_T and orientation (Euler angles φ, ϑ, ψ in RPY notation) of the body and three angles of leg servo drives α, β, γ. Transformation matrixes in eq. (1) includes nine parameters overall. As an example, the $\mathbf{T}_{65}$ is shown in eq. (2). The variables with "zero" index, e.g. x_0, is used for translational and rotational offset which differs for each individual leg of robot.

$$\mathbf{T}_{65} = \begin{bmatrix} c(\alpha+\alpha_0) & -s(\alpha+\alpha_0) & 0 & x_0 \\ s(\alpha+\alpha_0) & c(\alpha+\alpha_0) & 0 & y_0 \\ 0 & 0 & 1 & z_0 \\ 0 & 0 & 0 & 1 \end{bmatrix} \qquad (2)$$

The FKM is the direct approach and the validity of implemented method can be easily tested using model in SimMechanics [3].
There are two options how to implement the FKM: a) multiply the transformation matrixes symbolically in Maple and use resulting (and fairly complex) $\mathbf{T}_{81}$; or b) multiply the matrixes numerically in Matlab to obtain $\mathbf{T}_{81}$.

3. Inverse kinematical models (IKM)

The design of IKM is rather complex problem compare to FKM due to possible singular or multiple solutions and also various requirements to be achieved. The most substantial difference of IKM is given by iterative character of computational algorithm.
There have been formed two inverse models for the robot spatial control. In both models, the desired position of end effector *P* is defined but there are different ways and consequences how to reach it.
Further, we describe them in short and in the conclusion introduce possible applications. The name of model (e.g. 3-3) indicates the number of inputs **X** and output **Q**.

3.2. Model 3-3

This model computes the three servo drives angles □, □, □ for one of the four legs while the position and orientation of the body is fixed. Algorithm explicitly allows the separation for individual legs and can be formally written as:

$$[\alpha,\beta,\gamma] = f(x_P, y_P, z_P, x_T, y_T, z_T, \varphi, \vartheta, \psi) \quad (3)$$

Let us define the vectors and rewrite eq. (3):

$$\begin{aligned} \mathbf{Q} &= [\alpha,\beta,\gamma]^T \\ \mathbf{X} &= [x_P, y_P, z_P] \qquad (4) \\ \mathbf{Q} &= f(\mathbf{X}, x_T, y_T, z_T, \varphi, \vartheta, \psi) \end{aligned}$$

Iterative algorithm (5) minimizes the error $\mathbf{X}^* - \mathbf{X}$ of the position of *P*, where $\mathbf{X}^*$ is desired and $\mathbf{X}$ actual vector of position *P*.

$$\begin{aligned}\Delta\mathbf{X}_n &= \mathbf{X}^* - \mathbf{X}_n \\ \Delta\mathbf{Q}_n &= \mathbf{J}^{-1}\Delta\mathbf{X}_n \\ \mathbf{Q}_{n+1} &= \mathbf{Q}_n + \Delta\mathbf{Q}_n \\ \mathbf{X}_{n+1} &= fkm(\mathbf{Q}_{n+1})\end{aligned} \quad (5)$$

Jacobian matrix **J** is given analytically by forward kinematical model (1):

$$\mathbf{J} = \begin{bmatrix} \frac{\partial x_1^P}{\partial\alpha} & \frac{\partial x_1^P}{\partial\beta} & \frac{\partial x_1^P}{\partial\gamma} \\ \frac{\partial y_1^P}{\partial\alpha} & \frac{\partial y_1^P}{\partial\beta} & \frac{\partial y_1^P}{\partial\gamma} \\ \frac{\partial z_1^P}{\partial\alpha} & \frac{\partial z_1^P}{\partial\beta} & \frac{\partial z_1^P}{\partial\gamma} \end{bmatrix} \quad (6)$$

The manipulator is regular, so the normal inverse of matrix can be used for the computation apart from singular configurations. During the testing of implemented algorithm properties, various

3.3. Model 12-18

The algorithm remains the same as shown in (5), however, the meanings of vectors **X** and **Q** as well as Jacobian matrix is rather different:

$$\begin{aligned}\mathbf{Q} &= [4\times[\alpha,\beta,\gamma], x_T, y_T, z_T, \varphi, \vartheta, \psi]^T \\ \mathbf{X} &= 4\times[x_P, y_P, z_P]\end{aligned} \quad (7)$$

Now, the computation is not separable for each leg. Obviously, the manipulator is redundant open kinematical tree. Redundancy causes the impossibility to invert the matrix; one should employ e.g. the pseudoinverse or dampest least square method instead [1]. For the iterative algorithm (5) Jacobian matrix has relevant complexity:

$$\mathbf{J}_{12-18} = \begin{bmatrix} \mathbf{J}_{\mathrm{leg}_1} & & & & \mathbf{J}_{\mathrm{body}_1} \\ & \mathbf{J}_{\mathrm{leg}_2} & & & \mathbf{J}_{\mathrm{body}_2} \\ & & \mathbf{J}_{\mathrm{leg}_3} & & \mathbf{J}_{\mathrm{body}_3} \\ & & & \mathbf{J}_{\mathrm{leg}_4} & \mathbf{J}_{\mathrm{body}_4} \end{bmatrix} \quad (8)$$

The submatrix $\mathbf{J}_{\mathrm{leg}_i}$ remains as defined in (6) , the $\mathbf{J}_{\mathrm{body}_i}$ forms:

$$\mathbf{J}_{\mathbf{body}_{\mathbf{i}}} = \begin{bmatrix} \frac{\partial x_i}{\partial x_{T_i}} & \frac{\partial x_i}{\partial y_{T_i}} & \frac{\partial x_i}{\partial z_{T_i}} & \frac{\partial x_i}{\partial \varphi} & \frac{\partial x_i}{\partial \theta} & \frac{\partial x_i}{\partial \psi} \\ & & \cdots & & & \\ \frac{\partial z_i}{\partial x_{T_i}} & \frac{\partial z_i}{\partial y_{T_i}} & \frac{\partial z_i}{\partial z_{T_i}} & \frac{\partial z_i}{\partial \varphi} & \frac{\partial z_i}{\partial \theta} & \frac{\partial z_i}{\partial \psi} \end{bmatrix} \qquad (9)$$

4. Conclusion

The algorithms briefly characterized above have been processed in Maple. Symbolic modifications and simplification of algebraic expressions have been performed and result has been implemented in Matlab environment with the SimMechanics applied for verification.
The model 3-3 can be used for positioning of individual legs together with the positioning and orientation of the body. The model 12-18 requires and guaranties the positioning of legs only. So, the body state is out of control. In fact, such approach allows the less restricted manoeuvrability of the robot but also could produce inappropriate behaviour due to uncontrolled body. The resulting computational models have been successfully tested using VRML visualization in Matlab.

Acknowledgment

Published results were acquired using the subsidization of the Ministry of Education, Youth and Sports of the Czech Republic, research plan MSM 0021630518 "Simulation modelling of mechatronic systems" and the project GACR 101/06/P108 "Research of simulation and experimental modelling of walking robots dynamics".

References

[1] Buss, Samuel R.: Introduction to Inverse Kinematics with Jacobian Transpose, Pseudoinverse and Damped Least Squares methods, Department of Mathematics University of California, San Diego La Jolla, 2004
[2] Grepl, R.: Simulating modelling of quadruped robot gait control, Journal Engineering Mechanics, Volume 12, Nr. 4, ISSN 1210-2717, 2005
[3] Grepl, R.: Virtual prototype of autonomous locomotion walking robot, Simulation modelling of mechatronic systems I, ISSN 80-214-3144-X, pp. 29-37, 2005

Application of the image processing methods for analysis of two-phase flow in turbomachinery

M. Śleziak *

* University of Opole, Natural-Technical Faculty,
Technology Department, Opole, 45-365, Dmowskiego 7, Poland

Abstract

The aim of this research is an application of digital image analysis for working out the method, which will allow to evaluate irregularity rate of two-phase flow across various geometry of tube bundle in aspect of the shell - and - tube heat exchanger optimization. Visualization of liquid flow in the shell – side enables an analysis of flow parameters by the use of image processing and analysis methods.

The images of two-phase flow were obtained by the use of digital high speed CCD camera, then were analysed, to obtain information about hydrodynamics of flow with respect to tube bundle arrangement. Optical techniques of measurement, based on correlation algorithms, allow accurate determination of stabilization of velocity fields for the whole field of flow in the shell-side. The aim of work was also an estimation of the influence of geometrical parameters (tube arrangements, tube spacings, number of rows) on the homogeneity of two-phase flow in the heat exchangers along with complex analysis of flow area, especially the still zones.

1. Introduction

The image processing techniques have been used extensively in many different applications. In particular, in fluid mechanics, image processing has become a powerful technique to study the flow phenomena, the flow pattern and the flow characteristics of two-phase flow. One of the recent progress is development of a set of algorithms for spotting recognition and tracking objects in a time-varying image. These algorithms enable to realize an automatic annotation of video image capturing of moving objects. In various types of chemical apparatus two-phase gas-liquid flows occur very often. The two-phase flow takes place along and across tubes bundle in the heat exchangers. An application of the flow visualization and digital image analysis method allow to determine local and overall flow parameters, and investigation of hydrodynamics of liquid flow inside the tube bundle, basing on techniques of digital image processing. The classical shell – and – tube heat exchangers are commonly applied for heat transfer in flow apparatus [1]. In order to increase liquid flow velocity inside the shell – side, the flow is intensified by the use of sectional cross baffles, and minimizing superficies of heat exchangers. Two-phase mixture flow is characterized by significant fluctuations. Fluctucations of two-phase flow are result of complicated geometry and oscillatory nature of two-phase flow in the shell-side. The shell-side defines the flow area of two-phase mixture, where velocity of flow is maximum in small passages of the shell-side [2]. The areas behind the tubes surface exhibit little fluctuations and became named "still zones" (Fig.1). The velocity of flow and the gas void fraction in the still zones are considerably smaller than in passages of the shell-side. The shell-side decides about local disturbances of velocity fields, trajectories of gas flow, large area of all groups of gas particles concentration. The optical techniques as well as digital image analysis might be used for operation, estimation and optimization of the apparatus and two-phase flow setups.

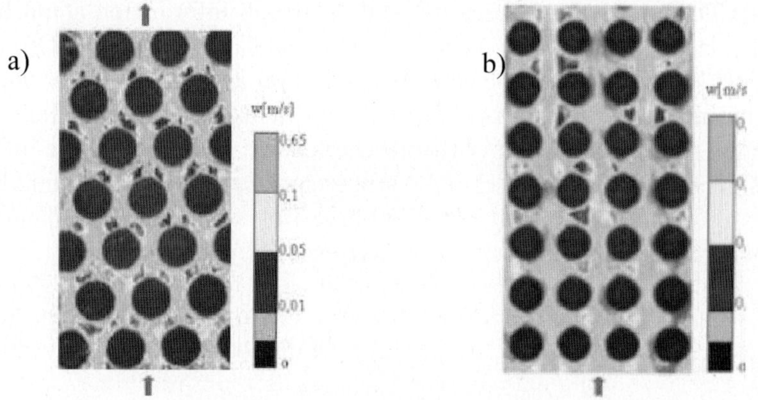

Fig.1.Still zone visualization behind row of tubes
a) triangular - staggered arrangement b) square in-line arrangement

2. Digital particle image velocimetry correlation algorithm

Digital image anemometry by the use of particle tracers called Digital Particle Image Velocimetry (DPIV) is a technique which allows determination of velocity vectors of liquid flows by means of the image correlation method [3]. Quantitative measurements of the flow and turbulence characteristics are obtained using DPIV. This is a non-intrusive optical technique that measures fluid velocity by tracking the displacement of tracer particles that have been added to the fluid. The particle locations are determined by capturing scattered light from the individual particles with a digital camera. The particle displacement in a small region of the image is calculated by comparing consecutive images. The velocity in the small region is then calculated by dividing the displacement distance by the known time delay between the images. The DPIV images were sub-divided into interrogation windows. The calculation of the correlation function to determine the displacement vector for each window can be performed either in the spatial domain or in the frequency domain. The digital spatial correlation function R_{II} required the evaluation of the following expression [4]:

$$R_{II}(x_1,x_2)=\sum_{i=-K}^{K}\sum_{j=-L}^{L} I(i,j)I'(i+x_1,j+x_2) \qquad (1)$$

where I(i, j) represents the intensity value for the (i , j) pixel. This function statistically measures the degree of correlation between two samples I (i , j) and I'(i , j) for a given shift (x_1,x_2). The shift position where the pixel values align with each other gives the highest cross-correlation value, and represents the average displacement of the particles in a given interrogation window. The algorithm used to evaluate the particle displacements for this experiment was based on the frequency-domain correlation method. The images were divided into 24 × 24 pixel sub-windows with 50% overlap. The correlation plane was created by transforming the intensity function from the spatial to the frequency domain using the discrete Fourier transform [4]. The cross-correlation was then obtained by using the Fast Fourier Transform (FFT) algorithm which is computationally efficient since it utilizes the symmetry property between the even and odd coefficients of the Discrete Fourier Transform. The FFT cross-correlation method resulted in a single peak on the correlation plane that represented the average displacement of the particles in the window during the time delay between the illumination pulses. The position of the correlation peak was then estimated to sub-pixel accuracy by using Gaussians - fit function [5]. The displacement vector was defined by the location of the peak with respect to the center of the window. The final algorithm filtered the vector fields. The vectors collected at a single spatial location were combined into a time record of velocity. The spatial correlation function is defined as:

$$f(r_j) = \frac{u_i^{'}(x+r_j)u_i^{'}(x)}{u^{'}_i(x)u^{'}_i(x)} \tag{2}$$

where x is the arbitrary location, $u^{'}_i$ indicates the fluctuation of the velocity component, and r_j is the distance in the j direction between the velocity measurements.

Visualization of liquid flow in the shell – side enables analysis of flow parameters by the use of image processing and analysis methods. Images were recorded with frequency 103Hz by the use of the digital high speed CCD camera. The bubbles of gas in shell-side are particle markers. The recording sequences of flow were conducted for two-phase gas-liquid flow across tube bundle, which were placed in triangular - staggered and square in-line arrangement for bubbles pattern. The velocities of phases were following for liquid V_L=(0,20 – 0,70) m/s and gas V_G=(0,50 – 7,0) m/s. The sequence of consecutive images was recorded during two-phase flow in shell-side, next were determined velocity fields of liquid using the correlation of images method (cross correlation DPIV) - (Fig.2). Dislocation of markers floating through liquid, in given interval of time, makes possible to receive the flow parameter which characterizes velocity of liquid flow. Distributions of velocity fields in dependence of geometry of tubes arrangement, staggered either in-line, have indicated values of the heat transfer coefficient.

a)

b)

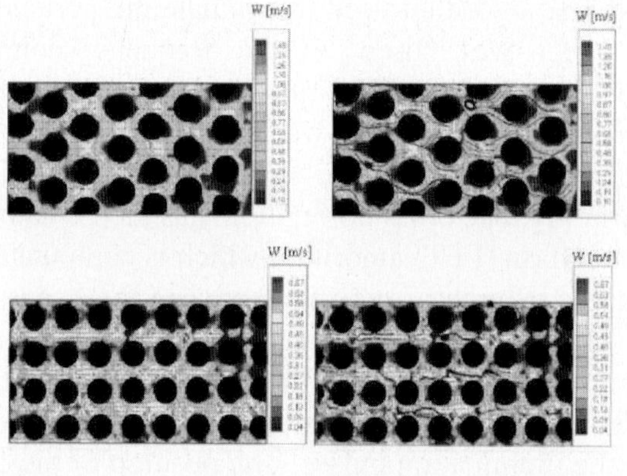

Fig.2. Liquid velocity fields and particles motion in the shell-side
a) Triangular - staggered arrangement b) Square in-line arrangement

Trajectories of particle markers motion were designed for the flow pattern estimation. Changes of flow direction were evaluated using motion of trajectory analysis, an area, where flow has stopped as well as still zones, where number of particle markers is not large (Fig.2).

The graphs showing velocity fields are presented in "figure 2", where area of local disturbances of velocity were located, caused by geometry of selected arrangement. Thus the pattern of tube arrangement has caused local turbulences. The consecutives of tube rows has an influence on change of trajectory of particles motion (Fig.2) as well as causes vortex generation behind tubes. Liquid flow around tubes is irregular for triangular – staggered arrangement also for this one there is a larger flow dynamics, where flux of liquid strikes in every row of tubes. It has enlarged the coefficient of heat transfer value for which arrangement is larger than other geometrical arrangements, but pressure loss is larger.

3. Conclusions

The results of flow visualization and digial processing methods allowed to draw detailed conclusions to the point of the hydrodynamics of two-phase flow in an area of shell-side. The conclusions are:

- Optical techniques of measurement, based on correlation algorithms, allow accurate determination of stabilization of velocity fields for the whole field of flow in shell-side.
- The two-phase flows characterized by large of turbulences and irregularitis of liquid flow around the tube bundle, for staggered arrangement especially.
- On the basis of distribution of velocity fileds there have been ascertained that staggered arrangements, which are applied in heat exchanger, are more efficient with regards to the flux of heat transfer. The triangular arrangement in actual heat exchangers is equivalent to the staggered arrangement.

References

[1] T. Hobler "Heat transfer and heat exchangers",WNT,Warszawa ,1986.
[2] K. Lam, C. Los "A visualization study of cross-flow around four cylinders in a square configuration, Journal of fluids and structures, 1992.
[3] R.J. Adrian "Imaging techniques for experimental fluid mechanics", Annual Reviews of Fluid Mechanics, 1991.
[4] A.Brathwaite "A novel laboratory apparatus for simulating isotropic turbulence at low Reynolds Number", Phd. Thesis, Georgia Institute of Technology, 2003.
[5] W.Suchecki "A visualization of flows by the used digital image anemometry", Engineering and Chemical Apparatus, Gda□sk, 2000.

Optoelectronic Sensor with Quadrant Diode Patterns Used in the Mobile Robots Navigation

D. Bacescu, H. Panaitopol, D. M. Bacescu, L. Bogatu, S. Petrache

"POLITEHNICA" University, 313, Spl. Independentei,
Bucharest, 060042, Romania

Abstract

In the last years, the authors' primary concerns were to design an optoelectronic family sensors. The sensors of this family are intended for the control equipments used in mobile robots navigation. The equipment is composed of two identical optoelectronic sensors. It works through the triangulation principle and it is not embarked on the structure of the robot. This fact makes possible the miniaturization of the mobile robot construction.
The authors have lately conceived and developed an optoelectronic sensor with photovoltaic quadrant. These optoelectronic sensors enable the equipment to have a rapid reaction when the target, with a embarked light source, changes its position in the working space.

1. Introduction

Many a time, in case of the minirobots, the sensorial equipment occupies a big volume comparable with their own gauge. As a result, the equipment cannot be mounted on the minirobot. Moving the sensorial equipment from the minirobot onto a follow-up and control installation independent with respect to this, the minirobot design becomes simpler and easier to be manipulated (actuated).
In this view, the authors were developed some equipments, which are based on the triangulation method and the comparison between the position of a minirobot into real workspace and the equivalent position of a virtual minirobot into virtual workspace. In this case, the real workspace has to be known a priori, with all the forms and dimensions of the obstacles. Furthermore, for the real workspace has to be created a digitizing equivalent map that represents the virtual workspace, memorized by a

computer. The equipment for tracking and adaptive control determines, in real-time, the minirobot position into the real workspace. Any command given to the minirobot at a certain moment of time is previously checked into the virtual workspace. For instance, if virtual minirobot movement into virtual workspace is valid then a command is send to the real robot for moving.
If virtual minirobot movement leads to an impossible situation into the real world (e.g. a stroke with an obstacle) the software will find the right solution into virtual workspace and will send the appropriate commands thus the real minirobot to be able to avoid the obstacle.

2. Family of optoelectronic sensors

A family of optoelectronic equipment for a mobile control in a laboratory workspace was created by the authors in order to be used at the mobile robots controlling. The first and the second optoelectronic sensors belonging to the same family are presented in fig.1. The disadvantages of these two models are the following: a small and fix stereo basis; low precision in measurement.

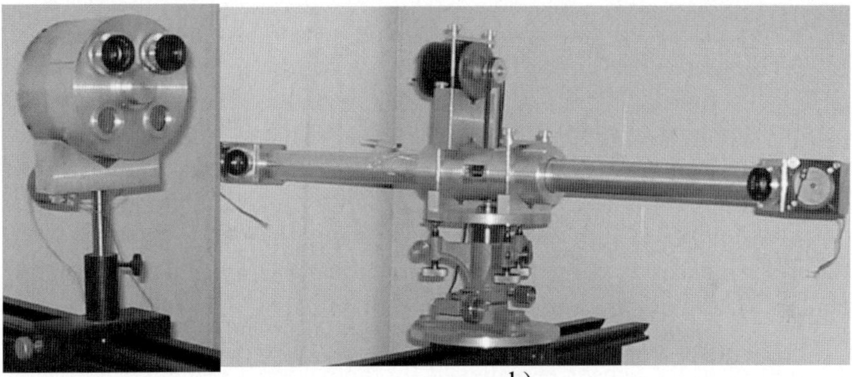

a) b)

Fig. 1. The first (a) and the second (b) optoelectronic sensors

The third optoelectronic sensor of the mentioned family determines the spatial position of a number of P_j light sources, placed on a movement object (e.g. on a minirobot that must be controlled). The photo of this sensor is shown in fig. 2. The optoelectronic sensor is in fact an installation based on three identical modules, named M_1, M_2 and M_3, which materialized a three-dimensional coordinate system at which the controlled displacement of a mobile object (minirobot – m_r) is referred. Because of the fix and relatively small stereo basis (but much more bigger than the models in fig.1), the working space controlled with an acceptable precision is limited at 2 m^2, a situation which is tolerated in laboratory condition.

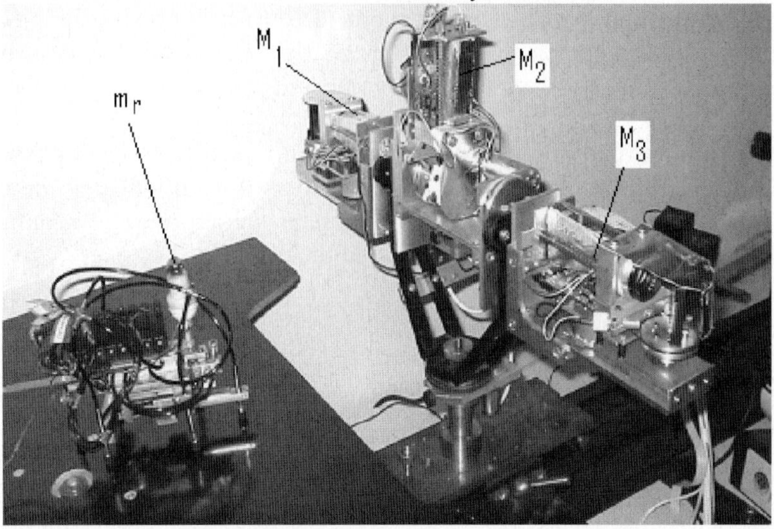

Fig.2. Photo of the optoelectronic equipment and the minirobot

The schematic of the fourth optoelectronic sensor of family is shown in fig. 3. The sensor has an objective Ob and a quadratic photosensitive sensor FE with four identically photoconductive zones. When the optical axis is on the direction of light source S, the signals from the four photoconductive zones are equal and if it is known the position of the axis is known the direction ObS. The tracking movement of light source S is produced by the stepping motors M_1 and M_2 by means of two warm-gear reducers m_1-rm_1 and m_2-rm_2. Measurement of the rotational angle α, is achieved by angle encoder $TIRO_1$, and measurement of rotational angle β, is achieved by angle encoder $TIRO_2$.
The control of the sensor is achieved by the personal computer by means of data acquisition board. The signals from the photoconductive zones give information about the orientation of deviation from direction of light source S and the motors are maneuvered so that to minimize this deviation. In the moment in which all the signals are equal, the axis of the sensor is on light source direction S, it is transferred in the memory of personal computer the content of the incremental counters of electronically module and is compute the angles α and β.
This optoelectronic sensor is able to tracking and determines, permanently the direction to a light source, without to determine the distance between sensor and the light source. To determine the distance to the light source needs two identical sensors mounted so that the triangulation principle to

be applied. In this case the system enable to determine accurately two coordinates of the light source irrespective of its position. The disadvantage is that the system has a large price.

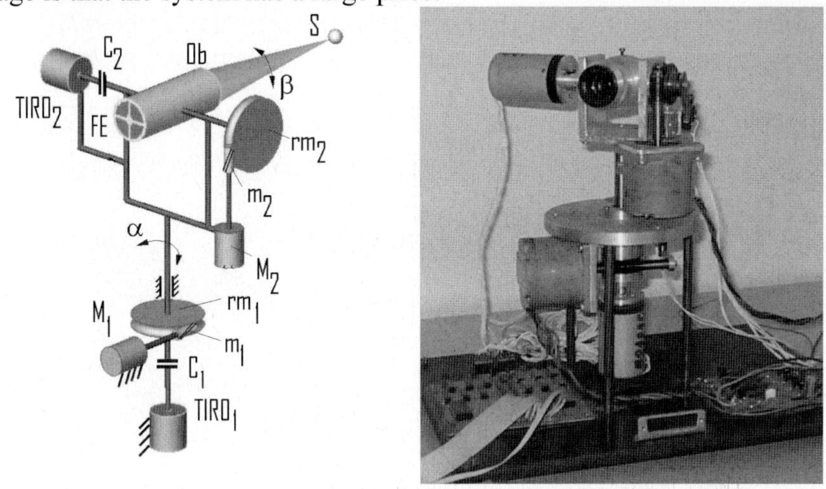

a) b)

Fig. 3. The first optoelectronic sensor with quadrant diode patterns
a) schema of the sensor; b) photo of the sensor

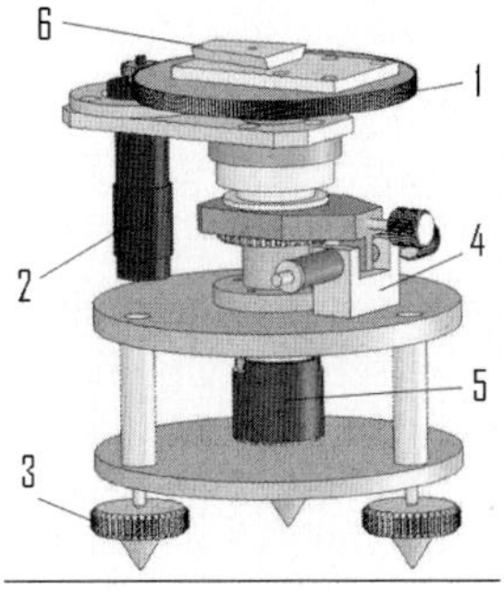

Fig. 4. The research platform

Fig. 5. Photo of the research platform with the quadrant diode and radiolocation sensors

After the studies and researches performed in the last time in this field, the authors propose the development of a research platform (figs. 4 and 5) and with a high flexibility. The modular construction of this one allows the endowment with one or several types of sensors at the same time, in order to study their behaviour. The sensors can be optic, radio waves etc.

Optic sensors, able to realize a sampling process-by means of which the signal is received and processed at equal periods and not continuously, have been tested on the platform, as well as the radiolocation sensors and the optic sensors with continous processing (quadrant diode patterns).
Some experimental data for the quadrant diode patterns are shown in the table 1. The quadrant diode sensor is shown in fig. 6. They refer to the repetitiveness of the information offered by the sensor with regard to the special position of the target (z - distance), for a 1,5 m stereo basis.

Table 1

	Basis stereo [mm]	X [mm]	Y [mm]	Z [mm]
1	1500	27,22	64,29	853,14
2	1500	26,96	64,48	853,00
3	1500	27,43	64,08	852,99
4	1500	27,13	63,99	853,43
5	1500	26,92	64,56	852,85

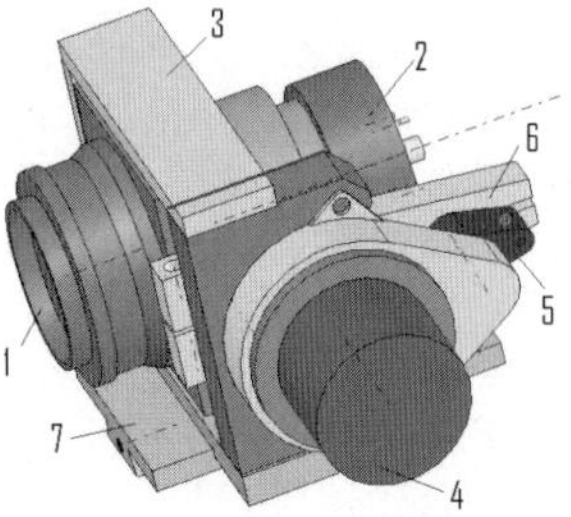

Fig. 6. Quadrant sensor: 1-lens; 2-quadrant; 3-chassis; 4-gearmotor; 5-eccentric; 6-crank guide; 7-adapter

3. Conclusions

The research platform that the authors propose in this paper enable the study of the behaviour of several types of sensors. Through its modular construction, it offers the possibility of studying one or several types of sensors at the same time. Among the different types of sensors which were studied so far, it is obvious that the quadrant diode sensors offered the best performances on the platform.

References

[1] D. M. Bacescu, D. Bacescu, H. Panaitopol, ”Aplicatii mecatronice bazate pe telemetria optoelectronica”, Medro, Bucuresti, 2005.
[2] D. Bacescu, H. Panaitopol, D.M. Bacescu, L. Bogatu., Mecatronica 2/2004, Romania, 16.

Mathematical Analysis of Stability for Inverter Fed Synchronous Motor with Fuzzy Logic Control

P. Fabijański, R. Łagoda
pawel@isep.pw.edu.pl; lagoda@isep.pw.edu.pl

Institute of Control and Industrial Electronics,
Warsaw University of Technology, Koszykowa 75,
Warsow, 00-662, Poland

Abstract

This paper deals with fuzzy logic control of inverter fed synchronous motor. In this article mathematical model of the motor drive system and stability condition of fuzzy logic control algorithm are represented. The essential fuzzy logic control configuration system consists of an internal current regulation loop and an external speed regulation loop. The speed and current loop in these systems are generally controlled by PI controller.

1. Introduction

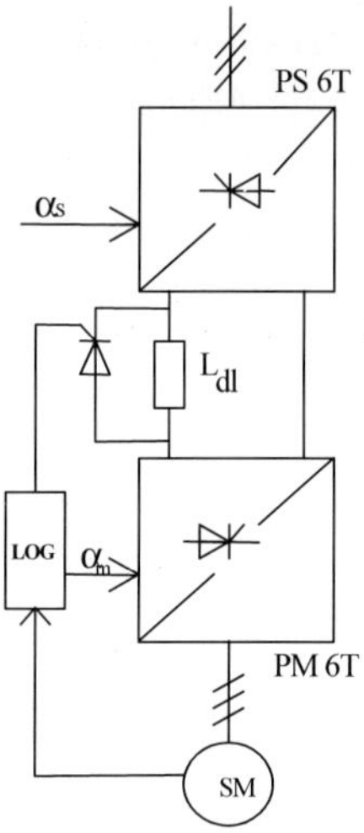

Fig. 1. General configuration of inverter fed synchronous motor

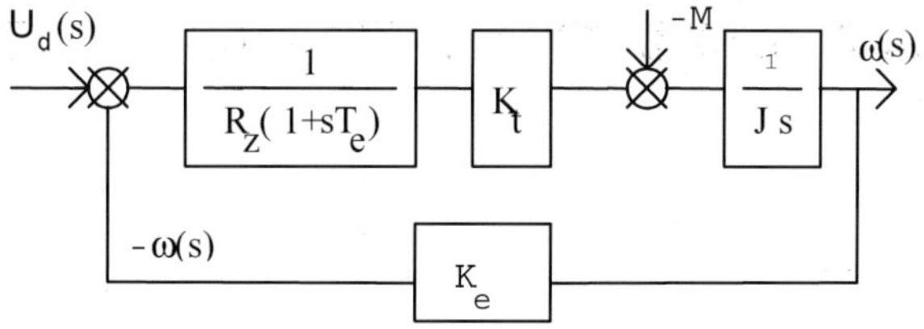

Fig. 2. Inverter fed synchronous motor-operating diagram

Our experimental drive system with fuzzy logic control of inverter fed synchronous machine consists essentially of a fully phase controlled six-pulse thyristor bridge converter operating essentially as a rectifier, dc intermediate stage with smoothing inductance and a six pulse thyristor converter working as an indirect type frequency converter (Fig.1). A fuzzy logic control and digital proportional and integral control method are proposed for electrical drive system with inverter fed synchronous motor. Block diagram for inverter fed synchronous machine is shown on Fig.2

2. Control scheme description

To control a invert fed synchronous machine a host computer with special microprocessor 1102 dSPACE. The host computer serves as a developing environment, it is in charge of the real-time program debugging, the execution code downloading, real-time parameter and state monitoring, and real-time gain auto-tuning. The fuzzy gain self-tuning algorithm is realized as follow: for the host computer, each cycle it synchronizes with the microcontroller and receives the sampling data in the precedent cycle, it then computes the sliding trajectory and the corresponding performance index. After this, suitable gains of the PI controller derived from the cycle information are calculated and send back to the microcontroller

3. Current and speed control

The current and speed control structure include the fuzzy logic controller. The speed control is achieved by self-tunning regulator. Inputs of the speed loop are the speed feedback and its reference. The output represents the references for the current loop. The inner control current loop is shown on Fig. 3.

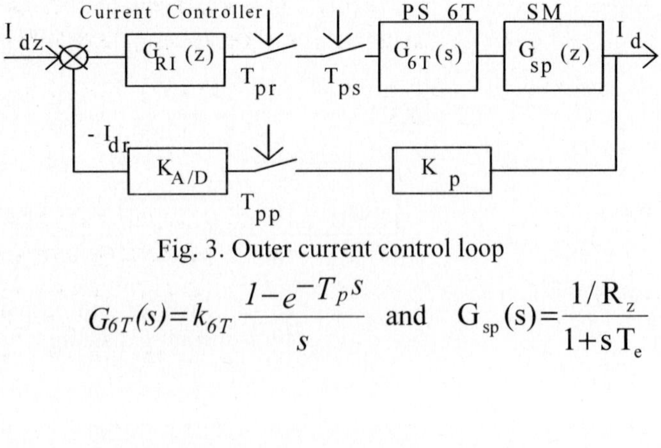

Fig. 3. Outer current control loop

where: $$G_{6T}(s)=k_{6T}\frac{1-e^{-T_p s}}{s} \quad \text{and} \quad G_{sp}(s)=\frac{1/R_z}{1+sT_e} \qquad (1)$$

We use a proportional-integral type regulator, defined by following function:

$$\alpha_S(nT_p) = k_p I_{de}(nT_p) + k_i \sum_{n=0}^{k} I_{de}(nT_p) \tag{2}$$

The recurring equation for angle of six pulse transistor bridge is:

$$\alpha_1(nT_p) = k_p I_{de}(nT_p) \quad \text{and} \quad \alpha_2(nT_p) = \alpha_2(n-1) + k_i I_{de}(nT_p) \tag{3}$$

and finally for PI current controller we have

$$\alpha(nT_p) = \alpha_1(nT_p) + \alpha_2(nT_p) \tag{4}$$

The inner loop of the control structure include the fuzzy controller. The basis of the fuzzy controller, the fuzzy rules, contain the relationship between the linguistic input variables and the linguistic output variable, which describe the conclusion to the state of the system.

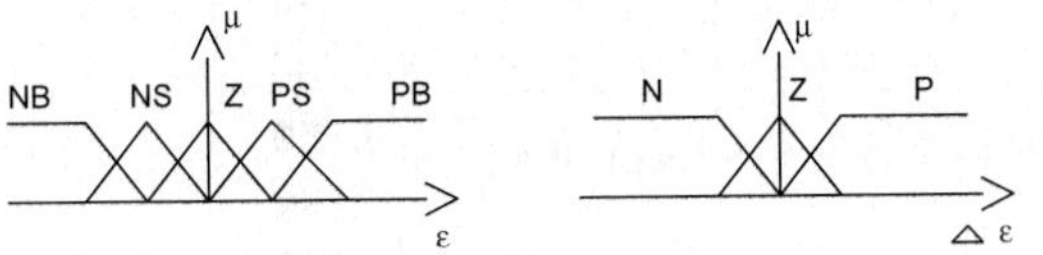

Fig. 4. Membership function for control current errors

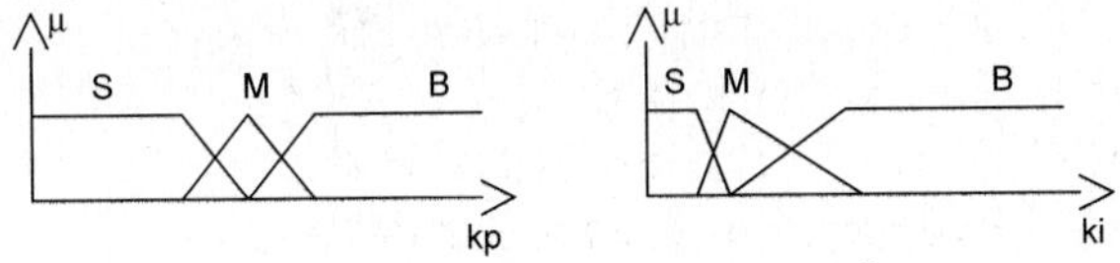

Fig. 5. Membership function for parameters of current controller

ε \ dε	N	Z	P
PB	S	S	M
PS	S	S	M
Z	S	M	B
NS	M	S	S
NB	M	S	S

ε \ dε	N	Z	P
PB	B	S	M
PS	B	S	M
Z	B	M	S
NS	M	S	B
NB	M	S	B

Tab. 1. Control matrix for kp and ki current controller parameters

The transfer function for inner current controller loop is:

$$G_1(z) = \frac{I_d(z)}{I_{dz}(z)} = \frac{G_{RI}(z)G_2(z)}{1 + K_{cl}\, G_{RI}\, G_2(z)} \quad \text{where:} \tag{5}$$

$$G_{RI}(z) = \frac{z(K_p + K_i) - K_p}{z-1} \quad \text{and} \quad G_{ob}(s) = k_{6T}\, \frac{1-e^{-T_p s}}{s}\, \frac{1/R_z}{1+sT_e} \tag{6}$$

Now we obtain the following transfer function for current regulation close loop:

$$G_1(z) = \frac{a_1 z + a_0}{b_2 + b_1 + b_0} \tag{7}$$

The stability condition for current controller are determined by using:

$$b_2 + b_1 + b_0 \rangle 0 ; \quad b_2 - b_0 \rangle 0 \ ; \quad b_2 - b_1 + b_0 \rangle 0 \tag{8}$$

The proposed control structure under application of fuzzy logic was tested with a 600W synchronous motor. In this case the stability condition for kp i ki current controller parameters are shown on Fig.7.

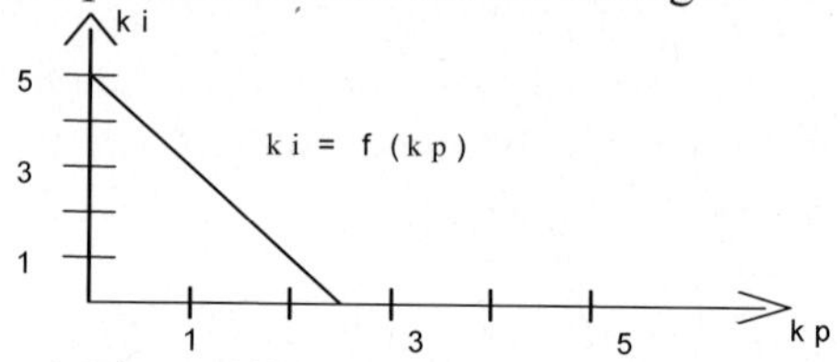

Fig. 6. The stability condition for current controller

The block diagram of speed controller and outer loop is shown on Fig.7, 8

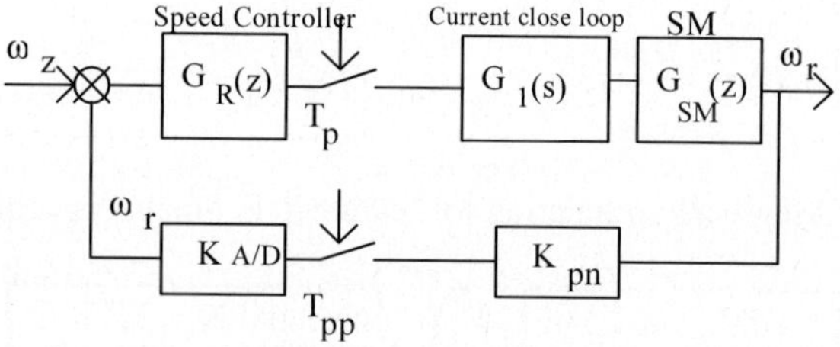

Fig. 7. Outer speed control loop, where: $G_{SM} = k/Js$

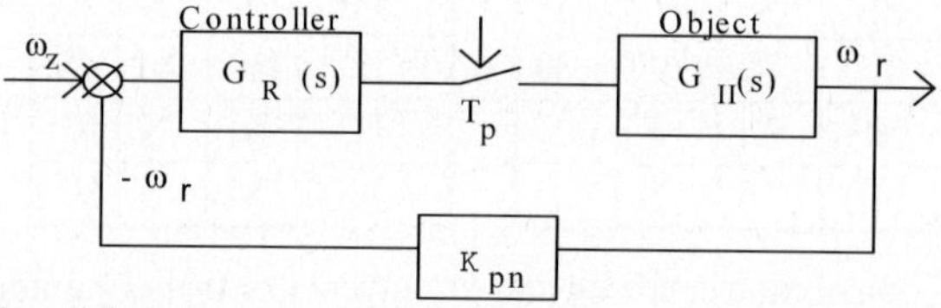

Fig. 8. Simplified diagram of speed regulator for outer regulator loop

$$G_{\omega}(z) = \frac{\omega_r(z)}{\omega\ (z)} = \frac{G_R(z)G_{II}(z)}{1 + K_p G_R(z) G_{II}(z)} \tag{9}$$

and then we have: $G_{\omega}(z) = \dfrac{A_1 z + A_0}{B_2 z^2 + B_1 + B_0}$ (10)

The stability condition are determined by using:

$$B_2 + B_1 - B_0 \rangle 0 ; \ B_2 - B_0 \rangle 0 ; \ B_2 - B_1 + B_0 \rangle 0 ; \ B_1 \rangle 0 \tag{11}$$

The stability condition for speed controller are shown on Fig.9.

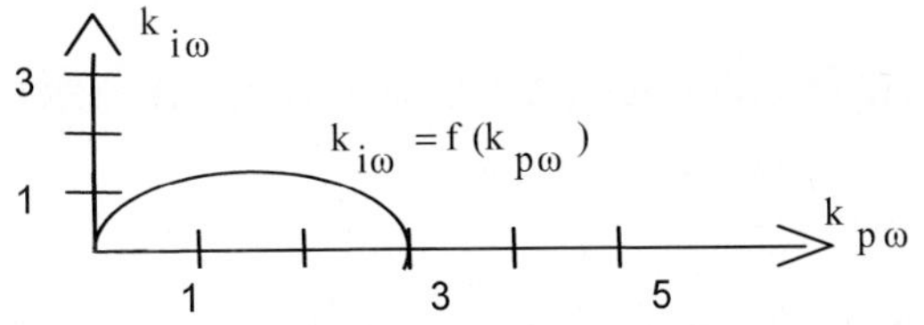

Fig. 9. The stability condition for speed controller

The laboratory test result on 600Wsynchronous motor was shown on Fig 10

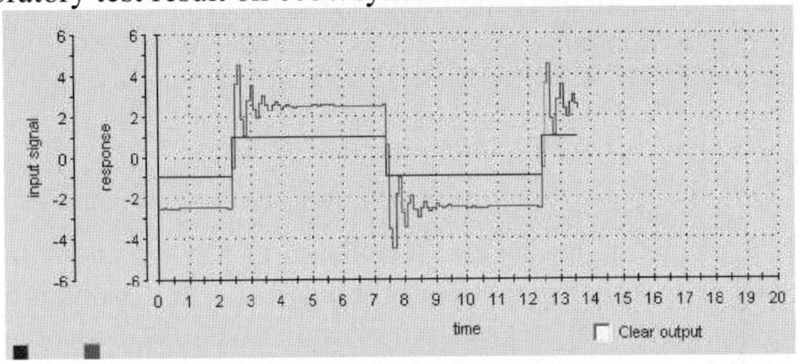

Fig.10. Output speed signal in case of change parameters of speed controller

4. Conclusion

The most important results of our investigation is description fuzzy logic control method of inverter fed synchronous motor. The stability condition are determined for current and speed controller. The simulated result were compared during the laboratory test on 600 W synchronous motor.
Another results of our investigation is description self-tuning fuzzy logic control method of inverter fed synchronous motor. The novel method is derived from a detailed analysis of the cycle information, it has been fully tested with a inverer fed synchronous motor drive. The experimental results show that the proposed algorithm has the feature of simplicity, versatibility and stability.

5. References

[1] A. Zajaczkowski, Indirect adaptive decoupling control of a permanent magnet synchronous motor, 20-th Seminar on fundamentals of electrotechnics and circuit theory, SPETO 1999

[2] S. Grundmann, M. Krause, V. Muller, Application of Fuzzy control for PWM voltage source inverter fed permanent Magnet Synchronous Motor Proceedings of the EPE 03, pp524 - 529

The influence of active control strategy on working machines seat suspension behavior

I. Maciejewski (a) *

(a) Koszalin University of Technology, Racławicka 15-17,
Koszalin, 75-620, Poland

Abstract

The paper presents the model and simulation of passive and active earth-moving machines seat suspension. The object of the simulation is the visco-elastic passive seat suspension, extended with the active control. In order to help the working machines operators against vibration, the active system with different control strategies is elaborated. Active system improves significantly the behavior of the seat suspension at low frequency excitation, with the most effectiveness obtained for the resonance frequency. As the results of the computer simulation, the power spectral densities of acceleration for a seat is presented in comparison with sample excitations on the earth-moving machines cabin floor. Additionally, the transmissibility functions for a passive suspension and the corresponding active suspension are shown.

1. Introduction

Truck drivers and operators of earth-moving machines during their work are running a risk of vibration, most often coming from surface unevenness [1]. The vibrations that occur in a typical earth-moving machines cabin ranges in 0 – 20 Hz [3, 4, 5]. This situation is very unfavorable, because a large majority of natural frequencies of human body organs are in the same range. This leads to the loss of the concentration, tiredness and decrease of the effectiveness of the work being performed.

The main dangerous range of vibration frequency for human body is between 4 – 9 Hz, because a large majority of human body parts and organs are in that range [1, 2]. Passive suspension of serial produced seats amplifies the low frequency vibrations, for the sake of resonance.

2. Inverse criteria of vibro-properties of seat suspension

The main problem of seat suspension control strategy is the two opposite criteria:

- minimization of absolute acceleration of a loaded seat,
- minimization of relative displacement of the seat suspension.

The correct control policy of the seat suspension should be done by minimizing the objective function [4, 6]:

$$J = c_1 \cdot E\left[\ddot{x}^2\right] + c_2 \cdot E\left[(x - x_s)^2\right] \rightarrow \min \qquad (1)$$

where c_1 and c_2 are weighting coefficients, depends on the significance of concurrent criteria.

On the one hand, the absolute acceleration on the seat should approach zero to protect driver's health. On the other hand, maximum relative displacement between the seat and floor of operator's cabin should approach zero as well in order to ensure the controllability of the working machines. The best compromise between concurrent criteria creates a large optimization problem.

3. Physical and mathematical model of passive and active seat suspension

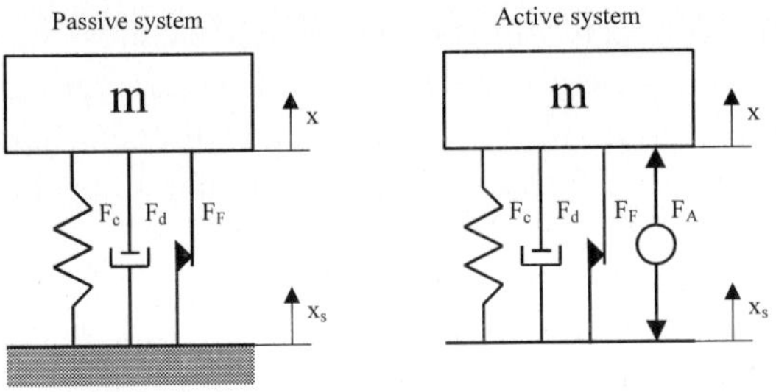

Fig. 1. Physical model of passive and active seat suspension

Simplified physical model of passive and active seat suspension are shown on fig. 1. Designations from physical and mathematical model are presented after equation (4).

Equation of motion for a passive seat suspension is given by:

$$m\ddot{x} = -d \cdot (\dot{x} - \dot{x}_s) - c \cdot (x - x_s) - F_F \cdot sign(\dot{x} - \dot{x}_s) \tag{2}$$

Equation of motion for an active seat suspension is given by:

$$m\ddot{x} = -d \cdot (\dot{x} - \dot{x}_s) - c \cdot (x - x_s) - F_F \cdot sign(\dot{x} - \dot{x}_s) + F_A \tag{3}$$

Active force generator is described as the first inertial element:

$$T_o \cdot \dot{F}_A + F_A = k \cdot U(t) \tag{4}$$

where: m – sum of mass of driver and seat, x – displacement of seat, x_s – displacement of cabin floor, d – viscous damping coefficient, c – stiffness, F_c – spring force, F_d – damping force, F_F – friction force, F_A – active force, T_0 – time constant of active force generator, k – proportional gain of active force generator, U(t) – control voltage of active force generator.

4. Control strategy of an active seat suspension

The active control systems used in simulation are presented on fig. 2a, 2b. The typical, proportional regulator for a sky-hook damper control strategy is used. The voltage signal from the acceleration sensor is fed to the input of the regulator and controls the active force generator. This system works by means of the "Sky-hook damper" algorithm [6] and ensures minimization of the absolute velocity of loaded mass. The second system based on double feedback loop control strategy, which ensures the selection between decreasing of acceleration acting on a driver and minimizing of relative displacement of seat suspension Computer model of passive and active seat suspension is performed using the simulation packet MATLAB – Simulink.

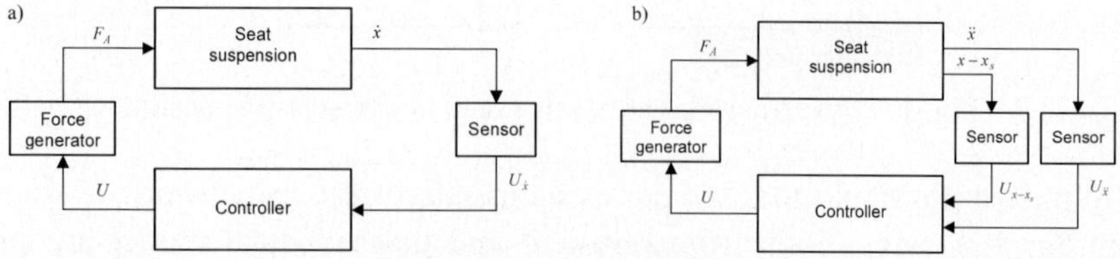

Fig. 2. General view of control system: a) sky-hook damper controller, b) double feedback loop controller

5. Comparison of simulation and experimental results

The experimental set-up consists of hydraulic shaker, vibration platform with mounted and rigid loaded seat suspension. For evaluation of seat suspension behavior, the white, band limited noise as excitation signal is used. During the test there were measured the following signals: acceleration of vibration platform, acceleration of loaded mass, relative displacement of suspension system and absolute displacement of vibration platform. Based on these signals, the Power Spectral Densities of acceleration and Transmissibility Functions are evaluated and shown on fig. 3.

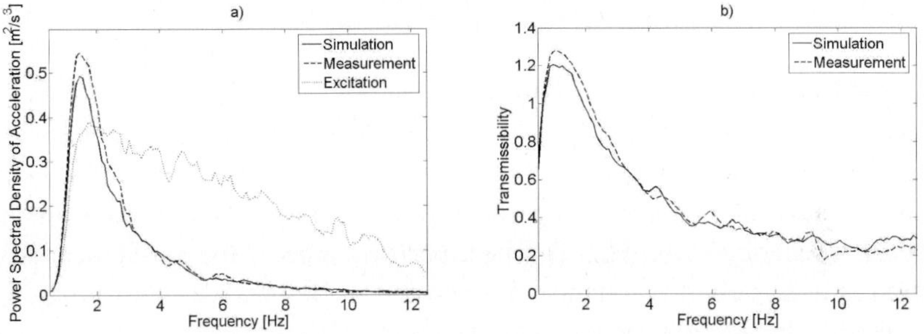

Fig. 3. Measured and simulated PSD (a) and transmissibility curves (b) of seat suspension

The results of simulated seat suspension are slightly better, than measured. The Seat Effective Amplitude Transmissibility factor (SEAT) [2] from simulation is 0,491 and from the measurement 0,504. The maximum relative displacement of suspension system from simulation is 68 mm and from the measurement 71 mm.

7. Simulation results for a different control strategy of an active seat suspension

The simulation investigations are performed for a different control strategy: sky-hook damper and double feedback loop control. The Seat Effective Amplitude Transmissibility factor (SEAT) [2] is lower about 34 % for sky-hook damper control in comparison with passive one, about 16 % for double-loop control. The maximum relative displacement of suspension system is lower about 6 % for sky-hook damper control in comparison with passive one, about 12 % for double-loop control. The PSD and Transmissibility for different control strategy are shown on fig. 4.

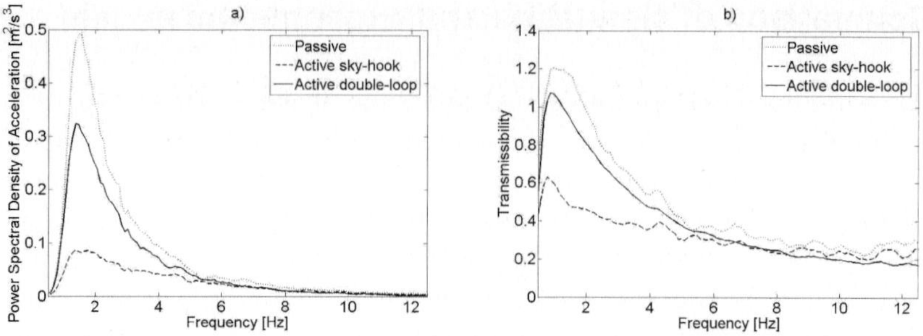

Fig. 4. Simulated PSD (a) and transmissibility curves (b) of seat suspension for passive and corresponding active system at different control strategy

8. Conclusions

The results of simulation shows, that active seat suspension with sky-hook damper control significantly improves vibro-properties of the seat in the range of frequency 0 – 6 Hz. The best performance of the active seat suspension is achieved at resonance frequency for the passive seat (about 1,5 Hz). In this case, the reduction of maximum relative displacement of seat suspension is on a low level. The active seat suspension with double feedback loop control improves both concurrent criteria: the acceleration on the seat and the maximum relative displacement. This control strategy allows the choice to be made for desired vibro-isolation properties of the active seat suspension.

References

[1] Engel Z.: Ochrona środowiska przed drganiami i hałasem. Wydawnictwo Naukowe PWN, Warszawa 1993.

[2] ISO 2631: Mechanical vibration and shock – Evolution of human exposure to whole-body vibration. 1997.

[3] ISO 7096: Earth-moving machinery – Laboratory evaluation of operator seat vibra-tion. 2000.

[4] Kowal J.: Sterowanie drganiami, Gutenberg. Kraków 1996.

[5] Krzyżyński T., Maciejewski I., Chamera S.: Modeling and simulation of active system of truck seat vibroisolation with biomechanical model of human body under real excitations. VDI Berichte Nr. 1821, 2004.

[6] Preumont A.: Vibration Control of Active Structures An Introduction, Kluwer Academic Publishers. London 2002.

Verification of the walking gait generation algorithms using branch and bound methods

V. Ondroušek, S. Věchet, J. Krejsa, P. Houška

Institute of Automatization and Computer Science, Faculty of Mechanics Engineering, Brno University of Technology, Technická 2,
Brno, 61669, Czech Republic

Abstract

The contribution is focused on generation of walking gates for quadruped robot using heuristic search state space algorithms. The efficiency of classical A* algorithm is improved by using branch and bound methods. Simulation verification shows reduction of number of states space nodes generated during the search.

1. Introduction

Automatic generation of robot walking gaits belongs to common requirements of mobile robotics. One of the possible approaches is use of algorithms based on state space search. Our team lately successfully tested A* and Beam-search algorithms. Efficiency of those algorithms can be theoretically improved by branch and bound methods. This paper describes our experiences with combining A* algorithm and branch & bound method used for automatic generation of quadruped robot walking gait. Tests were performed using simulation software, see [3,4]

2. Used Approach

Choosing the appropriate walking gait belongs to the set of planning tasks. The aim of such a task is to find the optimal path, in our case defined as a sequence of states and operators that perform transitions between states. Each state represents a particular configuration of the robot. The rule for robot´s configuration change represents the operator realization and such operator performance thus creates a new state. Therefore the whole task can be internally represented using continuous deterministic graph (tree). To find a solution for such task, informed methods of the state space search can be successfully used, for example A-star or Beam Search. Further improvement can be obtained by combination of those algorithms with branch & bound methods [6]; methods which refuse solution evidently worse than solutions already found during the initial stages of state space search – so called branching of the tree. To refuse the solution certain still acceptable limit evaluation of the node is used, so called bound. Among algorithms further developing branch & bound methods we can mention e.g. Futility cut off, Waiting for quiscence or Secondary search. Algorithm of A* with branch and bound:

1. Set bound to infinity
2. Set maximum depth for branching
3. Determine actual configuration of the robot
4. Using depth first search generate sub-tree whose tree represents actual configuration of the robot.
   a) Each newly created state evaluate using A* cost function.
   b) If the state currently evaluated is a leaf (i.e. it is located in maximum depth) compare its evaluation with bound value. If the bound value is higher, then

   Bound := leaf evaluation,

   Evaluated state note as temporarily the best one and remember its position

   Finish expansion of the parent of currently evaluated leaf and continue in expansion according to depth first search.
5. Based on remembered position of the best node from step 4 determine the rule which lead from actual configuration of the robot (root node) to the subtree containing the leaf with the best evaluation.
6. Use rule determined in step 5 on to actual configuration of the robot
7. Set newly created state as actual configuration of the robot
8. Repeat step 4 until actual configuration of the robot reaches goal state.

One can see from the description of the algorithm that bound value changes during the search, it is monotonously falling. Generated tree branching, which represents the main difference against classical A* algorithm appears in step 4b.

3. Implementation

The proposed walking gait generation algorithms needed to be tested in a financially and time undemanding way. That is why a virtual prototype of four-legged walking robot performing planary movement (constant distance of the robot body above the surface is considered) was designed. It is a software simulation developed in Borland Delphi 6. While designing the software simulation we need to take into consideration both the robot's behavior and its interaction with environment, as well as errors occurred during servo-motors positioning etc. With regard to the above stated requirements the virtual model comprises these main parts: [2]
- Module of simplified kinematic model of the robot in 2D
- Module of introduction of errors (environment simulation)
- Module of walking gait generation (AI algorithms implementation)
- Main simulation module
- Module of data representation

By errors we mean the errors in servodrives positioning which are unavoidable in real application. Such errors bring the necessity of gait replanning, when planned action can not be used due to the difference in expected and real state of the robot.

4. Obtained Results

Using above described software the tests were performed comparing results of A*, beam search and branch & bound algorithms To compare the results the cost function previously exhibiting the best results was used [2].

$$h^*(i) = k_1 d(i) + k_2 \Delta(i) + k_3 step(i) + k_4 move(i) + k_5 rot(i), \qquad (1)$$

where:

d(i) represents geometric distance between the *i* state and the target state,

$\Delta(i)$ gives us the deflection of the *i* state from robot's ideal direction,

step(i) is the number of leg movements of the robot on its path from the initial state s_0 to the *i* state.

move(i) is the number of translatory movements of the robot on its path from the initial state s_0 to the *i* state,

rot(i) is the number of rotational movements of the robot on its path from the initial state s_0 to the *i* state.

The constants k_i were defined experimentally: $k_1 = 10$, $k_2 = 50$, $k_3 = 3$, $k_4 = 3$, $k_5 = 5$.

Maximal depth of sub-tree expansion, (see algorithm description, step 2) was experimentally set to MaxDepth=3. Higher depth significantly increases algorithm response time; lower depth results in unacceptable walking gait. Quantitative comparison of walking gaits from [0,0] position to [0,9] position using A*, and "A* & Branch and bound" algorithms is shown on tab. 1.

Tab.1: Influence of introduced errors on the state space size,
RL = robot's length, RPN = number of path replanings

a) A-star algorithm

erros included	no of states in		RPN
	path	state space	
no errors	27	17540	0
0.02 [0.6%RL] 1.15 [deg]	26	23990	0
0.04 [1.3%RL] 2.30 [deg]	31	22165	0
0.06 [2%RL] 3.45 [deg]	34	20920	1
0.08 [2.6%RL] 4.60 [deg]	30	37290	2

b) A star algorithm with branch & bound

erros included	no of states in		improvement [%] of states in state space	RPN
	path	state space		
no errors	24	6674	62 %	0
0.02 [0.6%RL] 1.15 [deg]	24	6674	72 %	0
0.04 [1.3%RL] 2.30 [deg]	25	6895	68 %	0
0.06 [2%RL] 3.45 [deg]	28	7740	63 %	1
0.08 [2.6%RL] 4.60 [deg]	30	10130	72 %	2

Comparison of both algorithms used for walking gait generation shows that branch & bound method significantly reduces numbed of expanded states, around 68% in average and therefore reduces computational requirements and thus the algorithm response time. Resulting robot gait was comparable and does not contain any redundant leg movements. Translatory movements of the body are sufficiently large. Resulting walking gait can be considered as energetically efficient.

5. Conclusion

This paper shows the usage of branch & bound modification of A* algorithm on automatic generation of a four-legged robot's walking gait. This modification brings only slight increase the algorithm complexity together with substantial reduction of number of expanded states generated thus reducing overall computational requirements and corresponding algorithm response time, which is advantageous mainly in cases where replanning of the gait is necessary due to the servodrives position errors.

Published results were acquired using the subsidization of the Ministry of Education, Youth and Sports of the Czech Republic, research plan MSM 0021630518 "Simulation modelling of mechatronic systems".

References

[1] Mařík V., Štěpánková O., Katanský J., a kol., Umělá inteligence 1, Academia, 1993

[2] Ondroušek V.; Březina, T.; Krejsa J.; Houška, P.: Using Virtual Prototype for the testing of algorithms generating robot's walking gait, Simulation Modelling of Mechatronic Systems II, pp.121-130, ISBN 84-3341-80-21, (2006), VUT v Brně

[3] Ondroušek V., Březina T., The Automatic Generation of Walking Policies for a Four-legged Robot in a Nondeterministic Space, Sborník národní konference Inženýrská mechanika, Svratka, 2006.

[4] Ondroušek V., Březina T.,Krejsa J., The Walking Policies Automatic Generation Using Beam Search, in: Proc. of the 12th International Conference on Soft Computing MENDEL 2006, Brno, Czech Republic, pp.145-150.

[5] Pearl J., Heuristics, Intelligent Search Strategies for Computer Problem

[6] Rich E., Knight K.: Artifical Intelligence - Second Edition.McGraw-Hill, Inc., New York, 1991.

[8] Řeřucha V., Inteligentní řízení kráčejícího robota, Vojenská akademie v Brně, 1997

[7] Solving, Addison-Wesley, Reading, Mass., 1984

Control of a Stewart platform with fuzzy logic and artificial neural network compensation

F. Serrano, A. Caballero, K. Yen (a), T. Brezina (b)

(a) Florida International University, 10555 W. Flagler St. Miami, FL 33174, USA

(b) Brno University of Technology, Brno, Czech Republic

Abstract

The paper is focused on the analysis of the Stewart platform proposed for the development of a device that will be used in the determination of mechanical properties of materials that should substitute spinal segments of human bodies. Due to the nonlinear characteristics of the Stewart platform, classical control techniques, such as PID control, are difficult to design for this kind of mechanical system. Therefore, a fuzzy controller was developed to minimize the positional and rotation errors for the platform in the task space. An ANN compensator was designed to improve the performance of the fuzzy controller and a simulation was done to compare the two controllers.

1. Introduction

The Stewart platform is a parallel robotic device used in many applications; industrial, flight simulators and for this case, biomechanical applications. One application was developed at the Brno University of Technology by the group leaded by Dr. Tomas Brezina for experimental modeling of arbitrary load and movement on the spinal cord [2]. In this paper, a fuzzy logic controller is suggested in order to develop an alternative method to find the appropriate control signal to minimize the errors of the platform trajectory.

2. Kinematic and Dynamic Model

The kinematic model [4] is derived using a base frame located in the center of the platform. This model is represented by the vectors q and b, a translation vector denoted by t, and the platform vector p which is transformed respect to the base frame using the rotation matrix $\Re$ for the six legs. The vector S represents the length and position of each leg and is given by:

$$S_i = \Re p_i + t - b_i \tag{1}$$

To derive the dynamic equations of the Stewart Platform [4], the Newton-Euler equations must be applied to calculate the position t and orientation α.

$$J\begin{bmatrix}\ddot{t}\\ \alpha\end{bmatrix} + \begin{bmatrix} m(\omega \times (\omega \times R) - g) \\ \omega \times I\omega + mR \times ((\omega \cdot R)\omega - g) \end{bmatrix} = HF(t) \tag{2}$$

Where J is the inertia matrix, m is the mass of the platform, ω is the angular velocity, g is the gravitational acceleration, R is the position vector of the platform, and H is the transformation matrix for the input force F.

3. Design of the fuzzy logic controller

This fuzzy controller was done based on the standard PD like rules [1]. The whole system has 150 rules defined in table 1 (25 rules for each input). Due to presence of nonlinear and highly coupled terms, it's difficult to find the gain matrices using traditional methods and the linearization of the dynamic system introduces nonlinear approximation errors on the platform trajectory.
The fuzzy controller structure is given in [1] and two 6x1 vectors:

$e = [e_t \quad e_\alpha]^T$ and $\dot{e} = [e_v \quad e_\omega]^T$ where e is the trajectory error and $\dot{e}$ is the error derivative respectively; these two vectors conforms the 12 inputs for the fuzzy inference system. The output vector F is a 6 x 1 vector which represents the 6 actuators to move the platform. Each variable (input and output) has 5 membership functions NB, NS, ZO, PS and PB and the membership functions were done using the inductive reasoning algorithm.

Table 1. Rules for the fuzzy controller

F_i		de_i/dt				
		NB	**NS**	**ZO**	**PS**	**PB**
e_i	**NB**	*PB*	*NS*	*ZO*	*PS*	*PB*
	NS	*PS*	*PS*	*PS*	*PS*	*ZO*
	ZO	*PS*	*PS*	*PS*	*ZO*	*NS*
	PS	*PS*	*ZO*	*ZO*	*NS*	*NS*
	PB	*ZO*	*NS*	*NM*	*NS*	*NB*

4. ANN Reference Compensation Technique

Some authors have suggested several compensation techniques such as the H_∞ compensator for adaptive control [3]; the RCT is an on-line training method used to get accurate values for the error signal without modifying the rules from the fuzzy logic controller. The schematic diagram is shown in figure 1, the ANN compensator has two inputs; the control input and the delay input for online training to be used with the function v explained in the next paragraph. It's important to take in count the sampling rate and the computation time for real time application to avoid delays which can yield instability in the Stewart platform dynamic model.

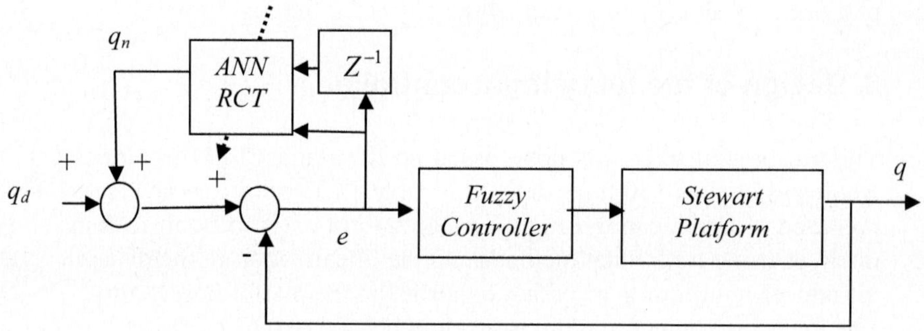

Fig. 1. Fuzzy Controller and ANN Compensator

Define three 6x1 vectors q_d, q, and q_n which represent the desired trajectory, actual trajectory and the output of the compensator. In order to find the target function to train the compensator, the target function v is computed using the estimated matrices k_1 and k_2 given by the fuzzy controller. This function is represented by:

$$v = k_1 \varepsilon + k_2 \dot{\varepsilon} \tag{3}$$

$$\varepsilon = q_d - q \tag{4}$$

5. Simulations

All the parameters used for this simulation such as mass, inertias and platform dimension are from Brezina's previous work [2]. The input was a sinusoidal deflection in ± 5 mm and a range of $\pm 10^{\circ}$ keeping the control signal in the range of 2000 N with frequency 2 Hz. The network architecture used was 6 input neurons (translations and angles) and 6 outputs (forces) and 12 hidden neurons. The back propagation algorithm was used and the learning rate was 0.6. In figure 2 there is a comparative graphic for the PD-like fuzzy controller with and without compensation. You can see the reduction of the elongation error of one leg using the ANN compensator for the fuzzy controller.

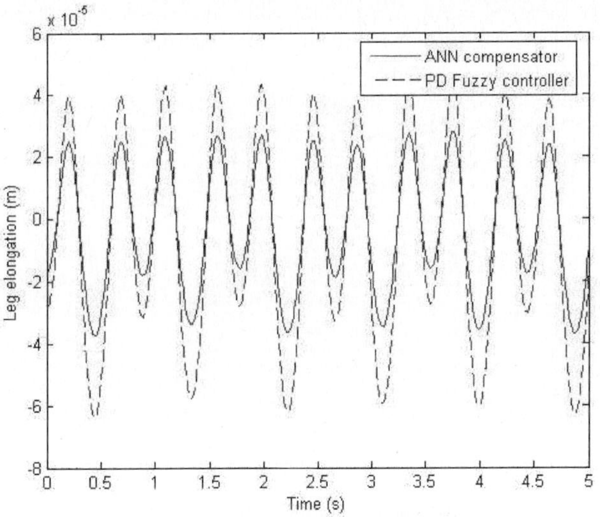

Figure 2. Error for the leg prolongation using a fuzzy controller

5. Conclusions

In this paper two alternatives are shown to compare the performance according to the error of the prolongation course of each leg. It can be noticed that in order to compensate some small changes in the rules of the fuzzy system due to perturbation, an ANN compensator helps to reduce this error in real time applications but it is very important to take into consideration the sampling time for the training of the RCT compensator. The RCT controller is very useful in this kind of applications, because it's not necessary to linearize the dynamic model, so there are no losses in the computation of the states when the system is used in real time.

References

[1] Huang Yongli, Yasunobu (2000), A general Practical design for fuzzy PID control from conventional PID control, *IEEE conference.*

[2] Brezina, T. (2005), Device for Experimental Modeling of Properties of Biomechanical Systems, Simulation Modeling of Mechatronic Systems, Brno University of Technology, Faculty of Mechanical Engineering.

[3] Se-Han Lee, Jae-Bok Song, Woo-Chun Choi and Daehie Hong Position (2003) Control of a Stewart platform using inverse dynamics control with approximate dynamics, *Mechatronics*, Vol. 13, No. 6, pp. 605-619

[4] Bhaskar Dasgupta and Mruthyunjaya (1998) *Mechanical and Machine Theory*, Vol. 33, No. 7, pp. 993-1012.

Mechanical carrier of a mobile robot for inspecting ventilation ducts[1]

M. Adamczyk

Department of Fundamentals of Machinery Design,
Silesian University of Technology, Konarskiego 18a
Gliwice, 44-100, Poland

Abstract

The focus of this paper is on design and development of a miniaturized mechanical carrier of a mobile robot for inspecting ventilation ducts. Work environment of this robot are ducts made of steel sheet. To enable this robot to easy move in this environment several problems have to be considered and then solved. The authors considered several conceptions. They chose two of them for further research. One of them was a four-leg walking platform. The other was a wheeled platform. 3D CAD systems were used for simulation of kinematics and dynamic behavior. They chose wheeled platform and then developed a real model of this platform. The platform is equipped with four magnetic wheels. Every wheel is independently driven by a step motor. The platform and all elements of the mechanical carrier are made from aluminum alloys. This platform can move on floor, walls and ceilings of round and square ventilation ducts using magnetic forces.

1. Introduction

Inspection robots are one of sub domains of Artificial Intelligence (AI) that are subject to rapid development in the last decade. New achievements in electronics enabled dozens of spectacular applications of mobile robots especially in such places and situations where operation of

[1] Scientific work financed by the Ministry of Science and Higher Education and carried out within the Multi-Year Programme "Development of innovativeness systems of manufacturing and maintenance 2004-2008"

humans is dangerous or even impossible: in small canals/ducts, in explosive or radioactive hazards atmosphere, in fire, etc. Autonomous inspection robots allowed exploring mysteries of deep sea, and penetrating areas inaccessible to humans [1].

The paper deals with a project of design and development of a miniaturized mechanical carrier of a mobile robot for inspecting ventilation ducts [5]. Project is made by a group of engineers from Department of Fundamentals of Machinery Design. There are many design, program and conceptual problems that are of special interests to researchers. One of the most important problems is climbing on vertical walls and ceilings in ventilation ducts. The proposed solution of inspection robot is equipped with magnetic wheels, eight actuators, several sensors [3] and specially designed control system [2].

The paper is composed as follows. In the next section research problems are formulated. Third section deals with design problems of a real robot focusing on the mechanical skeleton, actuators, low level controller and sensors. The paper ends with some conclusions.

2. Research problem

The main problem addressed by this project is to find an easy method of free movement in ventilation ducts of different but standardized diameter and shape. The authors set up some restrictions [5]. The robot in the first stage has to work only within square and round ducts. Diameter of round canals is limited from 300 mm to 700 mm and dimensions of square canals are limited from 250 to 600 mm. The minimal radius of elbows is 300 mm. Different sections of ventilation ducts can be connected by means of suitable adapters.

The inspection robot will have to drive in horizontal and vertical canals. To achieve this goal several problems should be considered and then solved. One of the ideas was to jostle opposite walls for climbing vertical sections. There are several robots using this conception but it is necessary that the ducts do not change the dimensions in large range. In our case it is impossible to use this idea. Other idea was to use some material properties such as magnetic or suction elements.

The authors assumed [5] that the speed of robot should be at least 0,02 [m/s] and the load capacity should be more than 0,6 [kg]. The robot should be capable of working autonomously; therefore it should be equipped with batteries and an energy-saving system.

The most essential problem concerns determination how the inspection robot should move in its environment. Two main approaches were considered for further research. One of them was a four-leg walking platform. The other was a wheeled platform. 3D CAD systems were used for designing and simulating kinematics and dynamic behavior. The authors chose wheeled platform and then developed a real model of this platform

3. Design

The first stage of design was to develop a mechanical body with actuators. Then, the robot was equipped with the respective control system [2]. Finally it was armed with sensors which allowed it sensing the environment [3]. Several concepts were created and then evaluated with respect to such criteria, as: high mobility by limited number of power transmission units, and ability to drive the robot through its inspected environment. Simultaneously, costs must not be too high and further development and modifications of the robot should be easily possible. Moreover, modifications and possibly repairs of the robot should be easy to carry out.

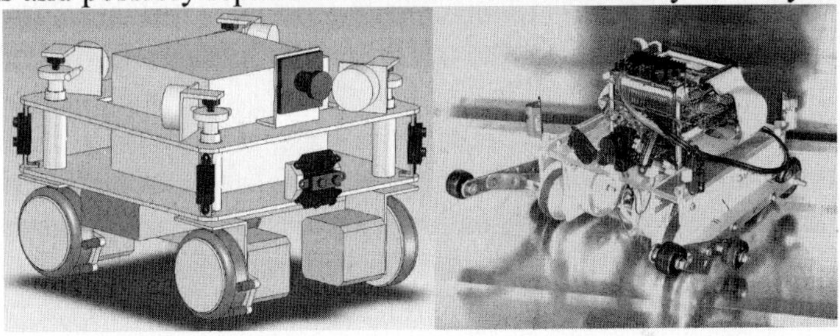

Fig.2. Model and testing version of inspection robot

The robot has been designed using CATIA 3D system. A virtual verification of concepts and a final design took place in simulation modules of CATIA. Fig.1. shows the virtual model of the inspection robot. All the elements were subject to FEM calculations to lower masses and inertia of the real object. Drawings obtained by means of CAD software were send to manufacturer that machined robot parts using CNC machining tool.

3.1. Mechanical skeleton

Elements of the skeleton have been obtained from aluminum and light material by CNC milling machine. Advantages of this solution are twofold. First, elements of complex shape could be machined precisely.

Then, aluminum elements allow obtaining low mass of the skeleton by relatively high durability features. The first testing version of the skeleton equipped with control and drives is shown in Fig.1.

Very important was climbing system of inspection robot. Two main approaches were considered: suction and magnetic foots. Author chose to use magnetic wheels. They use neodymium magnets in wheels for climbing vertical walls. This approach has several important advantages: most important is low mass and dimension if compared with magnetic forces they produce, low cost is also important.

3.2. Actuators

There are many possible solutions of drives for a welled robot. In this project only electric drives were considered, including step motors and permanent motors with control of rotations. The step motors were finally selected due to such features as small dimensions and weight if compared with torques that they produce. The complete drive system implements eight motors from two types with similar parameters. Every wheel has 2 DOFs – one for rotation and one for left/right movement.

3.3. Low-level controller

The low-level controller allows to drive step motors in many different ways. It allows one or two coil control and micro stepping (stepper). Motor currents limiter has selectable range from 0.5 to 3.0 Amps per winding in 0,1 Amp increments. Controller can be powered from any supply voltage from +12 to +50VDC. To work properly, the controller needs to be connected to a high level controller which is a source of digital control signals.

3.4. Sensors

Sensors allow the robot active operating in the environment [3], providing the robot with feedback that may be used to navigate and move through the environment. A variety of sensors may be applied, which require differing supply and control signals.

Infrared distance sensors were selected for this project for measuring the distance to obstacles. The sensor has integrated design that includes emitter and receiver diodes. These sensors have several important advantages, as: precise operation also in case of color or reflecting objects, large useful measuring range, low costs, and no external control unit required.

The author used also acceleration sensors for measurement acceleration of gravity. This sensor has also integrated design that includes all necessary parts for working. The advantages are: precise operation, large measuring range and low cost.

The last group of sensors is focused on assessing internal condition of the inspection robot, such as electric energy consumption.

4. Conclusions

Design and development of an inspection robot is connected with the necessity to solve many crucial problems. There are many issues concerning design and development of the complete system, and development of software that would be capable of intelligent moving the robot through complicate ducts installation. Construction enables further development and implementation of efficient algorithms for better navigation in more complex environments [2,4].

One of very demanding tasks is the ability to move in real ducts. Inside canals there can be a lot of dust and it can by really dirty. These contaminants can have very significant influence on efficiency the unit. A simulation in virtual reality gave the author a lot of information that was very useful in optimization the designed structure.

Sensing the environment is also very important. There is a variety of sensors that can sense environment on many ways, identifying location, direction and many more.

References

[1] A. Morecki, J. Knapczyk (Eds.): „Podstawy Robotyki. Teoria i elementy manipulatorów i robotów(in Polish)", WNT, Warszawa 1997.

[2] P. Przystałka, M. Adamczyk, "EmAmigo framework for developing behavior-based control systems of inspection robots" In the current proceedings of the 7th International Conference "MECHATRONICS 2007".

[3] M. Adamczyk, A. Bzymek, P. Przystałka, A.Timofiejczuk, "Environment detection and recognition system of a mobile robot for inspecting ventilation ducts" In the current proceedings of the 7th International Conference "MECHATRONICS 2007".

[4] W. Panfil, M. Adamczyk, "Behavior-based control system of a mobile robot for the visual inspection of ventilation ducts" In the current proceedings of the 7th International Con-ference "MECHATRONICS 2007".
[5] W. Moczulski, M. Adamczyk, P. Przystałka, A. Timofiejczuk „Mobile robot for inspecting ventilation ducts" In the current proceedings of the 7th International Con-ference "MECHATRONICS 2007".

The issue of symptoms based diagnostic reasoning

J. M. Kościelny, M. Syfert

Institute of Automatic Control and Robotics, ul. Św. A. Boboli 8,
Warsaw, 02-525, Poland

Abstract

In the paper the influence of the dynamics of the symptoms genesis on the correctness of generated diagnoses was shown. An approach of allowing for those dynamics to be taken into account in the algorithms of parallel and serial diagnostic reasoning was given. Finally, the algorithm of proper reasoning when the information about symptoms times is omitted was presented.

1. Problem formulation

The universal scheme of diagnostic system is shown on Fig. 1. Process models are used for fault detection. Generated residuals are the difference between value of measurement signals and calculated model outputs. Residuals or diagnostic signals arising as a result of evaluation (binary or multi-valued, crisp or fuzzy) of diagnostic test results are the input signals to fault isolation procedures. The faults pointed out in the diagnosis are the outcome of the isolation procedures.

The mapping of the space of diagnostic signals values S into the space of faults F is necessary for diagnosing.

Different forms of this representation are known. Structural and directional residuals are used most often [1, 3, 5]. The application of Fault Isolation System FIS [3] with using of multi-valued diagnostic signals is an another solution. All these representation methods have statistical nature. However, the diagnosed process are dynamical systems. Therefore, from the moment of fault occurrence to the moment when one can obtain measurable symptoms the particular time period passes. It depends, among others, on the dynamic characteristic of tested part of the process. If we consider the set of diagnostic signals which detect particular fault, at that moment after the fault appeared only part of this signals take a values that are the fault symptoms. Only just some period of time all signals take values which testifies about fault appearance.

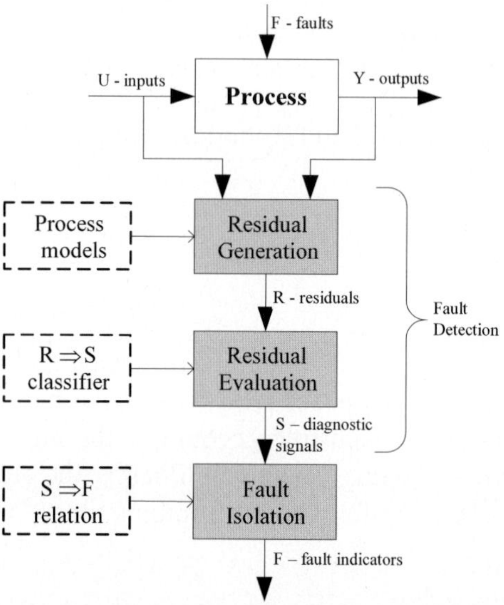

Fig. 1. Diagnosing scheme.

The wrong diagnosis could be generated if one didn't take the dynamic of symptom into consideration [2]. The following question appears: Is it possible to elaborate faults isolation algorithm insensitive to symptoms delays? It is the subject of this paper.

2. Symptom based diagnostic reasoning

Time instants when the symptoms appear depend on dynamic characteristic of tested part of the process, fault type (abrupt, growing) and development in time characteristic, applied method and detection algorithm parameters. It's possible to calculate analytically this times based on dynamic description (e.g. transmittance) of controlled part of the process (where the fault is an input and the process value is an output) and transient response of fault appearance. We make an assumption that limitation function parameters are known and there is no influence of the diagnostic test methods on the process operation.

In practice, it is very difficult to obtain the information about symptoms erasing times. So, it was decided to try to design an algorithm, which should assure of proper diagnosis formulation, for single faults, without taking into account the symptoms erasing times. It is described below.
Let us assume, that the value of "**1**" of the diagnostic signal denotes symptoms appearance, while the value "**0**" denotes its lack. In the known reasoning methods [1, 2, 3, 5] the fault symptoms as well as the lack of others are used parallel. The following rules of reasoning are used:

- The "0" value of the diagnostic signal testifies, that none of the faults controlled by that diagnostic signal had occurred:

$$s_j = 0 \Leftrightarrow \forall_{k: f_k \in F(s_j)} z(f_k) = 0 \,. \tag{1}$$

- The "1" value testifies, that at least one of the faults from the set $F(s_j)$ had occurred:

$$s_j = 1 \Leftrightarrow \exists_{k: f_k \in F(s_j)} z(f_k) = 1 \,. \tag{2}$$

while: $z(f_k)$ - is attributed to each of the faults f_k from the set F. It is defined in the following way:

$$z(f_k) = \begin{cases} 0 - \text{the state without fault } f_k \\ 1 - \text{the state with fault } f_k \end{cases} \tag{3}$$

It is easy to take into account, in the diagnostic reasoning, the symptom which can be observed. They can be immediately used in reasoning. To be able to take into account, in the reasoning, the lack of symptom, one must wait for the predefined period during which the symptom can occur. If the diagnostic reasoning takes into account only the symptoms that appeared (eq. 4) than the diagnosis modified after successive symptoms notifications would be proper. The achieved fault isolability can be lower due to not use of eq. (3) in comparison with the algorithms, which use both reasoning rules.
Serial diagnostic reasoning is based on the analysis of successive diagnostic signals. The diagnosis is formulated in several steps, in which the set of possible faults is gradually constrained [3, 4].

In the case of serial reasoning, the diagnostic relation R_{FS} is defined by attributing to each diagnostic signal the subset of faults detectable by this signal:

$$F(s_j)=\{f_k \in F : f_k R_{FS} s_j\} \tag{4}$$

The first steps of the algorithms are analogous to those ones in the case of parallel reasoning. The fault isolation procedure is started after the first symptom is observed. Its occurrence indicates that one of the fault from the set $F(s_x)$ of the faults detectable by that diagnostic signal had arisen. Such a subset of possible faults is indicated in the primary diagnosis:

$$(s_x=1)\Rightarrow DGN_1 = F(s_x=1). \tag{5}$$

The subset of diagnostic signals S^l useful for isolation of faults from the set F^l is created:

$$S^1 = \{s_j \in S: F^1 \cap F(s_j) \neq \varnothing\}. \tag{6}$$

It can be reduced by the signal s_x, which started the isolation procedure: $S^{l*}=S^l-s_x$.

When single fault occurrence is assumed, the following rules of reducing the set of possible faults indicated in the consecutive steps of diagnosis formulation are used: the value of „1" of the diagnostic signal causes the reduction of the set of possible faults by the faults undetectable by that signal. The new set of possible faults is a product of past possible faults and the set of faults detectable by that signal $F(s_j)$:

$$s_j=1\Rightarrow DGN_p = DGN_{p-1} \cap F(s_j). \tag{7}$$

During serial diagnostic reasoning the preliminary diagnosis is formulated after the first symptom is observed and then constrained when further, consecutive symptoms are taken into account. The diagnosis, in any reasoning step, is proper and points out such faults, for which observed symptoms are consistent with those ones defined in the signatures.

The diagnosis based on symptoms can be also formulated in a parallel way. Let S_p:

$$S_p = \{s_j \in S : s_j = 1), \quad p=|S_p| \tag{8}$$

denote the set of observed symptoms. The formulated diagnosis based on p symptoms has the following shape:

$$DGN_p = \left\{ f_k \in F : \underset{s_j \in S_p}{\forall} \left(r(f_k, s_j) = 1 \right) \right\}. \tag{9}$$

4. Conclusion

It was shown, that one can achieve proper diagnosis without the information about symptom arising times. In this case, only the symptoms are used, while equal to "0" values of diagnostic signals are not taken into account during reasoning. Nevertheless, it leads to decrease of fault isolability. Finally, in some cases, larger amount of fault is pointed out in elaborated diagnosis that in the case of taking the symptom times into account.

Acknowledgments

This work was supported in part by the Polish Ministry of Science and Higher Education under Grant no. 1527/T11/2005/29.

References

[1] J. Gertler "Fault Detection and Diagnosis in Engineering Systems", Marcel Dekker, Inc. New York - Basel - Hong Kong, 1998.

[2] J. M. Kościelny, Zakroczymski K. "Fault Isolation Algorithm that Takes Dynamics of Symptoms Appearances Into Account", Bulletin of the Polish Academy of Sciences. Technical Sciences, Vol.49, No 2, 323-336, 2001.

[3] J. Korbicz, J.M. Kościelny, Z. Kowalczuk, W. Cholewa „Fault Diagnosis: Models, artificial intelligence methods, applications", Springer, 2004.

[4] J. M. Kościelny, M. Syfert „On-line fault isolation algorithm for industrial processes", preprints of: 5th IFAC Symposium SAFEPROCESS, Washington D.C., USA, 9-11.VI, 777-782, 2003.

[5] R. Patton, P. Frank, R. Clark (Eds.) "Issues of fault diagnosis for dynamic systems", Springer, 2000.

The idea and the realization of the virtual laboratory based on the AMandD system

P. Stępień (a) *, M. Syfert (b)

(a), (b) Institute of Automatic Control and Robotics, Faculty of Mechatronics, Warsaw University of Technology, ul. św.Andrzeja Boboli 8, Warsaw, 02-525, Poland

Abstract

This paper presents the idea and realization of the virtual laboratory based on the AMandD system. This Computer Aided Control System Design (CACSD) environment can cooperate with external control systems and real devices with OPC Server. System installed at standard PC hardware is an important concept because it reduces the cost of experimental development and standardizes the computational engine. Virtual object–real controller configuration is presented, which demonstrates the capabilities and the performance of the AMandD environment.

1. Introduction

The virtual laboratory enables the user to implement and conduct experiments on models of controlled systems and their controllers. It is useful for new models design as well. That means two prospects: creation of virtual object, which is under control of real controller or creation of virtual control system for real object. Such laboratory stands allow designing in the same environment, from analysis phase to complete done project. The advantage of that approach is low cost of realization because it doesn't require purchasing any special and complicated devices. It's based on common software and hardware, and enables systems tests in laboratory.

Computer Aided Control System Design (CACSD) software tools allow prototyping of processes/objects and their control systems. To these design tools belong such environments as Matlab, LabView and the AMandD system, which is being developed at the Institute of Automatic Control and

Robotics, at Faculty of Mechatronics of the Warsaw University of Technology. They usually have graphical interfaces, which allow defining dynamic models as block diagram models. Various block-set libraries provide pre-configured blocks and connectors that can be incorporated into a model by simple drag and drop operations. The primary advantages of the proposed approach are as follows: 1) the required computer hardware is low cost, based on PC, 2) available plants of different authors can be supported under the same CACSD environment with no hardware modifications, 3) built block diagrams ca be utilized to prototype control strategies, eliminating the need for low level programming skills, 4) it's possible to use applications written in low level languages, too. CACSD environments have suitable tools for data processing acquired from process and ability to perform complex calculations. Additionally, they are able to perform calculations in real-time.

2. The AMandD system

The AMandD System is being developed at the Institute of Automatic Control and Robotics at Faculty of Mechatronics of the Warsaw University of Technology. It consists of components realizing various functions, in on-line and off-line mode. The AMandD system is a tool for creating measurements and automation applications that can be executed in real-time. The modules of the system exchange processing data and messages by its native communication server. It has to run for right system working. It doesn't require any user service. To receive and send messages by the system module it has to be connected with server. Modules can be divided into following groups: I/O Modules, Computational Modules, Utility Modules, Configuration Modules.

3. The virtual laboratory

The goal of the virtual laboratory was connection possibility and cooperation between real and virtual plants and regulators. Thanks to connection with real world, it's possible to validate designed models in real-time. The laboratory stands utilize standard PC hardware as its cost is considerably lower than the ones with industrial devices. There is also a central unit – an individual standard PC realizing control, monitoring and diagnostics of processes taking place in laboratory. All stands are connected by Ethernet using OPC Data Access Standard. Fig 1 shows the scheme of developed virtual laboratory. There are 3 physical objects in laboratory and each of them is connected with control system:

- TTS – Three Tank System (serial configuration) and Industrial^{IT} ABB system with built-in OPC Server, AC 800M controller and I/O modules.
- Boiler and PlantWEB (DeltaV) Emerson Process Management system with OPC Server, DeltaV M3 controller and I/O modules.
- AMIRA Three Tank System and Proficy HMI/SCADA CIMPLICITY system with VersaMax PLC and I/O modules.

and virtual laboratory stand with 2 models implemented in the AMandD System:

- Boiler-Turbine – model of processes taking place in the power boiler and the cooperating turbine.
- Controller – virtual digital controller with PID algorithms.

Structure of the virtual laboratory is flexible and can be developed in future.

Data from the all laboratory stands is directed to the central steering computer. It's a simple PC hardware with installed the AMandD system. Thanks to realized visualizations, the user can observe processes taking place in laboratory objects and interfere in them, changing their parameters. Also diagnostic tasks are possible to implement. Furthermore, every real object has its own model, implemented in the AMandD. All laboratory stands are connected with themselves by Ethernet and there is possibility to use every control system to control any real or virtual object. Designing of control or diagnostic system in the AMandD for every object is possible as well. To illustrate the realizability of the virtual laboratory the model of the processes proceeding in a boiler-turbine plant (virtual object) integrated with AC 800M ABB programmable logic controller (PLC) has been realized. Virtual object was implemented in the AMandD system, in the PExSim module. By the use of the AMandD components, the communication between the above object and the real stand was established, and the central visualization of proceeding processes was made.

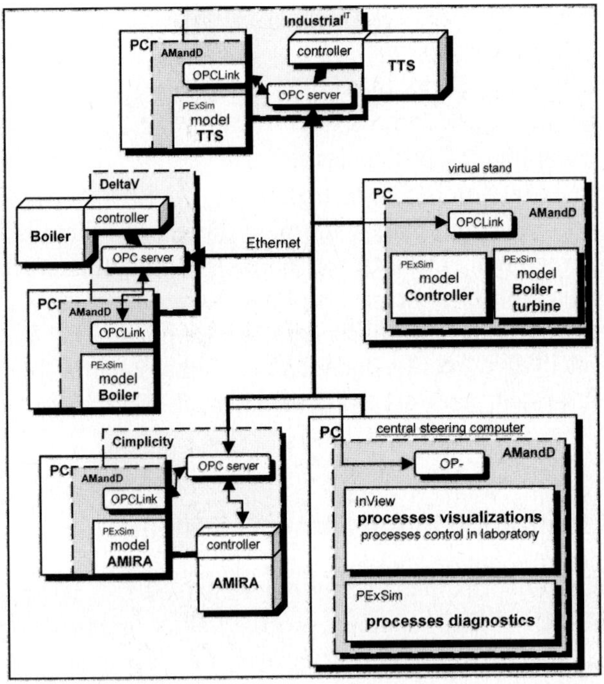

Fig. 1. Laboratory stands scheme in the virtual laboratory

4. Boiler-Turbine model

4.1. Process description

Virtual Boiler-Turbine object has been developed on the basis of electronic analog model realized at the Institute of Automatic Control and Robotics at Faculty of Mechatronics of the Warsaw University of Technology. It is a model of processes taking place in the power boiler with 380 t/h capacity cooperating with the 125 MW turbine.

Controlled values are:

- water level in the boiler drum – h,
- steam pressure in the boiler – Pk,
- superheated steam temperature – Tk,
- pure oxygen content in flue gas – O2,
- rotational speed of turbine – n.

Main disturbances in modelling processes are:

- fuel calorific value,
- fuel supplying boiler mass flow,
- air supplying boiler mass flow,
- water supplying boiler mass flow,
- steam extraction before turbine.

The Boiler-Turbine is the plant with many inputs and outputs. Besides interactions between main circuits, there are also through interactions, which cause automatic control systems coupling. While designing and analysing, theory of multivariable control system was applied. The Boiler-Turbine model consists of five interdependent subsystems. Each of them represents boiler-turbine system as a controlled system of appropriate process variable.

4.2. Model realization in the AMandD system

Boiler-Turbine model is declared as object named BT. It consists of five branches and each of them contains process values assigned to proper subsystem in accordance with controlled value. In order to send process variables outside or to acquire variables, they have to be assigned to topics. Variables are assigned to two topics - bt_procvar and bt_calcvar. Signals, which are coming from outside real controller belong to first topic and Boiler-Turbine process variables belong to the second. Next, connection with OPC Server was established. OPC Server is one of the Industrial^{IT} ABB system components. In OPC Server, OPC items are divided into two groups too. In particular group, process and control variables values are assigned.

Boiler-Turbine model was realized in PExSim module. It consists of five interdependent subsystems with through interactions. Model is realized in paths. Main path is named bt_BTSimulator and consists of five subpaths. Each of them is responsible for appropriate process variable modeling.

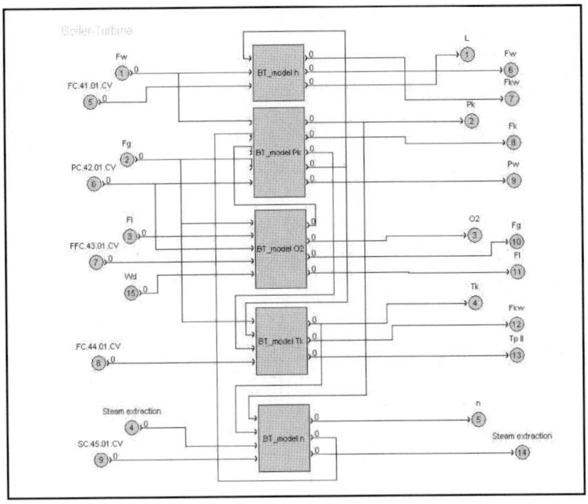

Fig. 2. Boiler-Turbine model structure

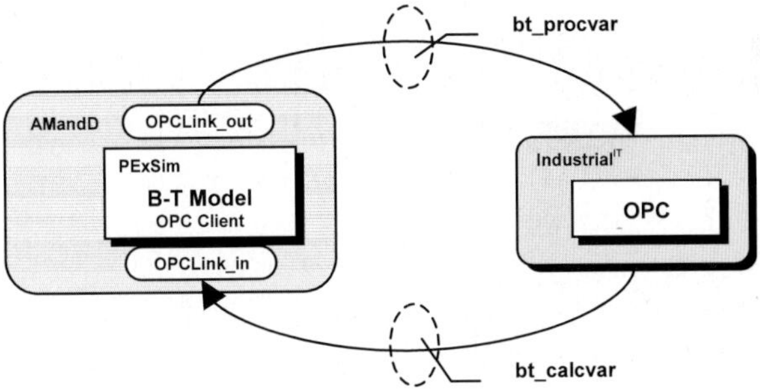

Fig. 3. Transfer data between Boiler-Turbine (bt_BTSimulator) and OPC Server

Additionally, visualizations were realized in InView module. They meet ASM (Abnormal Situation Management) requirements. Visualizations show Boiler-Turbine scheme with bar graphs, which inform about process variables values and mass flows of supplying boiler mediums. At this level the user can also change process parameters.

4.3. Control in Industrial^{IT} system

At the beginning the Industrial^{IT} system with cooperating controller and I/O modules was configured. When working with control projects you have to work in Plant Explorer, where the user has access to control projects via different views, called structures. New project was created in the Control Structure. There the user creates control networks, set the OPC data source definition aspects and connect demanded libraries. New variables were also created, which were automatically available via OPC Server. Then in application of Control Module the graphical objects of control systems were created and variables to the objects were connected.

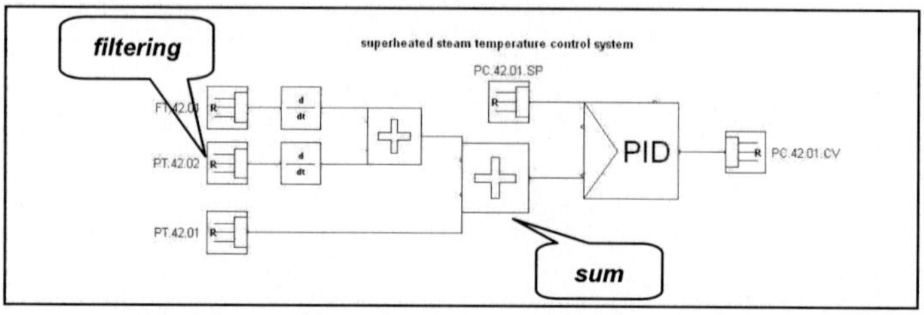

Fig. 4. Control system in Industral^{IT} – example of regulation loop

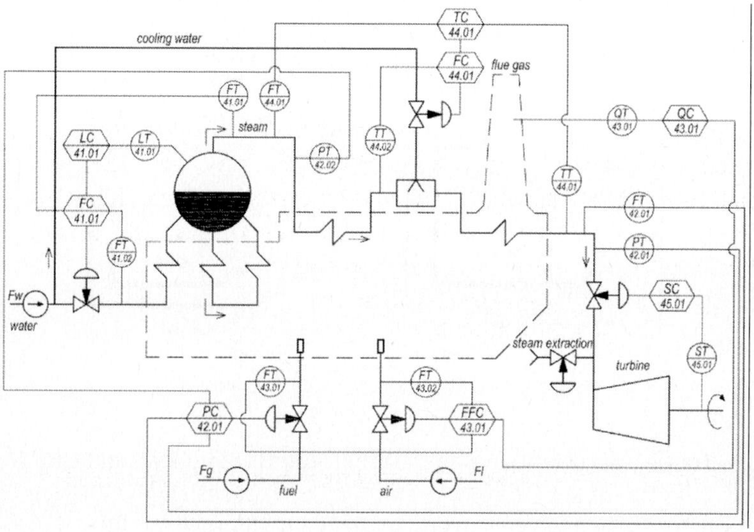

Fig. 5. Boiler-Turbine control diagram

Positive tests of communication between the virtual and real element of laboratory was carried out. The aim was to write and read online process data from PLC to AMandD. The best result of communication transfer was 0,5 s. It is satisfactory outcome when process proceeds not very fast.

As an example Fig.6 shows reaction of Boiler-Turbine model to set point change of superheated steam temperature (a) and water level in the boiler drum (b).

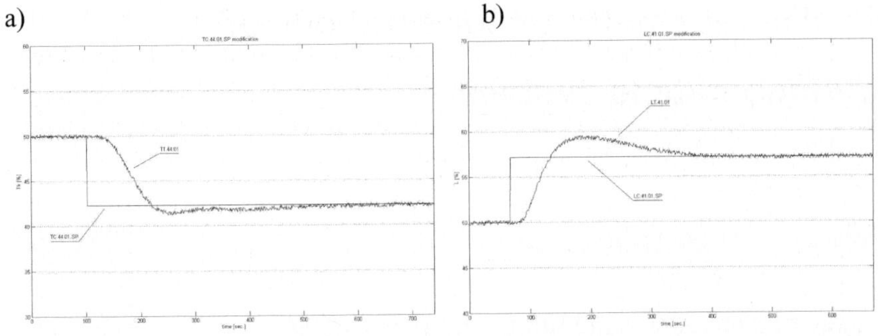

Fig. 6. Set point change of superheated steam temperature (a) and water level in the boiler drum (b) of the model

5. Conclusions

In this paper the idea of the virtual laboratory and example of its working has been presented. The proposed approach advocates the use of the AMandD system to modeling of plants using standard low-cost PC hardware, which can cooperate with real industrial devices. To illustrate this possibility one configuration controller-object was shown. Conducted experiments confirmed that described conception is suitable and efficient solution for further research work.

References

[1] P. Stępień, "The concept and realization of virtual laboratory" Master's thesis. 2006.

[2] A. Syrzycki, K. Cieślicki, "Electronic analog model of processes taking place in Boiler-Turbine plant", IAiR PW, Warsaw, 1986.

[3] W. Findeisen, "Automatic control engineering", PWN, 1978.

[4] J. Rakowski, "Automatic control of power station thermal equipment", WNT, 1976.

[5] W. E. Dixon, D. M. Dawson, B. T. Costic, M. S. de Queiroz, "Towards the Standardization of a MATLAB-Based Control Systems Laboratory Experience for Undergraduate Students", 2001.

[6] S. Persin, B. Tovornik, N. Muskinja, "OPC-driven Data Exchange between MATLAB and PLC-controlled System, 2000.

[7] The Abnormal Situation Management (ASM) Consortium, http://www.asmconsortium.com

The discrete methods for solutions of continuous-time systems

I. Svarc

Institute of Automation and Information Technology
Faculty of Mechanical Engineering - Brno University of Technology
Technická 2, Brno 616 69, Czech Republic

Abstract

The first part of this contribution deals with discretizing differential equations. Difference equations can also be obtained by discretizing differential equations. A first order differential is approximated by a first order difference, a second order differential by a second order difference, etc. The other way of discretization is discretization by Z transformation of transfer function *G(s)*. This contribution is concerned with the Euler's method and bilinear method. The contribution solves the link between *s* and *z*.

The last part of this contribution contains solutions of unit step response and impulse response of continuous-time systems by discrete methods that were introduced here.

The contribution shows the new possibility of how to solve continuous-time control systems by discrete methods.

1. Introduction

The traditional approach to designing digital control systems for continuous-time plants is first to design an analog controller for the plant, then to derive a digital counterpart that closely approximates the behaviour of the original analog controller. Techniques for designing analog controllers for continuous–time control systems are well established. A control engineer may have more experience in designing analog controllers and therefore may wish to first design analog controllers and then to convert them into digital controllers.

The other approach to designing digital controllers for continuous-time plants is to derive a discrete-time equivalent of the plant and then to design a digital controller directly to control the discretized plant. There are several methods of how to obtain discrete-time equivalents of continuous-time systems. These methods are as follows:

-numerical approximation of differential equations;
-discretization by Z transform of *G(s)* (Euler's method, bilinear transformation method, ...);
-discretization of continuous-time state variable models;
-numerical differentiation; ...etc.

This contribution presents a numerical approximation of differential equations and discretization by Z transform of *G(s)*. The problem of a continuous-time plant with a discrete-time plant can be viewed as converting the analog transfer function *G(s)* to a differential equation and then obtaining a numerical approximation to the solution of the differential equation or the direct discretization.

2. Numerical approximation of differential equations

Difference equations can also be obtained by discretizing differential equations. Here a first order differential is approximated by a first order difference, a second order differential by a second order diference, etc.

In order to discretize a differential equation, the following terms are used instead of the differentials (T is a sampling interval)

$$\frac{dx(t)}{dt} \approx \frac{\Delta x(k)}{T} = \frac{x(k)-x(k-1)}{T};$$
$$\frac{d^2x(t)}{dt^2} \approx \frac{\Delta^2 x(k)}{T^2} = \frac{x(k)-2x(k-1)+x(k-2)}{T^2}; \tag{1}$$
$$\frac{d^3x(t)}{dt^3} \approx \frac{\Delta^3 x(k)}{T^3} = \frac{x(k)-3x(k-1)+3x(k-2)+x(k-3)}{T^3} \quad ....$$

For example to discretize a differential equation of second order

$$a_2 y''(t) + a_1 y'(t) + a_0 y(t) = b_1 u'(t) + b_0 u(t) \tag{2}$$

We can insert (1) into (2) to obtain

$$a_2 \frac{y(k)-2y(k-1)+y(k-2)}{T^2} + a_1 \frac{y(k)-y(k-1)}{T} + a_0 y(k) = b_1 \frac{u(k)-u(k-1)}{T} + b_0 u(k) \tag{3}$$

The result for $T=1$ (4)

$$(a_2 + a_1 + a_0)y(k) - (2a_2 + a_1)y(k-1) + a_2 y(k-2) = (b_1 + b_0)u(k) - b_1 u(k-1)$$

3. Discretization by z transformation of G(s)

Euler's method. A transfer function *G(s)* that is to be discretized is given. The equivalent discrete-time transfer function can be obtained by replacing each *s* in *G(s)* [1] by

$$s \cong \frac{1-z^{-1}}{T} = \frac{1}{T}\frac{z-1}{z} \tag{5}$$

For the equation (2) the transfer function *G(s)* is as follows

$$G(s) = \frac{Y(s)}{U(s)} = \frac{b_1 s + b_0}{a_2 s^2 + a_1 s + a_0} \tag{6}$$

The discrete-time equivalent using Euler's method is

$$G(z) = \frac{Y(z)}{U(z)} = \frac{b_1 \frac{1}{T}\frac{z-1}{z} + b_0}{a_2 \frac{1}{T^2}\left(\frac{z-1}{z}\right)^2 + a_1 \frac{1}{T}\frac{z-1}{z} + a_0} \tag{7}$$

and the difference equation is (3) and for $T = 1$ again the equation (4).

Bilinear transformation method. The bilinear transformation method is also called a trapezoidal integration method or Tustin transformation method. By this method we approximate the left half of the *s* plane into the unit circle of the *z* plane [1]

$$s \cong \frac{2}{T}\frac{1-z^{-1}}{1+z^{-1}} = \frac{2}{T}\frac{z-1}{z+1} \tag{8}$$

Using the basic relation $z = e^{sT}$, where T is some chosen sampling interval, relations between the primary strip in the s and the z plane can be established. The inverse relation $s = \ln z / T$ is given as the series

$$s = \frac{1}{T} ln\, z = \frac{2}{T}\left(\frac{1-z^{-1}}{1+z^{-1}} + \frac{\left(1-z^{-1}\right)^3}{\left(1+z^{-1}\right)^3} + \frac{\left(1-z^{-1}\right)^5}{\left(1+z^{-1}\right)^5} + \ldots \right) \tag{9}$$

The corresponding *z*-transmittance of each of these s^{-1}, s^{-2}, ... , s^{-5} is also listed in table 1 – [2].

Consider our differential equation (2). The transfer function *G(s)* was expression (6). We have to rewrite *G(s)* as a ratio of polynomials in s^{-1} as follows

$$G(s) = \frac{Y(s)}{U(s)} = \frac{b_1 s + b_0}{a_2 s^2 + a_1 s + a_0} = \frac{b_1 s^{-1} + b_0 s^{-2}}{a_2 + a_1 s^{-1} + a_0 s^{-2}} \tag{10}$$

	Z - transform
s^{-1}	$\frac{T}{2}\cdot\frac{1+z^{-1}}{1-z^{-1}}$
s^{-2}	$\frac{T^2}{12}\cdot\frac{1+10z^{-1}+z^{-2}}{\left(1-z^{-1}\right)^2}=\frac{T^2}{12}\cdot\frac{1+10z^{-1}+z^{-2}}{1-2z^{-1}+z^{-2}}$
s^{-3}	$\frac{T^3}{2}\cdot\frac{z^{-1}+z^{-2}}{\left(1-z^{-1}\right)^3}=\frac{T^3}{2}\cdot\frac{z^{-1}+z^{-2}}{1-3z^{-1}+3z^{-2}-z^{-3}}$
s^{-4}	$\frac{T^4}{6}\cdot\frac{z^{-1}+4z^{-2}+z^{-3}}{\left(1-z^{-1}\right)^4}-\frac{T^4}{720}=\frac{T^4}{6}\cdot\frac{z^{-1}+4z^{-2}+z^{-3}}{1-4z^{-1}+6z^{-2}-4z^{-3}+z^{-4}}-\frac{T^4}{720}$
s^{-5}	$\frac{T^5}{24}\cdot\frac{z^{-1}+11z^{-2}+11z^{-3}+z^{-4}}{1-5z^{-1}+10z^{-2}-10z^{-3}+5z^{-4}-5z^{-5}}$

Table 1. Z-transmittance of $s^{-1}, s^{-2}, \ldots, s^{-5}$

By using table 1 for $T=1$ we obtain

$$G(z)=\frac{(0,5b_1+0,083b_0)+0,83b_0z^{-1}+(0,083b_0-0,5b_1)z^{-2}}{(a_2+0,5a_1+0,083b_0)+0,83z^{-1}+(0,083b_0-0,5a_1)z^{-2}} \tag{11}$$

and the difference equation is

$$\begin{aligned}(a_2+0,5a_1+0,083b_0)y(k)+0,83b_0y(k-1)+(0,083b_0-0,5a_1)y(k-2)=\\=(0,5b_1+0,083b_0)u(k)+0,83b_0u(k-1)+(0,083b_0-0,5b_1)u(k-2)\end{aligned} \tag{12}$$

4. Application

For example: Determine the step function response for the system

$$y''+5y'+6y=3u$$

We will first solve the system by continuous - time method:

$$h(t)=L^{-1}\left\{\frac{G(s)}{s}\right\}=L^{-1}\left\{\frac{1}{s}\frac{3}{s^2+5s+6}\right\}=0,5-1,5e^{-2t}+e^{-3t}$$

Then we will use the first numerical method (numerical approximation of differential equations). For $T=1$ we have the equation (4) and for numerical values we have

$$12y(k)+7y(k-1)+y(k-2)=3u(k)$$

Step function response by [2] $h(k)=0{,}59h(k-1)-0{,}08h(k-2)+0{,}25\eta(k)$

$\Rightarrow$ $k=0: \; h(0)=0{,}25; \; k=1: \; h(1)=0{,}398; \; k=2: \; h(2)=0{,}465;$ ......

Second numerical solution (discretization by Z transform of *G(s)* – Euler's method for $T = 1$, equation (7)):

$$G(z)=\frac{3}{\left(\frac{z-1}{z}\right)^2+5\frac{z-1}{z}+6}=\frac{3}{12-7z^{-1}+z^{-2}}=\frac{3z^2}{12z^2-7z+1}$$

$$h(k)=Z^{-1}\left\{\frac{z}{z-1}G(z)\right\} \rightarrow h(0)=0{,}25; h(1)=0{,}396; h(2)=0{,}465;...$$

Third numerical solution (bilinear transformation): transfer function (7)

$$G(z)=\frac{(0{,}5b_1+0{,}083b_0)+0{,}83b_0z^{-1}+(0{,}083b_0-0{,}5b_1)z^{-2}}{(a_2+0{,}5a_1+0{,}083b_0)+0{,}83z^{-1}+(0{,}083b_0-0{,}5a_1)z^{-2}}$$

Numerical $$G(z)=\frac{0{,}249z^2+2{,}49z+0{,}249}{3{,}749z^2+0{,}83z-2{,}251}$$

and the solution is similar as in the second numerical solution.

5. Conclusion

In the same manner (discrete methods) as the step function response of the system, the impulse response and other continuous-time systems can be solved.

References

[1] W. S. Levine, "The Control Handbook", CRC Press, Inc., Boca Raton, Florida, 1996

[2] I. Švarc , "Automatizace – Automaticke rizeni", CERM, Brno, 2005

Acknowledgements: The results presented have been achieved using a subsidy of the Ministry of Education, Youth and Sports of the Czech Republic, research plan MSM 0021630518 "Simulation modelling of mechatronic systems".

Control units for small electric drives with universal software interface

P. Houška, V. Ondroušek, S. Věchet, T. Březina

Institute of Automation and Computer Science,
Faculty of Mechanical Engineering, Brno University of Technology,
Technická 2,
Brno, 61669, Czech Republic

Abstract

This contribution deals with a design of software interface for control units of different types of electric drives. Major part of control units software can be uniform, only the power circuit differs, as the analysis of existing solutions has shown. Thus it is possible to design universal software interface, in terms of this analysis, consisting of interconnected cooperating modules. This paper describes objectives and implementation of such modules, and also provides description of proposed universal interface architecture. The method of employment is shown on a case of control units for drives with DC and stepper motors.

1. Introduction

We were solving many problems within last few years of utilization of different types of electric motors. The biggest problems were caused by control units used for controlling of the motors. The main problem was an incompatibility in operating with these control units. Another problems were different quality of the control process and limited possibilities of adjusting parameters of controllers implemented in control units.
In terms of these practical experiences the requirements for control units were specified. Existing control units was analyzed at the same time. The concept of control units with universal software interface for small electric drives arises from this analysis and given requirements. Universal software interface can be used over serial communication busses. The communication libraries were designed for the purpose of easy implementation and simple incorporation of power drive control into the applications.

2. Control units analysis

Above all the software of the control unit must solve task of regulation, sensing of feedback values and controlling of power circuit. These entire tasks are solved discretely in numeric form. It means that there are many inaccuracies originate from rotation speed sensing, current measurement and sensing other values. The action value computation is influenced by rounding errors and limited computation precision. The biggest inaccuracies arise from converting action value to the signal of pulse width modulation (PWM). The software of control units should be able to deal with these inaccuracies too.

In most cases a PSD controllers are used for automatic regulation [1]. Settings of these PSD are depended on operating conditions and a design of power drive [2]. It is possible to achieve a high accuracy of regulation with the PSD controllers, but at the price of increased "hardness" of power drive (too high gain in P-component of controller). The "hardness" manifests itself through increasing mechanical stress of a drive and whole frame structure of controlled device. Transient overshooting of a controlled signal is caused by an S-component of controller. Many publications deal with the possibilities of electric motors control by means of FUZZY controllers and/or neural networks [3]. Controllers utilizing reinforcement learning principle achieve very interesting results too.

The main problem of this solution is inefficient ability of controlling dynamic processes. An absence of operating standard is another disadvantage of commercial control units. Not even company standard often exists - various types of control units have various types of operating interface. Consequently the new control unit means learning of another type of operating with unit.

Complexity of control system is given by a computing power and a size of memory used by microcontroller, in which the whole control process, sensing and communication with environment is implemented. Many cost-effective control units are based on inefficient 8-bit microcontrollers or on 16-bit DSP processors in the case of better control units. New cheap microcontrollers based on ARM architecture are coming up on the market over last few years. These microcontrollers have more computing power, in consequence better potential to implement more sophisticated control algorithms.

3. Control units conception

Conclusions resulting from the above mentioned analysis are:

- Hardware of control units differs each other in the part of power electronics only,
- Software of control units differs each other in a part of power electronics control only,
- I.e. major part of hardware and software is identical

The purpose of this project is design a library of hardware and software units, which can be composed into the control unit with desired properties and unified behaviour. General purpose schema of control unit is on fig. 1. On this schema the small motor “depended” parts are dashed line enclosed.

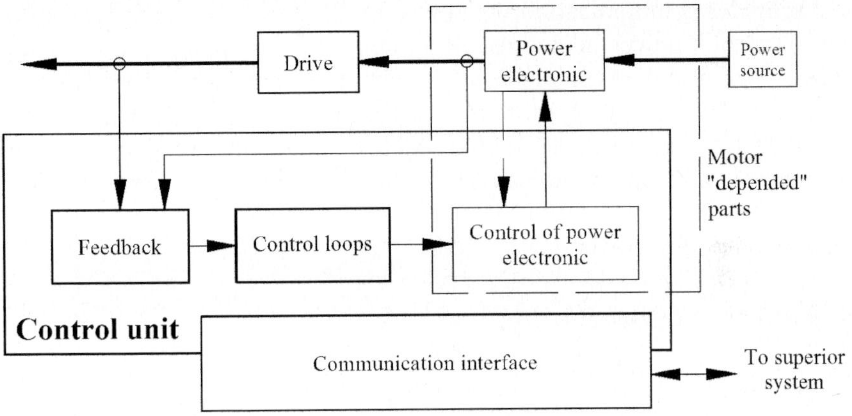

Fig. 1. Schema of control unit

3.1 Communication interface

Internal protocol [4] that is a laboratory standard for several years has been chosen as the communication protocol. This protocol solves addressing problem too. Minimal dependence on used type of bus and device identification are some of its advantages. Used busses are UART, I2C, SPI, USB and CAN bus. Communication commands result from the used protocol. Communication libraries are proposed for unit operation control. It is supposed to use the libraries for Microsoft .NET Framework 2.0 (Mono on Linux based operating systems), NI LabView and common microcontrollers.

3.2 Software conception

Software will be designed in order to keep maximal generality and minimal dependence on used hardware. Hardware service has to be solved on the lower level of software, usually marked as driver. Some applications require measuring not only kinematics values but also a current and a temperature, some other need to evaluate force or position from external sensors. It hast to be possible to adjust all of these possibilities.

3.3 Hardware conception

Power elements are produced as monolithic integrated circuits for motors power in view. Main part of the hardware conception is microcontroller that provides communication with master units, acquires data from sensors and controls power elements. Requirements on microcontrollers are defined by motor type, required sensors, communication interface and control type. The using of 8–bit microcontrollers (MCS52) by Silicon Laboratories (SiLabs) and 32–bit (ARM) microcontrollers by NXP are considered. The main advantage of SiLabs microcontrollers is bigger precision of AD converters. The advantage of NXP microcontrollers is higher computing power.

4. Realised control units

Purpose of already realized control units:

- DC motors, supply voltage from 6V to 48V, current to 6A,
- Unipolar and bipolar stepper motors, supply voltage from 5V to 24V, current to 1A.

SiLabs 8-bit microcontrollers with computing power of 20MIPS are used for these control units. Developed communication libraries are used for operating with the control units.

4.1 Control unit for DC motors

The control unit measures rotation speed, voltage, current, temperature and it is able to interpret logic signal from two switches (e.g. reference point). Control unit is manned with microcontroller C8051F006 and monolithic integrated power circuit TLE 6209 that provides an elementary diagnostic, as well as current protection and thermal protection. The frequency of output PWM signal can be set on 5 kHz or 20 kHz. Furthermore, parameters of motor, gearbox, control algorithm and output values are adjustable too. To cover precision of measurement it is necessary to calibrate all analogue measured values before putting the control unit into operation.

4.2 Control unit for stepper motors

Control unit measures only logic signal from two switches, on the other hand the rotation speed is evaluated from control signal frequency. Control unit is manned with microcontroller C8051F331 and integrated power circuit ULN2003A. Micro-stepping with resolution 64 micro-steps is used to obtain smooth rotor motion. Parameters of the motor, gearbox, control algorithm and output values are adjustable.

5. Conclusion

Eight DC motor control units and two stepper motor control units (described in chapters 4.1 and 4.2) have been finished. In present day we are focused on possibilities of torque/force control loop integration. The new revision of hardware is prepared for testing this new torque/force control loop. Consequently a problem of implementation of different robust control algorithms is solved. Realized control units have shown applicability of developed solution.

Acknowledgement

Published results were acquired using the subsidization of the Ministry of Education, Youth and Sports of the Czech Republic, research plan MSM 0021630518 "Simulation modelling of mechatronic systems".

References

[1] Kamalasadan, S., Hande,A.: A PID Controller for Real-Time DC Motor Speed Control using the C505C Microcontroller", 17th International Conference on Computer Applications in Industry and Engineering (CAINE), Orlando, FL, 2004, pp.34-39

[2] Caprini.G. C., Innocenti F., Fanucci L., Ricci S.: Embedded system for brushless motor control in space application, MAPLD International Conference, Washington, 2004, p151/5

[3] Marcano-Gamero C.R.: Synthesis and Design of a Variable Structure Controller for a DC Motor Speed Control, Modelling and Simulation – 2006, Montreal, Canada, 2006, pp26-30

[4] Houška, P.: Distributed control system of walking robot; Ph.D. Thesis; ÚMT FSI VUT v Brně; 2004

Predictor for Control of Stator Winding Water Cooling of Synchronous Machine

R. Vlach (a) *, R. Grepl (b) , P. Krejci (c)

(a) Institute of Solid Mechanics, Mechatronics and Biomechanics, Brno University of Technology, Technicka 2, Brno 61669,Czech Republic vlach.r@fme.vutbr.cz,

(b) Institute of Solid Mechanics, Mechatronics and Biomechanics, Brno University of Technology, Technicka 2, Brno 61669,Czech Republic grepl@fme.vutbr.cz,

(c) Institute of Solid Mechanics, Mechatronics and Biomechanics, Brno University of Technology, Technicka 2, Brno 61669,Czech Republic krejci.p@fme.vutbr.cz,

Abstract

This project is concerned with non-convectional direct stator winding slot cooling using water. The aim is to find optimal algorithm for control of water cooling. The control algorithms are tested on the experimental device, which is part of real synchronous machine with permanent magnets. The thermal model was built as a base for pump control algorithm model of a machine without thermal sensors. The Thermal model is possible used as predictor of machine heating in real time.

1. Introduction

The paper is concerned with computational simulations of stator winding heating of synchronous machine. The synchronous machine operates as high-torque machine with maximal torque 675 Nm at 50 rpm. The machine is used for the direct drive of the rotary or swinging axis, for example rotary tables of the machine tools.

The aim was to find predictor of synchronous machine thermal phenomena, so that the thermal model would be used for pump control of water cooling systems.
Software MATLAB was used for computational simulation of synchronous machine thermal phenomena. Computational simulations describe direct stator winding cooling by water.

2. Thermal model

The computational model geometry arises from real synchronous machine. It describes the heat of a part of synchronous machine mainly stator winding. The machine has 36 pair of winding slots and permanent magnets on the rotor. Rotor with magnets is not modelled, because the heat loss is only in the stator winding and rotor effect is negligible on the heating of stator. The brass tubes were comprised in the middle of each winding slots. Cooling water flows in the brass tube. Symmetry of machine was assumed so only one pair of winding slot is modelled.
The thermal network method [3] was used for description of machine heating. Thermal networks (Fig.1) consist from twenty-eight nodes. Last eight nodes (from 21 to 28) are used for description of cooling water heating. Thermal model describes transient state, because machine operates with varying load.
Thermal network is possible to be described by differential system equation:

$$C_i \frac{d\vartheta_i}{dt} + A \cdot \vartheta_i = b \quad . \tag{1}$$

where:

C_i	is thermal capacity concentrated in node i
A	is matrix of thermal conductivities
b_i	is heat loss in node i and heat flux to ambient

Temperatures of nodes describe heating of cooling water is given by:

$$\vartheta_i\left(a_Q + \sum_j a_{ij}\right) - \vartheta_{(i-1)}\left(a_Q - \sum_j a_{(i-1)j}\right) - \sum_j \vartheta_{ij} a_{ij} - \sum_j \vartheta_{(i-1)j} a_{(i-1)j} = 0 \tag{2}$$

where:

ϑ_i	is temperature of water node i
a_Q	is thermal conductivity of flowing water
a_{ij}	is thermal conductivity between nodes i (water node) and j (solid parts)

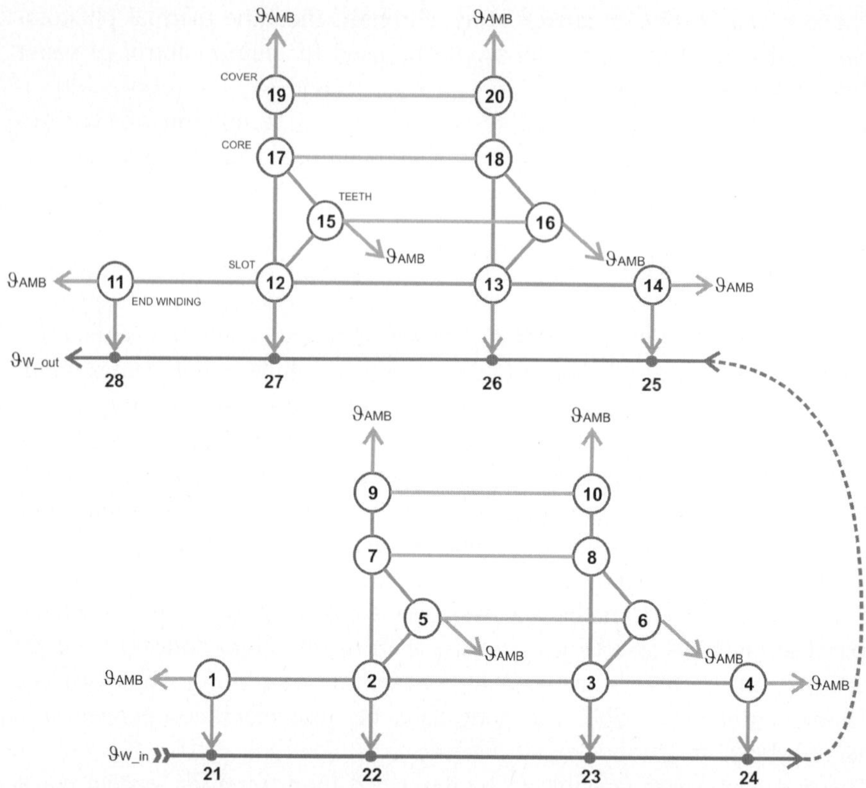

Fig. 1. Thermal network of synchronous machine

The measuring was used for verification of thermal model. Thermal network parameters were identified by using genetic algorithm, so temperature differences between measuring and simulation was minimal. The result of identification is summarized in figure 2.

	Heat losses (cold) [W]	Heat losses (heat) [W]	Winding temp. [℃]	Surface temp. [℃]	Output water temp. [℃]	Input water temp. [℃]
Measuring	55.4	69.8	90.3	85.9	27.7	31.2
Simulation	55.4	70.1	91.4	84.1	27.7	31.6
Error [%]	-	-1.19	-1.19	2.08	-	-1.01

Fig. 2. Result of thermal model parameters identification

3. PREDICTOR OF THERMAL PHENOMENA

Thermal model can be used for simulation of dynamic behaviour with respect time variable of heat load. Scheme of using thermal model as heating predictor is showed in figure 3.

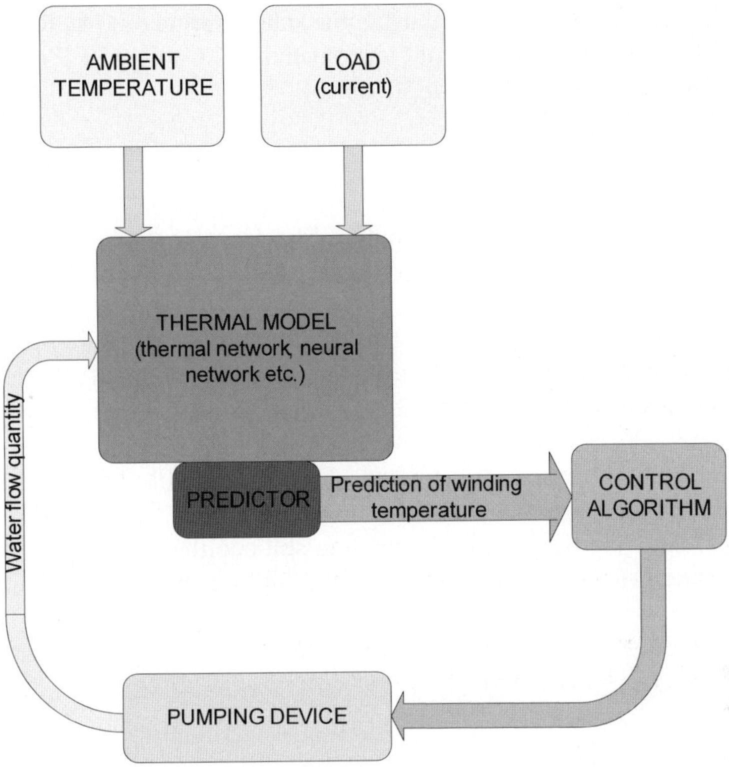

Fig. 3. Thermal model using as heating predictor

The pump capacity is determined on the basis of winding temperature from thermal predictor. Only ambient temperature and stator current are inputs to thermal model.

4. CONCLUSION

The idea is using thermal model for control of machine heating without thermal sensors in the machine. The thermal model is possible to be used

for prediction temperature of machine individual parts in real time, so algorithm of pump capacity control will be better.
Control quality depends on accuracy of thermal predictor, so more experiment will be realized.

ACKNOWLEDGMENT

Published results were acquired using the subsidization of GAČR, research plan 101/05/P081 „Research of Nonconventional Cooling of Electric Machines“ and research plan AV0Z20760514.

References

[1] R.Vlach, "Drive of Stator Winding Slot Cooling by Water", International Conference on Electrical Machines ICEM2006, Chania,Crete, 2.9-5.9. 2006

[2] R. Vlach, "Computational and Experimental Modelling of Non-convectional Winding Slot Cooling", International conference Mechatronics, Robotics and Biomechanics 2005, Ttřešť, 26.-29. 9. 2005, Czech Republic

[3] J. Hak, O. Oslejsek, "Computed of Cooling of Electric Machines" , 1.volume. VUES Brno 1973,CZ

[4] V. Holan "Non-convectional winding slot cooling of synchronous machine using heat pipe and water cooling", Diploma project, FSI TU Brno, 2006

[5] E. Ondruska, A. Maloušek, "Ventilation and cooling of electric machines" SNTL Praha 1985

The Design of the Device for Cord Implants Tuning

T. Březina (a), M. Z. Florian (b), A. A. Caballero (c)

(a) Institute of Automatization and Computer Science, Faculty of Mechanics Engineering, Brno University of Technology, Technická 2, Brno, 61669, Czech Republic

(b) Institute of Solid Mechanics, Mechatronics and Biomechanics, Faculty of Mechanics Engineering, Brno University of Technology, Technická 2, Brno, 61669, Czech Republic

(c) Department of Electrical & Computer Engineering, College of Engineering & Computing, Florida International University, 10555 W. Flagler St. Miami, Florida 33174, United States of America

Abstract

The most important moments of the design of the device for biomechanical components testing are described in this contribution. The device is designed in such a way that its movements are as close as possible to real physiological movements. Both general motion and effector action forces are reached by pair of robots with parallel kinematics. This structure is designed to meet the possibility of its use as an element of building architecture and to test then e.g. complete spinal segments. This fact is very convenient in the case of fixed vertebral bodies as the changes of mechanical properties of surrounding spinal elements occur.

1. Introduction

The end of the 20th century and the beginning of 21st century is characteristic on one hand by fast development of science and techniques, on the other hand by hurried life style that brings degradation of the human organism and subsequently considerable health problems. The spine and big

joints are degraded most often. Clinical solution is often based on application of fixator, eventually on complete endoprosthesis.
The demands on this device derive from the need of experimental modeling of arbitrary motion and load of the backbone (spinal segments) and of the joints (especially hip joints).

2. Requirements for the device

The following requirements have been defined which arise from the analysis of a pool of such tasks [1]:

- The device must be able to load up the test specimen with the assigned load force and to carry out any loading cycle in six DOF.
- Load forces and moments were estimated, with values approx. 2000 N and 10 Nm.
- Due to allowance of test specimen parameters, the accuracy of positioning is without special demands.
- Regarding experimental modeling of motion and load of the backbone the device has to affect individual spinal segments in given range.
- Due to assumed clinical application of the device an exploitation of electro – mechanical transmission is advised. The device dimensions should be as small as possible.

3. Basic considerations

3.1. Parallel mechanism concept

With respect to the determination it is suitable to conceive the device as two toroidal plates – interconnected by active elements - into which tested segments will be fixed. For backbone testing the device could consist of n, $n > 3$ stacked layers of such arrangement.
The concept of parallel mechanism called Stewart platform (hexapod) [2] naturally corresponds to a single layer of such device (Fig. 1). It provides wide range of motion and accurate positioning capability and large amount of rigidity, or stiffness, for a given structural mass, enabling the Stewart platform to provide a positional certainty.
Typically, the six linearly actuated legs are connected to both the base plate and the top (mobile) plate by universal joints in parallel located at both ends of each leg. The position and orientation in six DOF of the top plate depend on the lengths of the legs.

Therefore Stewart platform allows examination of e.g. total endoprothesis of hip joint (Fig. 2) eventually spinal element (Fig. 3). In the case of fixator application to spinal segment a surgeon is also interested in mechanical influence of neighbor segments (Fig. 4) e.g. correction of the scoliotic curve of vertebral column in various cases of scoliosis. Thanks to the possibility of modular arrangement of designed device we can achieve experimental solution of this problem.

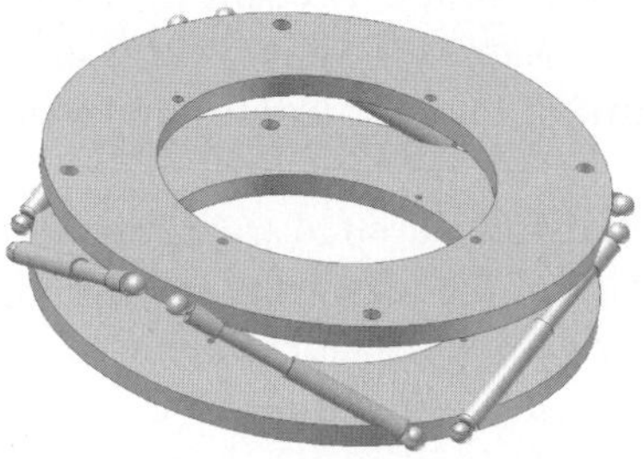

Fig. 1. Scheme of Stewart platform

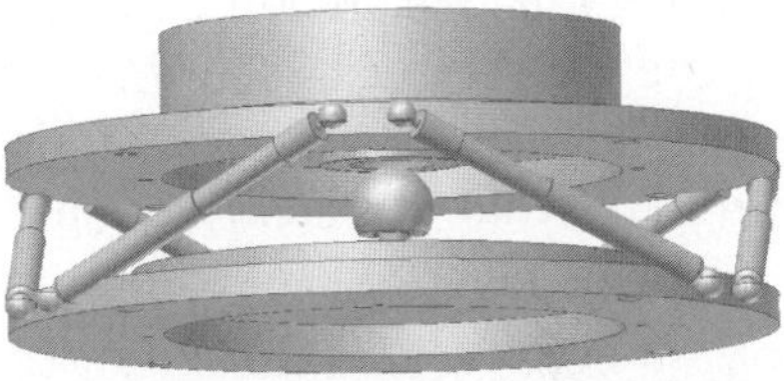

Fig. 2. Configuration for testing total endoprothesis of hip joint

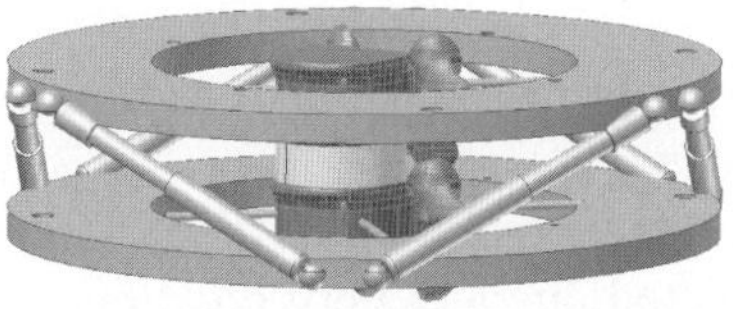

Fig. 3. Configuration for testing spinal element

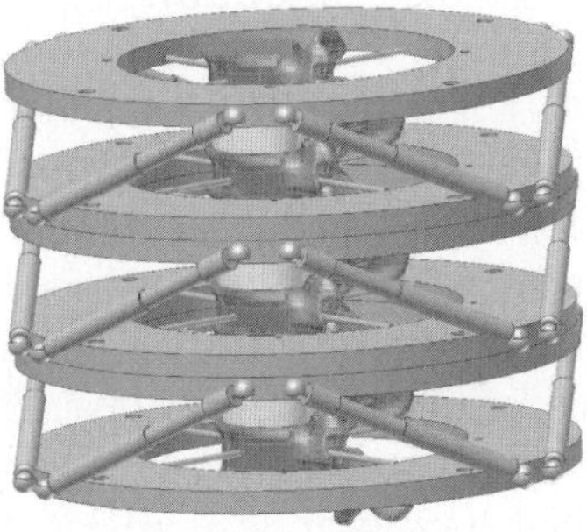

Fig. 4. Configuration for testing spinal segments

3.2. Computational models

Numerical simulations using analytical models [2] were first performed to find out the ranges and trajectories of prolongation of legs, velocities and forces in legs determining later actuator design. Harmonic trajectories of deflections and angular deflections along the coordinate system with respect to the equilibrium point (working layout) of top plate satisfying the device specification represented the courses of control variables in mentioned simulations.

3.3. Control concept

A control actuates the six leg forces to proper the deflections and angular deflections of the top plate of Stewart Platform given as desired trajectory over time. The six legs forces are the inputs into the mechanism while the lengths and velocities of the legs form the outputs. So the control is realized by actuating the six legs forces, sensing the legs lengths and velocities, and reading the desired trajectory.
The basic goal of such controller is to transform desired trajectory of the top plate into the corresponding trajectories of the legs using inverse kinematics. To avoid the computation of forward kinematics, a lower level controller for each leg assure the leg to keep its desired trajectory.

4. Spatial and functional integration

4.1. Drive unit selection

For this application the particular actuators must produce high forces in small velocities. This fact disposes the use of linear motors, with relatively small forces and high linear deflections in high speed and acceleration. So it is necessary to use drives based on rotational electrical motor followed by transmission of rotational movement into translational. The ball screw is used due to minimal backlash.
Finally, the Maxon drive unit which consists of DC motor RE35, single-stage planetary gear head GP32C was selected due to a very good ratio of the proportion power/weight, which is the most important one for this application. Possibility of a short-term overloading also brings in the possibility to use the motor with lower parameters than the nominal ones are and so to design the linear drive of smaller dimension.

4.2. Structural design

Energy transfer of linear actuator was reached by the chain of motor – gear head – spur gearbox – ball nut – ball screw (Fig. 5). To transfer the force, which is formed between the frame and the ball screw, spherical ball pin is used. Both ball screw and rear part of the frame are equipped with the thread for spherical ball pin screwing. The design of single layer of the experimental device is shown in Fig. 6.

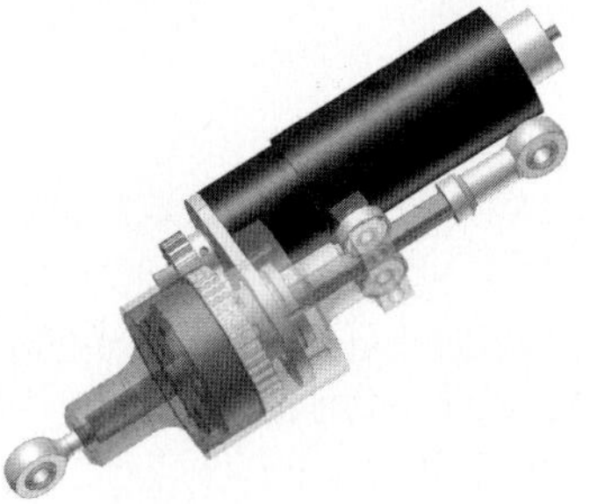

Fig. 5: Design of linear actuator

Fig. 6: Design of the experimental device

4. Conclusion

Design of the testing device was reached by mechatronic approach by relatively simple way. Thanks to the Stewart platform ability of general motion the device can be used as testing device for other clinical applications (e.g. dental prothesis) with motion, load forces and moments inside the original device requirements. Now a proof of concept of the linear actuator including control for proving suitability and verification of the real parameters is in preparation. After eventual redesign whole testing device will be built.

Acknowledgement

Published results were acquired using the subsidization of the Ministry of Education, Youth and Sports of the Czech Republic, research plan MSM 0021630518 "Simulation modelling of mechatronic systems" and the project 1P05ME789 "Simulation of mechanical function of selected segments of human body".

References

[1] Březina T. et al.: Simulation Modelling of Mechatronic Systems II, FME BUT, Brno, 2006.
[2] Merlet, J. P.: Parallel robots, 2nd Edition. Kluwer, Dordrecht, 2005.

Time Series Analysis of Nonstationary Data in Encephalography and Related Noise Modelling

L. Kipiński (a)

(a) Wrocław University of Technology, Faculty of Fundamental Problems of Technology, Division of Biomedical and Measurement Engineering, Wybrzeże Wyspiańskiego 27, Wrocław, 50-370, Poland

Abstract

In this report, statistical time series analysis of nonstationary EEG/MEG data is proposed. The signal is investigated as a stochastic process, and approximated by a set of deterministic components contaminated by the noise which is modelled as a parametric autoregressive process. Separation of the deterministic part of time series from stochastic noise is obtained by an application of matching pursuit algorithm combined with testing for the residuum's **weak** stationarity (in mean and in variance) after each iteration. The method is illustrated by an application to simulated nonstationary data.

1. Introduction

In brain evoked activity measured by means of EEG/MEG, one can observe time-dependent changes of its various characteristics like amplitude and frequency, as well as the contaminating noise. For this reason, it is necessary to use the analysis methods designed for nonstationary signals, since the standard EEG/MEG methodology based on signal averaging and simple spectral analysis is insufficient. Time-frequency estimation methods such as short-time Fourier transform, Wigner distribution, or discrete and continuous wavelet transform are very useful, yet, statistically inefficient. They also have some inherent limitations. Thus, the representation of the evoked-response generative process given by these methods is incomplete. In this research, EEG/MEG signal is investigated as a stochastic process which can be decomposed to a set of deterministic functions representing its nonstationarity and stationary residua. For modelling the stochastic EEG/MEG noise, statistical time series analysis methods are used.

2. Statistical time series analysis

A time series (TS) model for the observed data $\{z(t)\}$ is a specification of the joint distributions (or possibly of only the means and covariances) of a sequence of random variables $\{Z_t\}$, with a realization denoted by $\{z(t)\}$ [1]. In a short form, an additive TS model can expressed by the sum of deterministic $d(t)$ and stochastic $l(t)$ components:

$$z(t) = d(t) + l(t) \tag{1}$$

Off course, there are many possible examples for this kind of a model, i.e. $d(t)$ can be a linear trend, a seasonal (periodic) function, or a sum of them, and $l(t)$ can be a set of observations of any (stationary or nonstationary) random variable. Let us take times series generated by an additive stochastic process given by (2), which is the sum of $m \in N$ sine waves or other non-commensurable periodic functions (or commensurable but with a period much longer than the periods of its particular components) $s(t)$ and a stationary noise $e(t)$.

$$z(t) = \sum_{i=0}^{m} s^{i}(t) + e(t) \tag{2}$$

Modelling of a process requires all the deterministic functions to be removed at first. It can be achieved by the preceding estimation of these functions or by differencing the series. There is a lot of helpful examining tools (i.e. statistical tests and spectral analysis methods (spectral density, periodogram) based on the Fourier transform) and estimation techniques (i.e. maximum likelihood method). Next, stochastic, stationary residua can be diagnosed (on the basis of the sample autocorrelation function (ACF) and sample partial autocorrelation function (PACF), and using some statistical tests) and an adequate parametric model of them (autoregressive (AR) and/or moving average (MA) for example) can be constructed [1].

3. Time series de-trending by matching pursuit

Unfortunately, EEG/MEG signals are nonstationary, and so the characteristics of $s^i(t)$ components in Equation (2) varies in time (thus we can not assume that each $s(t)$ is periodic). In consequence, a typical time series decomposition, based on the Fourier transform, disappoints in this case. Therefore, it is required to construct the ongoing EEG/MEG noise model by using an effective approximation of the nonstationary components of the EEG/MEG signal.

In the first step of the iterative matching pursuit (MP) algorithm proposed by Mallat and Zhang [2], the atom g_{γ_0} that gives the largest product with the signal is chosen from the large, redundant dictionary D (usually composed of Gabor functions being cosines modulated in amplitude by a Gaussian envelope). In each consecutive step, the atom g_{γ_m} is matched to the signal $R^m z$ that is the residual left after subtracting the results of previous iterations:

$$\begin{cases} R^0 z = z \\ R^m z = \langle R^m z, g_{\gamma_m} \rangle g_{\gamma_m} + R^{m+1} z \\ g_{\gamma_m} = \arg\max_{g_{\gamma_i} \in D} \left| \langle R^m z, g_{\gamma_i} \rangle \right| \end{cases} \quad (3)$$

The possible stopping criteria for this algorithm are: 1 – fixing a priori the number of iterations m, irrelevant of the content of the analysed signal, 2 – explaining a certain percentage of the signal's energy, 3 – the energy of the function subtracted in the last iteration reaches a certain threshold. It is assumed that the residual vector obtained after approximation of m time-frequency waveforms is the noise, which converges to $N(0,\sigma^2)$ with increasing m [3].

Here, we propose to combine TS analysis with MP algorithm by a new model resulting from (2) and (3):

$$z(t) = \sum_{i=0}^{m} \langle R^i z, g_{\gamma_i} \rangle g_{\gamma_i}(t) + e(t) \quad (4)$$

In this model, the MP procedure is performed as long as the residuum is nonstationary. The accomplishment of a weak stationarity entails stopping the algorithm. Next, TS methods should be applied for modelling the residuum. Practical measures of the stationarity are the results of statistical tests: Kwiatkowski–Phillips–Schmidt–Schin (KPSS) test [4] for stationarity in mean, and the White test for homoscedasticity [5].

3. Application to simulated nonstationary data

In order to illustrate the idea of the proposed algorithm and to examine its properties, a simulated signal (Fig. 1) was constructed from a sum of *m* Gabors and a stochastic (but not Gaussian) noise generated by ARMA(5,3) process. This signal is stationary in mean, but heteroscedastic (covariance vary over time), so it is necessary to remove that nonstationarity by means of MP.

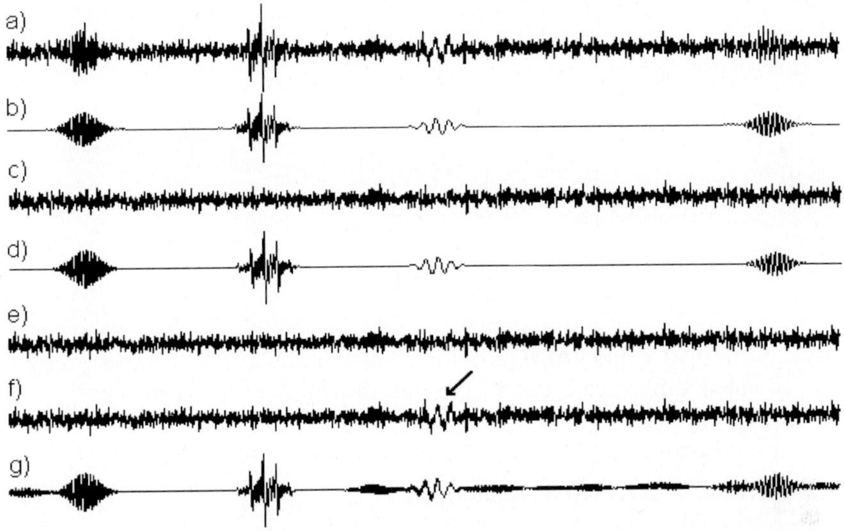

Fig. 1. a) Simulated signal constructed with a sum of *m* Gabor functions b) and autoregressive moving average noise c); d) the approximation after *m* iterations of MP algorithm stopped due to stationarity of residua e); f) nonstationary residua after *m-1* iterations (the arrow points at "the reason" for the White test's rejection of the null hypothesis); g) excessive approximation (for $n>m$ iterations, MP starts to explain the noise).

Residuum is tested after each iteration, and after the *m*-th one we have no reason to reject the null hypothesis of the White test about its homoscedasticity. This means that the residuum is weakly stationary, thus we can use statistical time series tools to describe it. Tests for independence and plots of sample ACF and PACF (Fig. 2b)) show that the residua generator is autocorrelated, and imply an autoregressive moving average model for the residua. The fit is obtained by the maximum likelihood estimators of the ARMA parameters, and minimum Akaike criterion. The values for all significant ARMA parameters are similar (into or near the confidence level) as the values used for the simulation. The noise error was examined too, and the result (Gaussian white noise) confirms correctness of the fit. In comparison, the "classical" MP algorithm looses information about the noise, or (if the number of iterations is large) it starts to explain the noise with deterministic functions (Fig. 1g)). The combined MP-TS algorithm is a clear improvement, as modelling of the noise is addressed explicitly.

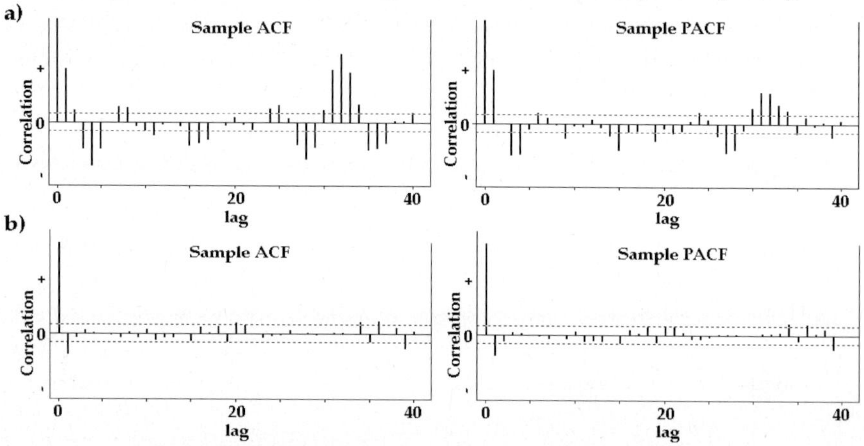

Fig. 2. Sample ACF and PACF plots for: a) simulated signal b) stationary residua. Evident difference is due to spectrum variability of simulated signal.

4. Summary and conclusion

In this paper, a combination of time series analysis methods with signal decomposition by matching pursuit is proposed. Although each of the two methodologies separately can be applied to build signal representation, their marriage seems very efficient, as their simultaneous use enables holistic description of both deterministic and stochastic components of a signal. It can be an effective method for noise description, prediction and filtering in any EEG/MEG measurement.

Acknowledgements: In this work, we used our modification of the MMP software written by A. Matysiak [6].

References

[1] P. J. Brockwell, R. A. Davies, "Introduction to time series and forecasting", Springer, New York and London, 2002.
[2] S. Mallat, Z. Zhang, IEEE Trans. on Sig. Proc. 41 (1993) 3397–3415.
[3] P. J. Durka, D. Ircha, K. J. Blinowska, IEEE Trans. on Sig. Proc. 49 (2001) 507–510.
[4] D. Kwiatkowski, P. C. B. Phillips, P. Schmidt, Y. Shin, J. Econometrics 54 (1992) 159–178.
[5] H. White, Econometrica 48 (1980) 817–838.
[6] A. Matysiak, P. J. Durka, E. M. Montes, M. Barwi□ski, P. Zwoli□ski, M. Roszkowski, K. J. Blinowska, Acta Neurobiol. Exp. 65 (2005) 435–442

Ambient dose equivalent meter for neutron dosimetry around medical accelerators

N. Golnik (a) *

(a) Institute of Precision and Biomedical Engineering, Warsaw University of Technology, św. Andrzeja Boboli 8,
Warsaw, 02-525, Poland

Abstract

A dose equivalent meter based on a recombination principle has been designed for routine measurements of ambient dose equivalent in mixed (gamma + neutrons) radiation fields outside the irradiation fields of linear medical accelerators. Two recombination chambers serve as detectors. Four different voltages are sequentially applied to the chamber electrodes and the ionization current values are measured for each voltage. The absorbed dose rate and ambient dose equivalent are calculated in real time, taking into account the dependence of the initial recombination of ions on linear energy transfer (LET). Tests at 15 MV Varian Clinac 2300C/D accelerator confirmed that ambient dose equivalent of mixed radiation in clinical conditions could be determined with accuracy of about 10%

1. Introduction

Radiation fields around medical accelerators, are slightly contaminated with neutrons, generated by photon-neutron nuclear reactions. This concerns practically all the accelerators operating at maximum photon energy of 15 MeV or higher. The photoneutron energy spectrum has a peak around 1 MeV, however, at the patient's plane, after the transmission through the accelerator head, neutrons have a distribution similar to that of the heavily shielded fission source.

The level of neutron production and its unwanted whole-body dose to the patient vary around different treatment units between 1 and 4.8 $mSvGy^{-1}$ (neutron dose equivalent per tissue dose at isocentre), depending on accelerator characteristics and the distance from the isocentre. The total neutron dose equivalent evaluated for a complete therapeutic treatment of 60 Gy photon dose is between 60 mSv and almost 300 mSv.

The International Electrotechnical Commission (IEC) recommended limits for the neutron absorbed dose in the patient plan [1], but practically almost no measurements are performed in radiotherapy departments, The main reason which discouraged medical physicists from making the measurements is the lack of convenient measuring equipment for the routine use.

A relatively simple measuring method, with a recombination chamber, has been recently proposed in our previous papers [2,3]. The methods are suitable for radiation protection measurements along the treatment couch outside the irradiation field. Among them, the method based on the determination of recombination index of radiation quality seems to be the most convenient for the routine measurements.

The paper presents a short overview of the method, a model of the device for automatic measurements of ambient dose equivalent meter and the results obtained at the medical accelerator.

2. Method

Recombination chambers are high-pressure, usually tissue-equivalent, ionization chambers operating under condition of initial recombination of ions. This kind of recombination occurs within tracks of single ionizing particles. It does not depend on the dose rate and depends on local ionization density within the tracks of ionizing particles i.e. on radiation quality.

The use of recombination chambers makes it possible to determine the total absorbed dose, which is proportional to the saturation current, and recombination index of radiation quality, Q_4 [4] in a phantom of interest.

The energy of neutrons generated by medical accelerators does not exceed several MeV. In such fields, the ambient dose equivalent H*(10) can be well approximated by the product of the ambient absorbed dose D*(10) and the recombination index of radiation quality Q_4 [5].

$$H^*(10) = D^*(10) \times Q_4 \tag{1}$$

where D*(10) is the absorbed dose in the appropriate phantom, simulating the ICRU sphere.

The method to determine the recombination index of radiation quality, Q_4 has been described earlier [4,5]. Here, only the main points of the method are briefly sketched. First, for a given chamber, the special voltage U_R has to be determined in a calibration procedure, as the voltage that ensures 96% of saturation in a reference field of gamma radiation. Usually, a ^{137}Cs

radiation source is used for this reason. Then, in the radiation field under investigation, one has to determine the saturation current and the ionization current at the voltage U_R. For radiation protection purposes, the saturation current can be approximated by the ionization current measured at a high voltage U_S, within the saturation range. Then, the Q_4 is determined as:

$$Q_4 = (1 - f_R)/0.04 \tag{2}$$

where $f_R = f(U_R)$ is ion collection efficiency measured at voltage U_R in the investigated radiation field.

A recombination chamber of F1 type [5] was used as a main detector in this work. The F1 chamber is a phantom, parallel-plate chamber with volume of 3.8 cm^3. It was filled with ethane (C_2H_6) up to a pressure of 0.7 MPa. The chamber has three TE electrodes, 34 mm in diameter. The wall thickness is 0.6 g/cm^2. The distance between electrodes is equal to 1.75 mm. The F-1 chamber is well sealed and its sensitivity usually does not change more than 0.5% per year.

The second chamber of REM-2 type [5] was used as a monitor of the radiation beam intensity. The monitoring chamber was supplied with the constant voltage of 300 V. The chambers were connected to the electronics by electrometric cables, type T3295 BICC (2 mm in diameter, PTFE insulation covered by graphite).

3. Dose equivalent meter

The measuring system consists of two ionization chambers (main detector and monitor) connected to a two channel automated electronic unit (ambient dose equivalent meter, ADEM) controlled from a PC computer [6]. ADEM contains four multiplex AD channels with resolution of ±15 bites Two of them are electrometric with measuring ranges of 2,5 nC and 25 nC for the measurements of the electrical charge and of 25 pA and 250 pA for the measurements of the ionization current. Two other channels are provided for the measurements of temperature and voltage in the range ±2,5 V and they have not been used in this work. AD conversion time for each channel is 25 μsek. The device contains also two stabilized high-voltage supplies. One is digitally controlled with resolution of 0,4 V in the range ±1638 V. The second one provides voltages in the range from 0 to 1600 V, with the step of 200 V and is controlled from the front panel. The device is controlled by a parallel port of a PC computer. With such solution, it can be used with all computer control systems.

4. Measurements

The values of the ionizations current of the main chamber have to be measured at saturation and at the voltage U_R. However, the results are much more precise if the measurements are performed for both polarities of the voltages and appropriately averaged [4,5]. The measuring system was calibrated at the Institute of Atomic Energy in reference radiation fields of ^{137}Cs, in terms of ambient dose equivalent. The calibration involved determination of the recombination voltage U_R. The value of $U_R = \pm 40$ V was used for the F1 chamber.
The sequential application of the positive and negative voltages was controlled by the PC computer through the AMED voltage supply.
The measurements were performed in the treatment room of the Varian Clinac 2300C/D at the Oncology Centre in Warsaw, with the accelerator producing 15 MV photons. The recombination chamber was placed on the treatment bed, at the distances of 50 cm and 100 cm from the isocentre. Two irradiation fields were used – with the photon beam collimated to the area of 10×10 cm^2 and 4×4 cm^2 at the isocentre. The beam intensity was 100 MU/min. The measurements were made on the treatment couch, at the distances of 50 cm and 100 cm from the isocentre. The chamber was placed on the PMMA phantom, but during the measurements with the larger irradiation field, the treatment bed was moved out of the beam in order to reduce the scattered photon radiation.
The time needed for the determination of H*(10) was usually about 20 -30 minutes, depending on the stability of the accelerator beam. The results were obtained on-line according to the equations (1) and (2).
Additionally, the neutron absorbed dose could be estimated after completion of the measurements. The method for such calculations is described elsewhere [2,3].

5. Results

The values obtained for the 10×10 cm^2 irradiation field are summarized in the Table 1. Increasing The value of the Q_4 increases with the distance from the isocentre. This clearly indicates that also neutron contribution to the ambient dose equivalent increases. In our conditions of irradiation the neutron contribution at 50 cm constitutes nearly 50% of H*(10), and at the distance of 1 m almost all the dose equivalent is due to neutrons.
The measurements of H*(10) at the irradiation field of 4×4 cm^2 resulted in the values of 60 mSvh^{-1} at 50 cm from the isocentre, 35 mSvh^{-1} at 100 cm

and 7 mSvh^{-1} at 300 cm The results are very similar to those obtained earlier with the laboratory measuring system [3].

Table 1. Basic experimental data obtained with F1 chamber and ADEM device. All the measurements were performed at the photon beam intensity of 100 monitor units (MU) and irradiation field of 10×10 cm^2.

Distance from the isocentre	On-line results		
	Q_4	$D^*(10)$ [mGyh^{-1}]	$H^*(10)$ [mSvh^{-1}]
50 cm	1.7	79	134
100 cm	5	7.6	38

6.Conclusions

The main idea of the present study was to create an automated measuring system with a recombination chamber, for the direct determination of the total H*(10 at medical accelerators outside the irradiation field. It was proved that the measurements could be performed in reasonable time of about 20 - 30 minutes and with accuracy of better than 10%. The significant advantage of using the recombination chamber is the direct reading of the result and relatively short time of the measurements.

References

[1] International Electrotechnical Commission "Medical electrical equipment - Part 2-1: particular requirements for the safety of electron accelerators in the range 1 MeV to 50 MeV", IEC 60601-2-1,1998.
[2] N. Golnik, P. Kamiński, M. Zielczyński, Radiat. Prot. Dosim. 110, (2004) 271.
[3] N. Golnik; M. Zielczynski; W. Bulski; P. Tulik; T. Palko Radiat. Prot. Dosim. (2007) doi: 10.1093/rpd/ncm125.
[4] M. Zielczyński, N. Golnik,. Radiat. Prot. Dosim. 52 (1994) 419.
[5] N.Golnik "Recombination methods in the dosimetry of mixed radiation" Institute of Atomic Energy Świerk (PL), Report IAE -20/A. 1996.
[6] Z. Rusinowski "Device for precision measurements of mixed radiation using recombination chambers" (in Polish) Institute of Atomic Energy Świerk (PL), Report B 41/98, (1998).

External Fixation and Osteogenesis Progress Tracking Out in Use to Control Condition and Mechanical Environment of the Broken Bone Adhesion Zone

D. Kołodziej, D. Jasińska-Choromańska

Warsaw University of Technology,
Institute of Micromechanics & Photonics, St. A. Boboli 8 street,
Warsaw, 02-525, Poland

Abstract

An external stabilization give a possibility to assure the right geometry of the fractured limb and safety load and proper unload the adhesion zone. In order to increase the possibility of a controlled medical interference in the broken bone healing process and fulfill the postulate of active healing it has to be known the present mechanical properties of the fractured bone. Applicate the fixator with right matching mechanic parameters is prelude in creation of a new fixator ability named "adaptive mode" described as a fluent time-changeable mechanical characteristic of the broken bone - external fixator system.

1. Introduction

External Osteosynthesis is a medical method which main assumption is assuring the right mechanical environment for the physically joined broken bone pieces. Mechanical stability issue is the main problem of the fixation systems and proper 3D configuration of the broken bone-fixation system can secure the young bone tissue before the crash As an exemplary model still can be taken an Ilizarov fixation system (Fig. 1) because its very high rigidity and assurance ability. It has been observed the relationship between the mechanical load of the adhesion zone that is located in the bone fracture, and the phenomenon of hardening and accelerating the bone remodeling process.

Fig. 1: Ilizarov fixation system

Reflection about this observation become a source of the next consideration about use the modern drive module and specify measuring systems to extend the possibilities of the existing external fixation systems. For some time now medics have been searching the method that allows them to direct influence on the treatment process. Present known methods are grounded on the observation of the consequences of the previous doings. The diagnosis still base on the visual quantitative subjective medical opinion. The medicine needs objective parameters that can describe in proper way the state of the young bone regenerates and also the mechanical condition of the whole limb.

2. Mechanical environment

Considering the structure of the bone mechanical environment, it can be differentiate an external from an internal mechanical environment. The External one is connected with the environment of the human body which gives high load impulses (forces, moments, etc.) that are shaping the internal environment. Loads are transmitted from the External by the fixator frame trough the bone screws to the Internal environment (Fig. 2) The adhesion zone can be in this way partially or fully relieved according to the mechanical profile of the fixator and its dump and carry loads ability.

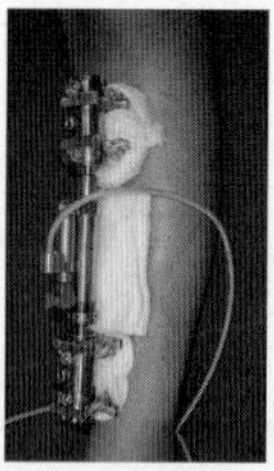

Fig. 2: Dynastab Mechatronics 2000 with measurement module [2]

The Internal one is directly connected with the closest surroundings of the adhesion zone and in this way this environment is shaping the future adhesion's mechanical profile. As common known the micro movements at the bone fracture can stimulate the growth process (Fig. 3) [3]. People should carry to shape these micro movements (range and loads) properly to assure that the adhesion growth and remodelling process goes in right way.

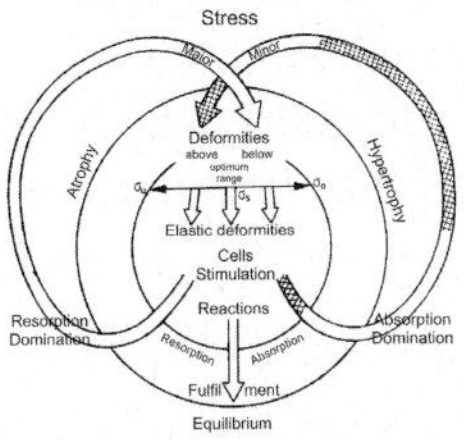

Fig. 3: Mechanical environment as a stimulation source in the broken bone tissue regeneration process [1]

In order to the time changeable mechanical loads that occurs in the bone fracture, the mechanical profile of the fixator frame should change its mechanical configuration. Tracking the occurred loads can be very helpful in the individual healing patient profile building process. Each information can be used in two ways. First of them is connected with the active mechanical crack zone securing. According to the occurring forces fixator should reconfigure itself and affect the proper shape in secure way (Fig. 4).

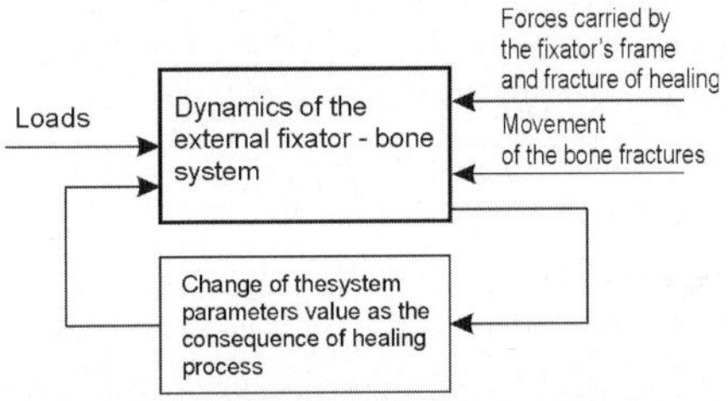

Fig. 4: Evolution of the mechanical environment of the fixation system [2]

The second one can be use in active bone stimulation proces, in which has to be firstly created the right bone loads and unloads profile. Only secure stimulation can properly accelerate the biological processes without any mistaken that can not be successfully retrieved in the future.

Healing progress tracking out

Heaving the proper knowledge about the present state of the mechanical properties of the bone regenerate enable to create new possibilities to influent in right way to the bone tissue. Algoritm of the crack zone unloading have to be strictly connected with the bone-fracture-bone system con-

dition. External osteosynthesis is used as a tool in the bone healing process. It can not be use as a medical replacement and in this way it always needs the human to supervise its adaptation.
Mechanical parameters of the adhesion can be taken using specify measurement module. Increasing or decreasing of the adhesion mechanical features give an informations about the healing progress (eq.1). These informations are limited to the features which can be described having force or pressure and the one direction bone pieces displacement data (Fig. 5). These limitations are dictated by the one of the active healing postulate, *do nor crush the adhesion and keep the right bone geometry, especially bones axis.*

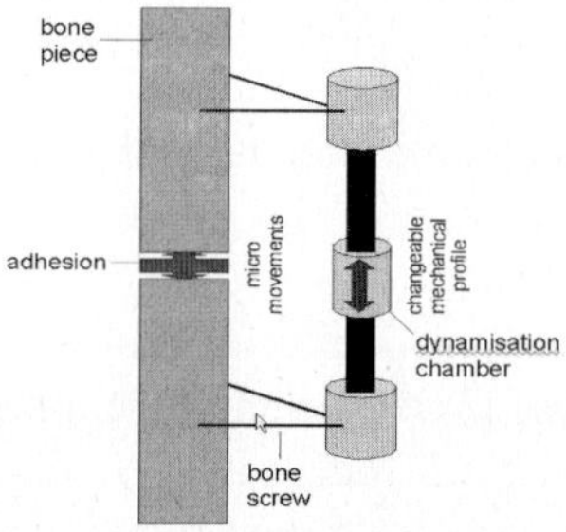

Fig. 5: Adhesion zone mechanical features measurement system

$$m(t)=\frac{F(t)_2}{F(t)}=\frac{F(t)_2}{F(t)_1+F(t)_2} \qquad \text{(eq.1)}$$

$m(t)$ – bone adhesion marker
$F(t)$ – total load
$F(t)_1$ – load carried by bone pieces through the adhesion zone
$F(t)_2$ – load carried by frame through the dynamisation chamber

An information about bone illness and course of healing process can be described using the bone adhesion makers [2]. This basic method of bone adhesion mechanical features approximating needs a special device but can give very important data that has direct connection with the functionality and mechanical limb ability.

Broken bone condition

Condition of the human bone system can be verified with a few ways. Lots of these means gives only a quantitative but not a qualitative information about the bone illness. The problem with the bones condition describing is connected with individual patient's illness and his disease history. It's common knowledge that the bone structure is not a homogeneous structure and the osteosynthesis phenomenon course is strictly connected with the general regeneration abilities of the human body. These abilities are being determined by the physical and psychological patients condition and these elements have an inseparable character.

Mechanical stimulation has very important property. Influence to the adhesion has the same kind like increasing body weight during the child period. Difference between child and a parent bone are very wide but the training of the bone structure method has the same nature. As common known each patient has its own illness history so therefore each case should be considered separately. Personal healing aiding module which can generate loads in controled way could be an additional information source about patients convalescence progress [4].

Summary

Question about abilities of the healed bone to carry the loads is still without the answer. Ignorance of the present load carry abilities causes that the patient is concerning for his limb for fear of pain and damage and do not load his limb in proper way. The direction of the researches is going to create modern external fixator which configuration profile is following according to the load changes and bone healing phase.

Acknowledgment

This work is a part of the Committee for Scientific Research project no. KBN 3 T11E 007 29.

References

[1] A. Morecki, *"Problemy Biocybernetyki i Inżynierii Biomedycznej - Tom 5 Biomechanika"*, Wyd. Komunikacji i Łączości, Warszawa, 1990.

[2] D. Jasińska-Choromańska, *"Modelowanie i symulacja w projektowaniu jednostronnych zewnętrznych stabilizatorów ortopedycznych"*, Wyd. Politechniki Warszawskiej, Warszawa, 2001.

[3] R. Będziński, *"Biomechanika inzynierska – zagadnienia wybrane"*, Oficyna Wyd. Politechniki Wrocławskiej, Wrocław, 1997.

[4] D. Kołodziej, D. Jasińska-Choromańska, *"Zintegrowana diagnostyka procesów zrostu kostnego"*, DPP, Warszawa, 2005.

[5] A. Morecki, J. Ekiel, K. Fidelus, *"Bionika ruchu – podstawy zewnętrznego sterowania biomechanizmów i kończyn ludzkich"*, PWN, Warszawa, 1971

Evaluation of PSG sleep parameters applied to alcohol addiction detection.

R. Ślubowski, K. Lewenstein, E. Ślubowska,

Institute of Precision and Biomedical Engineering,
Warsaw University of Technology, A. Boboli 8, Warsaw 02-525, Poland

Abstract

The results of detection of alcohol addiction based on the analysis of human sleep are presented in this paper. Sleep was described by numerical parameters calculated from the standard processed records of polysomnography (PSG) signals.

The database used in the experiments consisted of almost 200 examinations: 50% of healthy and alcoholic addicted patients, and 50% males and females, with normal age distribution.

We have used two different methods: statistical estimator and neural networks to evaluate the diagnostic value of the sleep parameters. We have proposed the set of 13 basic parameters to detect alcohol addiction. The differences in diagnostic value of these features are noticeable, but not very significant (the differences of the diagnosis correctness lie between +2% and –4%), but each of them improves the total quality of learning process.

Finally, we have obtained about 75% correctness of alcohol addition diagnoses.

1. Introduction

Polysomnography (PSG) is an overnight test used to evaluate abnormalities of sleep and/or wakefulness and other physiologic disorders that have an impact on or are related to sleep and/or wakefulness. A polysomnogram consists of a simultaneous recording of multiple physiologic parameters related to sleep and wakefulness. By international standards, a polysomnogram have to include at least 4 neurophysiologic channels: one electroencephalography (EEG), two electrooculogram

(EOG) channels and one electromyography (EMG) channel. The overnight recording is divided into epochs of approximately 30 seconds. According to standard procedure [4] predominant stage of sleep is assigned to the entire epoch on the basis of EEG, EMG, and EOG recordings. The total time of sleep and time spent in each of the 6 sleep stages are calculated.
Changes of some sleep parameters (sleep latency, total sleep time, stage REM, stages: 3, 4, REM latency) had been observed in most of researches, concerning an influence of alcohol addiction on sleep pattern. The findings were collected and briefly characterized in Kirk J. Brower's study [1]. In the paper [3] we have used neural networks for detection of alcohol addiction on the basis of sleep parameters, now we want to show diagnostic value of particular features calculated from PSG recordings.

2. Materials

We have used the database consisting of almost 200 examinations containing processed records of the polysomnography signals of alcoholics and age-matched healthy control subjects. There were 85 healthy and 87 alcoholic patients, 86 males and 86 females. Detailed description of the collected data was presented in the paper [3]. Twenty-six numerical parameters characterizing sleep saved in the database are presented in Table 1.
There are some general indicators (concerning the whole sleep) and those more detailed (referring to the isolated characteristic stages of sleep [4]). The recognition of alcohol addiction (i.e. medical statement if the patient is addicted) is an essential supplement to the collected data.

Table 1. Specification of sleep parameters and sets of parameters used in the experiments.

Parameter / Set of parameter	1	2	3	4	5	6	7	8	9	10	11	12	13	14	15	16	17	18	19	20	21	22	23	24	25	26
	min	min	min	min	%	%	min	%	min	%	min	%	min	%	min	%	min	%	min	%	min	min	min			min
	Total recording time	Total sleep period	Time of awaking	Total sleep time	Sleep maintenance	Sleep efficiency	Stage 1 NREM	Stage 1 NREM	Stage 2 NREM	Stage 2 NREM	Stage 3 NREM	Stage 3 NREM	Stage 4 NREM	Stage 4 NREM	Stage 3 plus 4 NREM	Stage 3 plus 4 NREM	Stage NonREM	Stage NonREM	Stage REM	Stage REM	Sleep latency	REM latency	Latency to stage 3 and 4	The number of awakes	The number of REM episodes	Average time of cycle
Set I	X	X	X	X	X	X	X		X		X		X		X		X		X		X	X	X	X	X	X
Set II			X	X			X		X		X		X						X		X	X	X	X	X	X
Set III			X	X			X		X		X		X						X			X		X		

3. Methods

We have used two different methods to study the data.
First, the data had been divided up into two groups: alcoholic and control one. Average and standard deviation were calculated for all of 26 parameters in these groups. To indicate the percentage differences between the groups the received values of averages were divided by the average from the control group and values of standard deviations were divided by the control group's average too. The differences are shown on Fig. 1.
Secondly, neural networks were used as a standard software tool in a multidimensional data analysis. In our experiments, networks were trained according to the strategy "with the teacher". The role of the teacher was played by the medical diagnosis of the subject's condition.
We have modelled a perceptron type neural network, with one hidden layer. The number of neurons in this layer was assessed experimentally as the lowest possible number enabling network training (to the almost zero training error). The input layer contained as many neurons as features describing the patient's sleep. The output (decision) layer contained only one neuron responsible for detecting alcohol addiction. According to the literature, as well as the author's experience [2], a network of such architecture has the greatest ability to generalize.
The Stuttgart Neural Network Simulator (SNNS - free-available software simulator) has been used in our studies to build and train networks used in the experiments.
We have implemented the "Quickprop" training algorithm with continuous sigmoid activation of neurons, because of the character of data. A classic four divided cross–validation method, multiple random initialisations and network trainings [3] have been used in order to check the correctness and repeatability of the results. The test's results of ten networks were averaged to get partial results. These partial results, after ultimate averaging for four divisions, provided us with the percentage of correct detected persons (Fig. 2).
In our neural network experiments we have used 19 parameters (Table 1, Set I) because some of 26 mentioned features were expressed both in minutes and in percents of total sleep time. There are also parameters, which can be calculated on the basis of the others from that set, for example "stage3 plus stage4". Therefore, we can simplify the analysed set to only 13 parameters (Table 1, Set II).

To evaluate an influence of particular parameters on result of alcoholic addiction detection, we checked how reducing of each parameter from training set (Table 1; Set II) changes the outcome. Fig. 2 shows the results of these experiments.
We have also calculated percentage of effective networks' initialisations as a tentative method of estimation the parameter's significance.

4. Results and discussion

The statistical analysis (Fig. 1) shows that some sleep parameters have small differences (<10%) of average between alcoholic group and control group. Standard deviation is also comparable. There are also some parameters with big difference (25%-100%) of average, but these parameters have very big (>100%) standard deviation in both groups. This means that detection of alcoholic addiction based on statistical methods would have too small correctness to be reliable.

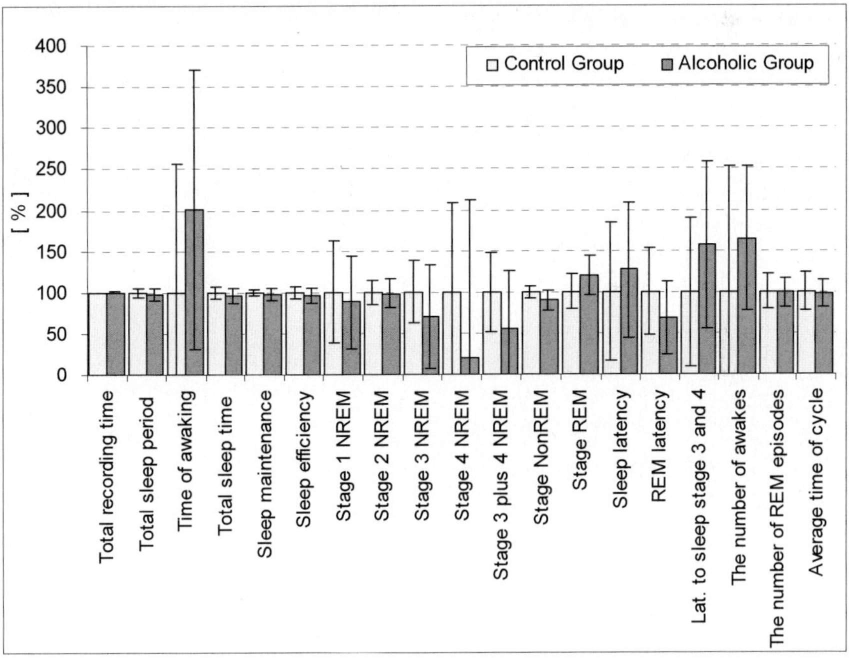

Fig. 1. Averages and standard deviations calculate proportional to averages for the control group.

The averaged results of neural network analyses are presented in Fig. 2. We have obtained the percentage of correct detections from 69,9% to 76,6% with the standard deviation of 0,8÷3,8% for all experiments. Effective initialisations were from 64% to 97,5%. Based on the results shown in Fig. 2 we can notice, that lack of some parameters in the Set II leads to better results (e.g. "latency to stage 3 and 4"). Simultaneously, a deficiency of the other ones causes noticeable deterioration of the detections' outcomes (e.g. "stage 4 NREM" and "stage 3 NREM"). The significance weight of parameters: "stage 3 NREM" and "stage 4 NREM" was confirmed by the values of the received numbers of the effective initializations and by substantial differences in the averages for the groups. Moreover, the big difference in the averages of "latency to stage 3 and 4" seems to be not correlated with alcohol addiction.

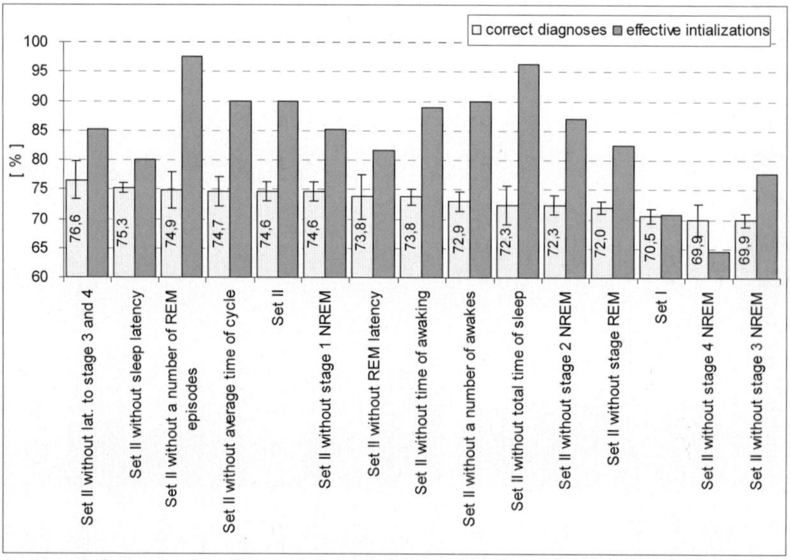

Fig. 2 Results of alcohol addiction's detection made by particular networks.

The noticeable differences of the results for the specified sets of parameters (Set I and Set II) indicate that the redundant information decrease reliability of the detections and optimization of the set's composition is required.

We have proposed the Set III (Table 1) to check if we can reduce a number of features from the Set II without worsening the correctness of diagnosing. We obtained 76,25% correctness of diagnoses with 51,6% effective initializations for the neuronal network with 5 hidden units.

Simultaneously, we achieved the 76,83% correctness of diagnoses with 96,4% effective initializations for the neuronal network with 7 hidden units. These results show that "weak features" are necessary to optimise the learning process' quality.

5. Conclusions

The research described in the paper leads to the following conclusions:

- Use of the NN and compressed features vector gives us a correctness of the diagnosis about 75%;
- The diagnostic values of features (from Set II) calculated from the PSG are noticeable, but not very significant (the difference of the diagnosis correctness lies between +2% and –4%). All of them are needed during learning process. We can eliminate these parameters, but we should add some hidden units to calculate the necessary values by the network;
- The most significant features are: "stage 4 NREM" and "stage 3 NREM", least significant seems to be "latency to stage 3 and 4".
- We think that the maximum of the diagnosis correctness could be under 80% using described PSG record and the optimal features vector.
- The length of particular sleep stages changes during the time of sleep. The parameters used in the experiments have been calculated as total sums of the six stages obtained from successive sleep cycles. We consider that it is necessary to take into account these changes to increase the reliability of diagnoses.

6. References

[1] Brower K. J.: Alcohol's Effects on Sleep in Alcoholics. Alcohol Research & Health, Spring, 2001 (3/22/01).

[2] Lewenstein K: "Artificial neural networks in the diagnosis of coronary artery disease based on ECG exercise tests." Warsaw 2002, Oficyna Wyd. P.W., Electronics vol. 140.

[3] Lewenstein K., Ślubowska E., Jernajczyk W., Czerwosz L.: "Detection of alcohol addiction based on PSG sleep patterns."; Polish Journal of Medical Physics and Engineering 2006 Vol.12 No.3; 121-130.

[4] Rechtschaffen A., Kales A: A manual of standardized terminology techniques and scorning system for sleep stages of human subjects. BIS/BRI. UCLA, (1968): Los Angeles.

Drive and control system for TAH application

P. Huták (a), J. Lapčík (b), T. Láníček (c)

(a), (b), (c) Brno University of Technology, Faculty of Electrical Engineering and Communication, Department of Power Electrical and Electronic Engineering, Technická 8, 616 00, Czech Republic

Abstract

The paper presents analytic design of active magnetic bearing PM synchronous motor for TAH (Total Artificial Heart) application and description of the drive and levitation electromagnet control. Pump system is double ventricle blood pump consisting of slotless PM axial flux synchronous machine and magnetic active bearings. The general electric machine theory is applied to the motor design. Power quality is ensured by the help of the rare earth PM and for good torque quality is used slotless, three phase winding and surface mounting PM. Control part deals with the synchronous motor and magnetic bearing control. Two structures, voltage and current, were design and tested for the position regulation. Mathematical model was successfully simulated. On the base of this model the regulation structure for positional feedback control was design. Design of the regulators is based on the mathematical model description of controlled system.

1. Realization of the magnetic bearing pump

In our solution Fig. 1 the drive system consisted of the permanent magnet synchronous slot-less motor and magnetic active radial bearings. These systems were tested both on the air and water environments. Problem of the magnetic bearings and electrical drive is very the same also for other type of the rotating pumps.

Fig. 1. Pump's drive system with the motor and magnetic bearings

2. General conception

At the first pump design, the base requirement was hydraulic output power. At this stage of design, properties and possible damage of blood elements were not be reflected. The pump contains 3-phase synchronous disk-type machine with slot-less winding (magnetic circuit doesn't contain any ferromagnetic materials), double sided rotor with permanent magnets and integrated turbine blades and two active axial magnetic bearings.

3. Motor and magnetic bearing control

Miniature controller "Easy 25" for brushless motors, mostly used for model motors, was found to be very useful for artificial heart motor supply. This controller is extremely simple to use – the easiest way it can be. Controller is ready for an immediate start without any prior settings, everything is done automatically. The controller is designed for the brushless sensor-less motors. Magnetic bearing has generally five control channels, four radial and one axial. For mostly used control systems each control channel has autonomous control and input signal is coming from the position sensors and systems have one degree of freedom. In the next text we will deal with one degree bearing. Dependency on if output signal is current or voltage the control can be current or voltage. Current control can be described by the equation of the second order. Voltage control can be described by the equations of the fourth order.
When we spoke about magnetic bearing control then we mean system of the linear, non linear, optimal and adaptive regulators. Into the control system is included also stability, transience and frequency characteristics of the bearings.

Mathematical model of the bearing and description depends on the chosen current or voltage control. Initial non-linear model of the bearing is described in the equation (1) and its non-linear analogy is described by the equations (2) and (3).

$$\left.\begin{aligned} m\ddot{y} &= \frac{c_L}{2}\left[\frac{i_1^2}{(\delta-y)^2}-\frac{i_2^2}{(\delta+y)^2}\right]+Q \\ \frac{c_L}{\delta-y}(i_1)^{\bullet} &+ \frac{c_L}{(\delta-y)^2}i_1\dot{y}+ri_1 = u_1 \\ \frac{c_L}{\delta+y}(i_2)^{\bullet} &- \frac{c_L}{(\delta+y)^2}i_2\dot{y}+ri_2 = u_2 \end{aligned}\right\} \quad (1)$$

$$m\ddot{y} = F_V + Q\ ; F_V = c_y y + h_{1i} i_v\ ; L(i_v)^{\bullet} + 2h_{1i}\dot{y} + ri_v = u_v \quad (2)$$

$$L\dot{I}_v + rI_v = U_v \quad (3)$$

Fig. 2: Disc with position sensors and control electronics

Initial non-linear bearing model is described in this case by the first equation in system (1), where controlling function have the currents i_1 and i_2. Linear model is described by the equation:

$$m\frac{dy^2}{dt^2} - c_Y \cdot y = h_i \cdot i + Q(t) \quad (4)$$

At conditions

$$-\delta < y < \delta \quad ; \quad -i_C \le i \le i_C \quad (5)$$

where i is the input current.

When we transform system of the equations (4) to the normalized form we have

$$x = \left(y, \frac{dy}{dt} \right)^T \quad ; \quad u = i \quad ; \quad f(t) = Q(t) \quad ; \quad C = D = (1,0)$$

$$A = \begin{pmatrix} 0 & 1 \\ C_Y / m & 0 \end{pmatrix} \quad ; \quad B = \begin{pmatrix} 0 \\ h_i / m \end{pmatrix} \quad ; \quad B_f = \begin{pmatrix} 0 \\ 1/m \end{pmatrix} \tag{6}$$

Using dimensionless variables then equation (4) can have two forms of the notation: using the dimensionless time $\tau = t / T_0$

$$\frac{dy^2}{dt^2} - x^2 \cdot y = x^2 \cdot i + Q(\tau) \tag{7}$$

Or using dimensionless time $t' = \tau \cdot x = t \cdot x / T_0$,

$$\frac{dy^2}{dt^2} - y = i + Q(t) / x^2 \tag{8}$$

at the conditions:

$$-1 < y < 1 \quad ; \quad -1 \leq i \leq 1 \tag{9}$$

4. Mathematical models of the bearing

Mathematical description of the bearing depends on the chosen control current/voltage, and on the mathematical model see Fig. 3.

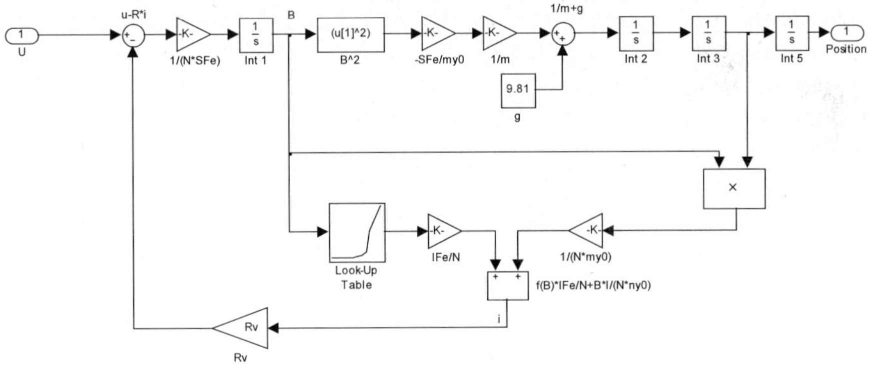

Fig. 3: Mathematical model of electromagnet with a non-linear magnetic circuit

5. Conclusion

The work was focused on the new pump design, its construction, mathematical simulation validation and the experiment as well. The centrifugal pumps with the smooth, spiral and double conical rotor that are driven by the electromotor and embedded on the magnetic bearings were designed, produced and their characteristics were experimentally obtained and were evaluated from the point of their future possible usage or corrections. Part of this pump design solution is the magnetic bearings stability control and solution. The pump development wasn't limited just for the construction solution, but also for the mathematical simulation of the flow in the pump bodies and the possibilities of new materials using, which could be the progress in the present biomedicine pumping technology.
PM machines are increasingly becoming dominant machines with the cost competitiveness of high energy permanents magnets. These machines offer many unique features. Motor efficiency is greatly improved and higher power density is achieved. Moreover, PM motors have small magnetic thickness which results in small magnetic dimensions. As for the axial flux PM machines, they have a number of distinct advantages over radial flux machines. They can be designed to have a higher power-to-weight ratio resulting in less core material. The noise and vibration levels are less than the conventional machines.

Acknowledgements

This paper was supported by Ministry of Education, Youth and Sports of the Czech Republic research grant MSM 0021630518 "Simulation modeling of mechatronic systems".

References

[1] LÁNÍČEK, T., LAPČÍK, J. PERMANENT MAGNET SLOTLESS SYNCHRONOUS MOTORS USED IN LABORATORY TAH APPLICATIONS, In Joint Czech - Polish Conference on Project GACR 102/03/0813. Low Voltage Electrical Machines. Brno-Šlapanice, FEEC BUT, 2005, s. 65 - 71, ISBN 80-214-3047-8

[2] LÁNÍČEK, T., LAPČÍK, J. ACTIVE MAGNETIC BEARINGS DESIGN FOR PERMANENT MAGNET SLOTLESS SM In XIV. International Symposium on Electric Machinery in Prague ISEM 2006. International symposium on electrical machines - ISEM 2006. Praha: Czech Technical University in Prague, 2006, s. 134 - 142, ISBN 80-01-03548-4

Acoustic schwannoma detection algorithm supporting stereoscopic visualization of MRI and CT head data in pre-operational stage

T.Kucharski (a) *,M.Kujawinska (a),K.Niemczyk (b)

(a) Warsaw University of Technology Faculty of Mechatronics, ul.Sw.A.Boboli 8.,Warsaw, 02-525,Poland
(b) Medical University of Warsaw Clinic of Otolaryngology, Postal address,

Abstract

In this paper the authors are focused on the development of novel algorithms for segmentation and extraction an acoustic schwannoma tissue (inner ear tumor) from MRI DICOM data The algorithm is meant to support the preparation of data for stereoscopic visualization system using Helmet Mounted Display (HMD) supporting pre-operational stage. The algorithm is based on histogram processing technique invoking to Otsu method.

1. Introduction

The goal of this work is to introduce simultaneous stereoscopic presentation of inner ear tumors and surrounding skull bones. Such visualization system may become in future a convenient tool supporting a surgeon decisions before inner ear operation or during monitoring of tumor growth. In this paper the authors focused on segmentation of acoustic schwannoma structures present in MRI DICOM images which were co-registered with CT by use of PMOD software. The specific features of data are:

- limited volume conditions. MRI series of images covers approximately 50mm of head volume,
- different MRI scanners spatial resolution in X,Y,Z directions,

- non uniform influence of contrast on patient

Therefore the existing methods of analyzing MRI data addressed to inner ear tumors recognition cannot cope with a whole variety of existing clinical cases. The technique like region growing applied in [1] treats a tumor as quasi-homogenous region what is not common. Author of this work decided to extract tumor according to a unique feature such as elongation of acoustic schwannoma in an inner ear direction, but this feature is not present in every case especially for small tumors. Other recognition techniques are based on probabilistic model of tissues [2]. This model is then adapted to every clinical case. The main disadvantage of this model is rigidness which means that intensity range and probability distribution of each tissue are known and constant. Other techniques dedicated for brain tumor recognition was examined also [3,4,5,6].

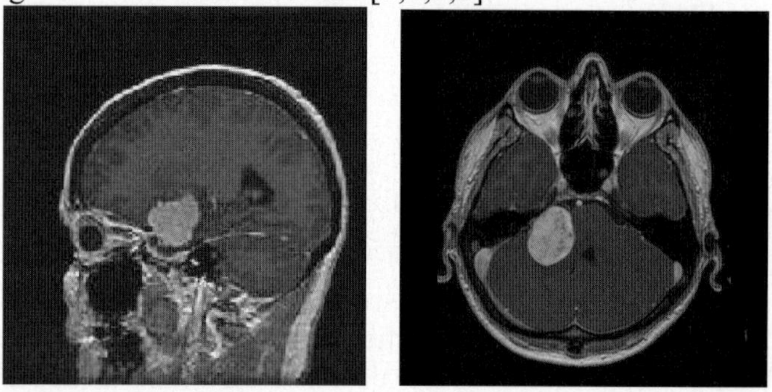

a) b)

Fig.1 MRI images containing: a) inner ear, b) brain tumor.

2. System architecture

The system for which image processing software is dedicated contains some crucial modules:

- Data acquisition units (MRI and CT scanners),
- Data processing unit (PC station),
- Data visualization unit (Stereoscopic Helmet Mounted Display HMD).

Medical data are acquired in Warsaw Central Medical University Hospital. They are captured in DICOM format which is common format to store such data. Information nested within DICOM is decoded, co-registered and processed according to presented schema (Fig2):

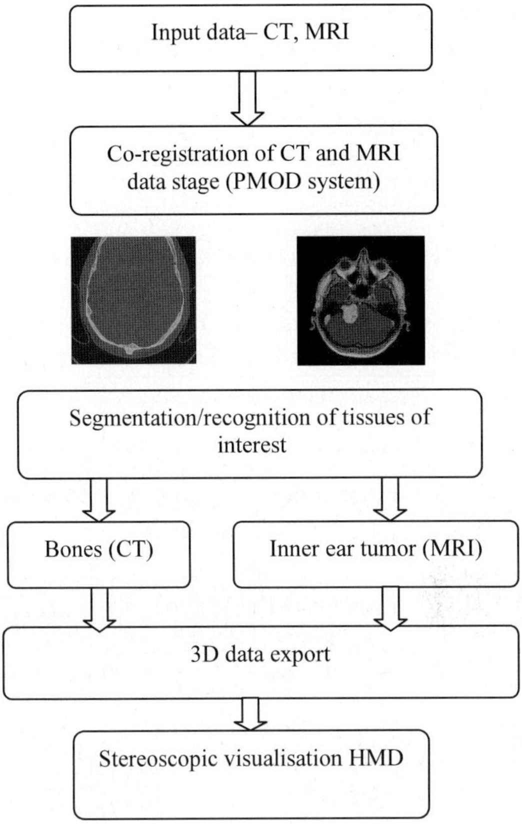

Fig2. Main data processing flow.

In this paper authors present algorithm of inner ear tumor segmentation and recognition from MRI series.

3. Inner ear tumor segmentation procedure.

As it was described in previous sections the character of analyzed (limited volume, various intensity conditions) MRI data made implementation of common used algorithms very difficult. One of the ways solving such problem may be a-priori knowledge of the place of potential occurance of acoustic schwannoma tissue. Such objects as seen at Fig.1a are localized in the neighborhood of head centre and inner ear. It is important information that can improve processing due to decreased number of potential objects that may be classified as tumor. For that issue volume of interest need to be created according to volume. This information supported by local and global data histogram processing may be useful in recognizing tissues of interest. Segmentation algorithm proposed followed by recognizing procedure is presented below (Fig.3):

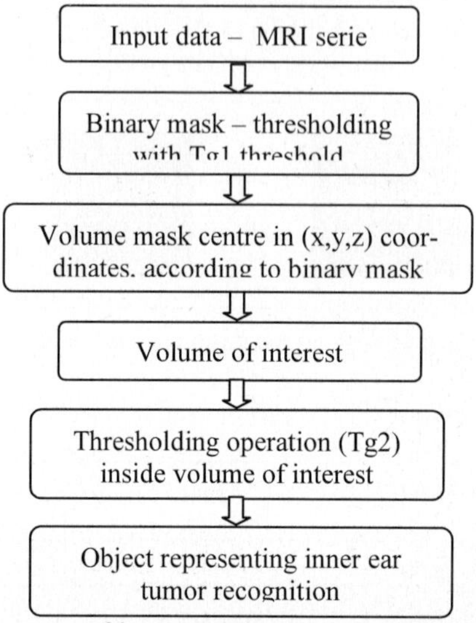

Fig.3 Diagram of inner ear tumor recognition procedure.

The diagram presented above takes a whole MRI images serie as an input. Based on 3D data set a global histogram is created (Fig.4a)

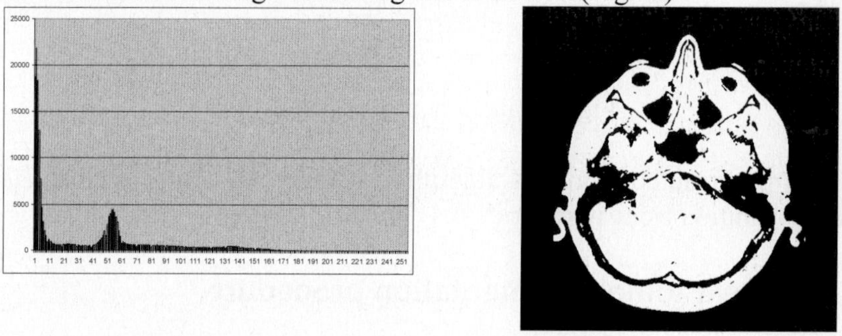

a) b)

Fig.4 Image of a) global histogram of entire 3D MRI data set, b) binary mask according to threshold via Otsu method.

The contrast injected into patient before MRI scanning changes intensity response within acoustic schwannoma tissue. However the intensity does not increase enough to create in a histogram a clear separate peak describing the mentioned tumor. This peak lies slightly beyond second peak that describes brain This observation allows to divide the segmentation process into 2 steps. In the first step the peak of brain, tumor and other objects possessing the same intensity range, are treated as a single peak. The global histogram (Fig.4a) can be treated as bimodal. Under such assumption the best solution would be dynamically adapting threshold value to every histogram. This demand is fulfilled by Otsu method [7]. The effect of thersholding is presented in Fig.4b. Local histogram is created in the next step (Fig.5b). The voxels of interest that are used to build new histogram are a part of binary mask and are placed inside a sphere with a centre and a radius defined empirically. Radius of a sphere delimities a position in relation to volume centre where tumors occurs (Fig.5a).

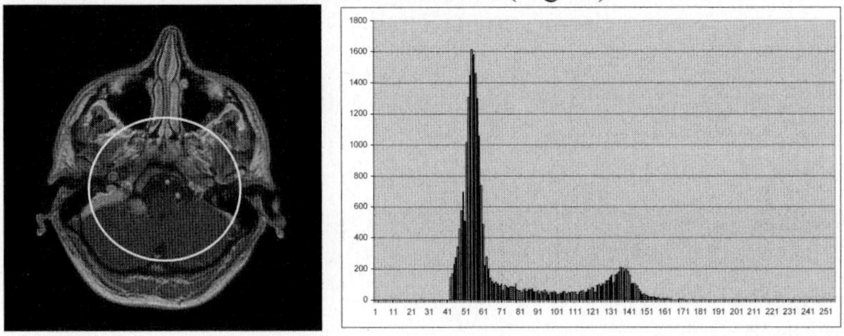

b)

Fig.5 Image of a) an exemplary MRI slice with a circle marked that simulates one of the sphere slices, b) histogram build form masked volume.

In the second segmentation stage the histogram shows bimodal structure also. Such bimodal character of local histogram emerges when huge background peak is removed. Huge peak seen in lower intensities represents brain and the smaller one represents tumor and other structures covering the same intensity range. Character of local histogram allows usage of Otsu method for second stage segmentation threshold. An exemplary MRI image consists of tumor segmented structures is presented below (Fig.6a):

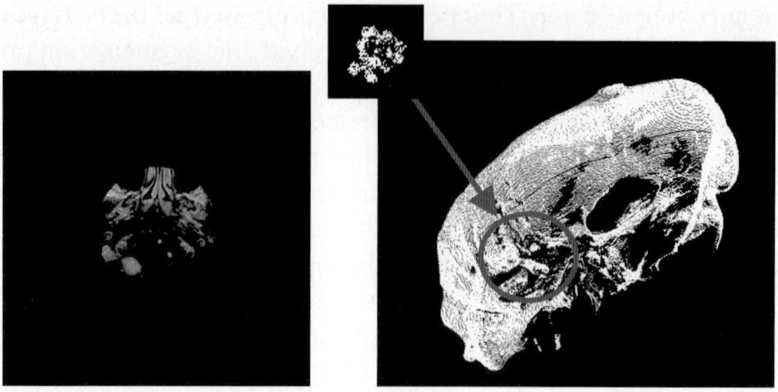

a) b)
Fig.6 Image of an exemplary MRI slice after second thresholding.

Local histogram thresholding operation finalizes fully automated inner ear tumor segmentation process. The next step is interactive and requires intervention of a user. The 3D visualisation of interactively marked inner ear tissue and head bone is presented below (Fig.6b).

References

[1] S.Dickson, B.T. Thomas, P. Goddard "Using neural networks to automatically detect brain tumours in MR images.", Int J Neural Syst. 1997 Feb;8(1):91-9

[2] M. Prastawa, E. Bullitt, S. Ho, G. Gerig *A brain tumor segmentation framework based on outlier detection*, Medical Image Analysis, ,275-283, vol. 8, Issue 3, September 2004

[3] E.M. Monsell, D.D. Cody, E. Spickler, J.P. Windham "Segmentation of acoustic neuromas with magnetic resonance imaging and Eigen image filtering.", Am J Otol. 1997 Sep;18(5):602-7

[4] C. Jiang, X. Zhang, W. Huang, C. Meinel *Segmentation and Quantification of Brain Tumor*, Proc. *IEEE VECIMS 2004,*Boston/MA(USA),61-6,2004

[5]. J.A. Sethian *Level Set Methods and Fast Marching Methods, Cambridge University Press, Cambridge, UK,* 1999

[6] A.S.Capelle, O.Alata, C.Fernandez, S.Lefevre "Unsupervised segmentation for automatic detection of brain tumors in MRI"

[7]. N.Otsu, *A threshold selection method from gray level histograms*, ,IEEE Trans. Systems, Man and Cybernetics,1979

Computer gait diagnostics for people with hips implants

D. Korzeniowski, D. Jasińska – Choromańska

Warsaw University of Technology, d.korzenioiwski@mchtr.pw.edu.pl
Warsaw, 02-525, Poland

Abstract

The whole diagnostics process should consist of the mathematics, clinic, quantity and subject analysis. The object movement can be described with use of parameters such as: angle, velocity and acceleration of the time function. It can also cause the change of the kinematics of the whole bone and muscle scheme of the patient after implanting. The tools analyzing the movement of the patient should include the analysis of the whole scheme and not its separate parts.

1. Introduction

In cases when the hip joint is somehow damaged, it automatically leads to impaired movement of the previously mentioned parts of the body. The rounded head of the thigh bone is perfectly centered inside (the hip socket / acetabulum).

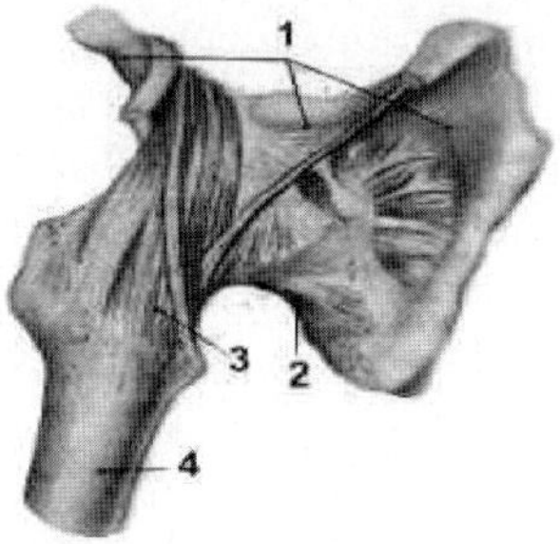

Fig. 1 The figure shows the structure of a hip joint [5]: 1) pelvic bone, 2)3) round ligament of the femur/ ligament of the femoral head, 4) femoral bone / thigh bone

The hip joint is composed of two parts – the hip socket (acetabulum) and the head of the thigh bone (femur). (Fig 1). The surfaces of these bones within the joint are covered with a layer of cartilage. This important surface allows the bones to smoothly glide against each other without causing damage to the bone. Cartilage works like a natural shock absorber.
Degeneration of the hip joint as well as different kinds of diseases are caused by a variety of diverse factors. Some of the reasons for the occurrence of diseases are:
- obesity - chronic and excessive load of joints (especially spine and lower extremities / lower limbs) causes cartilage to begin to show physical signs of wear and tear particularly in knee joint
- repeated loading (standing, sitting) or the lack of movement
- unbalanced diet – a lack of vitamin D3 and antioxidants (vitamins A, C, B6)
- genetic factors : (arthrosis / degenerative affection of joints)
- mechanical insults to the joint from an acute injury (sprain, dislocation) as well as microdamages
Professional sportsmen are especially at risk of arthrosis of a hip, knee, shoulder. They make up a higher proportion of the total population.
It is quite common that despite the natural factors that cause joint damages there are random factors such as injuries that can occur after car accidents or bone fractures.

2. Systems of analyzing and diagnosing patients' physical abilities after implantation

The whole diagnostics process should consist of the mathematics, clinic, quantity and subject analysis.
The object movement can be described with a use of parameters such as: angle, velocity and acceleration of the time function.
It can also cause the change of the kinematics of the whole bone and muscle scheme of the patient after implanting. The tools analyzing the movement of the patient should include the analysis of the whole scheme and not its separate parts.For diagnosis and analysis of a range movement and analysis of walk are used:
- *Clinic analysis* – consists of medical examination and visual diagnosis by an orthopedist
- *Quantity and subject analysis* – doing quantity and subject analysis leads us to many more or less advanced systems. Present researches were done with Vicon help / using Vicon and Poligon programme.
Within the scope of diagnostics:

- measurement of the ground reaction forces during walking (based on an averaged data of three series for each lower limb) in three directions
- using dynamic electromyography

Measurements of kinematic dimensions (regarding motion): pelvis position, angles in joints, in three dimensions and rotation

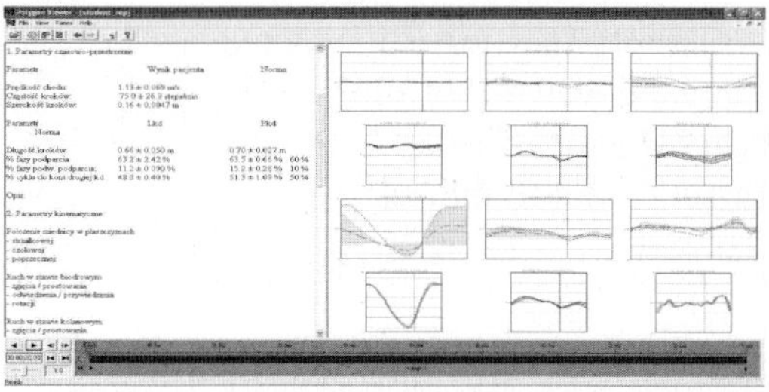

Fig. 4 The figure show the results of the research of the measurement of the kinematic pelvis and hip joint position in a frontal, sagittal, transverse plane

- the measurement of the time parameters of walking: walk cycles, speed, length and frequency of steps

Next system of quantitative analysis and objected is Sybar. It is an integrated system which co-operates with Kistler's platform and cameras VHS and EMG.

- *Mathematical analysis*- in case of mathematical analysis the systems consist of information processed with individual measures, by some from measurement systems

- *The static investigations* - in case of static investigations it consists of the measurements of angles, range of movement, speed of steps. The measurement of angles was conducted with the help of the programme WinAnalize.

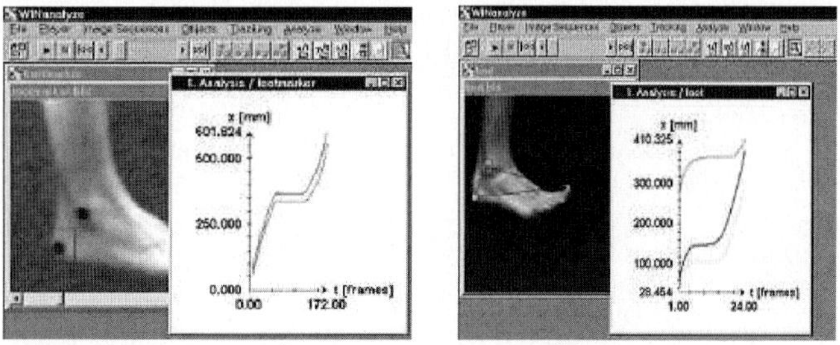

a) b)

Fig. 6 To follow object a) with use markers, b)without use markers

- *The dynamic Investigations* – in case of dynamic investigations it consists of Electromyography EMG - Measurement of muscles electrical activity, measurement of reaction force to foot – Ground reaction force vector

3. Research and results:

In order to define / state and measure the range of motion of implanted limbs the series of simple tests were conducted
- lifting the operated leg forward / lifting the leg with implanted hip joint
- lifting the operated leg backward
- lifting the operated leg out to the side

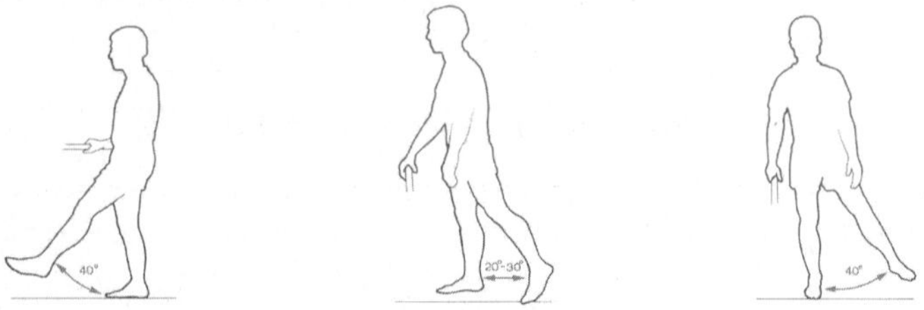

Fig. 7. Lifting the leg forward Fig. 8 Lifting the leg backward Fig 9 Lifting the leg out to the side

These exercises should be done under the physical therapist's control in order to prevent the dislocation of a joint. At the beginning of the rehabilitation process the patient lifted the operated leg / the leg with implanted hip joint/ forward maximum 40 degrees. It is extremely satisfactory because in case of a healthy person the same value reaches ca. 30 degrees.

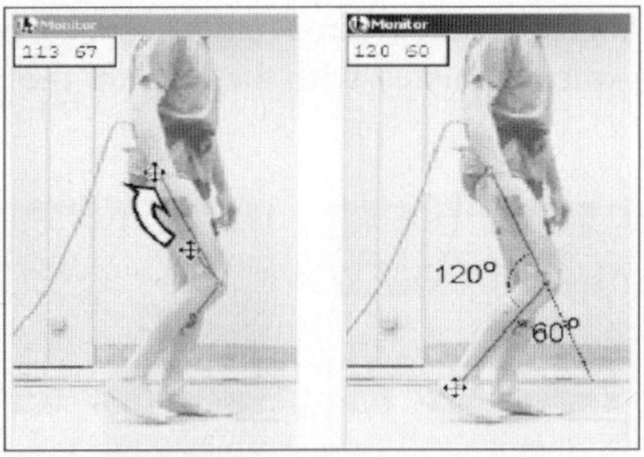

Fig. 10 The figure shows the research done by a motion analysis system – Sybar

An identical situation occurred when the patient lifted the leg backwards and the result was between 20 and 30 degrees. While the result of a healthy person was 35 degree. It is well seen that the outcome is proper / appropriate. The maximum range of motion in case of the prosthesis that are most

often used is about 126 degrees. This measurement from a physiological point of view enables the patient a full range of motion. The orthopaedic surgeons claim that implanted people cooperation between the ball, which replaces the spherical head of the femur and the cup, which replaces the worn-out hip socket is significant in case of long duration of the setting of prosthesis as well as their later mobility. In case of people after implantation many everyday activities becomes a real problem. One of such activities is standing up and sitting. This activity for a healthy person seems to be trouble-free.
They will be instructed by the physical therapist in a specific exercise program how to perform this action.

4. Conclusions

Nowadays, the average age of people having a hip joint implantation is still dropping and so far it seems to be important to deal with this topic.
As I mentioned earlier, so far there is not ideal and cheap and fulfilling all functions of diagnostic and analyzing instrument mentioned above.
Most of the equipment / devices available on the Polish market is even if extremely expensive do not fulfill the requirements.In most cases surgeons, orthopedics, or physical therapists facilitate their work simplifying diagnostics and movement analysis in order to use the methods described above selectively.In most cases it is not connected with any health hazard a professional incompetence. However, nobody knows if for instance a broad scope of joint movement will not lead to contractures of particular groups of muscles.The analysis of these methods and tests /researches will help to state / determine basic requirements needed to judge implanted people

Bibliography:

[1] Będziński R., Ścigała K., Physical and Numerical Model of the Tibia Bone, 12th Conference of the European Society of Biomechanics, Dublin, Irlandia, 2000.
[2] Bobyn J.D., Mortimer E.S., Glassman A.H., Engh C.A., Miller J.E., Brookes C.E.: Producing And Avoiding Stress Shelding. Clin. Orthop., 1992, T. 274
[3] Internet: www.stryker.pl.
[4] Marciniak W., Szulc A.: Wiktor Dega – Orthopedic and Rehabilitation, PZWL, Warsaw 2003.
[5] Internet: Biomexim.com.pl.
[6] Korzeniowski D., Jasińska-Choromańska D.: „Analysis of gait diagnostics for people with implants" Třešt, Czech Republic, 2005.

Time Series Analysis of Nonstationary Data in Encephalography and Related Noise Modelling

L. Kipiński (a)

(a) Wrocław University of Technology, Faculty of Fundamental Problems of Technology, Department of Biomedical Engineering and Instrumentation, Wybrzeże Wyspiańskiego 27, Wrocław, 50-370, Poland

Abstract

In this report, statistical time series analysis of nonstationary EEG/MEG data is proposed. The signal is investigated as a stochastic process, and approximated by a set of deterministic components contaminated by the noise which is modelled as a parametric autoregressive process. Separation of the deterministic part of time series from stochastic noise is obtained by an application of matching pursuit algorithm combined with testing for the residuum's weak stationarity (in mean and in variance) after each iteration. The method is illustrated by an application to simulated nonstationary data.

1. Introduction

In brain evoked activity measured by means of EEG/MEG, one can observe time-dependent changes of its various characteristics like amplitude and frequency, as well as the contaminating noise. For this reason, it is necessary to use the analysis methods designed for nonstationary signals, since the standard EEG/MEG methodology based on signal averaging and simple spectral analysis is insufficient. Time-frequency estimation methods such as short-time Fourier transform, Wigner distribution, or discrete and continuous wavelet transform are very useful, yet, statistically inefficient. They also have some inherent limitations. Thus, the representation of the evoked-response generative process given by these methods is incomplete. In this research, EEG/MEG signal is investigated as a stochastic process which can be decomposed to a set of deterministic functions representing its nonstationarity and stationary residua. For modelling the stochastic EEG/MEG noise, statistical time series analysis methods are used.

2. Statistical time series analysis

A time series (TS) model for the observed data $\{z(t)\}$ is a specification of the joint distributions (or possibly of only the means and covariances) of a sequence of random variables $\{Z_t\}$, with a realization denoted by $\{z(t)\}$ [1]. In a short form, an additive TS model can expressed by the sum of deterministic $d(t)$ and stochastic $l(t)$ components:

$$z(t) = d(t) + l(t) \tag{1}$$

Off course, there are many possible examples for this kind of a model, i.e. $d(t)$ can be a linear trend, a seasonal (periodic) function, or a sum of them, and $l(t)$ can be a set of observations of any (stationary or nonstationary) random variable. Let us take times series generated by an additive stochastic process given by (2), which is the sum of $m \in N$ sine waves or other non-commensurable periodic functions (or commensurable but with a period much longer than the periods of its particular components) $s(t)$ and a stationary noise $e(t)$.

$$z(t) = \sum_{i=0}^{m} s^i(t) + e(t) \tag{2}$$

Modelling of a process requires all the deterministic functions to be removed at first. It can be achieved by the preceding estimation of these functions or by differencing the series. There is a lot of helpful examining tools (i.e. statistical tests and spectral analysis methods (spectral density, periodogram) based on the Fourier transform) and estimation techniques (i.e. maximum likelihood method). Next, stochastic, stationary residua can be diagnosed (on the basis of the sample autocorrelation function (ACF) and sample partial autocorrelation function (PACF), and using some statistical tests) and an adequate parametric model of them (autoregressive (AR) and/or moving average (MA) for example) can be constructed [1].

3. Time series de-trending by matching pursuit

Unfortunately, EEG/MEG signals are nonstationary, and so the characteristics of $s^i(t)$ components in Equation (2) varies in time (thus we can not assume that each $s(t)$ is periodic). In consequence, a typical time series decomposition, based on the Fourier transform, disappoints in this case.

Therefore, it is required to construct the ongoing EEG/MEG noise model by using an effective approximation of the nonstationary components of the EEG/MEG signal.

In the first step of the iterative matching pursuit (MP) algorithm proposed by Mallat and Zhang [2], the atom g_{γ_0} that gives the largest product with the signal is chosen from the large, redundant dictionary D (usually composed of Gabor functions being cosines modulated in amplitude by a Gaussian envelope). In each consecutive step, the atom g_{γ_m} is matched to the signal $R^m z$ that is the residual left after subtracting the results of previous iterations:

$$\begin{cases} R^0 z = z \\ R^m z = \left\langle R^m z, g_{\gamma_m} \right\rangle g_{\gamma_m} + R^{m+1} z \\ g_{\gamma_m} = \arg\max_{g_{\gamma_i} \in D} \left| \left\langle R^m z, g_{\gamma_i} \right\rangle \right| \end{cases} \tag{3}$$

The possible stopping criteria for this algorithm are: 1 – fixing a priori the number of iterations m, irrelevant of the content of the analysed signal, 2 – explaining a certain percentage of the signal's energy, 3 – the energy of the function subtracted in the last iteration reaches a certain threshold. It is assumed that the residual vector obtained after approximation of m time-frequency waveforms is the noise, which converges to $N(0,\sigma^2)$ with increasing m [3].

Here, we propose to combine TS analysis with MP algorithm by a new model resulting from (2) and (3):

$$z(t) = \sum_{i=0}^{m} \left\langle R^i z, g_{\gamma_i} \right\rangle g_{\gamma_i}(t) + e(t) \tag{4}$$

In this model, the MP procedure is performed as long as the residuum is nonstationary. The accomplishment of a weak stationarity entails stopping the algorithm. Next, TS methods should be applied for modelling the residuum. Practical measures of the stationarity are the results of statistical tests: Kwiatkowski–Phillips–Schmidt–Schin (KPSS) test [4] for stationarity in mean, and the White test for homoscedasticity [5].

3. Application to simulated nonstationary data

In order to illustrate the idea of the proposed algorithm and to examine its properties, a simulated signal (Fig. 1) was constructed from a sum of m

Gabors and a stochastic (but not Gaussian) noise generated by ARMA(5,3) process. This signal is stationary in mean, but heteroscedastic (covariance vary over time), so it is necessary to remove that nonstationarity by means of MP.

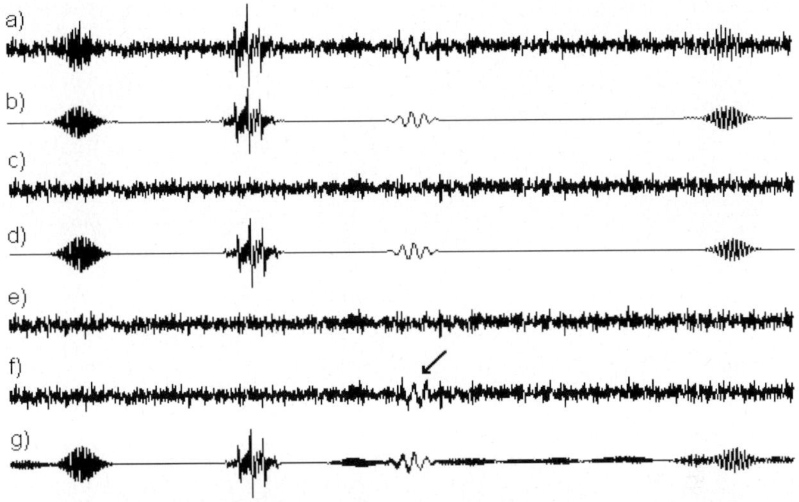

Fig. 1. a) Simulated signal constructed with a sum of *m* Gabor functions b) and autoregressive moving average noise c); d) the approximation after *m* iterations of MP algorithm stopped due to stationarity of residua e); f) nonstationary residua after *m-1* iterations (the arrow points at "the reason" for the White test's rejection of the null hypothesis); g) excessive approximation (for $n>m$ iterations, MP starts to explain the noise).

Residuum is tested after each iteration, and after the *m*-th one we have no reason to reject the null hypothesis of the White test about its homoscedasticity. This means that the residuum is weakly stationary, thus we can use statistical time series tools to describe it. Tests for independence and plots of sample ACF and PACF (Fig. 2b)) show that the residua generator is autocorrelated, and imply an autoregressive moving average model for the residua. The fit is obtained by the maximum likelihood estimators of the ARMA parameters, and minimum Akaike criterion. The values for all significant ARMA parameters are similar (into or near the confidence level) as the values used for the simulation. The noise error was examined too, and the result (Gaussian white noise) confirms correctness of the fit. In comparison, the "classical" MP algorithm looses information about the noise, or (if the number of iterations is large) it starts to explain the noise with deterministic functions (Fig. 1g)). The combined MP-TS algorithm is a clear improvement, as modelling of the noise is addressed explicitly.

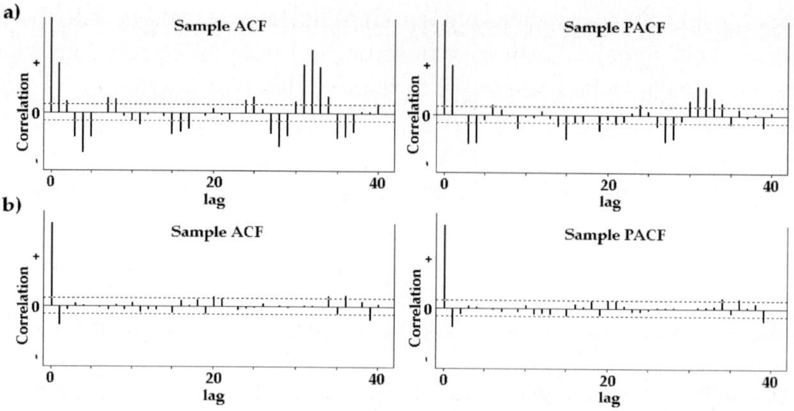

Fig. 2. Sample ACF and PACF plots for: a) simulated signal b) stationary residua. Evident difference is due to spectrum variability of simulated signal.

4. Summary and conclusion

In this paper, a combination of time series analysis methods with signal decomposition by matching pursuit is proposed. Although each of the two methodologies separately can be applied to build signal representation, their marriage seems very efficient, as their simultaneous use enables holistic description of both deterministic and stochastic components of a signal. It can be an effective method for noise description, prediction and filtering in any EEG/MEG measurement.

Acknowledgements: In this work, we used our modification of the MMP software written by A. Matysiak [6].

References

[1] P. J. Brockwell, R. A. Davies, "Introduction to time series and forecasting", Springer, New York and London, 2002.
[2] S. Mallat, Z. Zhang, IEEE Trans. on Sig. Proc. 41 (1993) 3397–3415.
[3] P. J. Durka, D. Ircha, K. J. Blinowska, IEEE Trans. on Sig. Proc. 49 (2001) 507–510.
[4] D. Kwiatkowski, P. C. B. Phillips, P. Schmidt, Y. Shin, J. Econometrics 54 (1992) 159–178.
[5] H. White, Econometrica 48 (1980) 817–838.
[6] A. Matysiak, P. J. Durka, E. M. Montes, M. Barwiński, P. Zwoliński, M. Roszkowski, K. J. Blinowska, Acta Neurobiol. Exp. 65 (2005) 435–442

Precision Electrodischarge Machining of High Silicon P/M Aluminium Alloys for Electronic Application

D. Biało (a), J. Perończyk (a), J. Tomasik (b), R. Konarski (a)

(a) Institute of Precision and Biomedical Engineering, Warsaw University of Technology, ul. Sw. A. Boboli 8, 02-525 Warsaw, Poland

(b) Institute of Metrology and Measurement Systems, Warsaw University of Technology, ul. Sw. A. Boboli 8, 02-525 Warsaw, Poland

Abstract

High silicon content aluminium alloys are preferred in electronic applications when low coefficient of thermal expansion and high thermal conductivity of material are require i.e. in housings for electronic packaging.
This material is difficult to machine due to the large volume fraction of hard Si particles dispersed in relatively soft aluminium matrix.
Electrodischarge machining (EDM) becomes very promising alternative to the traditional machining methods.
In the present study aluminium alloys with 6-40% Si particles was manufactured by powder metallurgy (P/M) route. During EDM holes of 0,4mm diameter were drilled. After EDM the following parameters were determined: drilling speed V_d, surface roughness R_a, oversize of the hole drilled Δ. Above mentioned parameters were correlated with EDM parameters.

1. Introduction

Materials for electronic packaging should possess certain physical properties to ensure high packaging density and reliability [1, 2]. High thermal conductivity for efficient heat dissipation, small thermal expansion tailorable to match that of adjacent components and low density for lightweight

are required of these materials. Materials developed so far for electronic packaging fall into the categories of metals (typically Al and Cu), alloys (such as Kovar), ceramics (such as Al_2O_3), sintered metals (such as W-Cu) and composites (typically Al-SiC) [3,4]. Al-Si based alloys with a high volume fraction of Si crystals, have the above mentioned properties and may be particularly suited for electronic and avionic systems.
This material is difficult to machine due to the large volume fraction of hard Si particles dispersed in relatively soft aluminium matrix. These particles caused rapid wear of cutting tools during machining and decreasing of accuracy and smoothness machined surfaces. Electrodischarge machining (EDM) becomes very promising alternative to the traditional machining methods.
In the present study aluminium alloys with 6-40% Si particles was manufactured by powder metallurgy (P/M) route. During EDM holes of 0,4mm diameter were drilled. After EDM the following parameters were determined: drilling speed V_d, surface roughness R_a, oversize of the hole drilled Δ.
Above mentioned parameters were correlated with EDM parameters.

2. Materials

Chemical compositions of Al-Si based alloys which were investigated are given in Table 1. Aluminium alloy (A/SO) without dispersed Si crystals was taken as a reference material.

Table 1. Code and compositions of the alloys.

Code	Elements (wt.%)					
	Si	Mg	Ni	Fe	Cu	Al
AS/0	-	1.6	1,1	1.1	2.5	bal.
AS/6	6	0,5	-	-	-	bal.
AS/20	20	-	2,0	5,2	-	bal.
AS/40	40	-	-	0,3	0,1	bal.

Alloys were prepared through a melt-spinning process, followed by pulverisation, warm compaction, degassing, hot extrusion and finally T6 heat treatment. Figure 1 shows microstructure of alloy with 40 % Si. It can be seen that alloy contains a very large volume fraction of fine Si particles that are homogeneously dispersed in the matrix.

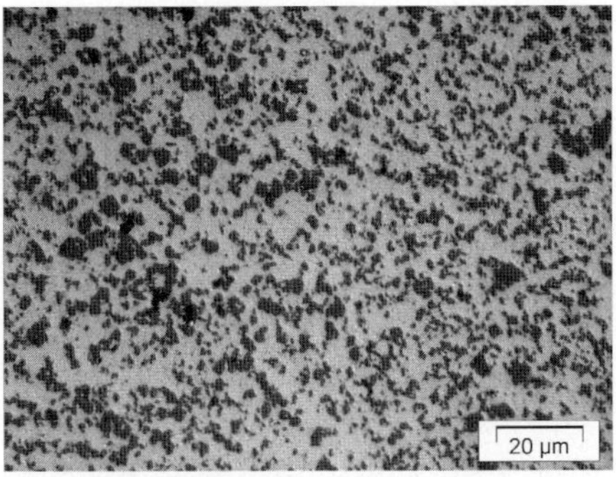

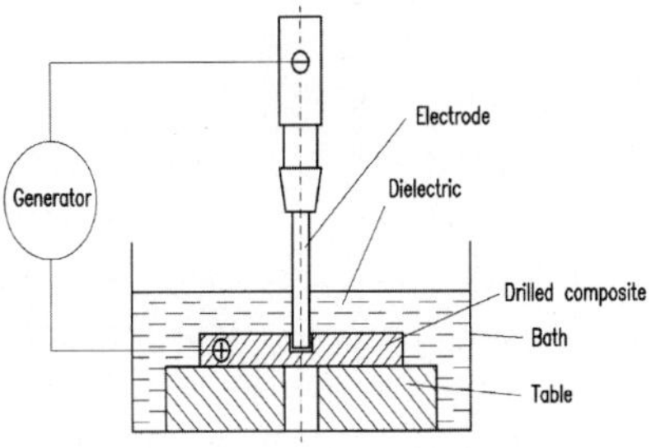

Fig.1.Microstructure of AS/40 alloy   Fig. 2. Scheme of EDM drilling process

3. Electrodischarge drilling

Electrodischarge drilling of microholes (EDM) was carried out on a drilling machine EDEA-25 provided with a generator RLC. The generator has several steps of adjustment of electrical machining parameters. In the experiments 4 steps of the energy of single discharge E_i were used: 0.05; 0.19; 0.84 and 1.83 mJ. Machining was done with an electrode of 0.4 mm diameter, made of M1E copper, in cosmetic kerosene environment, with supply voltage U_o = 240 V, closed-circuit voltage U_r = 195 V and straight polarity.
The scheme of the system for electrodischarge drilling is given on Fig. 2.
Typical EDM parameters were determined in the course of technological trials, as follows:

- average drilling speed V_d [µm/sec],
- roughness of the hole surfaces after machining,
- oversize of the hole drilled $\Delta = D - d_E$ [mm],

where: D – diameter of the hole, d_E – electrode diameter.

4.Research Results

Fig. 3 shows impact of single discharge energy on the average drilling speed.
As it can be seen on the figure, average drilling speed V_d grows along with the increase of single discharge energy for all material tested. Influence of silicon content on V_d is not clear at all. Generally, average drilling speed has tendency to groving for higher volume fraction of Si in alloys. But this principle is disturbanced by the different content of other components Mg, Ni, Fe, Cu in tested alloys.

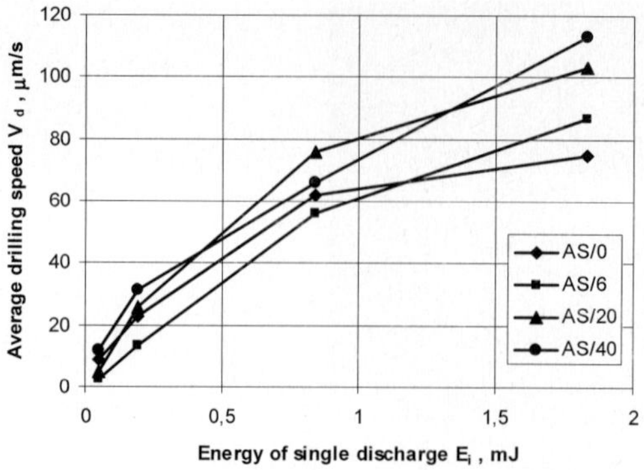

Fig. 3. Average drilling speed V_d vs energy of single discharge E_i

Impact of discharge energy E_i on oversize Δ of the hole drilled is shown on the Fig. 4. The higher values of E_i were applied, the bigger is oversize of the hole drilled. Influence of Si cristals is also visible. For the higher content of this phase in materials, lower hole oversize occurred.

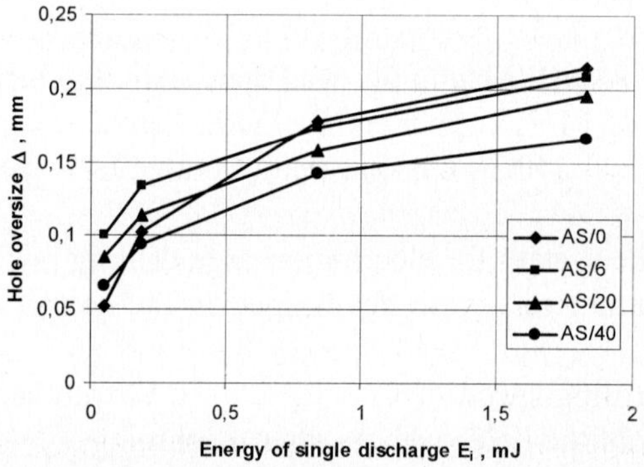

Fig. 4. Hole oversize Δ vs energy of single discharge E_i

Roughness changes expressed by the R_a parameter are shown on the Fig.5. Surface roughness is also influenced by presence of Si crystals. Their growing content in composites is accompanied by reduced roughness. One can conclude when comparing data from diagrams 3 and 5, that surface roughness is related with the EDM process output. The more effective is machining (higher drilling speed), the higher is surface roughness – there is simple relation with E_i. Higher energy of single discharge causing more protrusions, irregularities and craters to form on the surface during discharges, resulting in an increased surface roughness [5, 6].

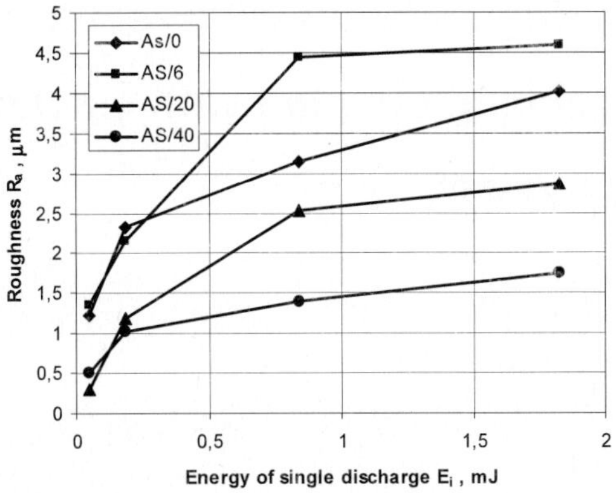

Fig. 5. Impact of discharge energy E_i on surface roughness R_a

5. Summary

It can be concluded as follows as regards the research carried out on electrodischarge drilling of microholes in aluminium alloys with Si content of 6-40%:

- Discharge energy E_i has substantial influence on microholes drilling process. Energy growth is accompanied by growing drilling speed (efficiency) and by increase of the hole oversize.
- As discharge energy grows, local micro-areas of erosion become wider and deeper, which – in effect – causes increased surface roughness of the holes drilled.
- Presence of Si crystals results in reduction of hole oversize and surface roughness.

References

[1] C. Zweben, JOM, July 1992, 15-23

[2] D. M. Jacobson , Advanced Materials and Processes, March 2000, 36

[3] R. M. German et al, The Int. Journal od Powder Metallurgy, 1994, vol. 30, No. 2, 205-215

[4] P. K. Mallick , Composites Engineering Handbook. Marcel Dekker Inc., New York, 1997

[5] K. E. Oczoś, Mechanics, 7, 1997, pp. 325-333 (in Polish)

[6] D. Bialo et al, EURO PM2005 Congress, Prague, Czech Republic, Oct, 2-5, 2005. Vol. 3, 225-230.

Modeling of drive system with vector controlled induction machine coupled with elastic mechanical system

A. Mężyk (a) , T. Trawiński (b)

(a) Silesian University of Technology, Department of Applied Mechanics, Faculty of Mechanical Engineering,
Gliwice, 44-100, Poland

(b) Silesian University of Technology, Mechatronics Department, Faculty of Electrical Engineering, Akademicka 2a,
Gliwice, 44-100, Poland

Abstract

The main aim of this work is formulation of the mathematical models of induction squirrel-cage motor, inverters with vector control, and mathematical model of elastic mechanical system coupled with induction motor. In mathematical modeling process of control system special attention is paid on good conformity between models of vector control system and its real industrial equivalent – Simovert VC. The numerous researches were carried out in order to examine the influence of elastic kinematic chain of mechanical system on control system stable work.

1. Introduction

In vector control systems during control process many information are needed about actual stator currents, voltages and rotor angular position and rotor speed. In order to prepare the proper control method many mathematical transformations must be made between different coordinates systems, and often observers must be used to estimate the rotor magnetic flux spatial position. So the induction squirrel-cage motor fed from inverters equipped with vector control system has comparable dynamic properties to separately excited direct current motor.

2. The mathematical model of mechanical system

Modelling of an electromechanical drive system as a system with feedback between its electric part and mechanical one is an example of mechatronic approach. A model of the mechanical part, taken as a discrete form, describes the following system of ordinary differential equations expressed as a matrix:

$$\mathbf{M}\ddot{\mathbf{q}} + \mathbf{C}_V\dot{\mathbf{q}} + \mathbf{K}\mathbf{q} = \mathbf{Q} \tag{1}$$

where: **M,** $\mathbf{C}_V$ and **K** - matrices of inertia, damping and stiffness, **q** - column matrix of generalized coordinates, **Q** - column matrix of generalized forces.

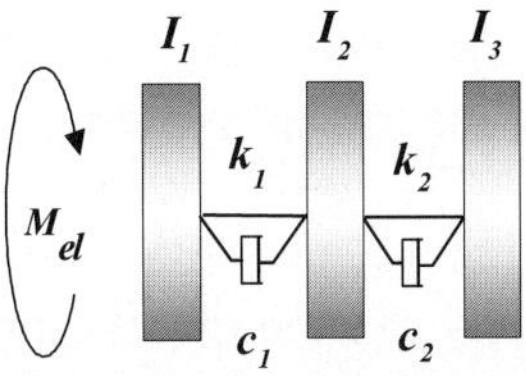

Fig. 1. Dynamic model of the drive system

The character of time courses of dynamic quantities is, to a considerable degree, conditioned by characteristics and power of a driving motor. Thus, the accuracy of the obtained solutions depends on the exactness of the assumptions made in a physical model of the mechanical system and in a model of the electric motor.

3. The mathematical model of induction squirrel-cage motor

Mathematical model of a squirrel-cage motor in natural coordinate is formulated under the following assumptions: the stator and the rotor windings are symmetrical, the rotor winding is considered as Q_r-phase winding (Q_r - number of cage bars), the skin effect phenomena in rotor bars are neglected, the air-gap is smooth and uniform, the magnetic circuit is linear. Under the above assumptions the mathematical model of a squirrel-cage motor neglects permeance and saturation magnetic field space harmonics. The mathematical model of a squirrel-cage motor in natural coordinates consists of the $3{+}Q_r$ voltage equations for stator and rotor windings and motion equation. Certain simplification of this model can be achieved by transformation of machine equations from natural coordinates to new coordinate systems [1-4]. If we assume that the fast stator current control is realized by inverter, the motor equation may be simplified too. The motor equation are simplified in significant way If we introduce the coordinate system "dq" (field coordinate system) which follows rotor flux space vector in such way that his real axis will align with rotor flux. In field reference frame the motor equation may be expressed by:

$$\tau_r \frac{d}{dt}\psi_{rd} + \psi_{rd} = L_m i_{sd}; \qquad L_m i_{sq} = (\Omega_{dq} - p\Omega_m)\tau_r \psi_{rd} \tag{1}$$

$$\frac{d}{dt}\Omega_m = \frac{1}{J}(pk_r \psi_{rd} i_{sq} - T_m) \tag{2}$$

where Ω_{dq}, Ω_m, k_s, k_r, τ_r, σ, τ_σ', r_σ, L_m, ψ_{rd}, i_{sq}, i_{sd} - rotor angular speed, angular speed of field reference system, stator leakage, rotor leakage, rotor time constant, motor leakage, modified stator time constant, modified motor resistance, magnetizing inductance, rotor flux, stator currents components.

Analyzing the equation (1) and (2) we may notice that they are similar to the ones describing separately excited direct current motor. If motor is excited, rotor flux arise ψ_{rd}, the electromagnetic torque " $pk_r\psi_{rd}i_{sq}$ " response is dependent only on current (stator current component i_{sq}).

4. Implementation of vector control of induction motor in Matalb/Simulink

The implemented model of vector control system as well the motor model, flux estimator and mechanical system is shown in Fig.2.

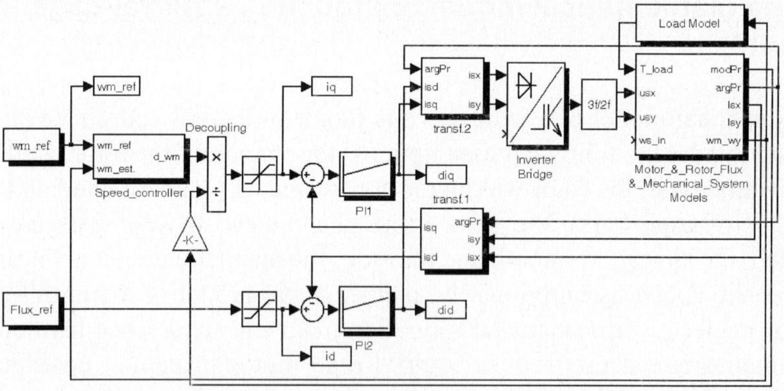

Fig. 2. Block diagram of induction motor vector control system implemented in Simulink

In the block diagram (see Fig.2) the part connected with speed and current controllers are referred to field reference frame. As a feedback signals for control purposes are used measured angular speed of the rotor coupled with mechanical system, estimated magnitude of rotor flux, flux angle and stator currents. Because measured stator currents are represented in the motor model in stator reference frame, they must be transformed to field reference frame with help of "transf.1" block (it is rotary transformation from stator to field reference frame). The reference currents becomes from currents controllers ("PI1" and "PI2") and must be also transformed, but from field reference frame into stator frame with help of "transf.2" block. After that block they are scaled in "Inverter Bridge" and compared with saw tooth signal in order to produce control signals for inverter keys (transistors which are modeled as ideal switches). The outputs of "Inverter Bridge" consist of three pulse width modulated voltage waves which are transformed to stator reference frame and fed the motor model. The block "Motor_&_Rotor_Flux_&_Mechanical_System_Model" includes motor model of rotor flux estimator and model of mechanical system. The general inputs of whole inverter are reference angular speed "wm_ref" and reference rotor flux "Flux_ref".

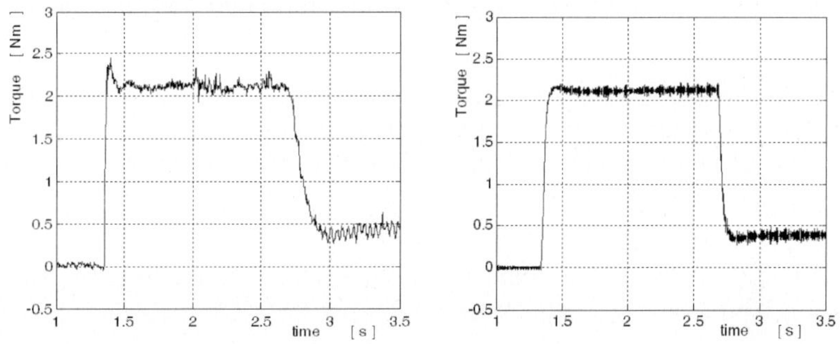

Fig. 3. Torque signal measured by torque transducer (left had picture) and torque signal from numerical simulation

5. Comparison between simulation and measurements

For comparison purposes the dynamic state of the motor was chosen in which the rotor and coupled elastic mechanical system accelerates to maximal angular speed during 1.2 s. In Fig.3. the comparison between measured torque (between motor and elastic system) and simulated torque is presented. From comparison of simulation and measurements data it follows that the conformity according torque time plots are quite good (see Fig.3). Some differences are observed at the beginning and end period of acceleration, but in the steady state conformity is excellent.

6. Conclusions

The developed dynamic model takes into account electromechanical couplings and enables computer simulation of dynamic phenomena occurring in kinematic pairs of the mechatronic drive system for various initial conditions. The results of numerical calculations prove that described model may be a helpful tool in the design process of vibrating mechatronic systems with vector control unit. The presented dynamic model, allows dynamic phenomena, occurring in kinematic pairs of the drive system both during starting and in the course of operation of the system loaded under steady - state conditions, to be computer simulated.

The investigations have been carried out within the project no 4T07C 00228, sponsored by the Ministry of Science and Higher Education

References

[1] A. M. Khambadkone, J. Holtz: ”Vector-Controlled Induction Motor Drive with a Self-Commissioning Scheme”, IEEE Transaction on Industrial Elektronics, vol.38, No.5, 1991,

[2] S. Wade, M. W. Dunningan, B. W. Williams: “Modeling and Simulation of Induction Machine Vector Control with Rotor Resistant Identyfication”, IEEE Transaction on Power Electronics, vol. 12, No. 3, 1997,

[3] J. Holtz: “Sensorless Control of Induction Motor Drives”, Proceedings of IEEE, vol.90, No.8, pp. 1359-1394, 2002,

[4] Trawiński T., Kluszczyński K.: Vector control of induction machine in presence of predominant parasitic synchronous torques, ICEM'98, Istambul- Turkey, vol.2/3, p.920., 1998.

[5] Mężyk A., Bachorz P., Świtoński E.: Dynamic analysis of mechatronic drive system with ansynchronous motor. Engineering Mechanics, Vol. 12, No. 3, s. 215-221, Brno 2005.

[6] Eds. Świtoński E: Modelowanie mechatronicznych układów napędowych. Red. nauk. Eugeniusz Świtoński. Wydawnictwo Politechniki Śląskiej, Gliwice 2004, 126 s., Monografia nr 70. (In Polish)

Method of increasing performance of stepper actuators

K. Szykiedans, Ph.D Eng.

Warsaw University of Technology, Institute of Micromechanics and Photonics, 8 Boboli Str. , Warsaw, 02-525, Poland

Abstract

The purpose of presented work was to analyze possibilities and work out new methods, allowing modifying characteristics of linear stepper actuators. A new method of steering was also considered, with increasing steering frequency of stepper different than uniform in relation to time. The method enables to attain higher maximum speed and in result spread area of functional characteristics. The results of this work, mainly the method of accelerating linear stepper actuators, may find wide areas of use.

1. Introduction

The purpose of this work was to analyze possibilities and work out new methods, allowing modifying characteristics of linear stepper actuators. It was assumed that the modifications will be possible to apply by their user. The subjects to analyze were two main areas where changes may be done. The first is a screw-nut gear with a bearing. Attributes which influence the functional characteristic the most strongly in this area are type and geometry (diameter, screw pitch) of a screw-nut gear and friction factor between a screw and a nut. The second area of possible changes is steering and power supply of an actuator. The subject to analyze was influence of a power supplying method, commutation type and steering mode (full- and microstep). A new method of steering was also considered, with increasing steering frequency of stepper different than uniform in relation to time.

2. New method of steering stepping actuators

As it is widely known and it is stated in many works, shape and range of any stepper motor or actuator depends on many factors stated earlier [1,4,7,9]. But it is not obvious that performance of stepper actuator also depends on frequency against time relation. This subject is now being examined more widely [2,3,5,8]. In most cases the aim is to improve actuator performance. There are different ways to achieve it, from usage of neural networks and fuzzy logic [5], to building prediction algorithms that works on a sophisticated hardware. Nevertheless these works are based on closed loop steering, what is negation of a main advantage of stepper motor - its possibility to work in open loop. During development of presented method of steering, it was assumed to keep possibility of work in open loop, and to keep it as simple as possible.
As it is predictable and was theoretically proven [6], stepper actuator pull-out characteristic will be the widest when control frequency will be changed in the way that will give dynamic position error as small as possible. From energetic point of view it means that force which is developed by actuator have to be equal to static and dynamic load.

$$F(f) = F_L(f) + M(d^2x / dt^2) \tag{1}$$

In case of stepper motor or actuator, any excess of energy put into the circuit will cause resonance effects. This will increase dynamic position error and loss of synchronism will be quicker. Analysis of equation (1) will give us for any set value of load F_L, value of frequency that is the maximum for this load. That will allow us to calculate frequency-time curve that shows most optimal acceleration. For this calculation stepper actuator characteris-

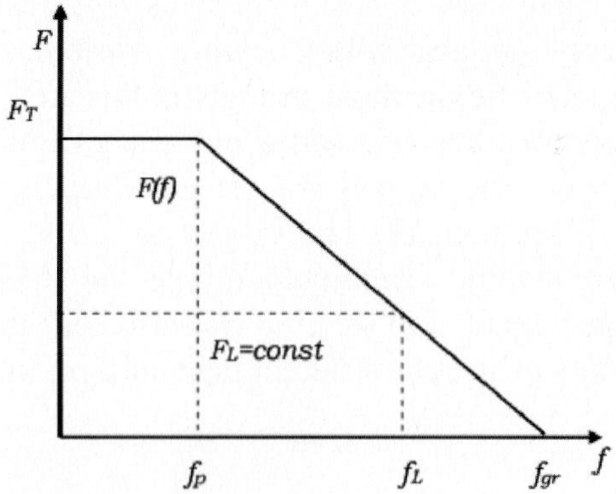

tic was simplified as it show on fig.1

Fig.1. Stepper actuator characteristic curve simplification

Some specific points of characteristic get their marking to be used in equations. Also time points of start and end of acceleration and deacceleration were properly marked.

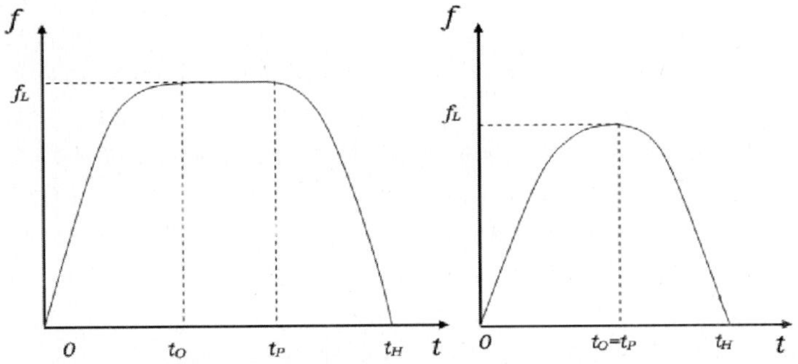

Fig. 2. Frequency against time in two cases of accelerating and deaccelerating machine, right with, left without, a period of work with a constant speed

To calculate equation of a curve described by equation (1) it was needed to find out what type of mathematic function will be suitable for it. Third degree polynomial was used as one that gives high convergence and most simple and elegant form. With all this assumptions, frequency against time during acceleration can be calculated as :

$$f(t)= f_L\left(\frac{t}{t_0}\right)^3 + 3(-f_L)\left(\frac{t}{t_0}\right)^2 + 3f_L\left(\frac{t}{t_0}\right) \tag{2}$$

Time t_o can be calculated from equation:

$$t_0 = \Theta \cdot h \cdot M \cdot A_R \tag{3}$$

where h – pitch of a thread, Θ – single step angle, M –load, and factor A_R is as follows:

$$A_R = \frac{1}{(F_T - F_L)} f_p + (f_{gr} - f_p)\left(-\frac{1}{F_T} ln|-F_T f_p + B| - \frac{1}{F_T} ln|-F_T f + B|\right) \tag{4}$$

and substitution B :

(5)

Any stepping machine that was accelerated into pull-out part of characteristic cannot be stopped incidentally. It has to be slowed down in a way that can be calculated similarly to acceleration.

Frequency against time during deacceleration has to be calculated as follows:

$$f(t)=\frac{f_L}{3t_P t_H(t_P-t_H)}t^3+\frac{f_L}{t_H(t_P-t_H)}t^2-\frac{f_L t_P}{t_H(t_P-t_H)}t+f_L\left(1+\frac{t_P^2}{3t_H(t_P-t_H)}\right) \quad (6)$$

time t_H :

(7)

factor A_H:

$$A_H=(f_{gr}-f_P)\left(\frac{1}{F_T}ln|-F_T f_L+C|-\frac{1}{F_T}ln|-F_T f+C|\right)-\frac{1}{(F_T+F_L)}f_P \quad (8)$$

and substitution C:

$$C=(F_T f_{gr}+F_L f_{gr}-F_L f_p) \quad (9)$$

It was tested how selected features of linear stepper actuator influence its functional characteristics. During experiments mechanical characteristics of tested actuators was measured. A comparison of actuator work in different modes was made. The experiments confirmed completely legitimacy of using proposed methods. At a basis of initial experiments, a computer simulating model of linear stepper actuator was created. Simulating experiments confirmed shortening time in which set speed is reached with nonlinear acceleration. Convergence of the simulation and experiments result was high. Tests shows that performance (range of its pull-out characteristic) of a stepper actuator can be increased from 25% even up to 88 %. These results were calculated in comparison to the best performance achieved on the same test bed, but frequency was changed uniformly in relation to time.

4. Conclusion

The results of this work, mainly the method of accelerating linear stepper actuators, may find wide areas of use. Linear stepper actuators are developing group of drives. They are used in new fields of technology, beginning with production lines automatization, through automotive, aeronautical and space industry, up to biomedical engineering. Proposed method of accelerating linear stepper actuators contributes to less consumption of energy, increasing efficiency of a drive. It is a reason to predestinate it as a more advantageous to use in devices with limited sources of energy. Obviously the method has also application to rotary stepper motors.

References

[1] Acarnley P.P.: *Stepping Motors: a guide to modern theory and practice.* IEE and Peter Peregrinus Ltd. London, New York, 1982

[2] Carrica D.O., González S.A., Benedetii M.: *A high speed velocity control algorithm of multiple stepper motors*, Mechatronics, vol.14, Elsevier Ltd, 2004, s. 675-684

[3] Ghafari A.S., Behzad M.: *Investigation of micro-step control positioning system performance affected by random input signals*. Mechatronics vol.15, Elsevier Ltd.,2005, s.1175-1189

[4] Kenjo T.: *Stepping motors and their microprocessor control*s, Clarendon Press, Oxford, 1984

[5] Melin P., Castillo O., *Intelligent control of a stepping motor drive using an adaptive neuro-fuzzy inference syste*m, Information Sciences, vol. 170, Elsevier Ltd., 2005, s. 133-151.

[6] Pochanke A.. Bodnicki M. *Badania symulacyjne stanowiska do wyznaczania charakterystyk częstotliwościowych silników skokowych*, Proceedings of XI Sympozjum Modelowanie i symulacja systemów pomiarowych, Krynica 17-21 września 2001, Kraków, 2001, s.73-80

[7] Phillips F. *How step-motor performance varies with type of control.* Control Engineering, vol. 7, 2001

[8] Szykiedans K.: *Doświadczalne badania charakterystyki granicznej silnika skokowego*, Proceedings of XIII International Symposium „ Mikromachines and servodrives", Krasiczyn, 2002, s. 395-399

[9] Yeadon W.H., Yeadon A.W.: *Handbook of small electric motors*, McGraw-Hill, 2001

Methods of image processing in vision system for assessing welded joints quality[1]

A. Bzymek, M.Fidali, A.Timofiejczuk

Silesian University of Technology at Gliwice,
18A Konarskiego Str., 44-100 Gliwice, Poland

Abstract

The paper deals with methods of image processing applied in a vision system for assessing welded joints quality. The vision system consists of two CCD and one infrared camera. The base of elaborated approaches to joint assesment is a set of images taken in infrared and visible light during welding process. Images taken in visible light are sources of information about outer conditions of the joint, while thermograms let us to obtain information concerning the joint interior. In the paper results of the application of processing of these two types of images are presented.

1. Introduction

Welded joints are widely used in vairous branches of industry and machinery assemblies. They have to meet quality standards specified in 0. Moreover, in many domains, like automotive industry, appropriate mechanical properties of welded joints and esthetic appearance are also recquired. One of the simplest and most popular methods of assessment of their quality is visual inspection. It is usually made after welding by a certified inspector with the use of conventional gauges or modern tools equipped with vision systems [3][8]. Such approach has some disadvantages – is time consuming and enables to examine only randomly chosen joints. In most cases visual inspection is the only way to decide whether joint is performed correctly. In some cases more sophisticated vision based systems are used for automated control of weld quality [2][9]. Methods of image processing and analysis applied in such systems are very efficient because they allow us to

[1] [1] Scientific work financed by the Ministry of Science and Higher Education and carried out within the Multi-Year Programme "Development of innovativeness systems of manufacturing and maintenance 2004-2008"

control weld condition and welding process parameters what correspond to joint quality. Because of the fact that visual inspection (manual or automated) allows us to asses only surface of welded joint and does not provide information about its internal quality, in some systems infrared detectors or cameras are used [4]. An infrared image analysis lets us to observe and control such weld parameters as bead width and depth size, temperature of weld pool and drops of melted base metal, it is possible also to track weld seam and assess its quality.

Research carried out by the authors is connected with development of a method of automatic inspections of welded joints with use of thermographic and visual images. Concept of observation system dedicated to MIG/MAG automata used in automotive industry was proposed [5]. In order to verify the assumptions made in the concept of vision system preliminary reasearch was carried out. Because of both visual and infrared images are used, appropriate methods of image processing had to be elaborated and applied. In the paper, results of the research dealing with observation and feature extraction from images acquired during welding process are presented.

2. Image acquisition

In the experiment a process of coated electrode welding was observed with the use of vision system consisting of two CCD cameras and one thermovision camera (Fig 1). Images (CCD and IR) were acquired simultaneously. Images from the second CCD camera observing the weld were acquired after the welding process.

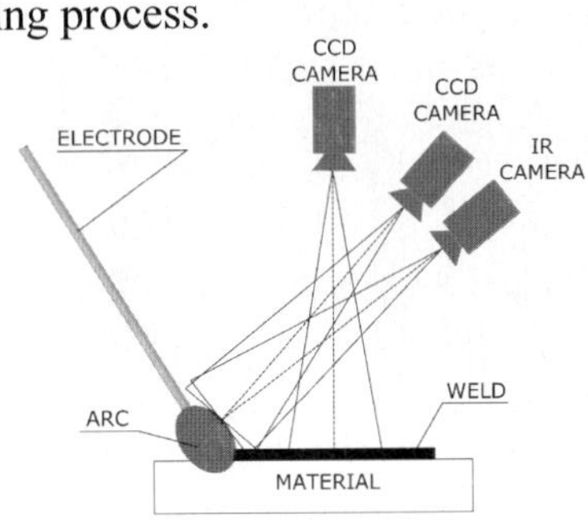

Fig 1. Schema of test stand.

3. Analysis of acquired images

Three groups of acquired images are sources of different information about the welding process and weld. Thermograms let us to observe thermal phenomena occurring during the process and to detect internal faults of the weld, whereas visual images (observation of welding area) gives some information about arc stability and environment – clouds of fume and particles of atrifacts. Moreover, such information correlated with thermograms enables us to eliminate this kind of nosie and helps to avoid incorrect conclusions. Images from the second CCD camera are helpful in identification of chosen features of a weld. In order to identify region of interest (ROI) parallel processing of visual images and thermograms has to be done.

3.1 Analisys of visual images.

Images acquired from the CCD camera observing the welding process (Fig.2a) give us information about process stability and presence of artifacts such as splinters. In order to extract a shape and dimentions of flame, fume and to localize artifacts, filtration and binarization were used Fig. 2b.

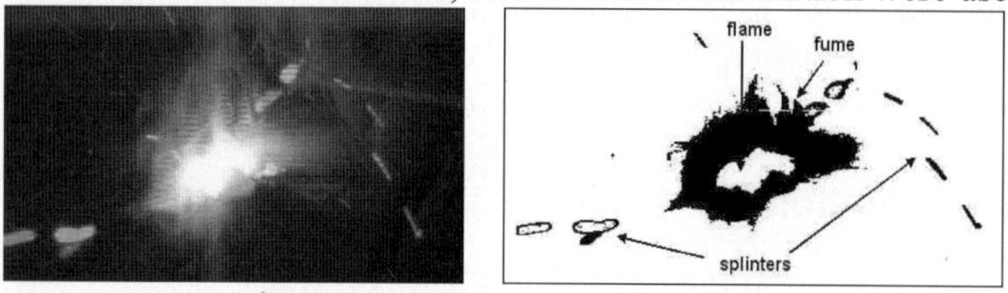

a) b)

Fig. 2. a) Photograph of welding region and b) processed image.

Regions of flame and fume were localized and their features were evaluated. After corellation with thermograms, artifacts such as splinters were identified and removed from the images before further processing.

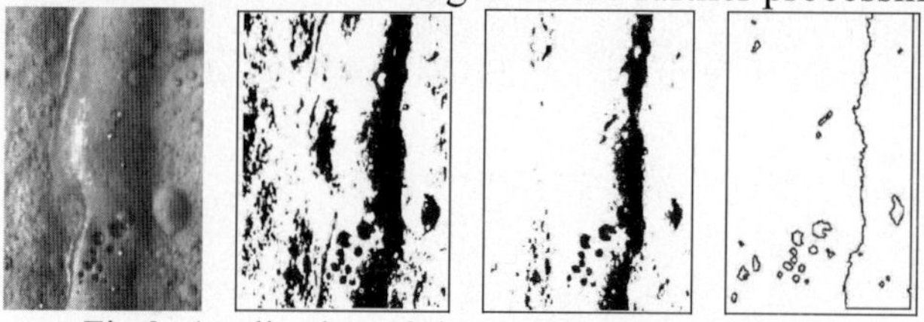

Fig 3. Application of shape detection algorithm.

Estimation of welded joint quality on the basis of images from the camera observing a hot welding region is very problematic. Reasons are changing condidtions of the process (unstable illumination of the weld). Therefore the second CCD camera was used to observe the joint. Features of the joint such as width, uniformity, scratches and porosity are possible to be estimated. In order to extract geometrical features methods of image preprocessing were used. To find an outer border of the joint, edge detection algorithm was applied. Relative position of borderlines of the joint was verified, and information about the uniformity was obtained. Another important factor of joints quality is porosity. In this case operations of cilcular shapes detection are applied. Exemplary images of welded joint after the cilcular shape detection algorithm are presented in Fig.3.

3.2. Analysis of thermographic images

In order to analyze thermographic images numerous methods can be applied [6]. The analysis should be preceded by the application of preprocessing operations which consist in selection of ROI, after that, methods like temperature profiles determination, thresholding can be applied.

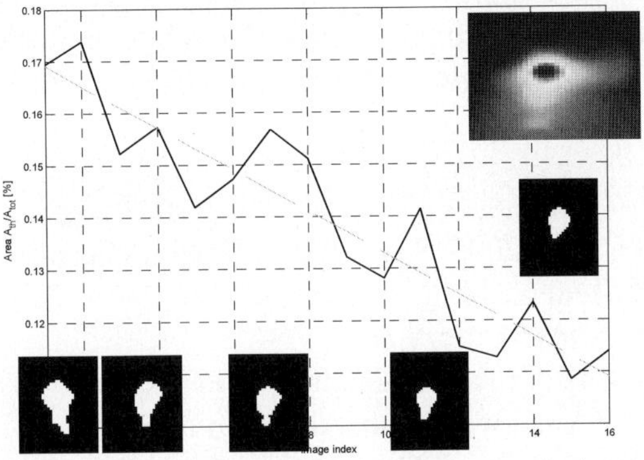

Fig 4. Changes of values of arc relative area versus image indexes and exemplary thermogam and thresholded binary images.

Thermograms acquired during the welding process give us information about distribution of temperature in arc region as well as about cooling area of a weld. It allows us to detect visible and hidden heterogeneity and defects in the weld structure. In this case methods of pulse active thermography can be used as well [7]. Analysis of arc area provides us with quantitative features which describe values of temperature in various points of arc region. In the paper a method consisting in analysis of a shape and size of arc area was proposed. A set of thermograms was thresholded and relative area of arc was estimated. A function of relative area of thresholded image versus indexes of sequence of recorded thermograms was estimated (Fig. 4). This function gives information about variability of welding proces in distinguished regions. Fig. 4 shows also exemplary thermographic image of arc area and selected binary images obtained as a result of thresholding. A linear decreasing trend of relative values of thresholded images is caused by decreasing of arc area as a result of a electrode moving away from stationary camera what was caused by arc instability due to hand welding by an inexperienced operator. In case of automated welding such results are not expected. This investigation is planned to be performed in the feature.

5. Conclusions.

On the basis of analysis of images acquired during the welding process information on outer condition of a welded joint and information about its interior can be obtained. Distinguished regions of temperature make it possible to identify the region of interest in images taken in visible light, and to limit analysis to this region. In case of images taken in the visible light the way of illumination plays important role and greatly influences quality of images. Strong light coming from the flame and lack of any other source of light causes that processing of images is not a trivial problem.

References

[1] PN-EN ISO 5817:2005 Spawanie - Złącza spawane (z wyłączeniem spawania wiązką) stali, niklu, tytanu i ich stopów – Poziomyjakości według niezgodności spawalniczych
[2] Barborak D. Castner H. "Automated weld inspection for shipbuilding".
[3] Czuchryj A. "Visual weld inspection" (in Polish)
[4] Fan, H., Ravala, N.K., Wikle, H.C., i Chin, B.A. „Low-cost infrared sensing system for monitoring the welding process in the presence of plate inclination angle". Journal of Materials Processing Technology vol. 140, no 1 (2003): 668-675.
[5] Fidali M., Timofiejczuk A., Bzymek A. „Koncepcja wizyjnego systemu oceny stanu procesu spawania i połączeń spawanych" DPS 2007, Słubice
[6] Gonzales, C. R., Wintz, P. "Digital Image Processing" Addison-Wesley Publishing Company (1987).
[7] Maldaque X. P. V "Theory and practice of infrareed technology for nondestructive testing". Willey-Interscience (2001)
[8] J. Noruk "Visual weld inspection enters new millennium" Sensor review Volume21 nr 4 (2001), pp.278-282
[9] Smith J.S, Balfour C. "Real time top face vision based control of weld pool size" Industrial Robot 32/4 (2005) pp.334-340.

Application of analysis of thermographic images to machine state assessment

M. Fidali

Silesian University of Technology, Konarskiego 18a,
Gliwice, 44-100, Poland

Abstract

Thermovision finds more and often the application in machinery and apparatus diagnostics. With the use of a thermographic camera one can carry out the non-contact simultaneous temperature measurements in many points of an object and record them in the form of a thermographic image. The thermographic image can be a source of diagnostic information, whose extraction requires that proper methods of the analysis of thermographic images need to be applied. In the article results of the application of selected methods of thermogram analysis are presented.

1. Introduction

Thermographic measurements find broad application in maintenance and technical state assessment of machinery and aparatus, industrial processes, as well as manufacturing [3],[4]. Thermographic inspections of technical objects are realized as a single or cyclic inspection and consist in controlling of a current technical state of an object or identification of a pre-failure state and assessment of damage size. Technological progress and price decrease of industrial termovision cameras make the application of such apparatus possible to monitor and assess a state of machines and devices.

Continuous thermo-diagnostics of objects is connected with conduction of necessary and systematic actions consisting in acquiring of diagnostic data which is decoded in recorded thermographic images. In case of acquiring of relevant diagnostic information proper thermographic image processing methods [2],[5] should be applied.

2. Analysis of thermographic images

During continuous object observation with the use of a thermographic device, a sequence of thermographic images in time t can be recorded. On the basis of acquired series of thermograms, multidimensional thermographical signal $ST(T(x,y),t)$ can be defined. If we consider a concept of conventional real time partition into "micro" (dynamic) and "macro" (exploatation) time [1], often applied in diagnostics, then a thermographic signal can be definied in these both domains.

Taking into account "micro" and "macro" time concepts, analysis process of thermographic signals can be divided into two stages. The first stage is connected with thermogram analysis and feature estimation. It enables determination of diagnostic signals in "micro" and/or "macro" time.

The second stage of analysis refers to analysis of diagnostic signals which were determined at the first stage. For this purposes classical signal analysis methods can be applied.

In the article the first stage of analysis of thermographic signal was presented. At this stage the most important task is analysis of thermogram series and acquisition of diagnostic features. Features are necessary for determination of diagnostic signals and thus a machine technical state.

Two simple methods of thermograms analysis were proposed. Common operation which was applied in both methods was the application of thresholding and estimation of binary images with the use of a measure of an area above the threshold level which was established experimentally. The measure area was treated as a diagnostic feature, and a diagnostic signal was built on the basis of its values. Thresholding was applied to two kinds of images: in the first method recorded thermograms were directly thresholded, in the second method an image of magnitude of Fourier spectra determined from recorded thermograms with the use of 2D Fourier transform were thresholded and estimated.

In order to verify proposed methods of analysis of thermogram series, an active diagnostic experiment was carried out. The aim of the experiment was acquisition of thermographic signals. An investigated object was a single-phase commutaotor motor, whose technical state was estimated as sufficient.

As a result of diagnostic experiments series of thermograms recorded during the object operation in different technical states were obtained. Differences in thechnical states were simulatated by changing of motor load and rotational speed.

In order to verify the first method of analysis recorded thermograms were thresholded with the use of different values of upper threshold and next relative area A_{th} of a region above the threshold level was computed for each image. A reference area A_{ref} was whole image. Functions presenting variation of this area versus index of recorded images for different thresholds were presented in Fig. 1. In figure binary images were shown. These images correspond to values of the maximum area.

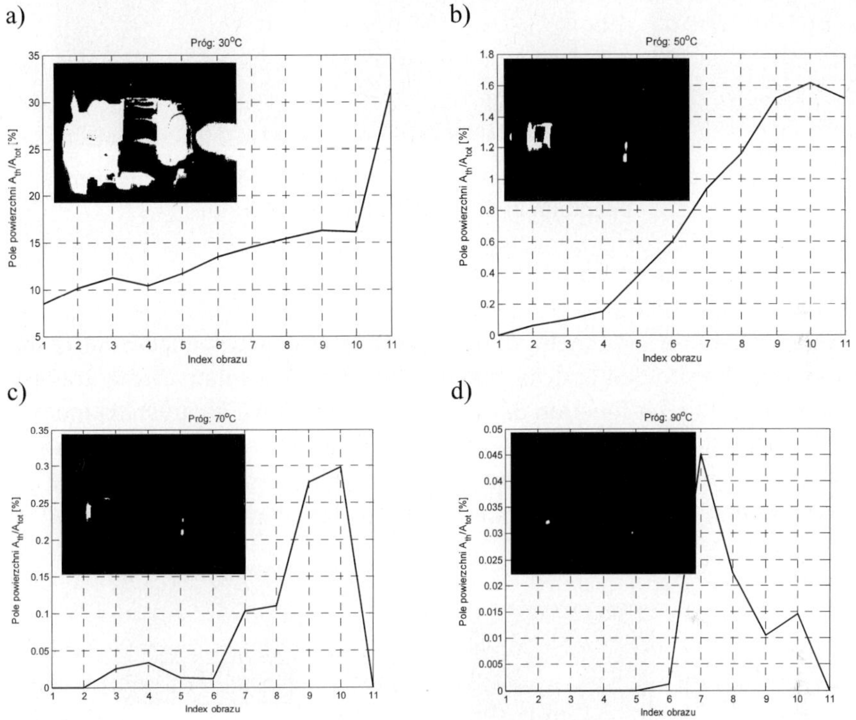

Fig. 1. Plots of a relative area values of thresholded thermographic images with thresholds a) 30°C, b) 50°C, c) 70°C, d) 90 °C and binary images correspond to maximum values of the area

The analysis of determined functions of an estimated diagnostic parameter indicates that it is possible to observe changes of thermal state of objects (Fig. 1b) and detect sudden thermal phenomena such as electric arc observed in image no. 7 (Fig. 1d).

In case of application of the second proposed method, images recorded during experiment were transformed to spectra with the use of 2D Fourier transform. In Fig. 2 there are presented exemplary magnitudes of Fourier spectrum estimated on the basis of thermograms recorded at the beginning of object operation (Fig. 2a), during operation in the moment of occurring of electric arc between commutator and one of carbon brushes (Fig. 2b) and at the end of machine observation when, in the bearing and commutator regions higher temperature caused by bearing seizing and commutation effect occurred (Fig. 2c).

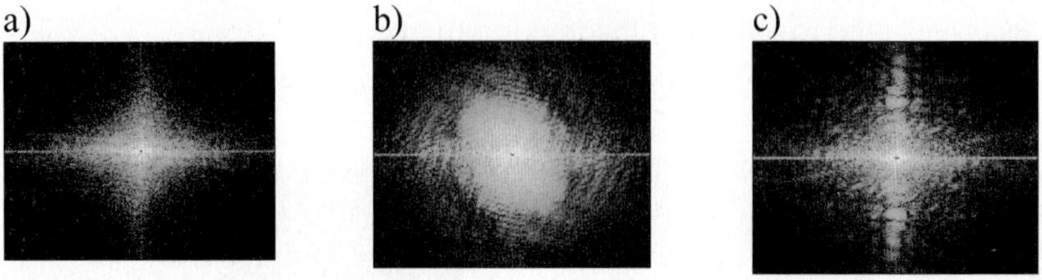

Fig. 2. Magnitudes of Fourier spectra of thermograms recorded during machine operation in different technical states.

Observations of determined Fourier images indicate differences as results of changes of machine technical state.

Similarly as in the first method, in the second one, determined Fourier images were thresholded and for each binary image, a relative area was determined. In Fig.3 a function of changes of the relative area versus indexes of binary images was presented. Determined function indicates that images created as a result of Fourier transform can be useful in a process of determination of changes of machine technical state during its operation.

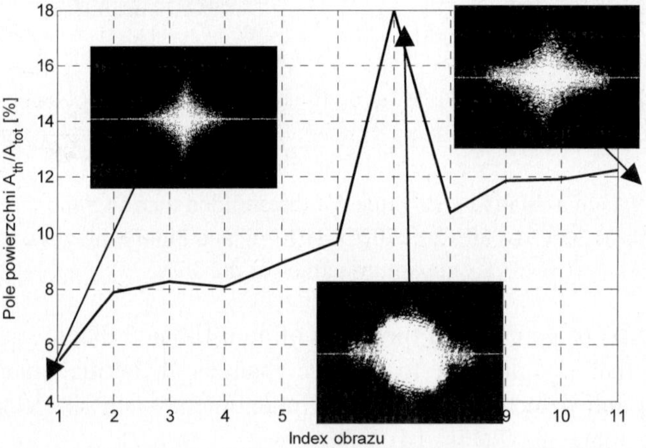

Fig. 3. Function of relative area values computed from binary images of magnitude Fourier spectra of thermographic images

Conclusions

In the article preliminary results of research whose aim was verification of a proposed concept of evaluation of a technical state of an object on the basis of results of analysis of sequence of termographic images were presented. Thermograms recorded during an active diagnostic experiment were analyzed. One of proposed methods was based on images computed with the use of 2D Fourier transform and such images were also processed. One stated that such kinds of images can be also a source of information about a diagnostic state thus can be processed with the use of different image processing methods. Thermographic as well as Fourier images were thresholded and such diagnostic features as the area of region above the threshold was used to determine diagnostic signals.

The analysis of determined diagnostic signals indicates possibilities of application of proposed methods of thermogram analysis to determination of one dimensional diagnostic signal. The proposed concept can be used for identification of changes of technical states during machine operation.

Results indicate that continuation of research in this area is necessary. Future research will be focused on determination of a set of diagnostic features which can be useful for classification of machine technical state.

References

[1] W. Cholewa "Method of machine diagnostics with application of fuzzy sets" Zeszyty Naukowe Nr 764. Politechnika Śląska, Gliwice 1983. (in polish)

[2] C. R. Gonzales,. P. Wintz "Digital Image Processing" Addison-Wesley Publishing Company 1987.

[3] H. Madura "Thermographic measurements in practice" Agencja wydawnicza PAK, Warszawa 2004. (in polish)

[4] W. Minkina "Thermovision measurements. Devices and methods" Wydawnictwo Politechniki Częstochowskiej. Częstochowa 2004. (in polish)

[5] Z. Wróbel, R. Koprowski „Thermographic image processing" Proceedings of VI krajowej konferencji Termografia i termometria w podczerwieni. Ustroń 2004. (in polish)

The use of nonlinear optimisation algorithms in multiple view geometry

Maciej Jaźwiński, Barbara Putz

Institute of Automatic Control and Robotics,
Warsaw University of Technology,
ul. św. Andrzeja Boboli 8, 02-525 Warszawa

Abstract

Search for optimal parameter set is a key point of stereovision algorithms and of other geometric computer vision algorithms performing scene reconstruction that use multiple views of a given scene. Optimisation algorithm must be robust and converge with high probability to one of the local minimum of a cost function. The paper discusses the use of nonlinear optimisation algorithms in viewing parameter estimation in reconstruction.

1. Introduction

Theory and practice of computer vision algorithms have evolved much during last decade. Scene reconstruction is now possible from pictures taken from uncalibrated cameras. It was not possible ten years ago, when process of camera calibration involved calculation of 11 parameters of each camera, from which epipolar geometry was computed. Because camera parameters changed during the robot was moving, it was not possible to perform dynamic reconstruction.

Comprehensive information on reconstruction using multiple view geometry can be found in monograph [4]. Stereovision systems are the most popular; basic reconstruction scheme with the use of data aquired from two views is pointed below :

1. Compute fundamental matrix F, representing the intrinsic projective geometry between two views and satisfying the relation $\mathbf{x}'^T F \mathbf{x}=0$ for any pair of corresponding points $\mathbf{x}$ and $\mathbf{x}'$ in the two images
2. Calculate camera projective matrix using epipolar constraint
3. For each of interest points in each image calculate its position in 3D scene.

In order to meet efficiency requirements, optimisation algorithm used in reestimation of fundamental matrix F, repeated many times from all point correspondences in step 1 must be robust. The well known basic optimisation methods like steepest descent method, Newton's method and Gauss-Newton methods are not efficient enough for solving reconstruction problems. To achieve fast and stable convergence, more advanced methods should be used.

2. Damped methods

In damped methods step length is controlled by damping parameter μ. Example of damped method is the Levenberg-Marquardt method [2], which is modification of Gauss-Newton method, introducing damping parameter. Damping parameter may μ may be given by the user, or calculated using some equation, most often in the form of $\mu_0 = \tau \max_i \{a_{ii}^{(0)}\}$, where τ is a parameter provided by user, and $a_{ii}^{(0)}$ are the elements of Hessian matrix. In the Levenberg-Marquardt method step length and step direction are calculated simultaneously, by solving equation:

$$(J^T J + \mu I)h_{LM} = -g$$

where g is the gradient of F(x), I is an identity matrix, J - Jacobian matrix, μ is the damping parameter, h_{LM} is the current step. By introducing Jacobian-based Hessian matrix approximation, the method requires only one-order partial derivatives.

For large values of μ factor μI dominates the left-hand side of equation and algorithm behaves like steepest descent algorithm which converges slowly; for small values of μ factor $J^T J$ dominates and algorithm behaves like Gauss-Newton algorithm.

The Levenberg-Marquardt method has become the standard of nonlinear least-squares routines due its simplicity and efficiency; see *Numerical Recipes* or [10]. It works very well in practise and is quite suitable for minimisation with respect to a small number of parameters, like stereovision based reconstruction [6,7].

Damped methods can be implemented as a model-trust regions metods described below.`

3. Trust region algorithms

The trust region methods [1,7] are characterized by two main concepts – a model function L approximating given cost function F, and trust region Δ. In the trust region methods it is assumed that model function is accurate in area of a trust region Δ. The step length is controlled directly by trust region radius, as opposed to damped methods. Quality of the model is evaluated by so called gain-ratio, dependent on parameter vector and the step in current iteration.

One of trust regions methods is the Dog Leg optimization algorithm, where the choose between the gradient descent step (if the Cauchy point lies outside the trust region), the Gauss-Newton step (if it lies inside the trust region) or combination of these - toward the intersection of the trust region with the line from Cauchy point to Gauss-Newton point - is performed, with the use of descent direction $J^T\varepsilon$ and Hessian matrix J^TJ. Important feature of Dog Leg version described in [7] is that normal equations can be computed only once for every successful iteration.

4. Specialized methods

Some methods may take advantage of problem's properties. In general case optimisation contained in this algorithms do not give better convergence, but when applied to some class of problems, they can make altered methods more efficient. Example of such method is sparse Levenberg-Marquardt method [2, 11], which uses sparse structure of parameters matrix in reconstruction problems from two views. In this case the Jacobian matrix has special form [4]:

$$J = \begin{bmatrix} A_1 & B_1 & \cdot & & \cdot & 0 \\ A_2 & & B_2 & & & \cdot \\ \cdot & & & \cdot & & \cdot \\ A_{n-1} & & & & B_{n-1} & 0 \\ A_n & 0 & \cdot & \cdot & 0 & B_n \end{bmatrix}$$

where A_i and B_i are partial derivatives of $\hat{\mathbf{X}}_i$ on **a** and $\mathbf{b}_i$ respectively, $\hat{\mathbf{X}}_i$ denotes the estimated value of *i*-th measured point with its parameter vector $\mathbf{b}_i$, **a** is a vector of camera parameters.

With the sparseness assumption, each iteration of algorithm requires computation time linear in n, the number of parameters. Without sparsness assumption the central step of algorithm has the complexity n^3 in the number of parameters. Analogously one can use sparse LM in the trifocal or quadrifocal tensor optimisation and in the multiple image bundle adjustment, taking advantage of the lack of interaction between parameters of the different cameras [4]. But Dog Leg algorithm can also benefit from sparse structure of Jacobian matrix in calculating descent direction and Hessian matrix. Thus when performing bundle adjustment in multiple view geometry, Dog Leg tends to be the best algorithm [6,7].

The interesting example of using sparse Levenberg-Marquardt method is presented in [9]. The reconstruction scheme of 3D NURBS curves is performed directly from its stereo images. The reconstruction of 3D curve is converted into control points and weights of NURBS representation of the curve, accordingly bypassing point-to-point correspondence matching. The Jacobian matrix has a sparse and simple form that allows efficient and stable Levenberg-Marquardt iteration.

5. Combined methods – new optimisation techniques

Some of more advanced algorithms tend to be hybrid algorithms, although this relation is not straightfoward. One example of hybrid method is algorithm presented by Madsen, and described in more detail in [8]. It combines Levenberg-Marquardt method with Quasi-Netown's method, starting with Levenberg–Marquardt method, and switching to Quasi–Newton, if algorithm detects that cost function is significantly non-zero. Research is performed also on methods that use other techniques. One example of such research is work of Heyden, Würtz and Peters [5]. Simple evolutionary algorithm used to back Levenberg-Marquardt optimisation gave improvement in the quality of reconstruction. Evolutionary algorithm may be used to perform optimisation data before or after optimisation by Levenberg-Marquardt algorithm.

Interesting result of research are published in [3], where evolutionary algorithm, which normally uses Gauss-Newton step, or gradient-descent step, was implemented to use both methods and choose better result. Test results show that this led not only to improvement in convergence time, but also in quality of optimisation result.

Tests performed by authors show, that approach used in [3] for selecting step length, applied to Dog Leg algorithm, don't give any significant improvement compared to Levenberg-Marquardt, or standard Dog Leg, in problems, where these algorithms should be used. For small scale problems enhanced Dog Leg algorithms gave better results then standard Dog Leg, but worse than Levenberg-Marquardt. For large scale problems (as BA) results were worse than obtained using standard Dog Leg algorithm.

6. Conclusion

An overview of available algorithms, with knowledge of each algorithm pros and cons, is required in order to choose the best optimisation algorithm for the given problem. For example, when performing bundle adjustment, Dog Leg tends to be the best algorithm, as it can too benefit from sparse structure of Jacobian matrix. For small scale problems, like stereovision based reconstruction, Levenberg–Marquardt seems to be the best algorithm. Supplementing optimisation algorithms with evolutionary algorithms may result in more precise or robust reconstruction.

References

[1] Berghen F. V.: *"CONDOR: a constrained, non-linear, derivate-free parallel optimizer for continous, high computing load, noisy objective functions"*, 2003-2004.

[2] Frandsen P.E., Jonasson K., Nielsen H.B. and Tingleff O.: *"Unconstrained Optimization"*, 3rd Edition, DTU, 2004.

[3] Gnosh A., Tsutui S., *"Advances in evolutionary computing. Theory and applications"*, Nat. Comp. Series, Springer-Verlag 2003, 45-95.

[4] Hartley R., Zisserman A.: *"Multiple View Geometry in Computer Vision"*, Second Edition, Cambridge Univ. Press 2006, UK.

[5] Heyden L., Würtz R.P., Peters G.: *„Supplementing bundle adjustment with evolutionary algorithms"*. International Conference on Visual Information Engineering - VIE 2006, 533-536, Bangalore, India.

[6] Jaźwiński M., Putz B.: *„Evaluation of the Levenberg-Marquardt and the Dog Leg optimization algorithms for small scale problems"*. Paper accepted for 13h IEEE International Conference MMAR 2007.

[7] Lourakis, M.L.A. Argyros, A.A.: *"Is Levenberg-Marquardt the most efficient optimization algorithm for implementing bundle adjustment?"*. ICCV 2005, Tenth IEEE Conf. on Comp. Vision, 1526-1531.

[8] Madsen K., Nielsen H.B., Tingleff O.: *"Methods for non-linear least squares problems"*. Technical University of Denmark, April 2004.

[9] Xiao Y.J., Li Y.F.: *"Optimized stereo reconstruction of free-form space curves based on a nonuniform rational B-spline model"*. J.of the Opt. Society of America A, vol. 22, no.9, Sept. 2005, 1746-1762.

[10] http://www.ics.forth.gr/~lourakis/levmar: levmar: Levenberg-Marquardt nonlinear least squares algorithms in C/C++.

[11] http://www.ics.forth.gr/~lourakis/sba: sba: A Generic Sparse Bundle Adjustment C/C++ Package Based on the Lev.-Marq. Algorithm.

Modeling and Simulation Method of Precision Grinding Processes

B. Bałasz (a) , T. Królikowski (a)

(a) Koszalin Universtity of Technology
Department of Fine Mechanics ul. Raclawicka 15-17
Koszalin, 75-016, Poland

Abstract

Grinding is very complex process depending on large number of correlated factors. In precise grinding it is very important to select optimal conditions and to preserve stable conditions during the process. The model of grinding process comprise usually a few elementary models: model of a grain, model of a grinding wheel topography, model of surface roughness, model of the process kinematics, model of a chip formation, forces and energy, thermal and vibration. The author of this paper undertook a study on developing algorithms and programs for complex simulation of grinding process. This paper presents assumptions, schemes, examples of models, and results of the advanced kinematic-geometrical model of grinding processes.

1. Introduction

The efficiency and quality of abrasive machining processes has a decisive influence on the costs and quality of elements produced as well as whole products. The machining potential of abrasive tools is used insufficiently. One of more important reasons for an insufficient use of the machining potential is a slow development of new abrasive tools – development work focuses more on the improvement of the known technologies and not so much on the creation of new abrasive tools. Also, due to high costs of research into tools from ultra-hard materials concerning new tools, such research has not made a sufficient progress. As a solution to the problem of second group of parameters a modeling and computer simulation of grinding process is one of the possible answer [1, 2, 3].

2. Fundamentals of the modeling and simulation

Models of the process geometry where developed based on the experimental results of microcutting process carried out with a single grain. The aim of the experiments was to obtain the conditions for chip formation according to grain shape, depth of cut and value of cutting speed for different type of the grains and machined material. To achieve that objectives the experimental stand has been developed, based on plane grinding machine equipped with grinding tool with a grip for mounting a single grain (cp. figure 1a), material sample mounted on dynamometer (cp. figure 1b), what enables the measurement of grinding forces during cutting. The results of each experiment where gathered in databases for father analysis.

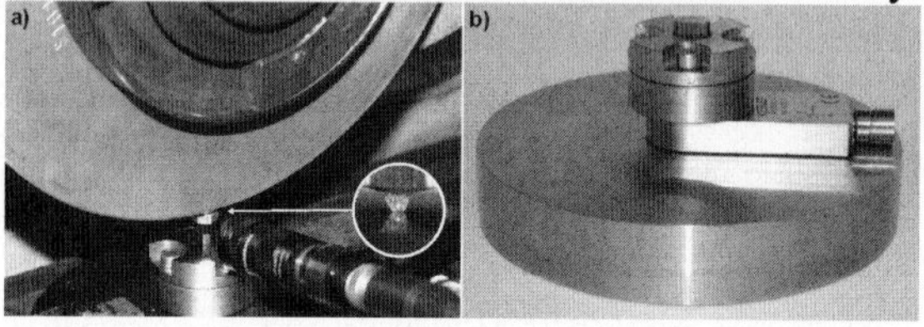

Fig. 1. Experimental stand for microcutting process: a) grinding tool with single grain, b) material specimen

Each material sample after cutting experiment was measured with profilometer in order to obtain the data on depth of cut, size of pile-ups and deviation of that on the length of the cut. A microscopic pictures of a grains and the scratches made by a cutting edges identified on the grains were also obtained with the use of scanning electron microscope (cp. figure 2).

Fig. 2 SEM-pictures of the grain and scratches made by single grain during microcutting

3. Modular simulation system

The objective of the research was to developed a modular system capable of providing simulation of machining processes with ability of flexible adaptation of different grinding process types regarding different grains type and size, models of grinding wheels (grain concentration, grains arrangement on surface), kinematics of processes. To achieve its flexibility simulation system was divided into four subsystems responsible for completion tasks connected with modeling, simulation computation, data management and simulation data analyzes. The diagram of simulation system module was presented on figure 3.

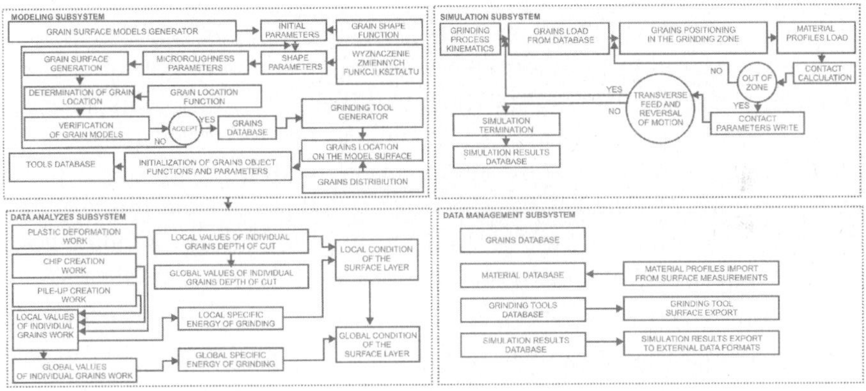

Fig. 3. Modules of the grinding simulation system

3.1. Modeling subsystem

The model of grinding process comprise usually a few elementary models: model of a grain, model of a grinding wheel topography, model of surface roughness, model of the process kinematics, model of a chip formation, forces and energy. Grains are generated with application of a two-dimensional elastic neuron network for the generation of the surfaces of abrasive grains with macro-geometric parameters set. In the neuron model developed, the output parameters are the number of the grain vertices, the apex angle and the vertex radius [4]. As a result of the work of the system, a random model of a grain with set parameters is obtained. The neuron model developed is used as a generator of the surface of the model of abrasive grains in the system of modeling and simulation of grinding processes. With every generated grain there is associated vector of grain parameters, describing temporal states of the grain during the whole process (e.g. number of contacts with workpiece material, volume of removed material, normal and tangential forces etc.). After grain generation, the working surface of the grinding wheel is generated by positioning a single grains with distribution of chosen model for grinding wheel type surface Thanks to that, the characteristic of behavior of contact during the process could be thoroughly discovered. On generated surface the model of the bond is placed on. As a completion to this task, models of grain displacement and removal and the dressing process are also elaborated.

3.2. Simulation subsystem

The structure of that subsystem is based on discrete time simulation. During the simulation grains moves over grinding zone with step of one micrometer and on the basis of grinding wheel velocity v_s and workpiece velocity v_{ft} time step is determined. In each time step the calculation of individual grains contact causing material profile modification take place and results are saved in the database. In order to reduce demand for computer memory, only grains and material profiles moving through grinding zone are read from databases and beyond the zone are released to databases.

3.3. Data management subsystem

The role of this subsystem is the manage the data created during the modeling, simulation and analyzes. The four databases were created for storing: objects of the grains, objects of the grinding tools, objects of material profiles before processing and simulation data, materials profile after processing and results of analyzes. A large number of data for simulation comes form outer sources, also the simulated data must be available outside simulation system , therefore a large number of procedure for importing and exporting data to different data formats (e.g. txt, csv, xml, sur) have been written as well.

3.4 Data analyzes subsystem

Data obtained during simulations are being analyzed with functions created in data analyzes subsystem. The most significant analyzes concern: the grain activity and its load, the average cut layers, flotation of single grain depth of cut along the grinding zone, and the influence of grains shape, size and arrangement on afore mentioned phenomena. Various data analysis methods have been implemented, range from dynamic sql-queries, statistical inference to data mining (eg. decision trees, clustering, time series, logistic regression).

4. Conclusion

The developed models of grinding processes reveals features which enables designing a new models of grinding tools with optimal grains shape and size, and its orientation on the grinding tool surface. The optimization process is feasible due to possibility of the models to carry out the simulation within a vast range of process parameters variability and exact gathering data concerning individual contact of grains. The innovation solution of presented models depends on isolation of individual grains during the simulation process and analyzing the phenomena in the grinding zone in relation to single grain. The most significant analyzes concern: the grain activity and its load, the average cut layers, flotation of single grain depth of cut along the grinding zone, and the influence of grains shape, size and arrangement on afore mentioned phenomena.

References

[1] Bałasz B., Królikowski T., Kacalak W.: Method of Complex Simulation of Grinding Process. Third International Conference On Metal Cutting And High Speed Machining, Metz, France 2001, pp. 169-172

[2] Bałasz B. Królikowski T.: Utility of New Complex Grinding Process Modeling Method. PAN Koszalin 2002, pp. 93-109

[3] Królikowski T., Bałasz B., Kacalak W.: The Influence of Micro- And Macrotopography of the Active Grinding Surface on the Energy Consumption in the Grinding Process. 15th European Simulation Multiconference, Prague, Czech Republic 2001, pp. 339-341

[4] Szatkiewicz T., Bałasz B., Królikowski T.: Application of an elastic neural network for the modeling of the surfaces of abrasive grains. Artificial Neural Networks in Engineering ANNIE 2005, ASME Press, New York 2005, pp. 793-800.

ACKNOWLEDGEMENTS

This work was supported by grant: KBN Nr 4 T07D 033 28 form Polish Ministry of Science and Higher Education

Determination of DC micro-motor characteristics by electrical measurements

P. Horváth (a)*, A. Nagy (b)

(a) Széchenyi University, Egyetem tér 1.
Győr, H-9026, Hungary

(b) Széchenyi University, Egyetem tér 1.
Győr, H-9026, Hungary

Abstract

It is generally difficult to carry out and measure breaking moments precisely in the mNm range. Instead of traditional methods using brake and additional mechanical elements to determine torque vs. angular velocity characteristics of micro-motors this paper suggest a new procedure based on purely electrical measurements. Theoretical background of the measuring procedure is discussed detailed. The suggested method is shown in the case of a RF300E DC micro-motor.

1. Introduction

The design process of a control system requires some knowledge about all parts of the system, including actuators. Small size DC motors are still often used in mechatronic systems as actuators. Their parameters must be determined experimentally. This process usually needs a brake to load the motor. Testing of even regular size motors can cause problems, because application of a brake needs additional mechanical parts (clutch, disc) to be fixed to the motor shaft. Fitting accuracy, additional weight and damping may all influence both the static and the dynamic behaviour of the motor. Measuring the braking moment with sufficient accuracy-especially at extremally low-power motors-can cause the core of the problem.
An ingenious idea can be found in [1], where eddy-current clutch and a DC motor with known characteristics, as a brake has been applied.

The aim of this paper is to present a parameter identification method without application of additional mechanical parts.

2. Modeling a DC motor

The dynamic model of DC motors is known well in literature [2] (Fig.1). The electrical part models the resistance *R* and inductance *L* of the armature winding. Motor constant *k* serves to calculate the back electromotive force owing to the motion of the coil in electromagnetic field.

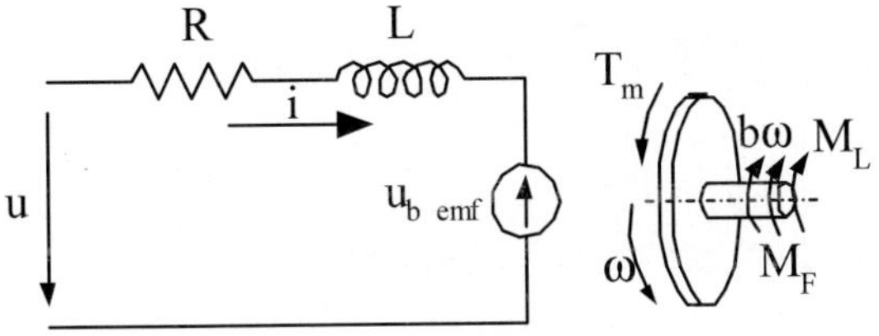

Fig. 1. Dynamic model of a DC motor

The free-body diagram of the mechanical part involves the mass- moment of inertia *J*, combined damping moment consisting of velocity-proportional $b\omega$ and constant Coulomb-type M_F friction parts, driving torque $T_m = k \cdot i$ and the external load M_L. On the basis of the model two equations can be written:

$$u = iR + L\frac{di}{dt} + k\omega \tag{1}$$

$$J\frac{d\omega}{dt} = ki - b\omega - M_F - M_L \tag{2}$$

3. The effect of Coulomb-friction

Damping law is the weakest point of the model, so the effect of the Coulomb-type friction on the dynamic behaviour must be analyzed theoretically. Let us consider a free run-out of the rotor without excitation and external load, assuming the initial condition $\omega(0)=\omega_0$. In this case electrical part has no effect to the motion of the rotor, so (2) becomes simple:

$$J\frac{d\omega}{dt} + b\omega = -M_F \tag{3}$$

Ignoring the details of the solution the following result occurs:

$$\omega = (\omega_0 + \frac{M_F}{b})\exp(-\frac{b}{J}t) - \frac{M_F}{b} \tag{4}$$

Fig. 2. depicts the shape of the theoretical run-out curve with Coulomb, velocity-proportional as well as combined friction moment.

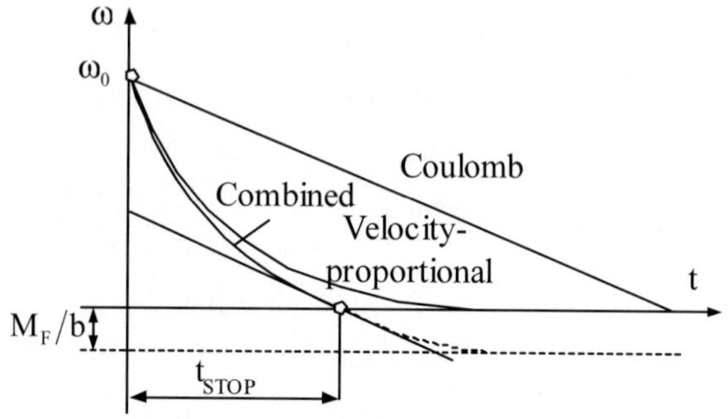

Fig. 2. Free run-out of rotor applying different type of damping

3. Parameter identification

The model describing the operation of a DC motor has the following unknown parameters: *R*, *L*, *k*, *b*, M_F and *J*. Some of the parameters, such as *R* and *L* can be determined by standard methods (R=11,52 Ω). Determination of the other parameters needs due considerations.

3.1. Motor constant

At stationary condition ω=constant the steady-state armature constant i_{SS} does not change, consequently the inductivity has no effect. Expressing motor constant from (1) we get

$$k = \frac{u - Ri_{S-S}}{\omega_{S-S}} \tag{5}$$

Measuring steady-state angular velocity $\omega_{S\text{-}S}$ happened by a commercial optical revolution-meter constructed originally to ball bearings. Motor constant has been calculated at various angular velocities and it really proved to be constant ($k \approx 0{,}00785$ Vs/rad).

3.2. Friction moment

Determination of the Coulomb-type friction moment occurs at starting phase of the unloaded motor. Motor current is increased from zero until the motor just begins to rotate ($i_0 \approx 0{,}0135$ A). At this instance the driving moment is equal to the friction moment ($M_F \approx 0{,}000106$ Nm):

$$ki_0 = M_F \tag{6}$$

3.3. Viscous damping coefficient

This measurement happens at stationary condition, when there is no inertia effect. Velocity proportional damping coefficient can be expressed from (2):

$$b = \frac{ki_{S-S} - M_F}{\omega_{S-S}} = \frac{k(i_{S-S} - i_0)}{\omega_{S-S}} \tag{7}$$

Viscous damping coefficient has been calculated at various angular velocities and proved to be constant ($b \approx 7{,}2 \cdot 10^{-8}$ Nms/rad).

3.4. Mass-moment of inertia

This is the most sophisticated task of the parameter identification process, to which dynamic measurement is necessary.We investigated the free run-out of the unloaded motor from $\omega = \omega_0$ to $\omega = 0$. Close to the stopping the viscous damping is negligible, so (2) can be written as

$$J\frac{d\omega}{dt} = -M_F \tag{8}$$

Instead of measuring the $\omega = \omega(t)$ time-history we apply the following procedure. The motor runs with stationary angular velocity ω_0 when we cut the input voltage. The motor starts operating as a generator and the oscilloscope with high input impedance measures the back emf which is proportional to the angular velocity. The run-out diagram of the investigated RF300E DC micro-motor can be seen in Fig.3. Even though the curve is noisy due to the commutation, fortunately near stopping the tangent of the curve dω/dt can be drawn precisely ($t_{STOP}\approx 1{,}03$ s). Applying (8) the mass-moment of inertia can be calculated ($J\approx 1{,}9\cdot 10^{-7}$ kgm^2).

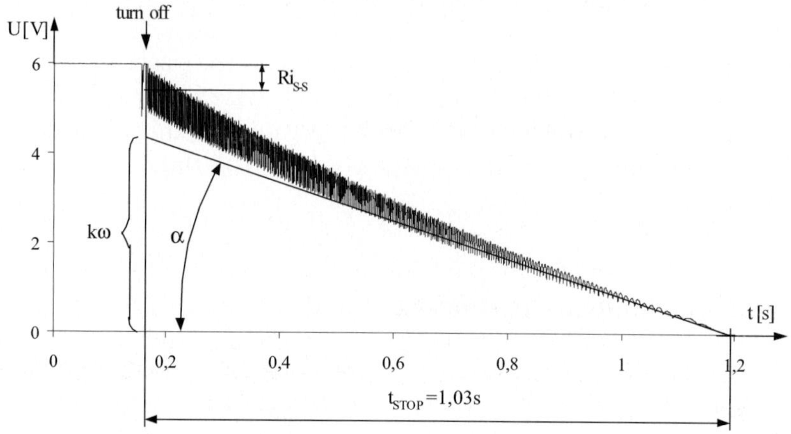

Fig. 3. Run-out curve for measuring mass-moment inertia

4. Conclusions

This paper presented a new method to identify parameters of DC micro-motors without applying additional mechanical parts. Results of the investigation showed (Fig. 3), that Coulomb-type friction is significant compared to velocity proportional damping, so it must be taken into consideration. The method outlined above can be applied at regular size DC motors too. By means of parameters determined above, static or dynamic characteristics of DC motors can be drown by known methods.

References

[1 A. Huba, A. Halmai: Berührunslose Drehmomentenmessung für extrem kleine Drehmomente, Periodica Politechnica, TU Budapest, 1987, Vol.31.pp.2-3

[2] G. F. Franklin, J.D. Powell: Feedback control of dynamic systems, Addison-Wesley, Stanford, 1992.

Poly-optimization of coil in electromagnetic linear actuator

Paweł Piskur, MSc (a) , Wojciech Tarnowski, Prof. Dr. Habilit., (b)

(a) (b) Control Engng Dept
Technical University, Koszalin, 75-620 Poland

Abstract

The main advantages of the electromagnetic linear actuators are the simple design structure, the fast response for input signal, a possibility to achieve a high linear acceleration and a low cost of maintenance [4][5]. Moreover, a linear motion is a natural output, so there is no need of any mechanical transmission.

On the other hand, the main drawbacks are: low energy efficiency and a need of the great current impulse source.

In the present paper an optimization of the actuator design is considered. The overall criteria are the maximal energy efficiency ratio and the minimal mass and volume of the actuator for the required kinetic energy of the core. These criteria may be transformed into design variables; in the present work we adapt two specific criteria:

- coil inductance (to be minimal);
- electromagnetic force (to be maximal).

A mathematical model for the two-criteria optimization is elaborated, and a poly-optimization is completed.

The final result is a set of Pareto optimal solutions [1], which makes possible to draw out some more general conclusions on a design of the actuator.

1. Introduction

Electromagnetic actuators are commonly used for various purposes [3], however the main drawback is the low energetic efficiency. The efficiency is here understood as a ratio of the kinetic energy of the core at the outlet to the electric supply energy delivered to the coil.

In the case of gun actuator, there is a set of coils, displayed by series. However, in this paper for a preliminary analysis there is only one coil system under consideration .

2. Object of poly-optimization

The system consists of one cylindrical coil and a ferromagnetic moving coaxial core (see Fig. 1), under normal atmospheric pressure. The coil is supplied by a constant voltage impulse of finite time. The goal of the system is to accelerate the core to a maximal kinetic energy.

3. Optimization criteria

The main functional criterion is a power efficiency η, what implies that for a demanded kinetic energy of the core, a system may be light and of limited dimensions, what concerns also an electric supply.
What more, an important parameter is the outlet velocity of the core: if one demands a high speed, also the current in the coil must quickly increase. To achieve it by a constant voltage, a small inductance is necessary (see Eqn. 1-3). Thus we decide to adopt the coil inductance L as another optimality criterion.

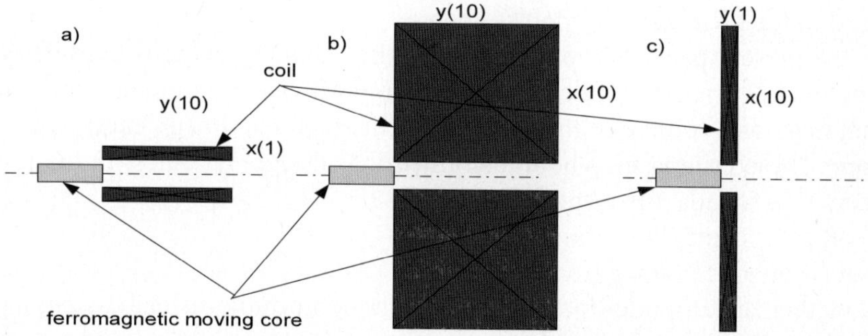

Fig. 1 Object of poly-optimization: the core and the coil; three exemplary design versions: a) x=1, y=10, b) x=10, y=10, c) x=10, y=1

4. Mathematical Model

Symbols

z - windings number [-]; u(t) - supply voltage of the coil [V];
i(t) - current in the coil [A]; d - core displacement [mm]
R - Ohm resistance of the coil [Ω]; R - Coil winding radius [mm];
$\mu_0\mu_r$ - relative permeability [H/m]; ρ - electrical conductivity [Ωm];

t – time [s]; v - velocity of the iron core [m/s]; m – mass [kg]
A - average area of the cross – sectional magnetic flux [mm^2];

Power effectiveness:

$$\eta = \frac{E_{KINETIC}}{E_{SUPPLY}} 100\% ; \quad (1)$$

where:

$$E_{SUPPLY} = \int_{t=0}^{t=t_1} u(t) i(t) dt ; \quad (2)$$

$$i(t) = \frac{u}{R}(1 - e^{-\frac{t}{\lambda}}) ; \quad (3)$$

$$\lambda = \frac{L}{R} ; \quad (4)$$

$$R = \rho \frac{l}{\pi r^2} ; \quad (5)$$

Inductance L

$$\mathrm{L} = \mu_0 \mu_r \frac{\mathrm{z}^2 \cdot \mathrm{S}}{\mathrm{length}} ; \quad (6)$$

where z and S are functions of x and y, and *length* is the y.

$$E_{KINETIC} = \frac{1}{2} m v^2 ; \quad (7)$$

$$v = \frac{1}{m} \int F dt ; \quad (8)$$

$$v = \frac{1}{m} \int \frac{1}{2} \mu_0 \mu_r A \frac{1}{d^2} z^2 i^2 (t) dt ; \quad (9)$$

$$E_{KINETIC} = \frac{1}{2} \int \left[\frac{1}{2} \mu_0 \mu_r A \frac{1}{d^2} z^2 i^2 (t) \right]^2 dt ; \quad (10)$$

$$F = \frac{1}{2} \mu_0 \mu_r A \frac{1}{d^2} z^2 i^2 (t) ; \quad (11)$$

where F is electromagnetic force, d and z are functions of x and y.

Decision variables

Arbitrarily was decided that dimensions of the coil *x* and *y* are to be the decision variables (see Fig. 1); remaining arguments of formulas are taken to be constant.

Constraints

Current density in a coil:

$$\rho \le \frac{i_{max}}{\frac{\pi \cdot d^2}{4}} = 3,66 \cdot 10^6 \quad \frac{A}{m^2}; \tag{12}$$

$$x \cdot y \cdot \rho \cdot 0,7 = i_{max} \cdot z; \tag{13}$$

Supply power $E_{SUPPLY} \le E_{NOM}$, where: E_{NOM} is a nominal (catalogue) power of the supply unit.

5. Solution of the poly-optimization problem

The main computation problem is, that the inductance *L* is a function of the coil parameters **and** of the current position of the core *s* and its parameters, as well. This relation is strongly nonlinear, and changing in time. An analytical form could be found as a rough approximation. So we have decided to use a MES method [6], and realize computations step by step, for discrete points of time and the core position *s*.

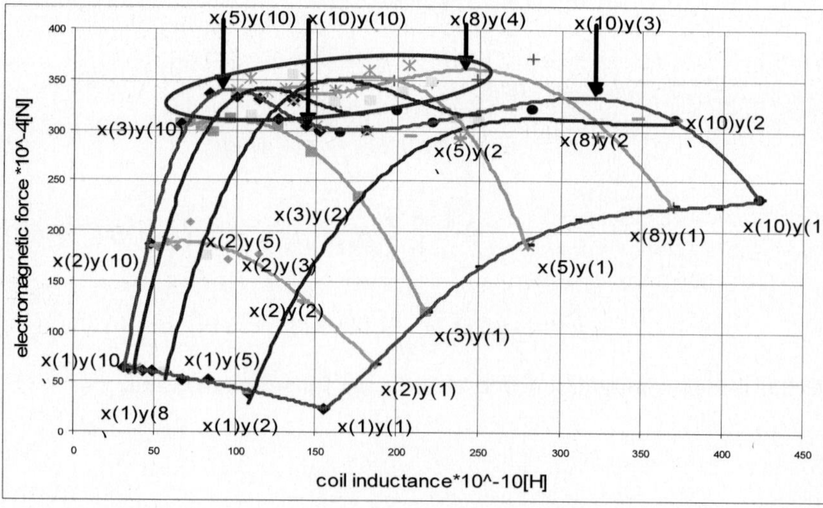

Fig. 2 All discrete solutions, the Pareto optimal solutions (left-upper edge) and recommended Pareto solutions (in the ellipse)

To find the optimal solution(s) we adopt a full decision space survey technique [2]. The length *x(i)* and the outer radius *y(j)* of the coil was varied in the range [5,..., 50 mm], with 5 mm step.
Results are shown in Fig. 2.

6. Conclusions and final remarks

Although the Pareto – poly-optimal solutions comprise larger set, a practical meaning have the solutions in the ellipse (Fig. 3), with the high value of the force.
As a continuation of the poly-optimization efforts, we include the voltage control signal as an important decision function.
The poly-optimal approach has been a powerful tool to examine the problem in a broad context.
Various optimization techniques should be tested, for example the Genetic Algorithms.

Acknowledgements

This work was supported by ZPORR (*Zintegrowany Program Operacyjny Rozwoju Regionalnego)*

References

[1] W. Tarnowski "Symulacja i optymalizacja w Matlab'ie" ,WSM, Gdynia, 2001.
[2] A. Jastriebow, M. Wyciślik „Optymalizacja – teoria, algorytmy i ich realizacja w Matlab'ie", Wydawnictwo Politechniki Świętokrzyskiej, Kielce, 2004.
[3] D.Howe "Magnetic actuators" Sensors and Actuators 81 (2000) p. 268-274.
[4] Comsol "Electromagnetic Module – User's Gide"

Characterization of Fabrication Errors in Structure Geometry for Microtextured Surfaces

D. Duminica (a), G. Ionascu (a), L. Bogatu (a), E. Manea (b), I. Cernica (b)

(a) "POLITEHNICA" University, 313, Splaiul Independentei, Bucharest, 060042, Romania

(b) Institute of Microtechnology, Str. Erou Iancu Nicolae 32 B, Bucharest, 077190, Romania

Abstract

The high accuracy, high resolution and freedom in choice of shapes make the etched silicon wafers an interesting alternative to study the effect of surface material and texture.
Well-defined surface textures were produced by UV photolithography and anisotropic etching of silicon wafers. The patterns included squares placed in a rectangular grid and parallel grooves of different sizes. The characterization of textured surfaces was also performed. The microstructure geometry errors were identified and statistically quantified.

1. Introduction

It is well known that the tribological performance of materials is highly dependent on the surface topography [1,2].
The monocrystalline silicon micromachining allows the obtaining of various shape cavities, using for this purpose the wet chemical erosion, selective, anisotropic and with shape influence by controlled doping of the processed material [3].
Silicon crystallizes in keeping with cubic system and it has the diamond structure. Its elementary cell is an octahedron limited by the family of planes {111}, explaining the angles formed by the tapered walls of the cavities with the planes (100)/(110)/(111) on which the wafers are cut.
Consequently, the anisotropic chemical erosion creates on a wafer surface cut on the plane (100) cavities having the contour of a square or rectangle

whose sidewalls form angles of 54,74° with the plane (100). The depth of these cavities is limited by intersection of the planes (111) which stop the action of the attack substance; that is why the width "*l*" and the depth "*h*" of a pyramidal cavity are dependent between them by the relationship $l/h = \sqrt{2}$.

2. Textures performing

p – (100) silicon wafers of 3 inch diameter and 375 μm thickness were thermally oxidized in wet oxygen atmosphere to obtain a silicon dioxide (SiO_2) layer of about 1 μm thickness, used as a protective layer (mask) during etching process. The oxide layer was patterned with arrays of quadratic openings and microgrooves aligned along the principal flat (<110> directions) of the wafers, using a standard photolithographic technique.
The silicon was then anisotropically etched in potassium hydroxide (KOH) (40 g/100 ml) at 80° C (etch rate of about 1.4 μm/min) to a depth varying with the time: 5 μm ($t_{etch\ 1}$ = 4 min.), 20-22 μm ($t_{etch\ 2}$ = 15 min.) and 50-80 μm ($t_{etch\ 3}$ = 40 min.). The remaining oxide was removed in an HF – solution: first in "Buffered HF" solution (NH_4F - HF) (6:1) at 32 °C (etch rate of about 0.1 μm/min.) and, finally, in DIP solution (HF:H_2ODI) (1:10) at 25 °C.
The patterns include squares placed in a rectangular grid and parallel grooves, as shown in figure 1.
The squares were manufactured to a width of 1.55 mm being disposed at a pitch of 3.1 mm. The grooves have a width of 30 μm and are placed at a pitch of 60 μm. In both cases the placing pitch is twice larger than the structure width.

3. Experimental results and conclusions

The measurements have shown that the size and lateral distribution of the structures at the wafer surface were defined with micrometer precision by the lithographic and etching processes.
The surface roughness measured on wafers was R_z=0.07..0.1 μm between the structures. In cavities R_z=0.3 μm (depth of 5 μm), R_z=0.2 μm (depth of 20..22 μm) and R_z=0.1 μm (depth of 50..80 μm).
Experimental results showed the presence of deviations from the ideal geometry due to processing errors. All the profile parameters were statistically quantified through histograms. Histograms of sample distribution were plotted for all the measured parameters.

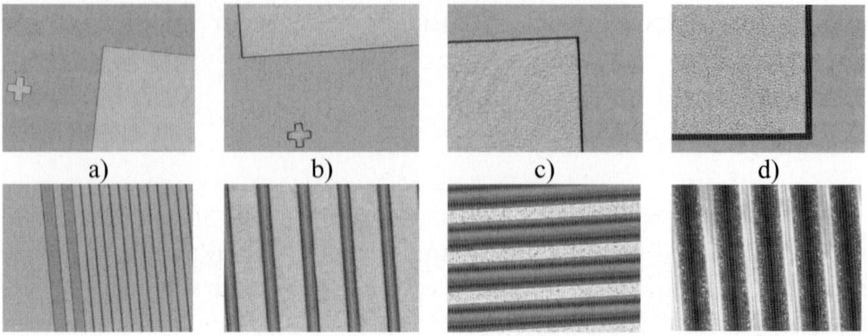

Fig. 1. Images of the Si wafers protected by SiO_2 mask, viewed at an optical microscope: a - quadratic openings, before the etching in KOH solution, b - quadratic openings, after the etching in KOH solution to a depth of 5 μm, c - quadratic openings, after the etching in KOH solution to a depth of 20 μm, d - quadratic openings, after the etching in KOH solution to a depth of 50 μm, e - microgrooves, before the etching in KOH, f - microgrooves, after the etching in KOH to a depth of 5 μm, g - microgrooves, after the etching in KOH to a depth of 20 μm, h - microgrooves, after the etching in KOH to a depth of 50 μm.

Fig. 2a) presents the obtained histograms of pitch “*p*” and cavity width “*l*” in the case of the rectangular grid for an etching time of 15 min, corresponding to a depth “*h*” of 20 μm. Fig. 2b) presents the obtained histograms of pitch “*p*” and cavity width “*l*” in the case of parallel grooves for an etching time of 15 min, corresponding to a depth “*h*” of 20 μm.

The histograms plotted for etching times of 4 min, corresponding to a cavity depth of 5 μm, and 40 min, corresponding to a cavity depth of 50 μm, presented a similar, bell-shaped pattern. A normal (Gaussian) curve was superimposed over the histograms, giving the first information of the type of distribution in the population of the error parameters.

In order to evaluate the normality of the distribution, the Anderson-Darling test for normality was used. The decision-making process was based on the probability value (*p*-value). Experimental results are presented in Table 1.

The results showed that fabrication errors presented a normal distribution centered about 0% in both cases (rectangular grid and grooves). A very good value of the standard deviation - approximately 0.15% (pitch) and 0.22% (width) - was obtained in the case of the rectangular grid. In the case of grooves, the fabrication errors distribute in a wider range of about 4.49% for the pitch and 8.45% for the width, proving less manufacturing precision that in the case of the rectangular grid. The interval [μ-3σ, μ+3σ], containing about 99.73% of the resulted values, was also established.

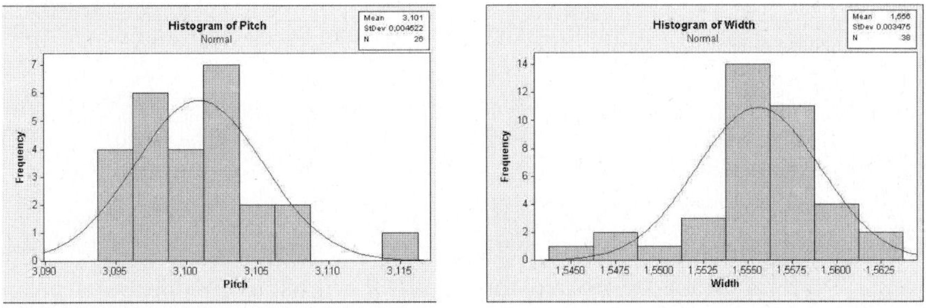

a)

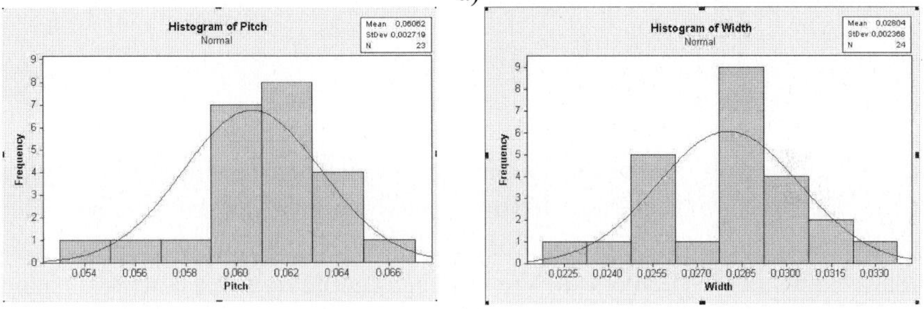

b)

Fig. 2. Histograms obtained in the case of the rectangular grid (a) and of parallel grooves (b) for an etching time of 15 min.

In other words, smaller dimensions result more imprecise than higher ones for the same grade of manufacturing accuracy.

It can be proved that, if the same ratio of dimensions and tolerances is maintained, it is necessary to increase the grade of manufacturing accuracy for smaller dimensions.

Table 1: Experimental results

rectangular grid etched during 15 min; "p"- pitch; "l"- width						
	μ	σ	σ/μ [%]	μ - 3σ	μ + 3σ	*p*-value
"p"	3.101	0.004522	0.15%	3.087434	3.114566	0.11
"l"	1.556	0.003475	0.22%	1.545575	1.566425	0.069
parallel grooves etched during 15 min; "p"- pitch; "l"- width						
	μ	σ	σ/μ [%]	μ - 3σ	μ + 3σ	*p*-value
"p"	0.06062	0.002719	4.49%	0.052463	0.068777	0.074

"1"	0.02804	0.002368	8.45%	0.020936	0.035144	0.074

The accuracy has to be increased with 5,7 grades in order to decrease tolerance interval as many times as nominal dimension (51,67 times), requiring a more exigent technology.

The width of a groove in photoresist layer (l_r) can be computed as:

$$l_r = l_s - \Delta l = l_s - 2e = l_s - 2h(1-A) == l_s\left[1 - \frac{2h}{l_s}(1-A)\right] \quad (1)$$

where: l_s – dimension at the interface with the substrate; Δl - overcorrosion; e – lateral attack; h – corrosion depth; A – anisotropy degree of processing:

$$A = 1 - e/h \quad (2)$$

The overcorrosion compensation has to be considered from the beginning of the mask design: the lines (distances between grooves) have to be wider and the gaps narrower in the photoresist layer. The dimension on the mask must besides be corrected with the dimensional deviation associated to the lithographic process. A supplementary correction of the dimension on the original drawing must consider the dimensional deviation associated to the mask manufacturing process. As integration density increases and the dimension l_s tends to the resolution limit of the lithographic process ($l_s \rightarrow l_{r\,\min}$), the anisotropy degree of the processing has to increase ($A \rightarrow 1$).

A high sensitivity technique, such as atomic force microscopy, will be used for the measurement of the surface roughness, in order to establish more precisely its influence and the influence of the surface textures on friction and wear behavior.

References

[1] A. Alberdi, S. Merino, J. Barriga, A. Aramzabe, Proceedings of the 14th International Colloquium of Tribology, Stuttgart/Ostfildern, Germany (2004), 51.

[2] G. Ionascu, C. Rizescu, L. Bogatu, D. Rizescu, I. Cernica, E. Manea, Acta Technica Napocensis, Series : Applied Mathematics and Mechanics, 49 Vol. III, Cluj-Napoca, Romania (2006), 765.

[3] G. Ionascu, "Technologies of Microtechnics for MEMS" (in Romanian), Cartea Universitara Publishing House, Bucharest, 2004.

Accelerated Fatigue Tests of Lead – free soldered SMT Joints

Z. Drozd (a) , M. Szwech (a) and R. Kisiel (b)

(a) Warsaw University of Technology, Institute of Precision and Biomedical Engineering, Division of Precision and Electronic Product Technology, sw.Andrzeja Boboli 8, 02-525 Warszawa/Poland
(b) Warsaw University of Technology, Institute of Microelectronics and Optoelectronics, Warszawa/Poland

Abstract

After implementation of EU Directive RoHS one of the main goals in development of electronics manufacturing technology, is the improvement of electronic interconnection systems reliability. Investigations of lead – free technology for small producers of electronic equipment (SME's) were performed in Warsaw University of Technology (WUT) in frame of EU GreenRoSE Project.
The methods and achieved results of accelerated thermal and mechanical cycling fatigue tests of PbSn and lead-free SMT solder joints are presented. Two PCB finishes: immersion tin and lead-free HASL made by ELDOS were tested. The samples were assembled and soldered by SEMICON on GreenRoSE pilot line. Thermal cycling tests were performed by WUT in two zone thermal shock chamber. For mechanical tests was used new laboratory stand developed by WUT. The failures occurred during the test were detected by resistance measurements and visual inspection..

1. Introduction

The main accelerating factors by investigation of life time are applications of appropriate range of test temperature and mechanical load of soldered components and PCB in following tests:

- temperature cycling
- thermal shocks
- mechanical cycling
- mechanical vibration test
- drop test
- electrical power cycling
- thermo - mechanical charges

The specimens for reliability testing were soldered with SnAg3Cu0,5 (SAC) alloy. For determining a reference level the specimens soldered with SnPb63 alloy were also tested.

For accelerated comparative reliability investigations of SMT soldered joints, performed in WUT, were applied thermal cycling, and mechanical fatigue tests. Achieved results are presented on Weibull plots.

2. Test methodology

In PW were designed simple and transparent test specimens enabling easy failures detection and obtaining the data for statistical analysis.

On the test PC board are soldered 35 jumpers type 1206 or 42 jumpers type 0805. For statistical analysis one jumper with two joints is considered as one sample. During the tests was verified the joints resistance stability. Resistance measurement circuit on the test board is shown on Fig.1.

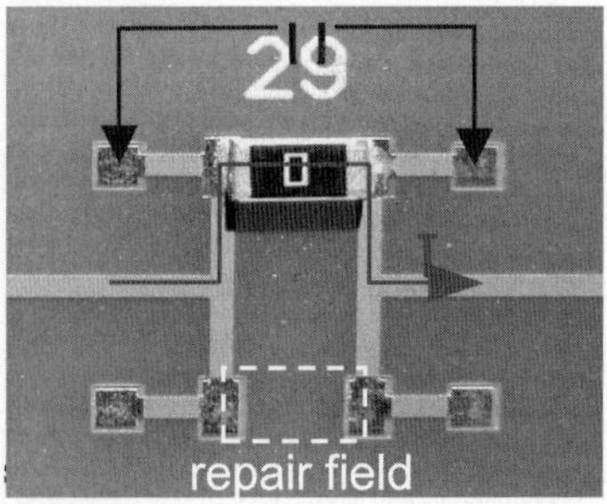

Fig 1 Four-point resistance measurement

The resistance of two joints and one jumper, measured during the test, is very reproducible during succeeding measurements. The only resistance changes were detected only when the soldered joints were damaged.

For resistance measurements of all components in series after breaks in certain joints is provided the shortening of damaged component by hand soldering of supplementary jumpers.

Fig.2. Temperature cycling chamber

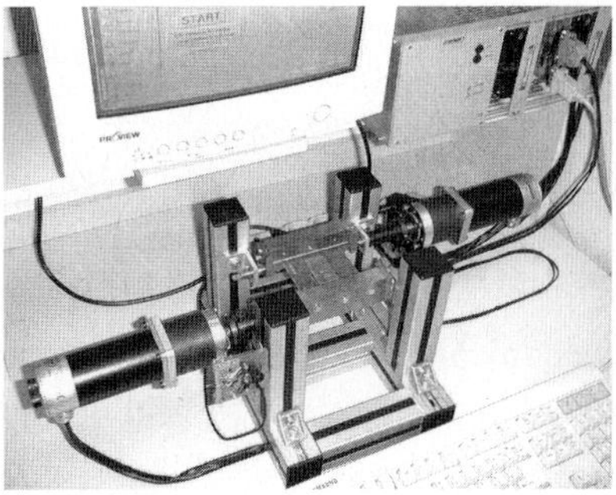

Fig.3. Mechanical fatigue test stand

Thermal cycling test was performed according to IPC 9701A and EN 62137, in dual zone test chamber shown on Fig.2. By appropriate samples configuration and temperature programming of the cold and heat zones can be achieved compatible temperature gradient. Mechanical cycling test was realized on the laboratory stand shown on Fig. 3. The PC controlled stand, developed in WUT, consists of bending system, two servomotors and motor controller.

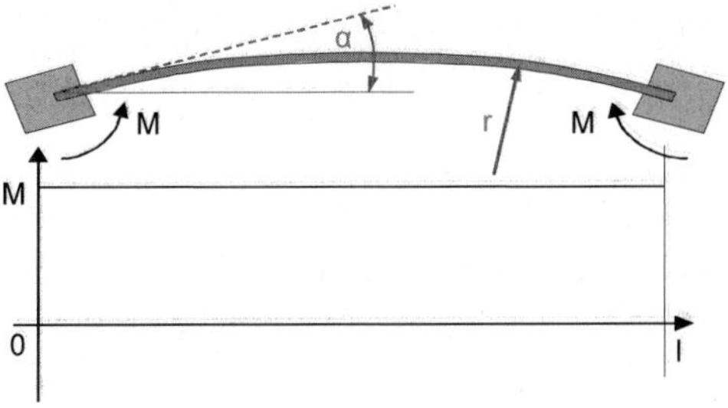

Fig. 4. Scheme of mechanical bend ing test

The scheme of bending system is shown on Fig.4. Because of constant moment M and bending radius r over the length of tested board the charge conditions are uniform for all components soldered on the PC board.

Tha main mechanical characteristics are expressed by following equations:

Bending angle:

$$\alpha = \frac{Ml}{2EJ}$$

where: E – Young module, J- Inertia moment of PCB

Bending moment:

$$M = \frac{EJ}{r}$$

Curvature:

$$c = \frac{1}{r} = \frac{2\alpha}{l}$$

where: l – PCB bending length

Strain:

$$\varepsilon = \frac{h\alpha}{l}$$

where: h – PCB thickness.

During the tests was applied curvature $c_{max} = 4{,}2\ m^{-1}$.

The plot of applied temperature cycle is shown on Fig. 5.

Mechanical test schedule by two-side bending is shown on Fig. 6. The parameters: t_r, t_d, α_{max}, α_{min} and number of cycles N are programmed by PC

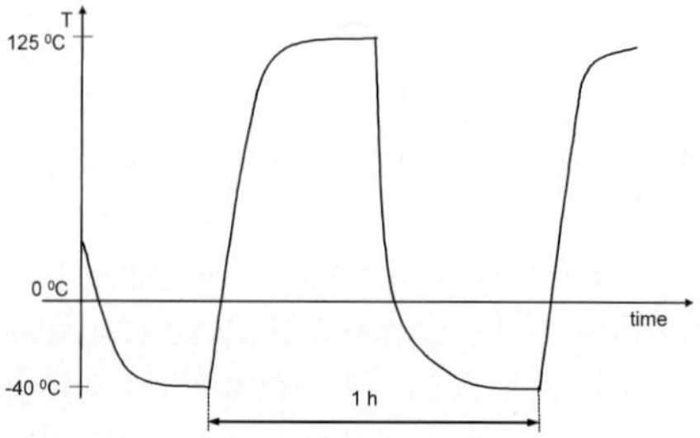

Fig.5. Temperature plot in cycling test

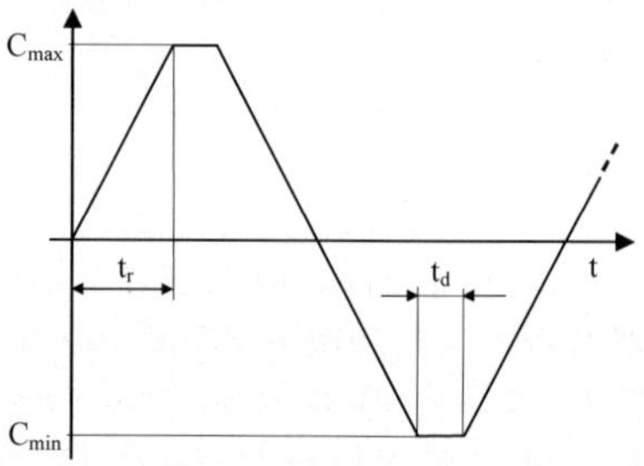

Fig.6. Mechanical tests schedule

Test results

During 4000 thermal cycles, in periods of 200 - 500 hours, was measured and registered the joints resistance value. The resistance changes more than 20mΩ from initial value and visible cracks were registered as failures. The failure probability for fatigue test can be described as Weibull distribution with reliability function R(t):

$$R(t) = e^{-\left(\frac{T-\gamma}{\eta}\right)^{\beta}}$$

where:β, η, γ parameters of Weibull distribution

On the beginning of mechanical cycling the joints resistance was constant, but after certain number of cycles were observed the resistance changes during the bending cycle. The PC bending can be also applied for early failures detection.

Selected measured characteristics of joints resistance R as function of board curvature c are shown of Fig. 7.

The cracks after thermal and mechanical cycling are shown on Fig. 8

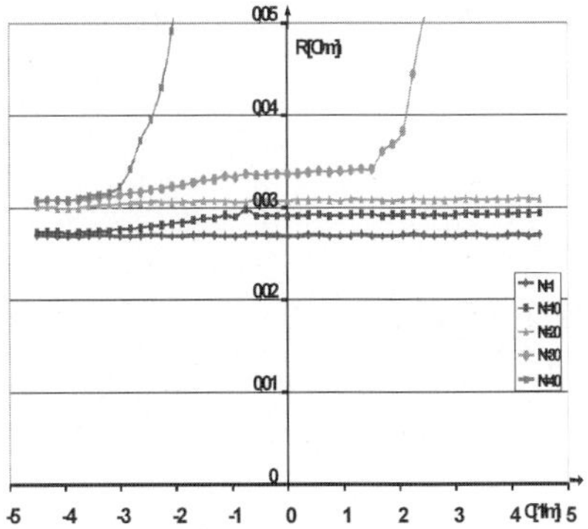

Fig. 7. Selected characteristics R(c)

Fig8. Selected cracks after thermal (top) and mechanical (bottom) cycling

Selected results of thermal and mechanical cycling test are shown as Weibull plots on Fig.9 and 10. It can be seen that general results of thermal and mechanical; tests are comparable.

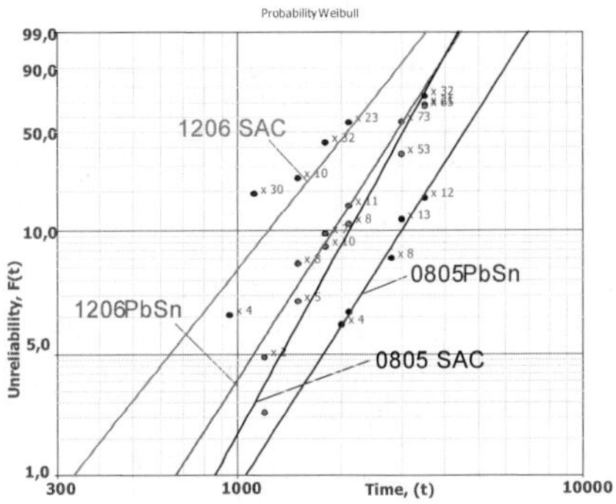

Fig.9 Weibull plots for thermal cycling

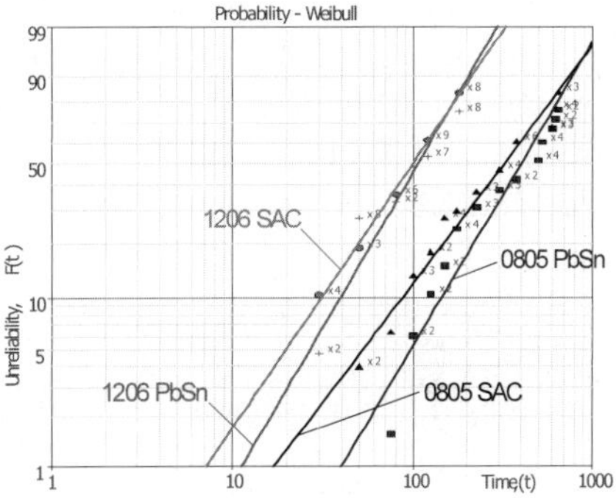

Fig. 10 Weibull plots for mechanical cycling

References

[1] Drozd Z., Szwech M.: Failure Modes and Fatigue Testing Characteristics of SMT Solder Joints. Proc. 1st Electronics Systemintegration Technology Conference ESTC 2006 Dresden 5 – 6.09.2006, pp.1187 – 1193

Early Failure Detection in Fatigue Tests of BGA Packages

R. Wrona (a) *, Z. Drozd (b)

(a) Telecommunications Research Institute, Poligonowa 30,
Warsaw, 04-051, Poland

(b) Warsaw University of Technology, Institute of Precision
and Biomedical Engineering, Division of Precision and Biomedical
Engineering, A. Boboli 8,
Warsaw, Poland

Abstract

The latest results of lead-free BGA solder joints investigations, performed in frame of GreenRoSE Project financed by EC, are presented. In our investigations were used test specimens BGA256r, developed in Warsaw University of Technology (WUT), and TopLine dummy components. Lead - free and lead - containing components were compared. Mechanical tests were performed by cyclic bending on mechanical stand developed in WUT. Thermal shocks are performed according to standards IPC 9701A and EN 62137.
In this paper are presented the results of mechanical tests.

1. Introduction

Resistance stability is one of parameters, deciding about reliability of solder joints. For early failure detection were used standard dummy components with modified daisy chain circuits for obtaining more measurement areas. This allows detection of areas where occur the first damages.
For more precise resistance measurements we apply the four point method. For this method were designed in WUT special BGA256r package and test board. All test boards and substrates for BGA256r were manufactured by ELDOS Company.

The assembly process was realized in WUT, Telecommunications Research Institute and in SEMICON Company on the GreenRoSE Pilot Line. For soldering was used SnAg3Cu0.5 alloy (SAC) and SnPb63 alloy to obtain reference level for lead-free components.

2. Resistance of Daisy Chain Circuits

For investigations were used TopLine BGA dummy packages, type CSP84 (pitch 0.5 mm), BGA100 (pitch 0.8 mm), BGA144 and BGA676 (pitch 1.0 mm), BGA272 (pitch 1.27 mm).
For obtaining more measurements points, original daisy chain were modified (Fig. 1).

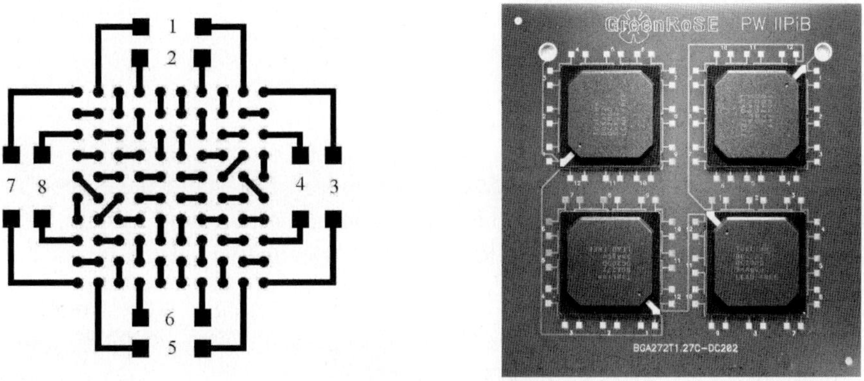

Fig.1. Modified daisy chain (left) and assembled test board with BGA272.

On each test board were assembled 4 components rotated 90° in relation to each other. This allows detection damage dependent on component position on the PCB.

3. Resistance of BGA Contacts

For precise reliability assessment and earlier failure detection was developed BGA256r package and PCB test board. For resistance measurement in this specimen the four-point method was applied. This method allows precise resistance measurements of single BGA contacts (Fig.3)

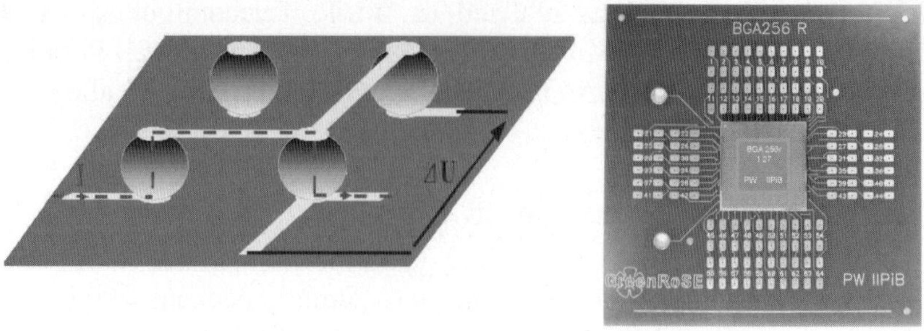

Fig. 3. Four-point scheme (left) and assembled specimen (right).

4. Mechanical Tests

4.1. Test stand

Mechanical tests by bending were performed on test stand developed in WUT (Fig. 4). The PC controlled stand consists of bending system, two servomotors and motor controller. The test specimen is fixed in special jaws and bent in two directions.

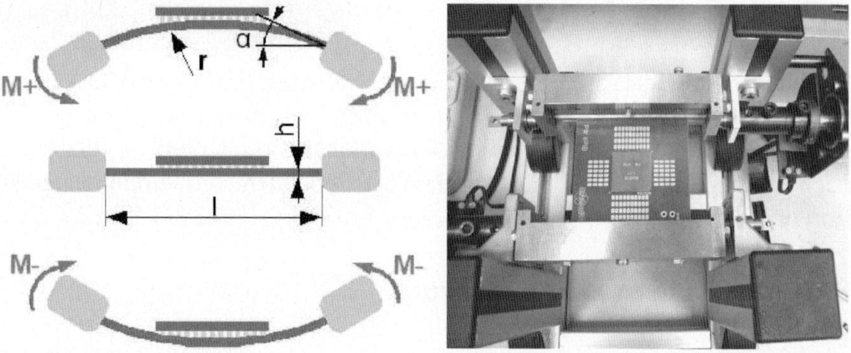

Fig. 4. Test stand scheme (left) and realized test stand (right).

4.2. Results of BGA256r

During mechanical tests each sample was bent by 200 cycles. One cycle time is 30 second. For BGA256r was applied the board curvature amplitude c=3.2 m^{-1} . After each 20 cycles the resistances of BGA solder joints were measured. Resistance distribution before the test is shown on Fig. 5. Weibull plot of failure probability are shown on Fig. 6.

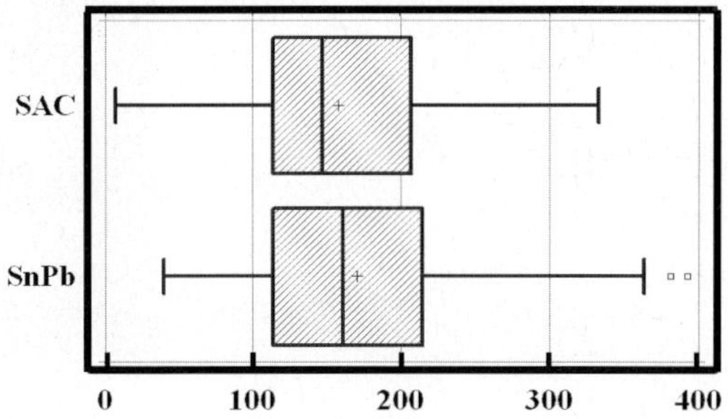

Fig. 5. Joint resistance (mΩ) before the tests.

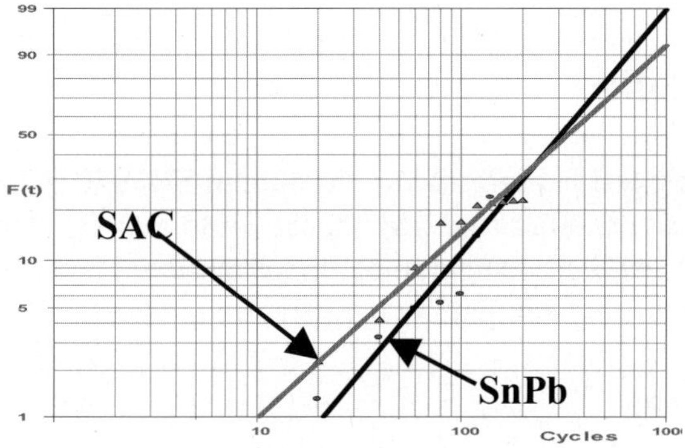

Fig. 6. Weibull plot of failure probability for BGA256r mechanical test.

4.3. Results of Daisy Chain Packages

During the tests the bending curvature 3.5 m^{-1} was applied. Other parameters were the same as by BGA256r.
Mechanical test for daisy chain test boards showed, that reliability of lead-free solder joints is slightly lower than lead-containing (Fig. 7).
The damage mechanism for both, lead-free and lead-containing solder joints are similar. First failures occur in solder joints on the components edge perpendicular to bending direction.

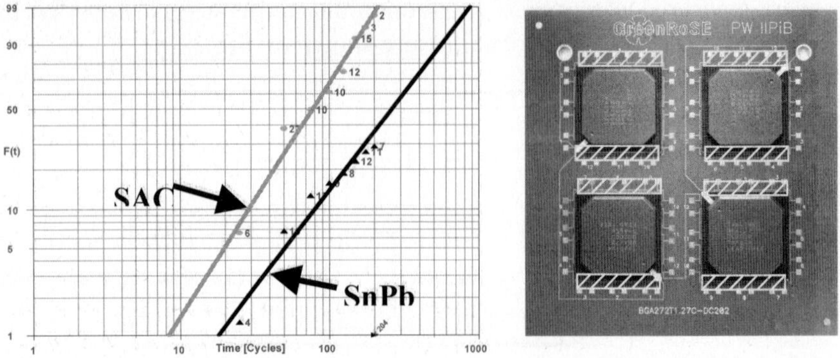

Fig. 7. Weibull plot of failure probability for daisy chain packages (left) and edges where failures occurs first (marked).

5. Thermal tests

Thermal cycling test is performed according to IPC 9701A and EN 62137, in dual zone test chamber (thermal shocks air to air). By appropriate samples configuration and temperature programming of the cold and heat zones can be achieved compatible temperature gradient.

6. Conclusion

BGA contact resistance may be good parameter for early failures detection, however further studies are necessary.

Reliability of lead-free solder joints is slightly lower than of lead-containing solder joints. Obtained results are satisfactory for consumer goods, but further investigations for reliable products are necessary.

In the future are planned mechanical deflection tests by twisting of the test specimens.

References

[1] Wrona R., Drozd Z., Szwech M.:Resistance measurement of BGA contacts during reliability tests. Nano Technologies for Electronics Packaging. 29th International Spring Seminar on Electronics Technology – IEEE Conference Proceedings, pp. 186-189.

[2] K.J. Puttlitz (Editor): Handbook of Lead-free Solder Technology for Microelectronic Assemblies. M. Detter Inc. N.Y. 2004. 1026 pages

Design and Fabrication of Tools for Microcutting Processes

L. Kudła

Warsaw University of Technology
Institute for Precision and Biomedical Engineering
ul. św. Andrzeja Boboli 8, 02-525 Warsaw, Poland

Abstract

Microcutting is one of the leading technologies for machining of precise components or patterned surfaces. For the execution of different microcutting techniques various tools are necessary. They are miniaturized versions of the tools used in a conventional range of dimensions or specially developed tools for the cutting in the micro scale. A design, materials and fabrication processes of such tools are diverse, but some common and specific problems become noticeable.

1. Introduction

Precise microstructures or patterned surfaces are key functional components of many micro-devices and micro-systems. Because the structures get more complex and smaller at the same time, a continuous progress in their machining is indispensable. One of the most versatile and effective processes is microcutting, consists of a mechanical material removal from a workpart, using tools with determined edge geometry. The microcutting means a very small (<10 μm) depth and thickness of cut (small value of feed) and therefore, minimal volume of removing material [1]. The realization of different microcutting operations needs a diversity of smallest tools. Design and fabrication of these tools is very important issue for the application of microcutting processes. The specific problems of technology of microcutting tools are a high requirements concerning the quality of cutting edges and next assurance of indispensable strength and wear resistance of the tool cutting part.

2. Microcutting techniques

In general, the microcutting techniques are similar to the typical cutting operations – Table. 1. Only a few of them are new technologies. They are fly-cutting, microgrooving and machining on atomic force microscopes (AFM machining) [2, 3]. The last technique is also denominated as nano-machining. The main optional realizations of kinetics are placed in the table in brackets. In some cases the rectilinear or X-Y motions are also used as the feed or infeed.

Table 1. Basic kinetics of microcutting techniques

Technique	TURNING		MILLING	FLY-CUTTING	DRILLING
Workpart	↷	⇦	⇦⇧	⇦	[↷ ⇩]
Tool	⇦⇧	↻ ⇩	↷ ⇩	↻ ⇩	↷ ⇩

Technique	REAMING	TAPPING	GROOVING		AFM MACHI-NING
Workpart	[↷ ⇩]	[↷ ⇩]	⇧	↷	⇦⇧
Tool	↷ ⇩	↷ ⇩	⇦	⇧	⇩

↷	↻	⇦ ⇧	⇦⇧
Rotation	Circulation	Rectilinear motion	X-Y motion

The tools for the microcutting have different size of dimensions. The geometry of the tool cutting part and the motions determined a form of the machined surface. By combining of more motions, the shapes of almost unlimited complexity can be produced. Well known advantage of microcutting technology is the possibility to machine 3D microstructures characterized by high aspect ratio. During the microcutting processes, a cutting force is highly concentrated on the fragile tool or on the workpiece. The relatively large elastic deformations of the miniaturized cutting tool and of the whole machining system could be generated. Therefore, the machining forces influence machining accuracy and limit machinable size of surfaces. Moreover, a very careful process execution is necessary to protect the cutting part of the tool against catastrophic destruction. Two main paths for design and fabrication of the tools for microcutting are possible. The first is downscaling of conventional tool forms and manufacturing processes. The next utilizes newly developed technologies.

3. Cutting tools technology

The microtools (turning cutters, drills, mills, reamers, tapping tools, hobbing cutters, tips etc.) are manufactured of high speed steels (HSS, HSS-E) like Cr4W6Mo5V2 or Cr4W6Mo5Co5V2, of sintered micrograin tungsten carbides (HM) like K10 (~WC95%Co5%) or K20/K30. Predominant tool materials are sintered tungsten carbides and monocrystalline diamond, but all the time persist the researches of new materials and coatings. The tools for microcutting have various design and variety of machining methods are necessary for their fabrication. The basics are fine abrasive techniques like grinding, lapping and polishing. In some fabrication processes micro-electrical-discharge machining (μEDM), laser beam machining (LBM) or focused ion beam machining (FIBM) have been investigated [4, 5]. Down-scaling of conventional tools is not unlimited. Therefore, the microtools have often simplified shapes and geometry differences of smallest drills, end-mills or reamers become slighter [6]. On the other hand a progress in manufacturing technology enables also very fine geometry executions. An example may be the modifications of cutting part in microdrills – Fig. 1.

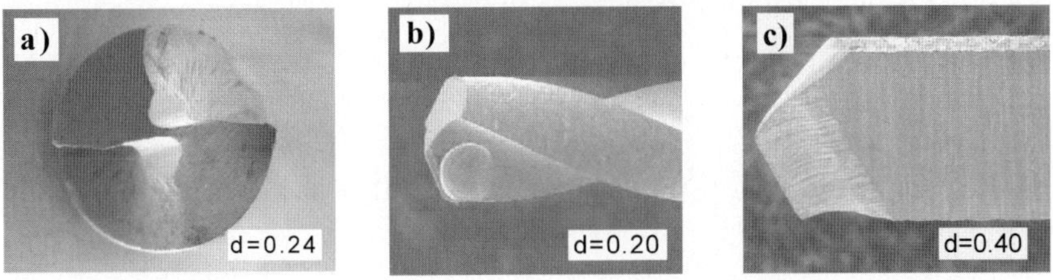

Fig. 1. Modifications of microdrills geometry: a) twist drill with undercut chisel edge, b) twist drill with shell-like cavity on rake surface, c) D-shape drill with groove parallel to major cutting edge [5]

For the cutting with a smallest thickness of cut, the radius of the tool edge and the waviness of the cutting wedge should be reduced to the values ranging of few microns or even considerably below. A maximal waviness W_{max} of the cutting edge is expressed by the equation

$$W_{\max} = \frac{1}{\sin\beta}\sqrt{R_{m\gamma}{}^{2} + R_{m\alpha}{}^{2} + 2R_{m\gamma}R_{m\alpha}\cos\beta} \tag{1}$$

where $R_{m\gamma}$ and $R_{m\alpha}$ represent the maximal roughness on rake and clearance surfaces and β is tool wedge angle. Self-evident conclusion is that the rake and clearance surfaces should be extremely smooth.

4. Non-conventional microcutting tools

The miniaturization of tools for microcutting creates an opportunity for the application of a very special design and fabrication processes. An example could be the fabrication of a particular micromilling (or micro-filing) tool [7]. The tool was made of Al disk with electroless Ni layer – Fig. 2. On the running track of the disk the microstructures have been shaped, using fly-cutting technique and diamond cutter. The microstructures were linear or prismatic with height of 25 μm and spacing of 177 μm. They have following geometry parameters: rake angle γ=–45° (negative), clearance angle α=10°, tilt angle λ=0°; 30°; 45°. After shaping, the microstructures were coated with a 1 μm diamond-like-carbon (DLC) layer. Another and very original solution for manufacturing of a thin milling cutter has been proposed in the work [8]. The whole disk of the cutter was produced from a gaseous phase by a chemical vapor deposition (CVD) of the diamond-like-carbon (DLC) layer, next separated from the template former and fixed by gluing with the shank – Fig. 3. The tool had 6 teeth with the clearance angle of α=13° and the rake angle of γ=8°. The machining tests showed a very good wear resistance of such a micromill.

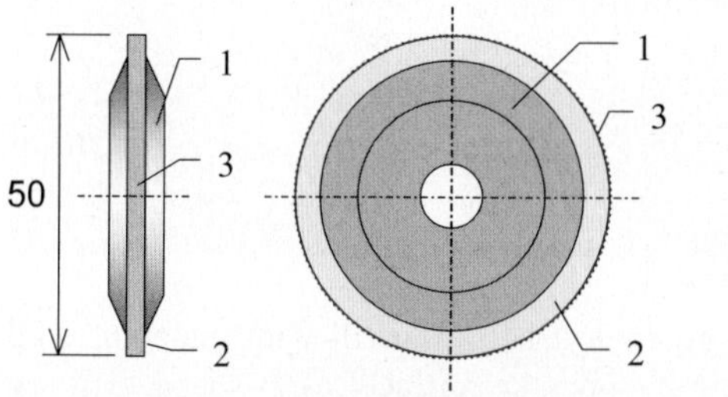

Fig. 2. Special disk tool for micromilling of flat surfaces: 1) Al-disk , 2) Ni-layer, 3) running track [7]

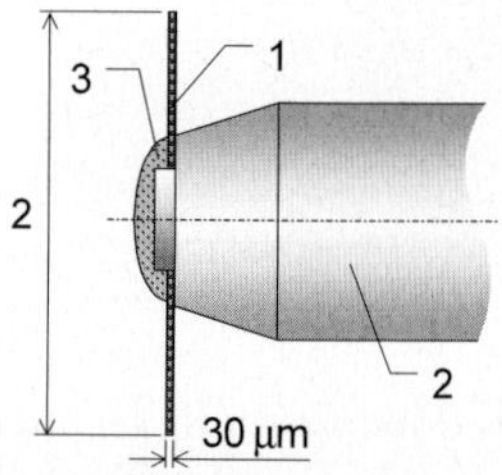

Fig. 3. Disk mill made of DLC layer by CVD method: 1) DLC disc, 2) steel shank, 3) glue [8]

Many tools for the microcutting are produced from the monocrystalline diamond. The geometry of the cutting part is formed by using of precise abrasive techniques. It is technology of relatively low effectiveness. An alternative may be the application of a single grain cutting tools, utilizing natural geometry of the diamond crystal habits. The diamond grains have very sharp corners and edges. Moreover, the grain tips have different values of the corner angle and may be individually chosen for a prospective application. The tool preparation procedure begins with a grain selection.

The grain having a shape of regular octahedral or cubic crystal (or its piece) is separated from the other grains and placed in a special instrument with grip. Next the grain is located in the holder, positioned and installed in the tool case – Fig. 4. The diamond tools with natural geometry grains have been successfully tested in microgrooving.

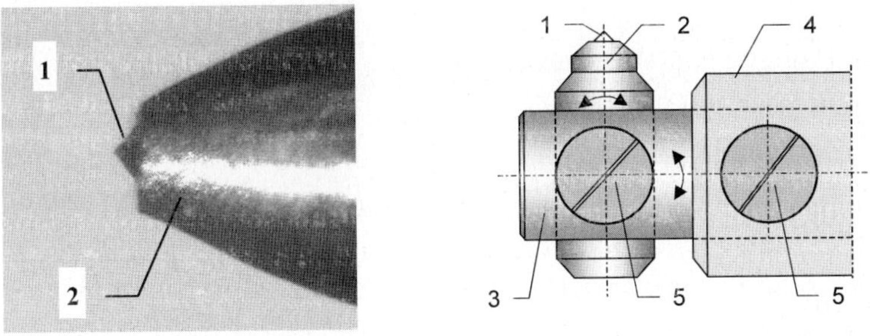

Fig. 4. Diamond turning tool with cutting edge having natural geometry of the crystal – grain fixed in holder and design of tool set (1 - diamond crystal, 2 - holder, 3 - adjustable arm, 4 - tool case, 5 - screws)

Presented selected problems and examples shown only general trends in technology of the tools for the realization of microcutting processes. The continuous progress in manufacturing techniques opened a challenge as for development of new tools as for new applications of microcutting.

References

[1] Masuzawa T., CIRP Annals, Vol.49/2/2000, 473.
[2] Kawai T., Ebihara K., Takeuchi Y., Proceedings of the 5th euspen* Int. Conf., 2005, Montpellier, France, Vol.2: 607.
[3] Ashida K., Morita N., Yoshida Y., Proceedings of the 1st euspen* Int. Conf., 1999, Bremen, Germany, 376.
[4] Picard Y. N. et all, Precision Engineering, Vol. 27(2003), 59.
[5] Kudła L., Proceedings of the 6th euspen* Int. Conf., 2006, Baden/Vienna, Austria, Vol. II: 160.
[6] Bissacco G., Surface Generation and Optimization in Micromilling. Ph. D. Thesis, 2004, Technical University of Denmark.
[7] Brinksmeier E. et all, Proceedings of the 4th euspen* Int. Conf., 2004, Glasgow, Scotland, 199.
[8] Wulfsberg J. P., Brudek G., Lehman J., Proceedings of the 4th euspen* Int. Conf., 2004, Glasgow, Scotland, 131.
* European Society for Precision Engineering and Nanotechnology

Ultra capacitors – new source of power.

Mirosław Miecielica (a), Marcin Demianiuk (b)

(a) Politechnika Warszawska, IIPiB
ul św. Andrzeja Boboli 8, 02-525 Warszawa, Poland

(b) OEM Automatic Sp z o.o.,
ul Postępu 2, 02-676 Warszawa, Poland

Abstract

Ultra capacitors & Super capacitors are emerging technology that promises to play an important role in meeting the demands of electronic devices and systems. Some people view it as the next-generation battery. Others view it as an independent energy source capable of powering everything from power tools to power trains. In this article we want present internal structure those components, advantages compared with batteries and conventional capacitors and the most interesting applications ultra capacitors in industry applications.

1. Introduction

Ultra Capacitors & Super Capacitors are emerging technology that promises to play an important role in meeting the demands of electronic devices and systems. Some people view it as the next-generation battery. Others view it as an independent energy source capable of powering everything from power tools to power trains. This kind of capacitors allowed to collect from few Farads to 2700 Farads in small volume (Fig.1.). Capacity 2700 F means that we can take 1A during 45 minutes, and voltage will fall down only 1 V. First association - this components work like a battery. However there are few meaningful differences. For example, we can charge these components during few seconds like standard capacitors.

Ultra capacitors offer a number of key advantages compared with batteries and conventional capacitors. Ultra capacitors deliver 100 times the energy of conventional capacitors and 10 times the power of traditional batteries. The duty cycles are up to one million recharge cycles, even in extreme environments. This technology reduces maintenance costs and adds value to other power sources. This is a non-toxic, environmental friendly solution and alternative to batteries.

Fig. 1 Ultra capacitors family from Maxwell [1].

2. Super capacitors technology.

In terms of energy density and access time to the stored energy, double-layer capacitors are placed between large aluminum electrolytic capacitors and smaller rechargeable batteries. The diagram (Fig.2) shows the domain occupied by double-layer capacitors in the power and energy densities space.

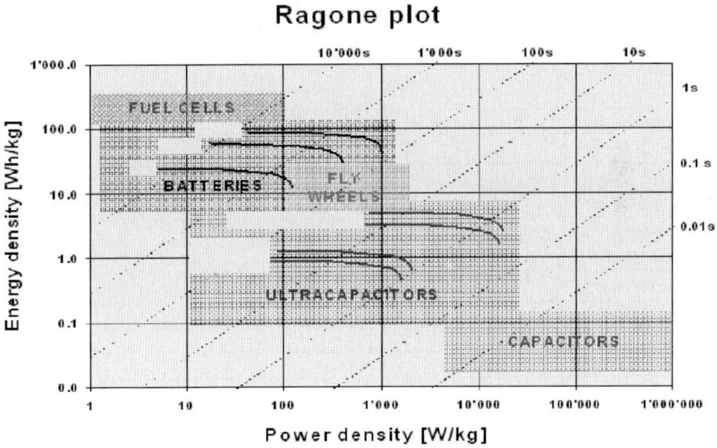

Figure 2. Ragone diagram, comparison of different energy storage and conversion devices [2]

Super capacitors consist of two activated carbon electrodes, which are immersed into an electrolyte (Fig.3). The two electrodes are separated by a membrane which allows the mobility of the charged ions but forbids the electronic contact. The organic electrolyte supplies and conducts the ions from an electrode to the other if an electrical charge is applied to the electrodes. In the charged state, anions and cations are located close to the electrodes so that they balance the excess charge in the activated carbon. Thus across the boundary between carbon and electrolyte two charged layers of opposed polarity are formed.

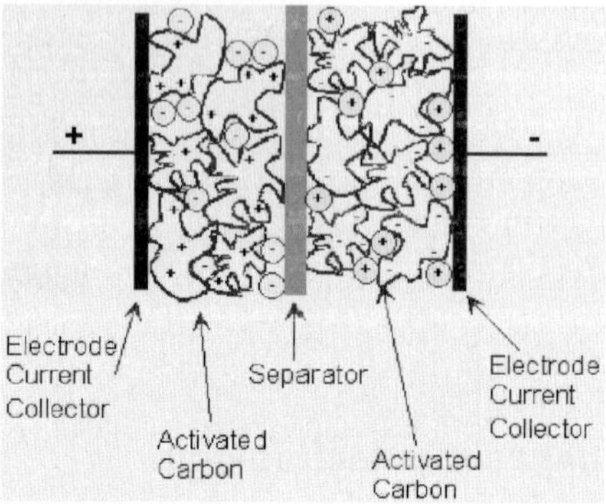

Figure 3. Electrochemical double - layer capacitor [2].

3. Super capacitors parameters.

The table present technical parameters one of the most popular model ultra capacitor – MC 2600 (2600F, 2,7V) from Maxwell. We can find the most important parameters like: capacity, voltage, short circuit current or internal resistance (Fig.4)

	Product Specification			
	MC2600	MC2600-W	Tolerance	Standard
Interconnects	Threaded Terminal	Weldable Post		(2X) M12 x 1.75 -6h x 14mm long
Capacitance, C_R [F]	2,600		+20%	
Voltage, U_R	2.7			
Internal resistance, DC [mohm]	0.40		Max.	Discharging at Constant Current (25°C)
Internal resistance, 100 Hz [mohm]	0.28		Max.	
Thermal Resistance, R_{th} [°C/W]	3			$\Delta T = D R_{th} I_c^2 R_d$ See Page 3 for Current Chart
Short circuit current, I_{sc} [A]	5,000			Caution, current possible with short circuit from U_R
Leakage current [mA]	5		Max.	72 hrs, 25°C
Operating temp. range [°C]	-40 to 65			
Storage temp. range [°C]	-40 to 70			
Endurance, Capacitance [F]	< 20% decrease from initial			1500 hrs @ U_R and 65°C
Endurance, Resistance [mohm]	< 60% increase from initial			
Maximum Energy, E_{max} [Whr/kg]	5.6			Full discharge from U_R
Peak Power Density [W/kg]	10,400			Matched load
Power, P_d [W/kg]	4,100			See additional technical information
Life Time	$\Delta C/C_R$ < 30%, ESR < 2.5 x increase			from initial spec after 10 years @U_R and 25°C
Cycle Lifetime	$\Delta C/C_R$ < 20% decrease, ESR < 2x increase			from initial spec after 1M cycles (U_R to 1/2 U_R) @ 25°C (I = 100A)

Figure 4. Technical specifications.

4. Applications.

Ultra capacitors are making a difference or better performance in a lot of areas, like automotive, industrial, traction and consumer electronic. Applications for ultra capacitors including technologies that require:

- burst power that can be charged in seconds and then discharged over a few minutes,
- short-term support for un-interruptible power systems,
- load-levelling for power-poor energy source such as a solar array,
- low-current, long duration power supply.

Transportation engineering

One of the main application of ultra capacitors are transportation area. The endless cycles of acceleration and braking of vehicles, buses, trains, cars and metro systems are ideal for this kind of technology. In those applications ultra capacitors are used for capturing regenerative breaking energy and reusable that energy to acceleration or supply of supplemental electrical systems. This kind of systems can be install on-vehicle or stationary designs (Fig 5a,5b).

Figure 5. Transportation systems: a) stationary, b)on-vehicle

In the automotive industry, due to the increasing power demand in future vehicles for comfort improvement, as well as ongoing public and governmental pressures for more environmentally friendly and fuel efficient means of transportation, automotive manufacturers are developing new vehicle subsystems and full hybrid systems. Super capacitors are ideal solution to supply additional energy for electric power steering, electromagnetic valve control, electromechanical braking, electric door opening or hybrid drive systems. The storage of braking energy can also be usefully applied for vehicles with internal combustion engines, especially for the improved alternators used as braking generators.

Industrial engineering

Ultra capacitor based energy storage and peak power solutions are key for countless industrial applications, where they store, bridge, deliver, ensure and smooth power and energy needs. They reliably bridge power in uninterruptible back up power and telecom network systems, assure around-the-clock power availability for wind turbine pitch systems, deliver peak power for drive systems and actuators, ensure peak shaving and graceful power-down of robotic systems, augment the primary energy source for portable devices such as power tools, smooth energy throughput from renewable energy storage sources like solar applications, and efficiently enable high power pulse forming in power generators. When appropriately applied, ultra capacitors represent an outstanding design option for advanced power systems design. Ultra capacitors are cost-effective, perform well, are very reliable and are first choice in terms of energy storage technologies in many electrical and electronic systems in the industrial domain.

Uninterruptible power supply (UPS)

Mission critical systems require high reliability and high availability bridge power backup for seconds to minutes. Ultra capacitors can provide high reliability, minimal to no maintenance, and highly available power backup to enable orderly shutdown or bridge to an alternative power source such as a generator, micro-turbine or fuel cell. Whether you are looking at portable wireless devices, low-earth orbiting communication satellites, cell towers or distributed, off-the-grid, high-quality electric power for commercial and residential building applications; premium power sources is key as it offers important benefits to a companies bottom-line.

5. Summary

In conclusion, ultra capacitors play a large part in revolutionizing transportation, automotive, UPS and also another domain industry. In this kid of industry which increasingly requires power technologies that respond to changing consumer demands for environmentally sensitive, high-performance and low-cost super capacitors are the best solution.

References

[1] Maxwell Technologies, Inc. – www.maxwell.com
[2] A.Schneuwly, G. Sartorelli, J. Auer, B. Maher: Maxwell Technologies, Inc. - Ultracapacitor Applications in the Power Electronic World 2006.

Implementation of RoHS Technology in Electronic Industry

R.Kisiel (a), K.Bukat (b), Z. Drozd (c), M.Szwech (c), P. Syryczyk (d) and A. Girulska (e)

(a) Warsaw University of Technology, Institute of Microelectronics and Optoelectronics, ul. Koszykowa 75, 00-662 Warszawa/Poland,
(b) Tele and Radio Institute, Warszawa/Poland
(c) Warsaw University of Technology, Institute of Precision and Biomedical Engineering, Warszawa/Poland
(d) Semicon Sp. Z o.o., Warszawa/Poland
(e) ELDOS, Wroclaw/Poland

Abstract

The goal of RoHS directive is to restrict the use of lead, cadmium, mercury, hexavalent chromium and two halide-containing flame retardants (PBB and PBDE). Engineers involved in design and manufacturing process are obliged to implement only RoHS compatible components and assembly process. There are two aims of the work. First, analyze of assembly process with applying halogen free laminate and RoHS compatible components in multizone reflow oven. On the base of performed experiments the oven parameters which give the proper SMT joints were selected. Second, the test samples for reliability investigation were done. The influence of RoHS compatible solders, component terminal finishes as well as PCB pads finishes on reliability of SMT joint were investigated and analyzed.

1. Introduction

On 2003, 13 February, European Commission had developed and imposed new regulations: WEEE(2002/96/EC and RoHS(2002/95/EC) to the electronics industry for the environmental protection. The main reasons for implementing new regulations were to reduce the excess hazardous materials which enter the landfill areas and reduce the influence of electronics waste to the environment as well as to increase the materials recycling ratio. The Waste of Electrical and Electronic Equipment (WEEE) directive requires manufacturers to reduce the disposal waste of electrical and electronic products by reuse, recycling and other forms of recovery. The manufactures had to be responsible to recycling of electronic waste from August 13, 2005. The Restriction of the Use of Certain Hazardous Substances in Electrical and Electronic Equipment (RoHS) directive restricts the use of six substances specifically within all electrical and electronic equipment traded in the EU member states. These substances are listed in Table 1. According to the RoHS directive, companies that are not in compliance will be unable to trade their products in member states of EU. Recognozing this global economy and the impact on the environment, other countries ought to implement EU directives too.

Tab.1 Restricted substances under the RoHS directive [1]

Substances	Symbol	Limits [ppm]
Lead	Pb	1000
Mercury	Hg	1000
Cadmium	Cd	100
Hexavalent Chromium	Cr (VI)	1000
Polybrominated Biphenyls	PBBs	1000
Polybrominated Diphenyl Ethers	PBDEs	1000

The implementation of lead-free technology to production status is still unsolved problem [2]. The assembly industry is still looking for answer which soldering materials, component termination metallurgies and printed circuit board finishes are optimal. The use of Pb-free materials and technology has prompted new reliability problems, as the result of different alloy metallurgies and higher assembly process temperatures relative to SnPb technology. The selection of RoHS compliant alternatives should be based on compatibility between materials used as component terminations, PCB pads finishes and flux-solder systems.

Investigations of Pb-free technology for small producers of electronic equipment (SME's) were made in Warsaw University of Technology (WUT) in frame of EU GreenRoSE Project. The results of performed experiments are presented in this paper.

2. Design and assembly with RoHS

Engineers and designers have to take extra precautions when they are involved either in manufacturing components or final systems. They are oblidged to implement only RoHS compatible components and assembly technology. RoHS no compliant materials and parts should be replaced with RoHS compliant alternatives that are selected based on availability, manufacturability, reliability, and cost consideration. System designers need to maintain proper documentation for four years after the product has been released to market to prove RoHS compliance all the way down to the subcomponent level.

2.1 RoHS compliant components and material selections

The main considerations for Pb-free component selection include terminal finish, moisture and thermal sensitivity and material and process compatibility. At present, pure Sn is the most widely adopted finish materials for leadframe, followed by Ni-Pd-Au plating. For array components SnAgCu solder ball metallurgy were widely adopted. For connectors, SnCu and SnAgCu finishes are employed as replacement of SnPb contact finishes.
Considerations for Pb-free board design include PCB pad finish and laminate material selection. The primary Pb-free alternatives to SnPb HASL are immersion Sn, immersion Ag, electroless Ni/immersion Au, SnCu HAL or organic solderability preservative (OSP). PCB finish selection must be based upon the finish wetting characteristics with lead-free solders, shelf life, pad planarity and cost. PCB laminate material must withstand multiple reflows and rework at the appropriate Pb-free processing temperature without thermo-mechanical damage. The laminate should be free from polybrominated biphenols (PBB) and polybrominated diphenyl ethers (PBDE). These halide-containing flame retardants are prohibited by RoHS legislation. The research institutes from many countries recommended Sn-Ag-Cu eutectic as the most promising Pb-free solder. Sn(3.0-3.9)Ag(0.5-0.7)Cu appears to be the leading choice adopted by industry both for reflow and wave soldering. Sn0.7Cu or Sn0.7CuNi is the low cost alternatives for wave soldering recommended by some researches.

2.2 RoHS compliant soldering process

During Pb-free soldering process the components are exposed to higher temperatures compared to SnPb reflow. Thermal sensitivity of components and PCBs creates the need for precise control of reflow temperature profile. Soldering process must be done in multizone convection reflow oven with temperature range 300-350°C in reflow zones. Such reflow zones are capable of heating the boards and components to the temperature range (245-255°C) required for lead-free soldering of most products. Finding the proper multizone reflow oven set up to perform proper soldering process is the big challenge for small companies.

3. RoHS compliant assembly of test boards

The halogen free test boards with immersion Sn and SnCu HAL as well as 0805 and 1206 components with Sn terminal finish were used for investigations. The SAC305 Pb-free solder paste and Sn37Pb solder paste (as a reference) were used for reflow soldering. The reflow soldering was done using multizone convection oven ERSA HOTFLOW 2/14 (7 top and 7 down heating zones and 2 top +2 down cooling zones). For selecting the optimum reflow temperature profile the Taguchi methodology was used [4]. During performed series of experiments the soldering temperatures were change in the range 235°C÷281°C and soldering time from 38 to 92s. The SMT joint resistance and shear force of components were measured after each test run. The methodology for SMT resistance and shear force measurements were described in [5].The results of experiments were shown in Fig.1. On the base of performed experiments the proper reflow oven set up for 0805 and 1206 components was establish (T_{max}=255°C, t=47 s) and series of test boards for reliability tests were manufactured.

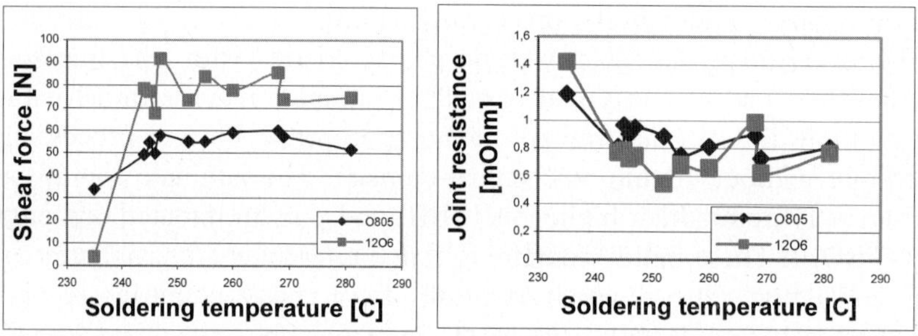

Fig.1 Influence of soldering tempetarure on shear force (left) and SMT joint resistance (right)

4. Reliability testing of SMT joints

For reliability testing the thermal cycling and mechanical fatigue tests were applied. During performed reliability tests the solder joint resistance was measured after established time intervals. The resistance changes of soldered jumpers more than 20 m□ from initial value or visible cracks were registered as failures. Thermal cycling test was performed in dual zone test chamber (-40°C and +125°C). The comparison of the Weilbull plots for solder joints created by SAC and SnPb alloys on test samples with Sn and SnCu HAL finishes are shown in Fig.2. There is significant difference in reliability of SAC and SnPb solder joints with Sn finish. The SAC alloy is

worse. For SnCu HAL finishes the SAC and SnPb alloys had comparable reliability.
Mechanical cycling test was realized on the laboratory stand developed in WUT [4]. The comparison of the Weibull plots for solder joints created by SAC and SnPb alloys on test samples with Sn and SnCu HAL finishes are shown in Fig.4. The SAC joints had slightly lower reliability than SnPb joints with Sn finish.

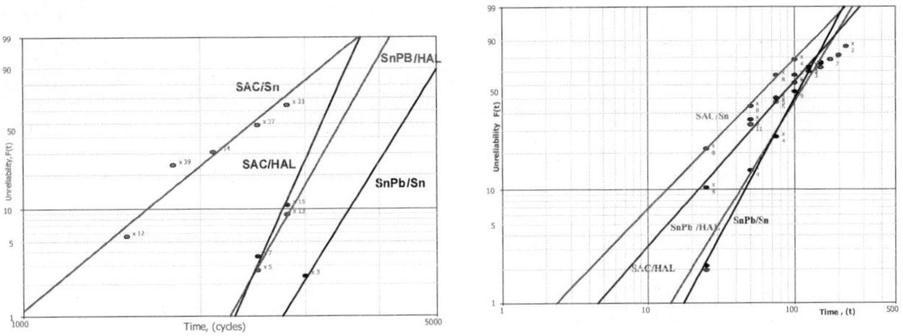

Fig.2 Weibull plots for solder joints created by SAC and SnPb alloys: after thermal cycling test (left) and mechanical cycling test (right)

5. Conclusions

In the paper there were presented the problems of implementation the RoHS directive in SMEs. The Pb-free reflow soldering process was applied for assembly halogens free PCB with RoHS compliant components. The test samples were reliability tested in thermal cycling and mechanical fatigue tests. The obtained results showed that the component size as well as PCB pad finishes had the influence on solder joint reliability. It was found that in some applications the reliability of SnPb solder joints are better than SAC solder joints.

References

[1] Eveloy V., Ganesan S., Fukuda Y., Wu J., Pecht M.G. IEEE Trans. on Components and Packaging Technologies, vol. 28, no 4, 2005, p.884-893

[2] Drozd Z., Szwech M. Proc of ESTC 2006, Dresden, Germany, p. 1187-1193

[3] Kisiel R., Syryczyk P. Proc. of 4th EMPS, 22-24 May 2006, Terme Catez, Slovenia, p.295-299

[4] Kisiel R., Gasior W., Moser Z., Pstrus J., Bukat K., Sitek J.: Journal of Phase Equilibria and Diffusion, vol.25, No 2, 2004, p.122-124

Simulation of Unilateral Constraint in MBS software SimMechanics

R.Grepl

Institute of Thermomechanics, Academy of Sciences of the CR, branch Brno, Czech Republic, email: grepl@fme.vutbr.cz

Abstract

This paper deals with implementation of simple unilateral constraint model in SimMechanics environment. The point – line segment contact is modelled using linear viscous – elastic model, the tangential component is considered with Coulomb friction. The block has been successfully tested on the case of circuit breaker problem.

1. Introduction

Unilateral constraints are relatively difficult to be modelled, but in exist in reality and are important in many industrial applications, e.g. cam-valve models, circuit breakers and others.

SimMechanics has powerful capabilities for mechatronic: it allows to simulate mechanical system (MBS) together with other domain models (electrical, control) very easily. However, it is lacking for unilateral constraint modelling for understandable reasons [1]. Many researches are interested in similar problem for different currently used MBS software [6].
This paper briefly describes the work related to:

- design of simulation blocks for simple unilateral contacts in SimMechanics
- with as fast as possible behaviour.

Particularly, in this paper we demonstrate the point – line segment contact element.
There are basically two classes of methods dealing with the unilateral constraints: the Newton impulses and Hertz continuous approach. We used the second one in our implementation, the linear viscous-elastic model is:

$$N = N_e + N_v = k\delta + b\dot{\delta} \qquad (1)$$

Contrary to Newton method, the continuous method provides the forces during the impact, which is very important for e.g. dimensioning of the parts.

The contact modelling implementation

Fig. 1 shows considered situation when the point E penetrates into flexible surface of edge (line segment) AB. The solution can be summarized in following steps: 1) obtain coordinates of points A, B and E, compute relevant vectors, 2) test if point E is in contact; 3) if so, compute depth of contact δ and relative speed in normal direction $\dot{\delta}$ and next; 4) compute normal force $N = N_e + N_v$ and the moment compensation; 5) and optionally compute tangential friction force T and finally; 6) apply resulting force to "point" body (point E) and apply force and compensatory moment to "line" body - particularly to point A.

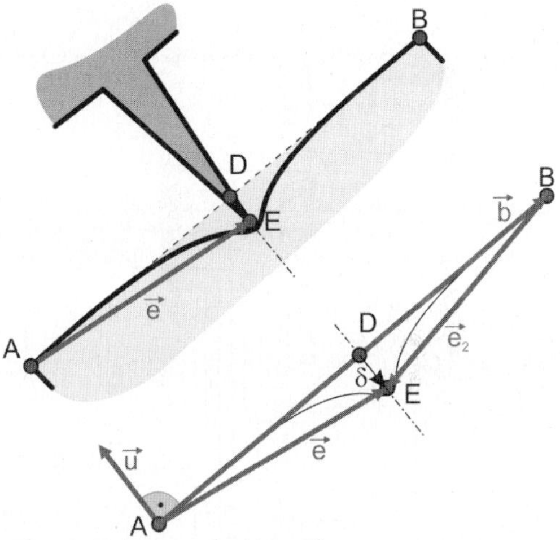

Fig. 1. Schema of point - line segment contact task

The global coordinates of points A, B and E can be read easily using Body Sensor block connected to relevant CS port of Body as shown in Fig. 2.

The testing of possible contact (relative position of point and line) can be accomplished by the vector product. Obviously, the resulting vector of vector product is oriented in z direction; therefore just following scalar formulation can be used:

$$\begin{aligned} &\vec{w} = \vec{u} \times \vec{v} \\ &w_z = u_x v_y - v_x u_y = f_s(\vec{u}, \vec{v}); \end{aligned} \qquad (2)$$

The tests can be written this way:

$$\lambda_1 = f_s(\vec{b}, \vec{e}),\ \ \lambda_2 = f_s(\vec{u}, \vec{e}),\ \ \lambda_3 = f_s(\vec{u}, \vec{e}_2) \quad (3)$$

The complete test criterion is

$$\lambda = \lambda_1 \leq 0 \wedge \lambda_2 \leq 0 \wedge \lambda_3 \geq 0 \quad (4)$$

If $\lambda = 1$, then point E is in contact with edge AB and we compute contact force, otherwise contact force is zero.

The deformation (depth) in contact δ can be computed as the distance of point to line using

$$\delta = u_x(x_E - x_A) + u_y(y_E - y_A) \qquad (5)$$

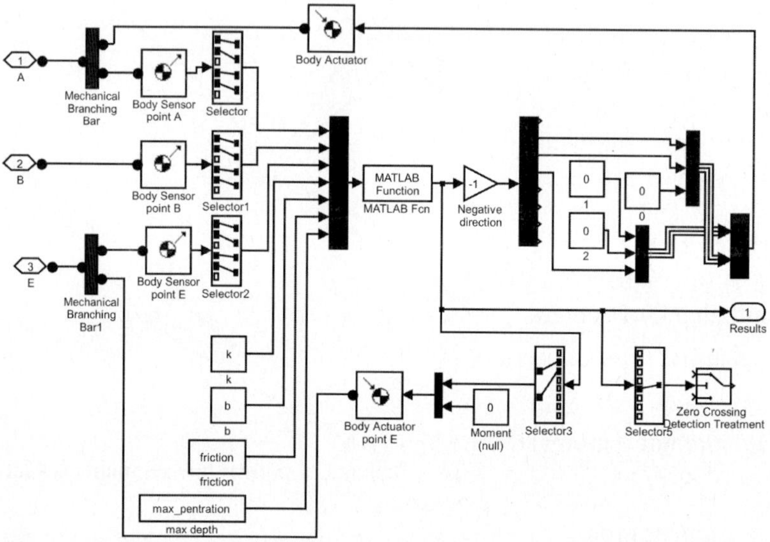

Fig. 2. Simulation schema of point – line segment contact block in Sim-Mechanics

Further, we can obtain the elastic component of normal force N_e according to eq. 1. Viscous component of normal force requires the relative velocity between point E and point D, which is the projection of contact point on line AB. We assume, that AB is part of the rigid body, than the principles of absolute and relative velocities of points on body are valid and we can formulate:

$$\vec{v}_{BA} = \vec{v}_B - \vec{v}_A, \; \vec{v}_{DA} = \frac{|\vec{e}\,|}{|\vec{b}\,|}\vec{v}_{BA}, \; \vec{v}_D = \vec{v}_A + \vec{v}_{DA}$$

$$\dot{\vec{\delta}} = \vec{v}_E - \vec{v}_D \qquad (6)$$

$$\dot{\delta} = \frac{1}{|u|}(\dot{\delta}_x u_x + \dot{\delta}_y u_y)$$

After that, the viscous component of force and complete vector of normal force are:

$$N_v = -b\dot{\delta}$$

$$\vec{N} = \frac{N_e + N_v}{|u|}\vec{u} \qquad (7)$$

If required, we can also model the Coulomb (dry) friction using well known formulation:

$$T = Nf \qquad (1)$$

where f is coefficient of Coulomb friction. Similarly can be modelled viscous friction or other more advanced (and complicated) models. As a minimal practical improvement, it is useful to consider the dependence of f on the velocity (kinematic and static friction coefficient – sliding and sticking mode).
Tangential friction force can be written in vector form thus way:

$$\vec{T} = -sgn(v)\frac{(N_e + N_v)f}{|b|}\vec{b} \qquad (8)$$

Finaly, we add both vectors and obtain total contact force in vector form

$$\vec{F} = \vec{N} + \vec{T} \qquad (9)$$

Computed resulting force F acts on "point" body in E and in opposite direction on "line" body in point D. In SimMechanics, one can apply any force and/or moment to fix defined point only, while the D is floating. Therefore, we apply the force to point A and to ensure the statically equivalent load, we add the compensating moment

$$\vec{M} = \vec{e} \times \vec{N} \qquad (10)$$

Practically in SimMechanics, the force and moment can act on body using Body Actuator block.
The computation described above is implemented in Simulink and m-Function block. All relevant blocks are included into masked subsystem which provides user friendly interface. User connects the three points A, B, E and defines parameters k, b, f. Resulting schema of block is shown in Fig. 2. The solution with relatively slow Matlab-Function can be replaced by S-Function in the future.

2. Conclusion

In this paper, we described the implementation of simple type of unilateral constraint problem in MBS software SimMechanics. The continuous Hertz

approach has been used with linear viscous and elastic component. There are still open questions about the parameters of contact problem which should be carefully selected; the rather suitable explanation is in [2].
From the practical point of view, the used might be informed, that the stiff solver has to be used for this problem, the best results has been obtained by ode23. The developed simulation blocks have been successfully tested on the problem of circuit breaker contact making.
In the future, the technique described here can be used for the building of more advanced constraint block such as point – curve or even curve – curve contact problem. The extension in spatial contact is also possible, but surely significantly more demanding.

Acknowledgment

The published results have been acquired be support of project AV0Z20760514 and GA□R 101/06/0063.

References

[1] Wood, G.D.: Simulating mechanical systems in Simulink with Sim-Mechanics, The MathWorks Inc.,www, 2002
[2] Pogorelov, D.: Universal Mechanics, manual, part 2,http://umlab.ru/, p.35–37, 2006
[3] Faik, Witteman: Modeling of Impact Dynamics: A Literature Survey, 2000 International ADAMS User Conference, 2000
[4] Tasora, A., Righettini, P.: Sliding contact between freeform surface in three-dimensional space, GIMC 2002, 'Third Joint Conference of Italian Group of Computational Mechanics and Ibero-Latin America Association of Computational Methods in Engineering, 22-24 June 2002, Giulianova - Italy, 2002
[5] Tasora A.: An optimized lagrangian-multiplier approach for interactive multibody simulation in kinematic and dynamical digital prototyping, VIII ISCSB, International Symposium on Computer Simulation in Biomechanics, 4-6 July 2001, Politecnico di Milano, Italy, 2001
[6] Bottasso, C.L., Trainelli, L.: Implementation of effective procedures for unilateral contact modeling in multibody dynamics, Mechanics Research Communications, Vol. 28, pp.233–246, 2001
[7] Wasfy, T. M., Ahmen, K. N.: Computational procedure for simulating the contact/impact response in flexible multibody systems, Computer methods in applied mechanics and engineering, Vol. 147, pp. 153–166, 2006

Fast prototyping approach in design of new type high speed injection moulding machine*

K. Janiszowski, P. Wnuk

Institute of Automation and Robotics,
Warsaw University of Technology, A. Boboli 8, Warsaw 02-525, Poland

Abstract

In paper is outlined an application of PExSim - a package for simulation and research dynamic control systems, for investigation a solution of very fast, hydrostatic positioning of a form in high speed moulding machine (HSIM). This construction, based on new locking mechanism, shall attain a 1s dry cycle time for form mass of 3000 kg and 0.5m distance of positioning. Complex mechanism of highly nonlinear transmission ratio, is moved by a set of hydraulic plunger cylinders, that are supplied by a constant stroke, small inertia pump, driven by very dynamic electric motor. Different limitations of final construction have induced many constraints at control. The design of locking mechanism and final control program was developed by many experiments performed by SimulationX and PExSim. Some expected transients of the mould control will be presented.

1. Introduction

Introduction of new types of electric AC-servo motors has induced small revolution in construction of different high power drive systems. In case of HSIM machines an electric direct drive in comparison of classic solution - hydraulic cylinder supplied by proportional, throttling valve and variable volume pump driven by electric cage motor, has increased power efficiency and deleted unwanted leakages. However pure electric drive has shown some disadvantages in use – transmission mechanisms have shown wear effects. HSIM machine makes usually 10000 heavy strokes, (eg. 3000 kg mass with acceleration of 2-4g) per day, hence construction has to be very endurable. A construction with intermediate hydraulic circuit, containing hydraulic accumulators, can significantly reduce the stroke effects. Application of AC-servo (working only in time of movement) with low inertia, constant stroke pump, without throttle valve, will remarkable increase power efficiency. Position control was very important – form has to be stopped at final position with velocity less 20 mm/s, when its maximum within movement was appr. 2m/s. Modelling of the entire drive system and development of proper control algorithm was quite complex problem.

2. Fast prototyping of construction

Fast prototyping approach was focused on two main aims: decrease of costs and time of project. To avoid unnecessary expenditure of building alternative prototypes (for choice of the best final machine), cooperating with different AC-servo drives, valves systems etc. all main mechanical components are modeled by simulation package SimulationX ™. This results were used for determination of extreme values of electric and mechanical components, that have determined power and dimensions of necessary components: motors, pumps, accumulators, switching valves, mechanism, etc. Next cooperating partners: machine builder (1 team), drive and melting extruder suppliers (2 teams), control engineers (2 teams) have to investigate its capabilities and deliver proper components. Each team was working individual and parallel. Institute of Automatic Control and Robotics of Warsaw University of Technology was involved in design of fast control of moulding form. Results of simulated (in SimulationX™) [2] responses of drive components, in form of parametric model were next used for development of two control concepts: modified state space control and adaptive open loop control for positioning of form. A last phase – design and testing of position control was implemented on PExSim platform.

3. PExSim package

Software simulation of dynamic systems is common used technique for investigation of system behavior in different operation conditions. All modern SCADA systems have some tools for modeling dynamic blocks, but they are adopted to a platform used by this system and are not standard. Other approach to simulation is based on special written tools packages, that can be used to modeling or simulation with demanded accuracy, when the models of process are known, e.g. SimulationX [2], designed for simulation of fluidic and electro-mechanical structures with built in large libraries of components of elaborated integration techniques. These tools are however limited in area of control design possibilities and are quite expensive. Other tools, like e.g. general purpose system - Matlab package [3] with Simulink, used in academic institutions are based on assumption, that system dynamic is well known. This approach, creating all possible effects from basic components, based on blocks with very simple nature, yields an uncomfortable result: a simulation of more complex phenomenon, e.g. nonlinear friction force or PID controller induces a quite complex block diagram with 30 - 40 or more basic blocks. User effort and time, necessary to creating, testing and verification of all used components, is high and in effect this tools is used for simulation rather simple systems.

The package PExSim (Process Explorer and Simulator), baased on plugin architecture can be used for emulation of complex dynamic systems, composed of predefined dynamic blocks or multivariable models (estimated with own procedures), can cooperate in real time with industrial environment and will be easy and flexible to extension by the user writing its own plugin objects, in C++. PExSim contains following operators and components, shown on Fig.1,

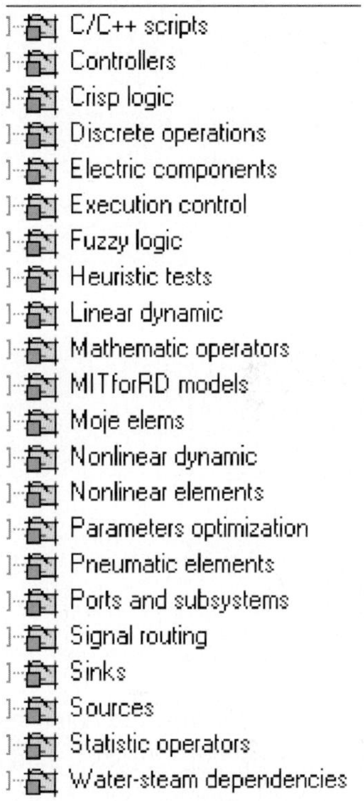

Fig.1.PExSim options

Signal "sources" contains deterministic or stochastic time functions, imported signals from real time interface or data files. As sinks are short described outputs, that can be plotted, stored on file or transferred to real time process or project subsystems. Components named as mathematic operators, discrete operations, crisp and fuzzy logic are used for signal processing. General purpose dynamic components – linear dynamic (transfer functions and state space representation) together with "nonlinear dynamic" – a solver of nonlinear defined equations show simulation abilities within deterministically desribed dynamic systems. MITforD models is a class of fuzzy MIMO models of TSK (Takagi, Sugeno, Kanga) form [1,4]. For construction of block diagram for simulation the user has different options gathered in: execution control, signal routing and nonlinear elements. Option "Ports and subsystems" enables construction of sub-pathes, representing dynamic subsystems of simulated structure in form of one block with defined inputs and outputs nodes. PExSim contains sets of founded specialized libraries too, like: controllers, electric components, pneumatic elements, water steam dependencies. Own user procedures can be incorporated into package by C/C++ scripts option. For current investigation properties of supervised values can be used statistic operators and heuristic tests. Option called "parameter optimization" can be used for minimization of defined quality index, e.g. sum of squares errors of simulated structure in respect to variable parameters of this model.

4. Modelling of HSIM machine

A drive system was divided into blocks: electric AC servo motors with pumps, creating pressures in hydraulic cylinders, cylinders with moved mass, mechanism for creating clumping force, controllers and displays, Fig.2.

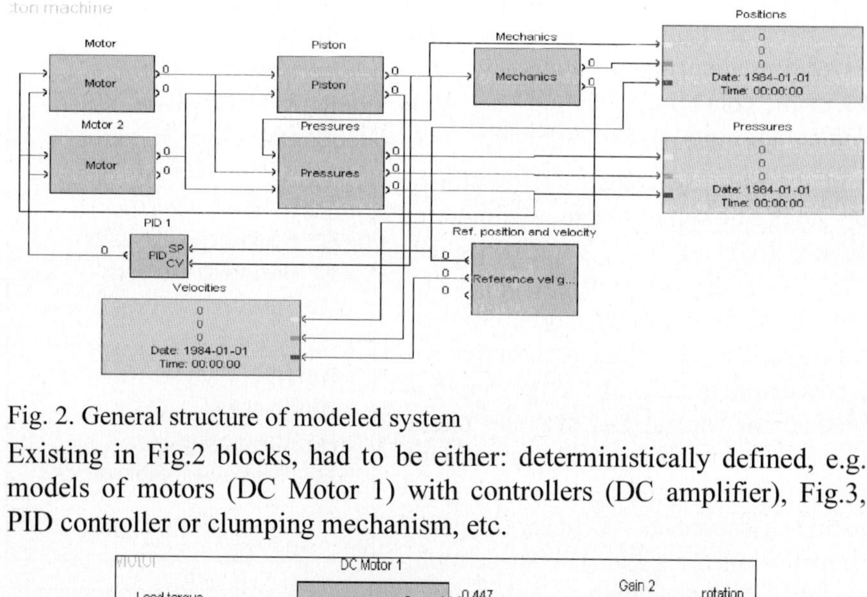

Fig. 2. General structure of modeled system

Existing in Fig.2 blocks, had to be either: deterministically defined, e.g. models of motors (DC Motor 1) with controllers (DC amplifier), Fig.3, PID controller or clumping mechanism, etc.

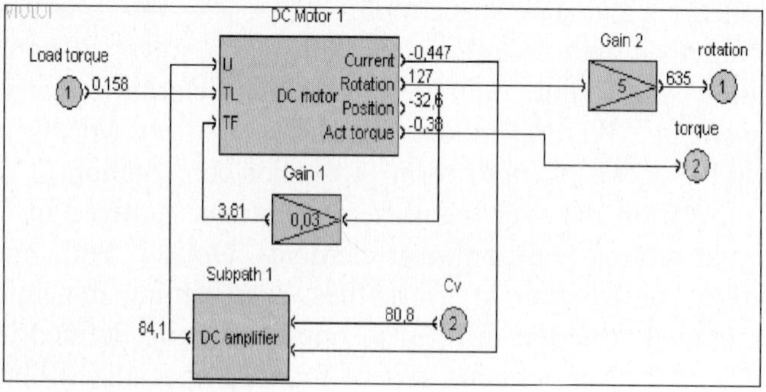

Fig. 3. Structure of motor "subpath" called Motor1 on Fig.2.

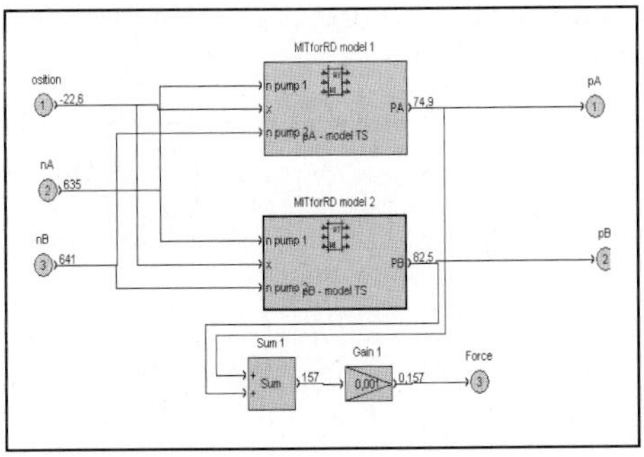

Fig.4 Model of pumps for Fig.2

or statistically estimated, using MITforD package, Fig. 4.

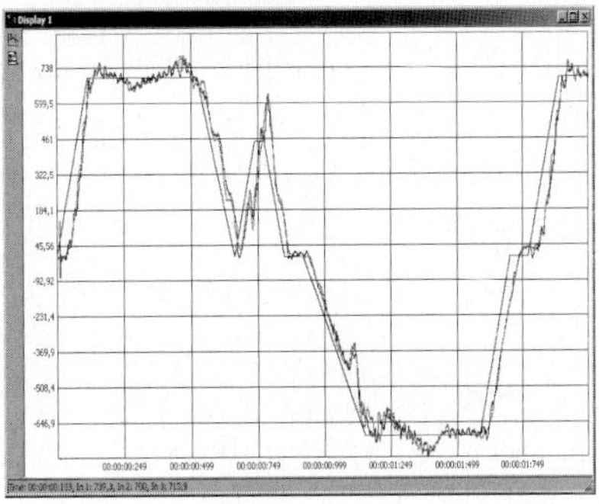

Fig. 5. Modeled and measured mould position

A block called mechanics, in Fig.2, was expressing a strongly non-linear transmission of force, developed in special toggle system. This mechanism was invented for amplification a clumping force in ratio 1:300 for the locking phase of clumping movement. Its action was practically static, variable ratio force transmission, dependent on mould-form position. It was modeled in form look-up table plugin, contained in non-linear elements option.

Different algorithms for mould form movement control were tested using PExSim package. A demand of very short time of complete mould movement: to the form and withdraw, has resulted in decision – drive as fast as possible, the whole movement time and control, with position measurement, only at closing mould to the form. At open control of form, only limitations of machine were important: maximal pressures in hydraulic pipes, maximal current of motors, maximal velocity of pumps and maximal acceleration of drive. At final position control, the form velocity and positioning error were important. Then two approaches were used: a modified state space control with feedback of position, velocity and deceleration of form and open loop adaptive control for minimal time of form stop. The proper instants of AC-servo time switching were determined adaptively within 4-6 first dry cycles of mould. This last approach was developed after investigations and experiments made in PExSim, with different adaptively estimated parameters. Finally a simple model of mould movement was derived, that had only two tuned parameters. This approach was next tested in SimulationX too. The transients of SimulationX were a common platform for all cooperating teams, so the corresponding results can be easy transferred and verified by co-partners of HSIM project. A sample of movements, for a 2000 kg form and displacement of 45 cm, plotted in SimulationX are presented on Fig. 6.

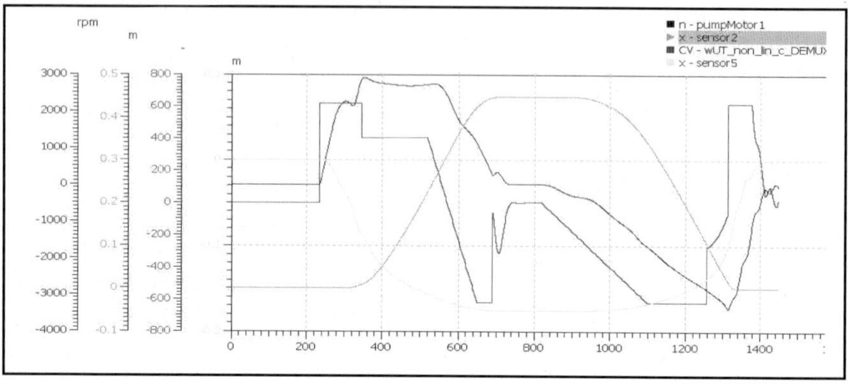

Fig.6 Transients of adaptive control movement of form in HSIM machine

6. Conclusions

A fast prototyping is very efficient and time saving approach in design of complex, new constructions. It should be mentioned, that within this project were tested three different configuration of hydraulic systems (with different number of motors, pumps and auxiliary valves), were investigated 2 different construction of clumping force mechanisms, 3 algorithms for injection and 3 algorithms for clumping form movements and 4 different versions of hydraulic supply, oil accumulator and cooling systems. Finally developed construction seems to be very economic, in component number, but of high quality and good parameters, has involves small energy loses and can be so fast as demanded, after getting together experience and knowledge of different cooperating teams. An elaboration of such effect by one team – machine constructor, will take much more time and will never resulted in so good final parameters of machine.

7. Aknowledgements

The paper was partially sponsored by grant 004/ITE/3/2005

8. References

[1] Janiszowski K. *Parametric models identification*, Exit 2002, Warsaw,

[2] System documentation, http://www.simulationx.com/
[3] System documentation, http://www.mathworks.com/
[4] Wnuk P. *Algorithms for fuzzy model structure identification*, Ph.D dissertation, Warsaw University of Technology 2004

Ultra-precision machine feedback-controlled using hexapod-type measurement device for six-degree-of-freedom relative motions between tool and workpiece

T. **Oiwa** (a) *

(a) Shizuoka University, 3-5-1 Johoku,
Hamamatsu 432-8561, Japan

Abstract

This paper describes a precise machine concept based on compensating for six-degree-of-freedom (6-DOF) motion errors between the tool and the workpiece. A hexapod-type parallel kinematics mechanism installed between the tool spindle and the surface plate measures the 6-DOF motions regardless with temperature fluctuation and external forces because the mechanism has a compensation system for the elastic and thermal errors of the joints and the links. Therefore, the tool position and orientation are compensated by using the measured 6-DOF errors. This paper describes the conception of the system. Moreover, a passively extensible strut with the compensation device for the joint's errors was tested.

1. Introduction

To realize an ultra-precise machine system with nanometer-order positioning resolution and positioning accuracy less than 100 nm, a mechanism, which generates highly accurate relative motions between its cutting tool/touch probe and the workpiece, is required as well as the accuracy improvements of each element of the machine. In an actual machine, however, internal and external disturbances noticeably cause positioning errors. Thus, much improvement in guide element accuracy and structural stiffness has been achieved to decrease the motion error and the elastic deformation. However, increased mass along with such improvements has dynamically caused further motion error and elastic deformations. Further, no machine structure can be made infinitely stiff. On the other hand, to compensate for the thermal deformation, prediction methods have been investigated using temperature sensors and thermal deformation analysis. However, the thermal deformation is hard to predict precisely. This study[1] presents a machine system equipped with a feedback sensor system for six-degree-of-freedom (6-DOF) motion errors between the tool and the workpiece. A hexapod-type parallel kinematics mechanism (PKM) measures the 6-DOF relative motions. This paper describes the conception of the system. Moreover, the strut with a compensation device was tested.

2. Machine tool with PKM measurement device

Figure 1 shows the principle of proposed machine system based on a measurement device for the 6-DOF motions between the tool and the workpiece. This system consists of a hexapod-type PKM, a conventional machine structure, and a controller. To measure the relative 6-DOF motions, the base platform and the moving platform of the PKM are mounted on the surface plate and the machine spindle, respectively. Both platforms are connected through six extensible struts with prismatic joints. Since each strut has no actuator, the moving platform of the PKM is passively moved in three-dimensional space by the conventional machine. Because change in the length of each strut is measured by a displacement measurement unit, the 6-DOF motions can be calculated by the forward kinematics of the hexapod-type PKM. Consequently the controller compensates for the motion errors and accurately actuates the tool. In the coordinate measuring machine, the coordinates of the probe tip are directly measured by the PKM.

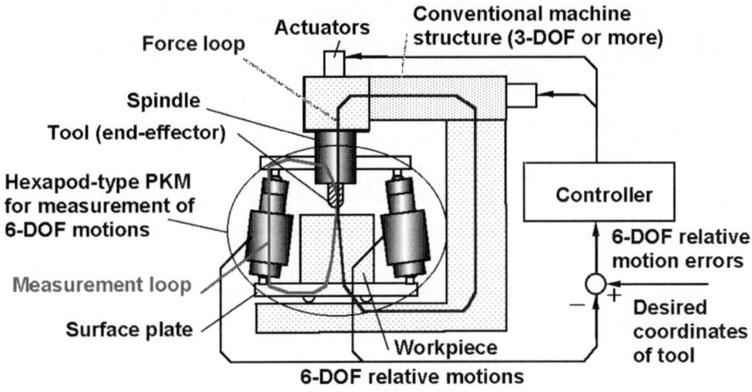

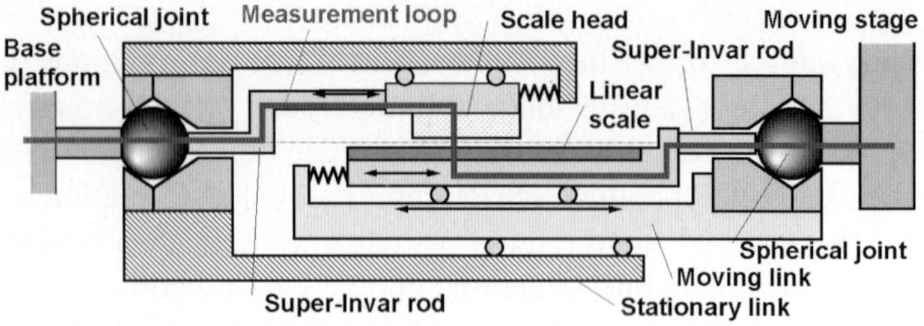

Fig. 2. Extensible strut with compensation device for both joint errors and link's elastic and thermal deformations

Fig. 1. Fundamentals of proposed machine system using a hexapod-type measurement device for 6-DOF motions between tool and workpiece

3. Extensible Strut of PKM

Figure 2 shows an extensible strut of the PKM. Each strut has a mechanical compensation device for both joint errors and link deformations [2]. Two rods connect the scale and the scale head of the linear scale unit to the two spherical joints. The scale head and the scale are guided by some linear bearings so that they can move only in the longitudinal direction. Thus, the scale unit can measure not only the displacement change of the prismatic joint but also the spherical joint errors and the link deformation in the longitudinal direction because each rod end is in contact with the master ball of the spherical joint. Further, because the rods are made of Super-Invar (thermal expansivity: approximately 0.5 ppm/K), each distance between the scale unit and a spherical joint is not influenced by any temperature change. Additionally, the distances are not influenced by any external forces because no external force is applied to the rods and the scale unit. To put it briefly, even if the strut is thermally or elastically deformed, the scale unit can accurately measure the length change of the strut.

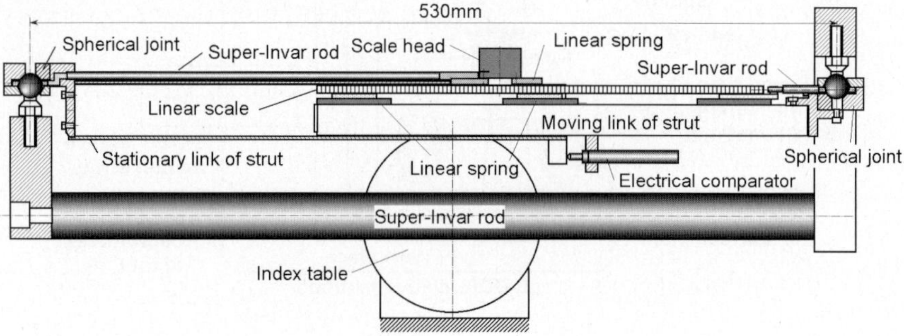

Fig. 3. Experimental setup for testing strut with improved spherical joints

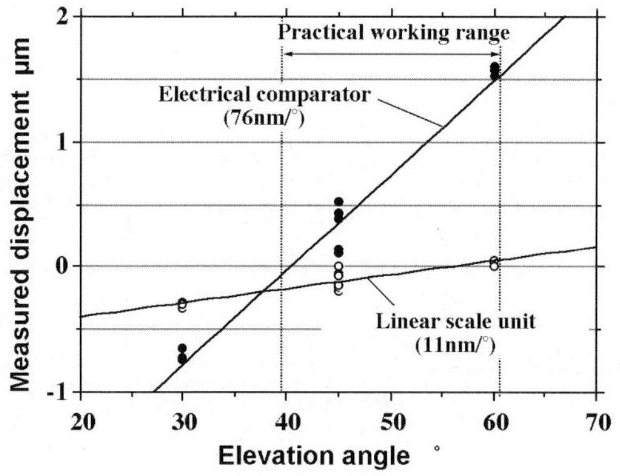

Fig. 4. Influence of elevation angle of strut on displacements measured by two measurement systems

4. Experiments for strut

Figure 3 shows the strut and its test bed. The test bed was made of low thermal expansion cast iron (expansivity: 0.8 ppm/K) and the Super-Invar. The strut is mounted on the bed by the spherical joints. A distance between both spherical joints is fixed, which is set to be approximately 530 mm. An electrical comparator (Mahr 1201IC+P2004M) measures the relative displacement between a stationary link and a moving link of the strut. This displacement measured represents deformations of both links. However, since the distance between the spherical joints is constant, a displacement measured by the linear scale unit must be zero, ideally.

First, to investigate an effect of the gravity on measured displacement of the scale unit, an index table tilted the test bed. The elevation angle was changed between 30° and 60° to horizontal. Figure 4 depicts an influence of the elevation angle on measured displacements. The elastic deformation of the strut reached to 2.3 µm when the strut tilted at a 30 to 60° angle.

On the other hand, the displacement measured by the linear scale unit was less than 0.33 µm then, 14% of that measured by the comparator. This proves that the linear scale unit with the compensation device accurately measures the distance change between both ball shanks of the spherical joints regardless of the elastic deformation of the links and the joints.

Next, to investigate the influence of the temperature fluctuation, the strut was heated for 7 min until its surface temperature rose several degrees. Subsequently, the strut was allowed to cool. The strut was lengthened so that it might be strongly influenced by heat. Figure 5 shows the strut's

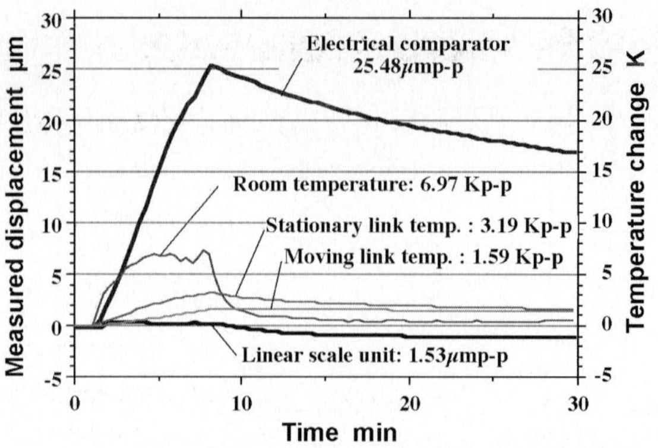

Fig. 5. Displacements measured by two measurement systems and strut temperatures

temperature change and the displacements. When the surface temperatures of both links rose 3.19° and 1.59°, respectively, the relative displacements reached 25.48 μm. Provided the average temperature of the strut is 2.38 degrees, calculated thermal expansion coefficient of the strut 530 mm long is 20.2 ppm/K, almost equal to a coefficient of aluminium.

On the other hand, the displacement measured by the linear scale unit was less than 1.53 μm. Thus, the scale unit with the compensation device accurately measures the distance change between both the spherical joints regardless of the temperature change. Moreover, this implies that the struts expansion coefficient has improved from 20.2 to 1.21 ppm/K (6.0 %).

5. Conclusion

An ultra-precise machine system, which compensates 6-DOF relative motion errors between the tool and the workpiece, has been described. A passive hexapod-type PKM consisting of six extensible struts with linear scale units is installed between the tool spindle and the surface plate of conventional Cartesian-coordinate-geometry mechanism. The passively extensible strut with the linear scale unit and the compensation device accurately measured the length change of the strut regardless of the strut's orientation change and temperature change.

References

[1]Oiwa, T, *Proc. KSME-JAME Int. Conf. Manufacturing, Machine Design and Tribology*, DLM305(2005)1-4.

[2]Oiwa, T, *Int. J. Robotics Research*, 24, 12(2005)1087-1102.

Mechatronics aspects of in-pipe minimachine on screw-nut principle design

M. Dovica (a) *, M. Gorzás (b)

(a) Technical University, Faculty of Mechanical Engineering, Department of Instrumental and biomedical engineering, Letná 9, 042 00, Košice, Slovak Republic

(b) Technical University, Faculty of Mechanical Engineering, Department of Safety and Quality of Production, Letná 9, 042 00 Košice, Slovak Republic

Abstract

This paper deals with the utilization possibility of kinematical couple screw-matrix in mechatronics concept of in-pipe minimachine which is assigned to move in the pipes with inner diameter less than 25 mm. We use the minimachine for inspection of inner surface defects. The motion principle is based on the transformation from the rotary to the linear motion by the screw and nut which creates a change of the distance between front and rear line of bristles. The motion of the minimachine insures the difference friction between the bristles and the pipe surface in the working stroke of minimachine.

1. Introduction

Nowadays the mobile machines for motion in the thin pipes (less than 25mm) represent a suitable area for research. Their utilization is oriented to the detection of defects on the inner pipe surface, the repair of localized defects, monitoring and maintenance of pipes and last but not least their utilization is oriented to the ability to draw new cables into the old and already unused pipe systems.
The utilization of motion principles by means of classic wheel and crawling traction for design of in-pipe minimachine is dimensionally limited. In the view of this reason for positioning and motion the bristles in the form

of flexible beams which are orientated under the precise angle towards the pipe surface are used and they use the difference friction. At the minimachines realization, the actuators that are designed and manufactured with different approach, are used. That is because of the efficiency of power fields which generating forces are need for the initiation of the minimachine in the motion decreasing with dimensions. [1], [2], [3]
The in-pipe minimachine, that will be analyzed next, is designed to motion in the relative straight pipe with approximately 25mm inner diameter in the view of the possibility of another extension by the monitoring system in the form of CCD cameras or surface defects sensor.

2. Concept of Minimachine

The minimachine is used for moving in the pipe with inner diameter approximately 25 mm. It is made of three modules, besides the module in the front and at the back are made of the rotary electromagnetic actuator, screw, nut and bristles. The middle module, which uses the own move of the minimachine, is made of the rotary electromagnetic actuator, screw and nut. The principle of the movement works as the transformation of the moving of rotary actuator through the screw and nut to the straight motion. It makes change in the distance between the minimachine modules in the front and at the back. The direction of its movement in the pipe is determined by pressing of bristles of the module in the front or the one at the back. Pushing forward of the bristles and their press to the inner wall of the pipe is also ensured by the screw and nut. Control of the minimachine ensures the change in the direction of the linear movement and cyclic repeating.

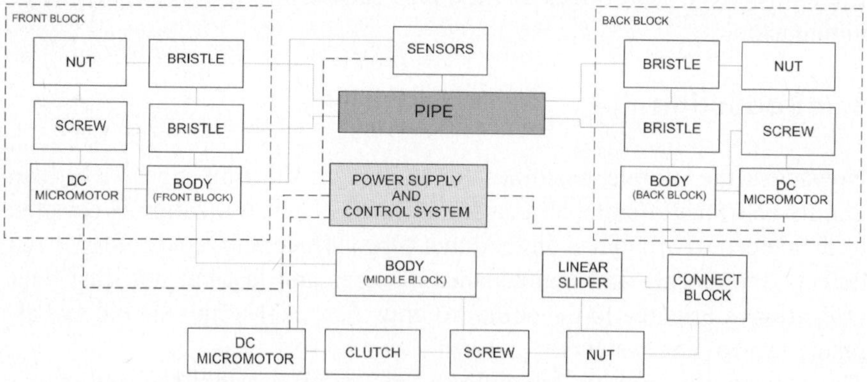

Fig.1 Block diagram of the in-pipe minimachine

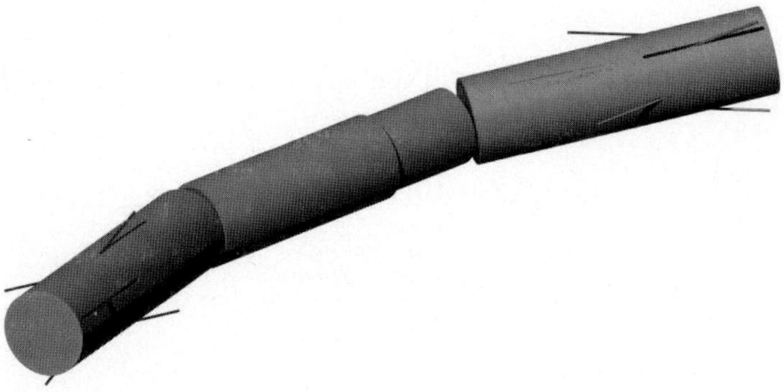

Fig.2 3D model of the in-pipe minimachine

3. Kinematics Analyse of the Minimachine

Regarding to the requirements which are defined in the phase of the design kinematics analyze process of the in-pipe minimachine is made. Kinematics scheme (Fig.3) represents the middle module and includes an actuator (M) whose motion is defined by the angular velocity ω_M through the clutch transmitted to the screw (PS). The movement of the ended component of the mechanism expressed by the parameter x is resulted by the nut (PM).

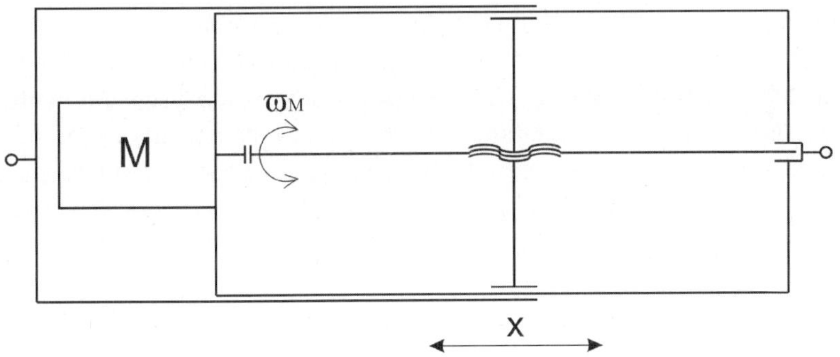

Fig.3 The kinematics scheme of the middle block of the in-pipe minimachine prototype

After substitution and conversion it is possible to move the ended component of the mechanism and express as follows:

$$v = \frac{dtg\alpha}{2} \cdot \varphi \tag{1}$$

where φ is the angular rotation of the screw, α is the spiral angle of nut thread[4].

For expressing the velocity of ended component of the mechanism it holds:

$$v = \frac{dx}{dt} \cdot \frac{d\varphi}{d\varphi} = \frac{dtg\alpha}{2} \omega_M \tag{2}$$

where ω_M is the angular speed of the actuator.

For expressing the acceleration of ended component of the mechanism it holds:

$$v = \frac{dv}{dt} = \frac{dtg\alpha}{2} \alpha_M \tag{3}$$

where α_M is the angular acceleration of the actuator.

Conclusion

In this article the particular stages of the design mechatronics concept of the in-pipe minimachine developed on the authors` department were described. The design is related to the elaborational requests and the parameterization, the stages of concept design and the prototype realization.
The result is the realized prototype. In the next research we will begin with the experimental verification of the parameters for purposes of the motional optimization in the limited area, the study of analyses and synthesis of the deflection and the tolerances of precision mechanisms of the in-pipe minimachine and the active control for compressive force of smart bristles.

Acknowledgement

The work has been supported by the Slovak Grant Agency for Science. Grant project VEGA No. 1/3159/06: The research of the nature inspired the motional principles and their application at the minimachine design.

References

[1] R. Isermann "Mechatronic Systems" Springer Verlag, 2003
[2] S. Iwashina, N. Hayashi, K. Nakamura "Development of In-Pipe Operation Micro Robot" Proc. of IEEE 5th Int. Symp. Machine a Human Sciences, 1994, pp. 41-45.
[3] A. Gmiterko, "Mechatronika: Hnacie faktory, charakteristika a koncipovanie mechatronických sústav" Emilena, Košice, 2004
[4] U. Fischer, a kol. "Základy strojnictví" Europa – Sobotáles, 2004

Assembly and soldering process in Lead-free Technology

J. Sitek (a), Z. Drozd (b), K. Bukat (a)

(a) Tele and Radio Research Institute, 11 Ratuszowa Street, Warsaw, 03-450, Poland

(b) Warsaw University of Technology, Institute of Precision and Biomedical Engineering, Division of Precision and Electronic Product Technology, 8 Sw. Andrzeja Boboli Street, Warsaw, 02-525, Poland

Abstract

Continuous miniaturization of electronic products & components and parallel increase their functionality in modern applications together with the EU RoHS directive restrictions delivers everlasting technological difficulties.

The problems with materials, assembly and soldering equipments selection, printing solder paste both on very small and big PCB pads, optimization of soldering processes for lead-free complex board were shown at the article.

1. Introduction

The 1st of July 2006 on the European Economic Area came into force the RoHS Directive [1]. According EU regulations lead-free materials have to be used in the manufacturing processes of electronic products. But not all electronic products are covered by the RoHS directive nowadays. Some electronic manufacturers, particularly those in the high performance and reliability sectors, are working to maintain Pb-base materials and processes. As the supply of Pb-base components is reduced, manufacturers using Pb-based processes are faced with the decision of reprocessing Pb-free components. This is a reason that some companies have to use lead-free technology even they are not obligated for this. More over the continuous miniaturization of electronic products and components delivers additional technological difficulties. Example, presence on the same PCB both CSP, BGA or other fine-pitch components together with very small passive components causes great challenge for assembly technologists.

The problems with materials, assembly and soldering equipments selection, printing solder paste both on very small and big pads, optimization of soldering processes for lead-free complex board were shown at the article. This work was the continuation of investigations connected with lead-free technology realized by authors as a part of the GreenRoSE-Collective Research Project – 2004-500225 funded by the EC under the 6FP.

2. Materials and equipment for tests

The double sided PCB tests with immersion tin or Ni/Au finishes contain SMD: passive, active, CSP and BGA components with different sizes and pitches and also THT passive and active components were used for investigations. The most of components had pure Sn coating on terminations except CSP and BGA components which had SnAgCu (SAC) balls and one THT integrated circuit with Ni-Pd coating on terminations. The lead-free solder paste with SnAg3.0Cu0.5 (SAC305) solder powder and ROL0 type flux inside was used for reflow soldering. The SAC305 was used for wave soldering also together with less and more active VOC-free fluxes.

The whole manufacture process of the complex PCB was divided on three main parts. First was assembly process of different SMD components including CSP and BGA parts and reflow soldering of the first side of the PCB. Next SMD active and passive components have picked and placed onto glue on the PCB second side and cured at reflow oven. The third stage was assembly process of THT components and wave soldering.

The assembly processes were carried out using production equipment for SMT and wave soldering. Solder paste was printed via 127 or 100 µm steel stencils using automatic printing machine - PBT MOTOPRINT V. Components were picked and placed using JUKI assembly system type KL2050L or FUJI AIM pick and place machine. The reflow soldering was done using convection oven ERSA HOTFLOW 2/14, which has 7 heating zones and 2 cooling zones or the VIP70A - 5 zones, convection, reflow oven, made by BTU.

3. Assembly and soldering difficulties

The PCB test contains passive components sizes from 0805 to 0201 and BGA components with ball numbers from 676 to 84 with pitch from 1.27 to 0.5 mm. Such differentiation of components is great challenge for assembly technologists.
The first technical assembly problem was observed during solder paste printing process. The complex PCB requires correct quantity of solder paste both on large pads as well as on very small pads [2]. It was observed wrong, conical shape of overprints made by the stencil 127 μm thickness and too small quantity of solder paste on pads for CSP with pitch 0.5 and for resistors size 0201 or smaller (Fig.1).

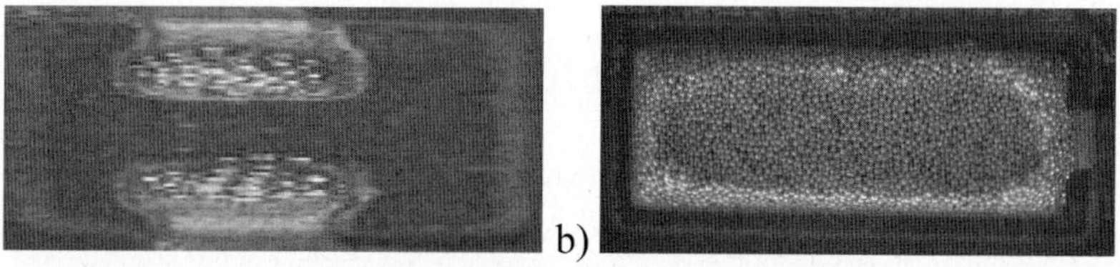
a) b)

Fig. 1. Examples of the solder paste overprints: a) print defect for pads R01005; b) correct print result for SO16 1.27 pitch pad.

The printing problems were resolved mainly by special design of a stencil. The area ratio (AR) of stencil windows for the smallest pads was changed from 0.45 to 0.7 (for CSP 0.5 pitch) and to 0.8 for RC0201 and smaller pads (AR = surface of stencil window divided by surface of stencil walls which create window). The thickness of stencil was reduced also from 127 □m to 100 □m.
The next observed technical problem was connected with pick and place very small components (R0201). The JUKI assembly system type KL2050L has not nozzle, as standard equipment, for such small components. The purchase of dedicated nozzle for RC0201 resolved this problem for this machine. The assembly process was carried out without problems using the FUJI AIM pick and place machine which has possibility pick and place both large and small components.

The presence on a PCB different component types (different size, pitch and materials) requires optimization of soldering profile, especially in lead-free soldering technology. The linear profile of soldering was used to experiments (Fig.2). The linear type of profile was used to minimize degradation of coatings on PCBs at high temperature during reflow soldering process because these boards had to pass though also via reflow oven during glue curing and wave soldering.

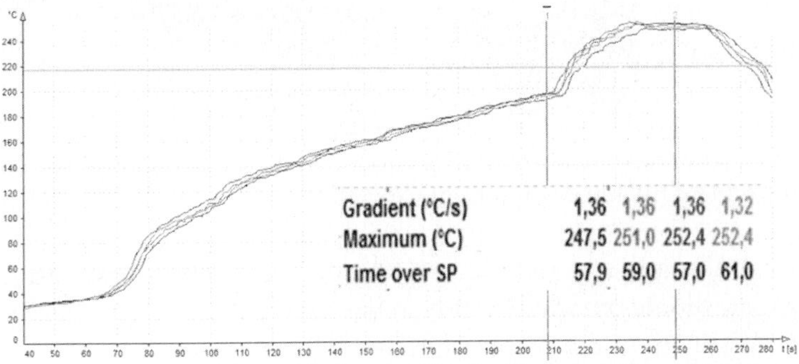

Fig. 2. Example of the soldering profile for Pb-free reflow soldering.

The highest soldering problem was concerned with the CSP84 - 0.5 pitch and R0201 components (Fig.3) during reflow soldering.

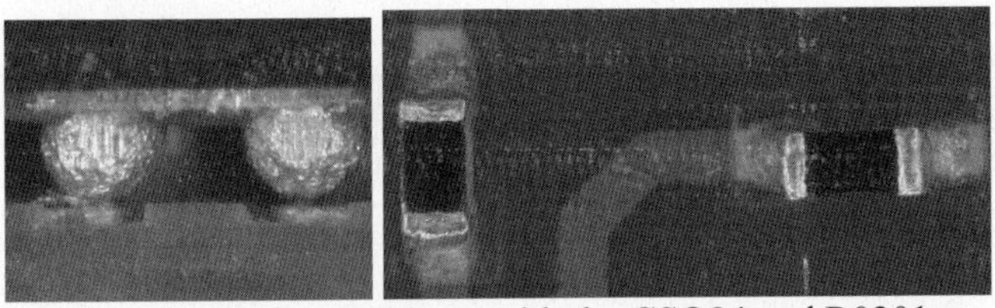

Fig. 3. Examples of soldering problems with the CSO84 and R0201 components.

The reason of above situation was first of all solder paste printing problems. The next reason was high value of ΔT (~13°C) on a complex PCB which effected reduction of soldering window. The soldering defects were reduced by improving printing process as was described above. After such correction acceptable reflow soldering results for all components were obtained using the 5-zones reflow oven.

The second part of soldering problems was concerned with wave soldering. The wave soldering process was carried out after pick and place process of active and passive components and after glue hardening in convection oven at the 120°C. All thermal processes before wave soldering decreased wetability of the Sn PCBs coating (See SOT23 components Fig. 4), the Ni/Au coating was not destroyed visible. The more active flux had to be used in air atmosphere to minimize soldering defects but they still were present at the QFP64 pitch 0.5 components (Fig.4).

a) b)
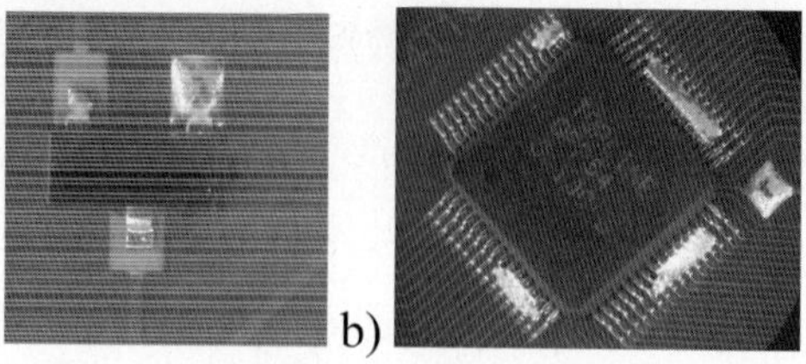

Fig. 4. Examples of soldering problems: a) the SOT23 – Sn wetability problems, b) the QFP64 pitch 0.5 component – small pitch and pad design problems.

Probably a nitrogen atmosphere and improve pads design near the QFP64 components should improve wave soldering results in this situation. The further investigations are planning in this subject.

Summary

Printing process is the most critical during assembly of complex PCBs in lead-free technology.

The dedicated nozzles for pick and place machine are essential for assembly components sizes: 0201 or smaller as also BGA and CSP.

The complex PCB in lead-free technology requires soldering profile with about 250°C peak temperature and precise convection oven with 5 or more heating zones.

The minimalization of thermal processes previous a wave soldering, good quality of PCB finish, adequate PCB pads design and more active flux are recommend for lead-free wave soldering of complex boards.

Acknowledgements

This work was realized as a part of the GreenRoSE - Collective Research Project – 2004-500225 funded by the EC under the 6FP.

References

[1] Directive 2002/95/EC on the restriction of the use of certain hazardous substances in electrical and electronic equipment, Official Journal of the European Union, 13.02.2003, pp. 19-23

[2] J. Klerk, „Large & Complex Boards“. ELFNET at SEMICON Conference, 5-6th April 2006, Munich, Germany

Applying Mechatronic Strategies in Forming Technology Using the Example of Retrofitting a Cross Rolling Machine

R. Neugebauer, D. Klug, M. Hoffmann, T. Koch

Fraunhofer Institute for Machine Tools and Forming Technology
Reichenhainer Straße 88, Chemnitz, 09126, Germany

Abstract

A machine for manufacturing bulk metal forming parts (a cross rolling machine with flat jaws) will be used as an example to demonstrate the idea of retrofitting metal forming machines that is based on linking single mechatronic components to one overall mechatronic system. The purpose of retrofitting was boosting the machine's performance range while expanding its functionality. Ideas for applications will be developed supported by simulation and based on an analysis of the machine's system to evaluate how they fit into the overall system. The foremost features of the machine under analysis are the cascade-shaped regulation of the main hydraulic machine and secondary axes using decoupled physical factors and the decentralized control structure distributed physically over several levels. Selected experimental findings will demonstrate that this overall mechatronic system can be used in practical applications.

1. Introduction

As with all modern production machines, metal forming machines are constantly called upon to be increasingly productive, flexible and efficient. Furthermore, the wide variety of materials to be worked and geometries to be produced spells out increasing demands made upon the complexity of the metal forming process. All of these requirements made of metal forming machines means they have to be considered overall mechatronic systems. This not only applies to coming up with new metal forming machi-

nes. It also figures prominently in renewing and upgrading existing metal forming machines (i.e., retrofitting).
Retrofitting not only has the purposes of modernizing metal forming machines and boosting process and machine reliability, but also upgrading the range of the machine's applications. This can be done by using the machine's performance ranges to a greater extent or integrating new functionalities. Meeting this target calls for including and efficiently taking advantage of the opportunities offered at the intersection of mechanical engineering, electrical engineering/electronics and information technology for the technological functioning of each metal forming machine system. Making actuator and sensor technology a part of the control and regulation design and implementing them are major factors in translating the targets of retrofitting the machine into reality. The foremost technological factors are the position and motion of the tool and work piece as well as the process forces.

2. Problem Description

Cross rolling is a continuous metal forming process for manufacturing graduated work pieces that are mostly symmetric to rotation with a high degree of dimensional, shape and mass accuracy. Components range from preform parts to finished shapes made of hollow or solid material. These iron and non-ferrous materials are cold-, semi-warm or warm-formed while cross rolling is done on cross rolling machines with round tools or, as in the case under consideration, cross rolling machines with flat jaws. Their characteristic feature is pressure forming the work piece by means of tools moving opposite one another that roll on the surface of the work piece and put it into rotational motion. In addition to the classical field of bulk metal forming technology (mass production with tools with a high degree of shape storage), cross rolling with flat jaws is a flexible forming technique for small and medium parts numbers with partially meshing and partially low-shape storage tools. While the machine structure on existing cross rolling machines with flat jaws does not have any substantial means of enhancing the production outcome in terms of quality, productivity and economical efficiency, this can be brought about by using control engineering to impact the complex interaction between the machine, tool and work piece as shown by the subsequent example of retrofitting a cross rolling machine with flat jaws. The reason for retrofitting this machine is not only to upgrade the usable performance range (rolling force or sled speed) and improve the control accuracy of the main axes (sled axes), but also to extend machine functionality (pendulum sled stroke) and integrate modular function axes (mandrel axes) into the higher-level control system. The mandrel axes designed as modular function axes are used for rolling the hollow components on the mandrel.

3. Strategy Development

The point of departure for meeting the target of retrofitting cross rollers with flat jaws was an analysis and description of process factors relevant to the forming technique that can be impacted by this machine structure such as forming force, forming speed and forming path. They are used to calculate the controller variables of speed, position and pressure applicable to the specific hydraulic linear actuators in conformity with the basic physical laws. The next step is upgrading hydraulic linear actuators to single hydraulic axes or single mechatronic components including or applying the necessary sensor technology (path sensor or pressure sensor) and basic regulation functionality (position and pressure). They were combined into functional groups (combined axes) as the overall mechatronic system of the cross roller with flat jaws in the way they interact in forming process factors and finally by including the entire machine structure. Even if the single hydraulic axes involved in the metal forming process have basic regulation functionalities, they are not sufficiently free of reactions among one another. This meant that it was necessary to study the effects that the various control circuits had on one another to draw conclusions on suitable higher-level control strategies and, in the final analysis, on control structures.

Suitable scenarios for the control structures were studied and analyzed using the means and methods of dynamic simulation on a complex component-oriented simulation model that can also replicate useful adapted control and process models. Matlab/Simulink was used as the simulation tool focusing on modeling the hydraulic drive system consisting of a total of four hydraulic cycles, although only two are of significance for implementing process factors. Two synchronization cylinders are used as the hydraulic linear actuators for the main function of the rolling sleds and two differential cylinders are used for the added function of the mandrel axes. Each of these hydraulic axes is impacted via one control circuit with an orthogonal effect (sled axes) or two control circuits with an orthogonal effect (mandrel axes). They are then combined to functional groups for the higher-level control functions (such as synchronization) as required by the rolling process. This overall system model broken down into control system, drive and process was used to simulate and determine an enhancement tool for the hydraulic drive system to come up with suitable control strategies for reliable process guidance. The basic regulating strategy shown in Figure 1 proved to be the one that best meets requirements under actual conditions.

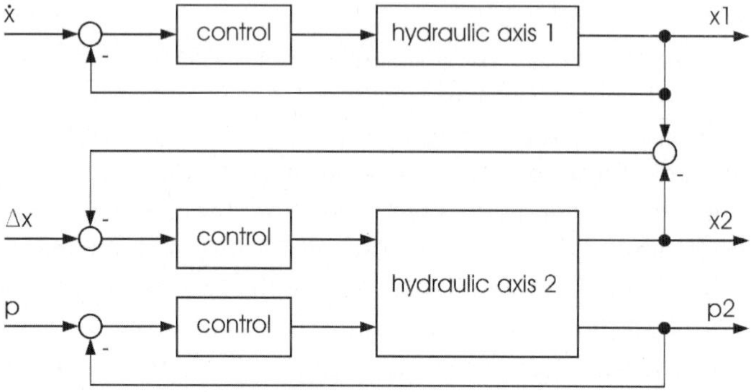

Fig.1. The basic regulating strategy

4. Strategy Implementation

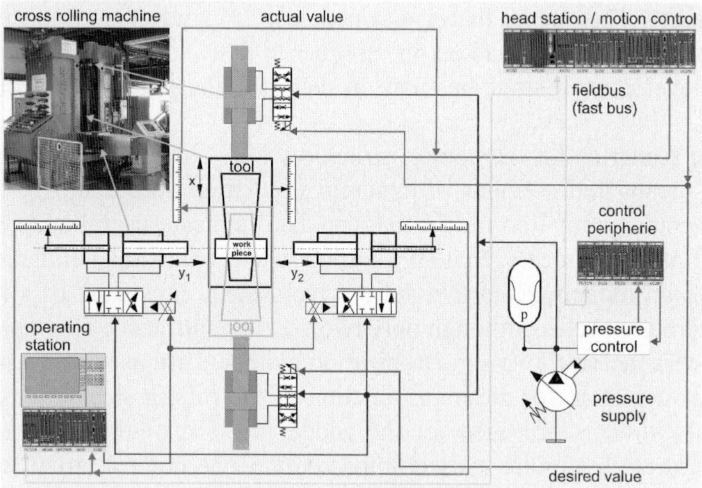

Fig. 2. The control structure

The control strategy for the cross rolling machine with flat jaws (including the visualization strategy needed) was developed and adapted to applications based upon the overall developed drive strategy and its basic regulating strategy. This control strategy was premised upon a uniform control platform with decentralized structures linked via bus system that are also used for implementing control functions. Then a motion control system was used as the control platform due to the major requirements that the technological properties make of regulating the axis functions of the metal forming process and the necessity of building the control system of freely configurable, structurable and scalable units because of the machine structure. This would not only have the advantage of combining the benefits of NC and PLC technology. This option also offers the possibility of implementing complex regulation functions in sufficient real-time with determined and reliable data communication even with control structures built decentrally such as the cross roller with flat jaws under consideration. Figure 2 shows the control structure as it was developed and built.

5. Results and Conclusions

Analyzing and upgrading the system to an overall mechatronic system aligned with the metal forming process meets the requirements of retrofitting the cross roller with flat jaws because the control strategy is sufficiently quick at regulating the complex structures of the hydraulic drive axes via decentralized bus structure at 0.5 ms of cycle time and at a maximum of 5 µs signal running time.

The subject matter of subsequent studies will be expanding and upgrading the process of the existing control strategies and directly integrating the cross roller with flat jaws into further process chains of bulk metal forming as overall mechatronic systems.

References

[1] R. Neugebauer, D. Klug, S. Noack, "Simulation of energy flow in hydraulic drives of forming machines" Australian Journal of Mechanical Engineering Vol. 2 (2005) No. 1, pp. 51–63

[2] R. Neugebauer, D. Klug, M. Hoffmann, "Mechatronical drive concepts for forming machines with electrical and hydraulic axes" 9th Scandinavian International Conference on Fluid Power (2005), Linköping (Sweden), Volume

[3] M. Hoffmann, T. Päßler, H. Koriath, A. Haj-Fraj, "Motion Steuerung in Umformmaschinen – Simotion-Applikation in einer Ziehpresse" Accuracy in Forming Technology (2006), pp. 399-408

Simulation of Vibration Power Generator

Z. Hadaš (a), V. Singule (b), Č. Ondrůšek (c), M. Kluge (d)

(a) Institute of Solid Mechanics, Mechatronics and Biomechanics, Faculty of Mechanical Engineering, Brno University of Technology, Technicka 2, Brno, 616 69, Czech Republic

(b) Institute of Production Machines, Systems and Robotics, Faculty of Mechanical Engineering, Brno University of Technology, Technicka 2, Brno, 616 69, Czech Republic

(c) Department of Power Electrical and Electronic Engineering, Faculty of Electrical Engineering and Communication, Brno University of Technology, Technicka 8,
Brno, 616 69, Czech Republic

(d) EADS Innovation Works, Sensors, Electronics & Systems Integration, Munich, D-81663, Germany

Abstract

This paper deals with the simulation of a vibration power generator that has been developed in scope of the European Project “WISE”. The vibration power generator generates electrical energy from an ambient mechanical vibration. The generator is a suitable source of electrical energy for wireless sensors which operate in vibration environment. When the generator is excited by mechanical vibration, its construction produces a relative movement of a magnetic circuit against a fixed coil. Thereby the movement induces voltage on the coil due to Faraday’s law. This paper describes the modelling of the vibration power generator in Matlab/Simulink.

1. Introduction

The aim of our work is the development of a vibration power generator, which generates electrical energy from an ambient mechanical vibration.

This generator shows an alternative for supplying wireless sensors with energy without the use of primary batteries. The parameters of the vibration generator are tuned up to the frequency and amplitude of the excited vibration. The design of the vibration power generator is tailored to the excited ambient vibration [2] and the appropriate designed vibration power generator can produce the required power. As the generator is excited by ambient vibration, the resonance mechanism produces a relative movement of the magnetic circuit against a fixed coil. This relative movement induces a voltage in coil turns due to Faraday's law.
The simulation modeling of this mechatronic system is very useful for optimization of the generator parameters. The generator model can be excited by sinusoidal, random or real vibration data and the expected generated output power and voltage are simulated in time domain.

2. Electromagnetic Vibration Power Generator

The model of electromagnetic vibration power generator consists of:

- **The Resonance Mechanism.** It is tuned up to the frequency of excited vibration and it provides a relative movement of magnetic circuit in relation to a fixed coil.
- **The Magnetic Circuit.** It provides a magnetic flux through the coil.
- **The Coil.** It is placed inside the moving magnetic circuit and it is fixed to the frame of the generator.
- **An Electrical Load.**

The individual parameters of mechanical and electromagnetic parts of this mechatronic system are in interaction. The design and parameters of resonance mechanism must be set up with dependence on required output power [3]. The electromagnetic parameters of the generator affect the behaviour of the resonance mechanism due to the dissipation of electrical energy from the oscillating system. The simulation model of this device can be used for setting up the generator parameters in dependence on required output power, overall size, weight etc.

3. Model of Vibration Power Generator

Design and parameters of the vibration power generator model are published in PhD. thesis [1] and the complex model of this generator is used for simulating modelling. The CAD model and real product of this generator is shown in Fig. 2. The resonance mechanics of the vibration power generator is tuned up to the stable resonance frequency 34 Hz of the vibration. The generator is excited by vibration with amplitudes in the range of 50 – 150 μm, i.e. the level of vibration 0.2 – 0.7 G. The generator is capable of generating electrical energy with an output power of around 5 mW and an output voltage of 2 Vrms for an average vibration level of 0.4 G. The generator dimensions are 50 x 32 x 28 mm and the generator uses a self-bonded air coil for harvesting of electrical power.

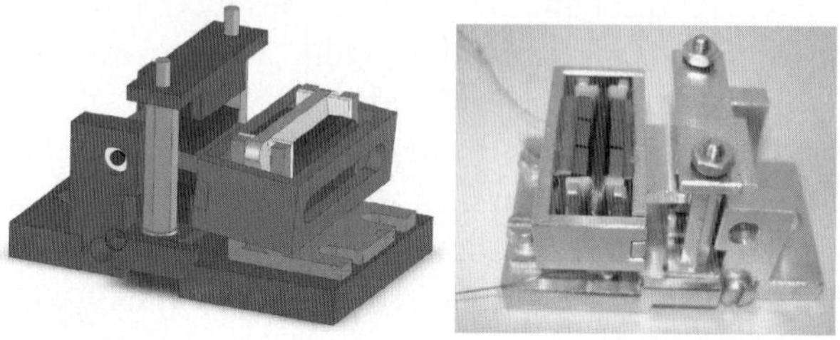

Fig. 1. CAD model and real product of the vibration power generator

4. Simulation of Vibration Power Generator

The Simulink environment was used for modelling of the generator as mechatronic system. The model consists of the resonance mechanism with models of the mechanical damping force, the electromagnetic circuit (magnetic circuit and coil) and the electrical load. The model is excited by sinusoidal vibration and the response of system is analysed.

The generated power depends on the quality of resonance mechanism. This parameter is represented by the mechanical damping force (primarily friction forces) represented by parameters F_0 and F_2. If the electromagnetic damping force in the generator, which generates useful electrical power, and the mechanical damping force in the resonance mechanism are equal, the generator harvests the maximal electrical power [3]. The electromagnetic damping depends on the magnetic circuit (B_x), the coil parameters (l, N, R_c) and the resistance of electrical load R_z. The magnetic flux and active length of the coil depends on the generator design. Others parameters are chosen optimally for generating of required output power and voltage. The number of coil turns and resistance of connected electrical load affect electromagnetic damping force and the number of coil turns is optimized to the appointed electrical load or inversely.

The complex model of the whole generator was created in SIMULINK and it is shown in Fig. 2. On the base of non-linear model [2] the friction coefficients are estimated for the actual design of the generator. The results of the simulation modelling and the measurement of real vibration power generator output are shown in Fig. 3. The electrical load 1 kΩ is used for both simulation and measurement. This model of vibration power generator corresponds with the real vibration power generator in the range from 0.2 – 0.7 G. The vibration power generator was excited with a resonance frequency of 34 Hz. The model provides the generated output voltage and power for a given time series of vibration data.

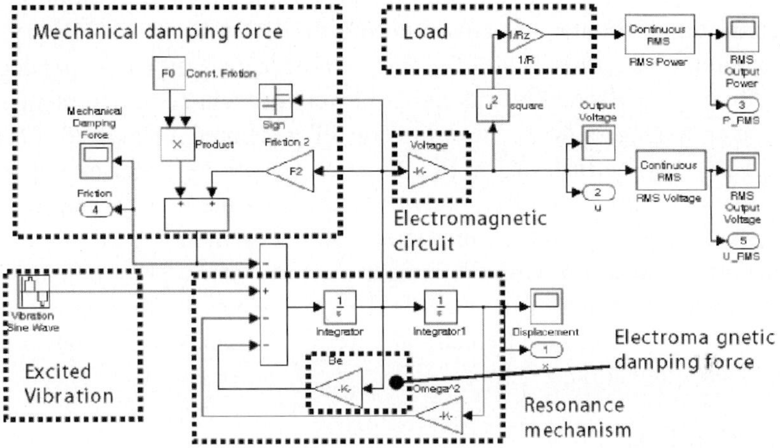

Fig. 2. Simulation Modelling of Vibration Power Generator

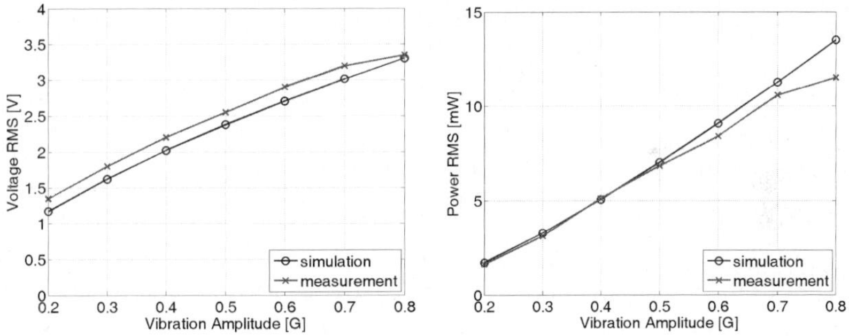

Fig. 3. Simulation and Measurement of Output Voltage and Power

The model of the vibration power generator shown in Fig. 2 can be excited by random vibration or real measured data of vibration. The model of the Grätz bridge (diode rectifier) and capacitor can be included in this simulation model too. As follows this model can simulate excitation by random or real vibration and monitor amplitude of the relative movement, rectified output voltage and actual output power. This process is very useful for design of optimal generator parameters.

5. Conclusions

The simulation of the vibration power generator is important for the design/tuning up of the generator parameters to its exciting vibration. This generator model is can be used for an optimization and minimization study too. The advantage of the simulation modeling of vibration power generator is the possibility to excite it with real vibration data and to monitor the expected output voltage and power during the excitation. It is very useful for designing the real product of vibration power generator in dependence of its vibration environment and output power and voltage requirements.
The power management for stabilization of generated voltage to the required value can be included in the vibration power generator model. The waveform of output voltage and power depends on the level of the applied vibration and the variation of vibration amplitude in time.
The development of the vibration power generator has a great potential as an inexhaustible source of the electrical energy. The vibration power generator can provide sufficient electrical energy for some wireless sensors in aeronautics applications.

Acknowledgement

The results published in this paper have been developed in the frame of the European FP6 Project "WISE - Integrated Wireless Sensing" (www.wise-project.org).
Additional subsidization has been received by the Ministry of Education, Youth and Sports of the Czech Republic, research plan MSM 0021630518 "Simulation Modelling of Mechatronic Systems".

References

[1] Z. Hadaš "Microgenerator – Micromechanical System" PhD. Thesis, Brno University of Technology, Faculty of Mechanical Engineering, Brno.
[2] Z. Hadaš, V. Singule, Č. Ondrůšek "Overall Tuning up of Vibration Generator and Choice of Energy Transducer Construction" 5th International Conference on Advanced Engineering Design 2006, Prague, 2006.
[3] V. Singule, Z. Hadaš, Č. Ondrůšek "Analysis of Generic Energy Harvesting Vibration Generator" 5th International Conference on Advanced Engineering Design 2006, Prague, 2006.

An Integrated Mechatronics Approach to Ultra-Precision Devices for Applications in Micro and Nanotechnology

S. Zelenika (a)*, S. Balemi (b)*, B. Roncevic (a)*

(a) University of Rijeka – TFR, Vukovarska 58, 51000 Rijeka, Croatia

(b) SUPSI – DTI, 6928 Lugano-Manno, Switzerland

Abstract

An effort to optimise both mechanical and electronic/control components of ultra-precision devices is presented. The considered mechanics is compliant, which overcomes the non-linearities of conventional devices. Design guidelines for hinge optimisation are given and a preliminary consideration of the scaling effects is performed. The developed control system is based on a rapid controller prototyping platform consisting of a Compact-PCI system running under the Linux RTAI real-time extension.

1. Introduction

Mechatronics is seen as the combination of mechanics, electronics, computer science and control. The focus of a mechatronics approach lies on the overall system behaviour, while the different components are seen as instrumental for obtaining the desired performances. In practice, the fact that the whole system is as good as its components is often forgotten. When considering dynamic behaviour as the most important issue, the impression that the model obtained from the identification procedure is valid in absolute terms often tends to prevail over the fact that different working conditions may produce unexpected results.

---

* The work was performed within the project "*Ultra-high precision compliant devices for micro and nanotechnology applications*" of the Croatian Ministry of Science, Education and Sports and the project "*A stronger Europe with micro and nanotechnologies (SEMINA)*" of the Swiss National Science Foundation.

This work follows an approach aimed at overcoming these limitations via the optimisation of all system components. The mechatronics device considered here and shown in Fig.1 is based on optimised compliant mechanical structures for ultra precision positioning (e.g. for handling and assembly of microcomponents or for STMs or AFMs). In fact, given the absence of mechanical non-linearities [1], compliant mechanisms are advantageous in high precision applications, allowing simple control typologies to be applied. The architecture of a single degree-of-freedom (DOF) optimised integrated mechatronics device is hence described.

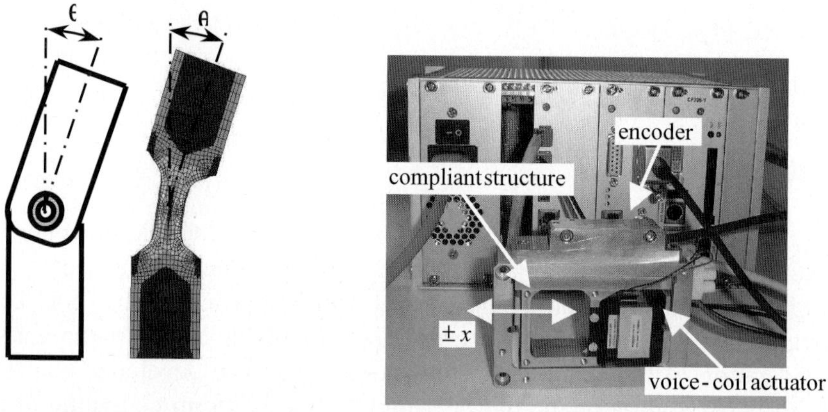

Fig. 1. Compliant joint and mechatronics device optimised in this work

2. Optimised mechanical structure

Mechanical aspects considered in the design process were the optimisation of the flexural hinge shapes (Fig. 2) in terms of compliance, strength and parasitic motions, as well as the scaling effects on the mechanical properties. Several hinge shapes were considered: the prismatic beam (P shape), the conventional right circular (RC) hinge, the optimal shapes obtained in classical mechanics (based on the authors indicated as the Grodzinski (G), Baud (B) and Thum & Bautz (TB) shape [2]), the optimised shapes obtained by coupling non-linear parametric optimisation algorithms with automatic FEM meshing and spline function generators like the optimised circular shape (OC shape), the optimised pure elliptical shape (OPE), the elliptical shape with $r_y = h_{min}/\pi$ (OEB) or the freeform optimised shape (FFO). Compliances around the primary hinge rotation DOF φ_z, as well as the transversal flexural (φ_y) and axial (x) directions were taken into account. It was thus established that the FFO and TB shapes will be the preferred choice when the goal is compliance maximisation along φ_z (Fig. 3a), while the G, B, OC and OEB shapes will be preferred when the parasitic shifts and the stress concentration in the axial and transversal directions are also important (Fig. 3b) [3].

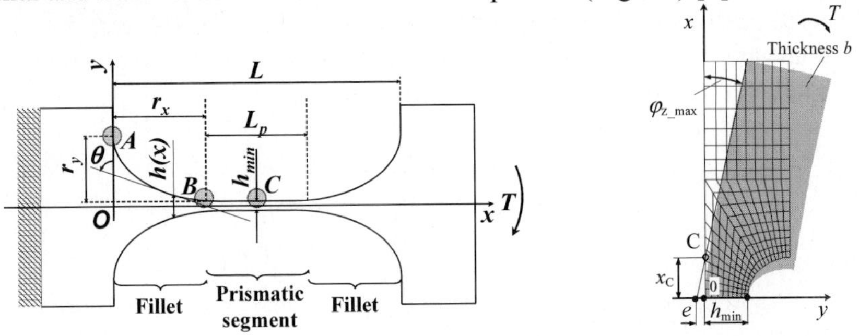

Fig. 2. Geometry of flexural hinge and parameterised shape for optimisation

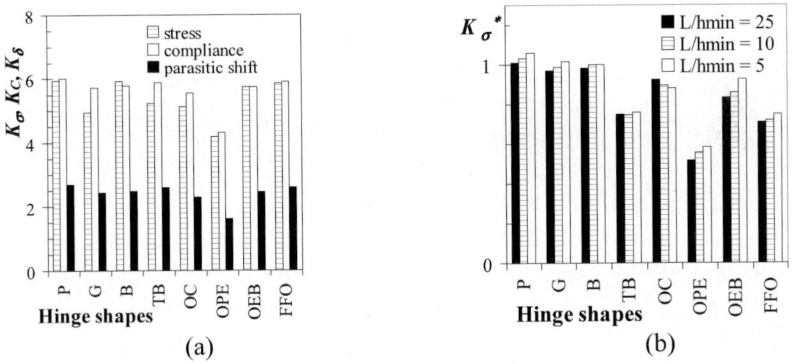

Fig. 3. Hinge behaviour along φ_z (a) and normalized stresses along φ_y (b)

When the considered applications are such that the dimensions of the mechanical structure must be minimised to nanometric levels, scaling effects on the entity of the mechanical characteristics must also be taken into account. In fact, it has been established that in the submicrometric domain the value of Young's modulus E can vary up to 70% with respect to its conventional value [4]. It is also known that the value of the Poisson coefficient ν is seldom known with an accuracy better than 20% [5], but its estimation at these dimensions has not yet been performed. An innovative methodology for determining ν is thus proposed here. The method is based on the calculation of the dynamic flexural response of Euler-Bernoulli-type cantilevers coupled with Von Kármán equations used to determine the variation of flexural stiffness of rectangular plates. In fact, the latter is a non-linear function of the deflection of the beam, varying from the value of E for plane strain (small loads) up to $E/(1-\nu^2)$ for plane stress (large load) conditions [6]. A seismic excitation of the cantilever with varying amplitudes will then result in an increment of the flexural stiffness, and thus of the frequency at which the response amplitude is maximal (Fig. 4). A suitable dimensioning of the cantilever of known E allows then a straightforward accurate determination of ν.

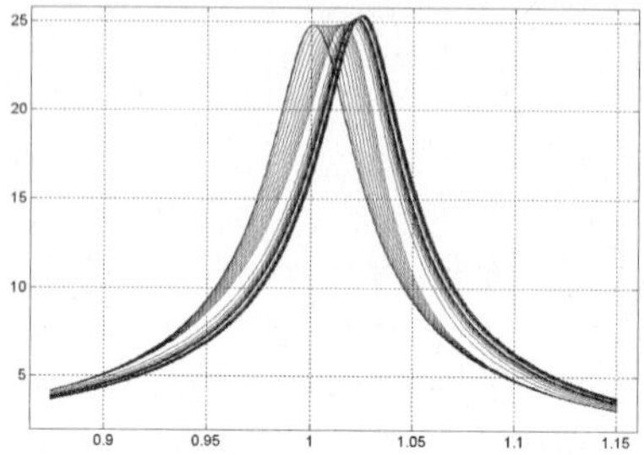

Fig. 4. Dynamic response of a micrometric silicon cantilever with $\nu = 0.22$

3. Actuators and sensors

Various actuating (DC micromotors, stepper motors, voice-coils, PZTs, inchworms, ultrasonic and inertial actuators) and feedback systems have been considered for the foreseen applications. Given the needed resolutions, accuracies and precisions, as well as the needed travel ranges, power requirements, ease of bidirectional control and large dynamic ranges, voice-coil actuators have been chosen. On the other hand, high resolutions and accuracies, excellent dynamic performances, large travel ranges and an easy integration with the compliant structure made optical encoders preferential over capacitive sensors, LVDTs and interferometric-based displacement measurement systems.

4. Control system

The control system is contained in a Compact-PCI rack with a power supply unit and a standard X86 processor computing board. The system exploits the results of the RTAI project (www.rtai.org), which offer real-time extensions of the Linux OS and interfaces with various CACSD tools (Matlab/Simulink or Scilab/Scicos). Within the same environment a graphical model can be prepared to feed the process with excitation signals and to retrieve data for the identification; the real-time application can send data to a remote PC, where the data is stored, displayed and analyzed.

The heart of the control system consists of two specially developed boards: a sinusoidal encoder signal interpolation board and a driver board for voice-coil motors. The three channel sinusoidal encoder interpolation board is built around a commercially available IC and it processes 1 Vpp sinusoidal signals. It is able to sample the inputs at a frequency of 500 kHz and to resolve 13 bits within a signal period. The three channel driver board has an output of up to 3.5 A per channel with a 16 bit resolution.
Other interface boards can be used as well: AD boards for various measurements or other Compact-PCI compatible boards. Their usage is immediate if the board is supported by the Comedi project (www.comedi.org); otherwise the drivers have to be written.
The initial integration of the control system with compliant mechanical structures allowed excellent performances with high flexibility and reliability at a limited cost. In fact, nanometric positioning accuracies (less than 15 nm) have been achieved in millisecond range time spans after a long (1 mm) range positioning step.

5. Outlook

The improvement in the design of the hinge shapes will be assessed with experiments. The objective is to accurately determine the stiffness of the structure as a function of the angle. Intuitively one would measure statically the dependence between motor currents and the resulting displacements and obtain the stiffness. However, this dependence is affected by the position-varying current-to-force characteristics of the actuator or by deviations due to the sensors' mounting inaccuracies. The tests on the structures will thus be based on the analysis of the resonance frequency at different positions. Precise estimates of the frequencies will be obtained using periodic excitation signals and the FFT analysis of the resulting data.

References

[1] L. L. Howell "Compliant Mechanisms" Wiley, New York, 2001.
[2] R. E. Peterson "Stress Concentration Factors" Wiley New York, 1974.
[3] S. Zelenika et al., Proc. 7th EUSPEN Int. Conf. (2007).
[4] B. Bhushan (ed.) "Springer Handbook of Nanotechnology" Springer, Berlin, 2004.
[5] M. J. Madou "Fundamentals of Microfabrication" CRC Press, Boca Raton, 2002.
[6] P. Angeli et al., Proc. XXXV AIAS Nat. Conf., (2006).

Conductive silver thick films filled with carbon nanotubes

Marcin Sloma (a), Malgorzata Jakubowska (b), Anna Mlozniak (b), Ryszard Jezior (c)

(a) Warsaw University of Technology, Faculty of Mechatronics, PhD student, Sw. Andrzeja Boboli 8 street, Warsaw, 02-525, Poland

(b) Institute of Electronic Materials Technology, 133 Wolczynska Street, Warsaw, 01-919, Poland

(c) Warsaw University of Technology, Institute of Precision and biomedical Engineering, Division of Precision and Electronic Product Technology, Sw. Andrzeja Boboli 8 street, Warsaw, 02-525, Poland

Keywords: thick film technology, carbon nanotubes,

Abstract

Search of new, better materials for conductive paste for thick film compositions deposited by screen printing was the main aim of this work as well as influence investigations of carbon nanotubes addition to functional phase of thick film paste. Different types of carbon nanotubes such as single-walled nanotubes (SWCNT), multi-walled nanotubes (MWCNT) and non purified non segregated nanotube clusters were mixed with conductive silver pastes. Obtained mixtures were screen-printed and fired or cured. The measurements of resistivity were carried out and changes in resistivity of obtained CNT/silver thick films in comparison to common silver layer were shown. The influence of different conditions of firing process were also investigated especially to avoid high temperature degradation and oxidation of carbon nanotubes. Obtained mixtures with optimal conductivity will be used for further experiments.

1. Introduction

Recently, carbon nanotubes (CNTs) have been demonstrated to possess remarkable mechanical and electronic properties, for example in field emission applications [1,2], chemical sensors [3] or as reinforcing material in composites [4,5]. CNTs can be synthesized by various methods such as arc discharge [6], laser vaporization [7] and chemical vapor deposition (CVD) [8,9]. In this case, purpose of carbon nanotubes addition is to create active centers in thick film layer, which can be used for instance as field emission source or adapted to very high frequency radio communication systems. However, there exists difficulty in fabricating high quality thick films with proper conductivity and containing reasonable amount of nanotubes. The paper presents first results of preparing thick film compositions containing carbon nanotubes. The authors encountered a lot of difficulties arise during the mixing process of the compositions Specially prepared thick film composition with selected nanotube material and silver nanopowder was screen printed and fired to obtain conductive layer. Direct measurement or conductivity parameter was carried out in reference to standard silver thick film conductive layer.

2. Materials and preparation

Three types of carbon nanotubes material were taken under investigation: single-walled nanotubes, multi-walled nanotubes, and nonsegregated material from synthesis process. The authors found that SWCNT caused a lot of difficulties in mixing process, and no suitable thick film composition was obtained. So to minimize time and costs of experiment, MWCNT material from synthesis was selected for further experiments. The material was taken directly from synthesis process without any purification or segregation, so it contained other carbon nanostructures. However SEM observations of used material showed that investigated material mostly contained multi-walled carbon nanotubes with small amount of amourphous carbon and some ferrous grains.

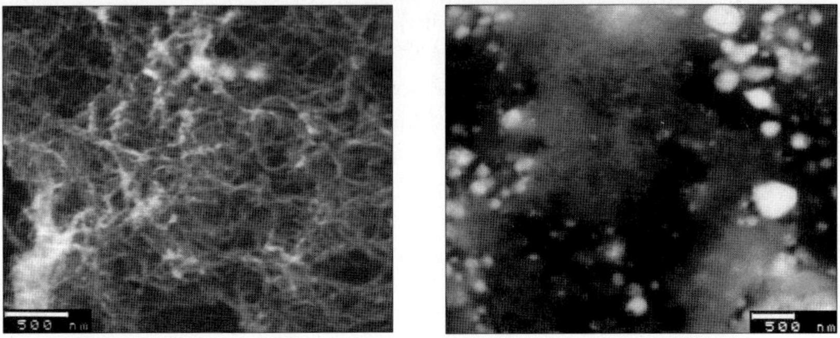

(a) (b)

Fig. 1. SEM images of carbon material: (a) Raw carbon nanotubes material. (b) The same material after firing in 650°C in air, showing ferrous grains and incineration residues.

Estimated mean dimension of nanotoubes was approximately 50nm in diameter and 1000-2000nm in length. Ferrous grains were estimated to be in size under 1000nm and were only observed after additional firing process in 650°C after degradation of nanotubes and other carbon structures.

Silver nanopowder used for Ag-CNT composition was obtained by chemical precipitation process by the authors and it was classified to be in range of 100-300nm dimensions with some addition of larger grains.

(a)

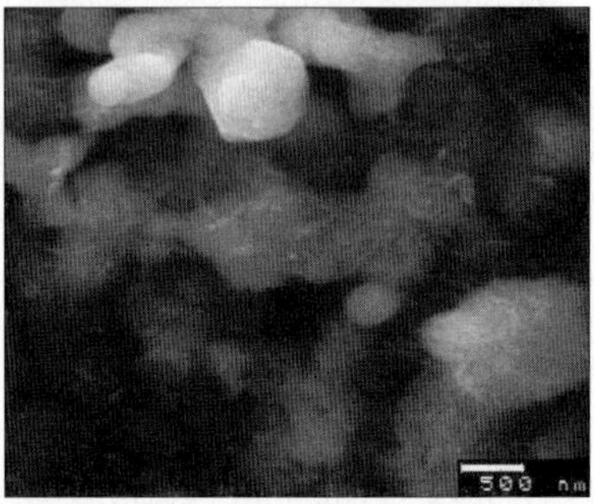

(b)

Fig. 2. SEM images of Ag nanopowder used for Ag-CNT composition (a) and Ag-CNT mixture (b).

As reference, standard silver thick film paste P-120 (ITME) was used, to prepare screen printed and fired conductive layer.

Compositions of nanomaterials were fabricated in specially adapted ultrasonic stirring process to obtain homogeneous mixture, which is essential to obtain well produced thick film layer with appropriate electric parameters. Two kinds of Ag-CNT compositions were prepared, regarding firing process. The compositions contained differend organic resins suitable for firing process conducted in air atmosphere (ethylcellulose resin) and for firing in protective nitrogen atmosphere (acrylate resin). Compositions were screen printed on alumina substrates in adequate pattern. Firing process for standard silver paste was conducted in 650°C in air, and for Ag-CNT pastes in 470°C in air and 650°C in N_2 respectively.

3. Results and discussion

Obtained thick film layers were well sintered in both cases with better results form specimen fired with higher temperature in nitrogene. Tracks fabricated from Ag-CNT material were practically indistinguishable from each other, whats proofs good selection of materials conducted at early stage. Unfortunately, against assumptions there where no nanotubes observed at surface of sintered layer. In spite of many observations it was impossible to detect them in any sample or region. It was caused by low amount of nanotubes in thick film composition. Only observations of scrapped away layer revealed that carbon nanotubes unmodified by the firring but hidden beneath main silver layer.

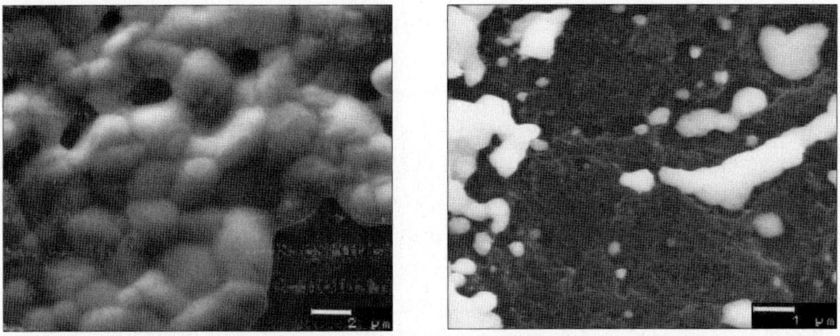

Fig. 3. SEM images of fired Ag-CNT layer, top view (on left), bottom view "scrapped away" (on right)

Resistance measurement was conducted with Keithley multimeter 2001 for both examined samples and for reference sample on the same length pattern, which allows to directly compare obtained parameters.

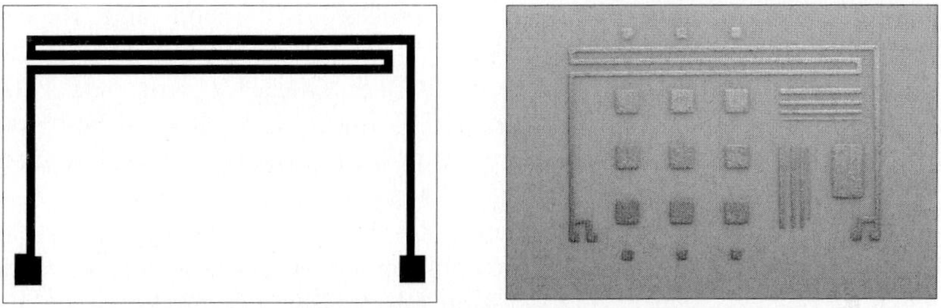

Fig. 4. Pattern of measured samples (drawing and object picture)

Obtained values of resistances for Ag-CNT compositions are varied from resistance for standard Ag thick film layer, but differences are not so significant. Main reason for that is that fabrication and firing processes are not yet optimal, and obtained layers are not ideal. But for instance taking under consideration very low temperature applied for air firing process, this layer have very fair conductivity and N_2 fired layer is in the same order of magnitude.

	Ag 650°C air	Ag-CNT 470°C air	Ag-CNT 650°C N_2
R [ohms]	0,4	1,8	0,8

Fig. 5. Comparison of measured resistances for examined samples

This results will be utilized in the further step of experiments where goal is to obtain thick film specimens with carbon nanotubes placed on surface of the layer.

4. Conclusion

One of the key issues in the development of thick film materials containing nanomaterials such as carbon nanotubes is controlling agglomeration process and aim for homogeneous mixture. It is crucial for obtaining well sintered layer which allows to gain proper electric parameters. The results presented in this paper demonstrate that specific selection of materials and fabrication procedures allows to overcome many inconveniences both during process and in resulted properties. Examined samples represented them selfs as very promising materials. Both layers have very fair conductivity which can be improved by optimization stirring and firing processes. After some major improvements materials will be used for further studies.

References

[1] T-Y Tsai, N-H Tai, I-N Lin "Characteristics of carbon nanotube electron field emission devices prepared by LTCC process", Diamond and Related Materials 13 (2004) 982–986

[2] N.S. Lee, D.S. Chung, I.T. Han, J.H. Kang, Y.S. Choi, H.Y. Kim, S.H. Park, Y.W. Jin, W.K. Yi, M.J. Yun, J.E. Jung, C.J. Lee, J.H, You, S.H. Jo, C.G. Lee, J.M. Kimb "Application of carbon nanotubes to field emission displays", Diamond and Related Materials 10 (2001) 265-270

[3] Kunjal Parikh, Kyle Cattanach, Rashmi Rao, Dong-Seok Suh, Aimei Wu, Sanjeev K. Manohar "Flexible vapour sensors using single walled carbon nanotubes", Sensors and Actuators B 113 (2006) 55–63

[4] A. Esawi, K. Morsi "Dispersion of carbon nanotubes (CNTs) in aluminum powder", Composites: Part A 38 (2007) 646–650

[5] A.R. Boccaccini, D.R. Acevedo, G. Brusatin, P. Colombo "Borosilicate glass matrix composites containing multi-wall carbon nanotubes", Journal of the European Ceramic Society 25 (2005) 1515–1523

[6] T.W. Ebbesen, P.M. Ajayan, Nature 358 (1992) 220

[7] A. Thess, R. Lee, P. Nikolaev, H. Dai, P. Petit, J. Robalt, et al., Science 273 (1996) 483.

[8] W.Z. Li, S.S. Xie, L.X. Qain, B.H. Chang, B.S. Zou, W.Y. Zhou, et al., Science 274 (1996) 1701.

[9] Z.W. Pan, S.S. Xie, B.H. Chang, L.F. Sun, W.Y. Zhou, G. Wang, Chem. Phys. Lett. 299 (1999) 97.

Perspectives of applications of micro-machining utilizing water jet guided laser.

Z. Sokołowski, I. Malinowski
Warsaw University of Technology,
Institute of Precision and Biomedical Engineering
Św.A. Boboli 8, 02-525 Warszawa, Poland, soko@mchtr.pw.edu.pl,

Abstract

The machining using laser energy stream concentrated within the water jet, having a length of several centimeters and the width comparable with the size of focus in traditional methods of machining, gives the ability of working on three dimensional, spatial elements. The cooling effect of water and its participation in removing the debris material, allow for attaining significantly higher quality of the work-piece surface and minimizes the harmful impact on environment and work safety. The capabilities of this method open for it applications, in which the traditional method of laser machining does not prove itself valid. In the article given is the principle of the method and the examples of its implementation. Determined are attained parameters of machining for various materials: silicon, gallium arsenide, ferrites, acid resistant steels, super hard materials. Given also are examples of application water jet guided laser machining in manufacturing technology of mechatronics.

1.The idea of laser machining in water jet.

General scheme of the laser beam concentrated in water jet is shown on the figure 6. Under pressure p the water flow out from the nozzle with velocity V. For small values of the diameter (d = 50…150 μm) and pressure >1MPa the typical equations of fluid mechanics need some corrections, however the water velocity V can be expressed as:

$$V \approx \sqrt{\frac{2p}{\rho}}, \quad \text{where } \rho \text{ is the water density} \qquad (1)$$

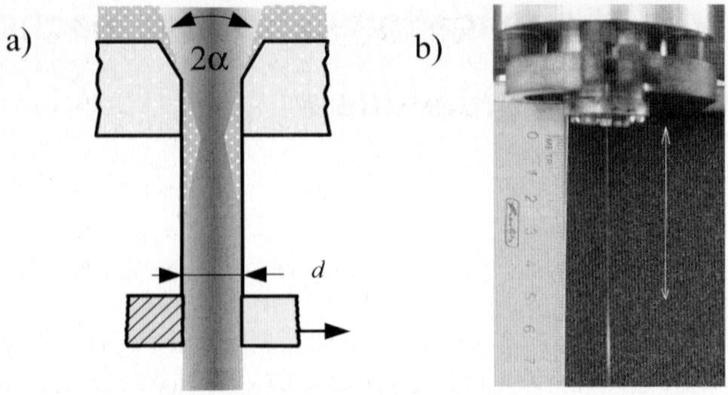

Fig. 1. a) Laser beam concentrated in water jet , where α is the angle of total internal reflection, b) the line work segment of laminar flow .

Water jet guided laser is a technique utilizing delivery of laser beam via total internal reflection of laser beam in water. This technique produces a water jet containing internally reflected and guided laser beam, which can be utilized for variety of purposes. The great advantages of water jet guided laser are:

1. Ability to maintain uniform beam width on a prolonged distance (approx 1000 times water-laser beam width) without divergence
2. Lack or minimum of thermal damage to cut material due to efficient cooling simultaneous with the laser ablation- "cold laser" – produces highly biocompatible cuts
3. Lack or minimum of impurities generated in water-laser jet process due to efficient material removal with water (efficient washing out of molten material)
4. Ability to produce high quality cuts or trenches with parallel walls, and of any shape
5. Minimum force exerted by water jet (approx 0.03-0.1 N)
6. Ability to utilize variety of laser wavelengths – most popular 1064 nm, (best results for green lasers such as Nd:YAG at 532 nm wavelength –where water absorption is minimum)
7. Ability to stop the laser light sideways movement with respect to work material without continued ablation – no need to retract or withdraw the beam
8. Small cutting radius (14 μm) and fine tolerance (1-3μm)
9. No burrs in the cut zone, no charring or contamination
10. No recrystallization, oxidation or micro cracks. Additionally the material is not distorting or warping. The tolerances of the final products are very small

11. Multi-layer, super-hard materials such as CBN (cubic boron nitride), PCD (polycrystalline diamond), PCBN (polycrystalline cubic boron nitride), or silicon nitride and porous material can be cut

The above advantages of water jet guided laser have caused it to be applied in medical device manufacturing [1], [2], [3]. Blades, scalpels, micro-tubes, stents and chips can be made using water jet guided laser machining.

Blades or scalpels are mainly used in surgical operations; they are also used by dentists, veterinary surgeons as well as by professionals in some other industries. A large amount of blades are produced for different applications. Blades are frequently made of steel, carbon steel, tungsten carbide, ceramics or silicon can be produced with a perfect cutting edge in spite of the complexity of the shapes and very economically . The main manufacturing tasks are: cutting out of slots, cutting of complex shapes, edging, drilling and grinding.

Tubes include a large array of devices such as cannulae, needles, or endoscopes. They are usually used for intra vascular operations so they tend to be less traumatic and more minimally invasive. Tubes are normally made of steel, stainless steel, titanium, or nitinol (NiTi). The main manufacturing tasks are: cutting out of slots, complex cutting, micro drilling of holes, edging. Tube wall thickness usually varies from 30 μm to 600 μm and require and excellent surface finish.

Microchips are integrated in sensors, transducers, pacemakers, auditory devices, or flex circuits. They use chips that have to be manufactured (e.g. wafer dicing, singulation). Wafers are typically made of silicon and their thickness vary from 20 to 1500 microns. The main manufacturing tasks are: wafer processing includes a large array of applications such as slicing, dicing, cutting edges, grinding, slotting and drilling. Multi-project wafers can be manufactrured. Complex shapes and grooves in PZT (piezoelectric ceramics) can be made using water jet guided laser for use in medical ultrasonics. Ferrite cores can be cut with uniform gap width (15μm wider than the width of water jet) for use in electronic circuits and advanced medical devices.

Cutting of medical stents Stents are used in medicine to open up, and maintain patent, congested blood vessels. They support and dilate the vessels' tissue and allow unimpeded blood flow. Usually they are made out of stainless steel or nickel-titanium. The contours, which have to be cut in these thin tubes, are very small and complex. Up to recently mostly classical laser cutting or chemical etching have been used. Classical laser cutting of stents however requires important post treatment by mechanical and chemical cleaning and polishing steps. The water jet guided laser inhibi-material changes – an important property for this application as product failure may be harmful to patient. The Laser Microjet cut stent-tubes are already sufficiently clean that the post treatment steps can be considerably shortened (see Figure 2).

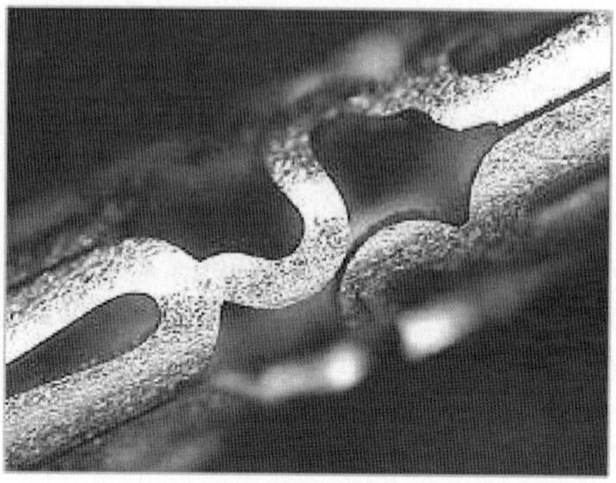

Fig. 2. 200 μm thick and 120 μm width Nitinol stent cut with the Laser Microjet (no further process)

The small radius of only 14 microns enables fine contours. Cutting speeds of more than 12 mm/s allow high throughput. The important advantages of the water jet guided laser in this application are: no heat influence, small beam radius, high speed, the smallest forces as well as the absence of contamination. The water jet guided laser is well adapted to medical devices; it can cut, drill, groove, mark, scribe or dice with a high degree of precision, speed, cleanliness and reliability. The problems related to heat damage, post-treatment, debris, deposition, or focal point, are totally removed. The future applications of water jet guided laser may include its wide use in biomedical engineering, as a surgical tool or technology for manufacturing of medical device elements.

Technical data of water jet process:

Wavelength: 1064 or 532 nm (solid state)
Pulse rep. rate: 25 kHz; 1000 Hz
Pulse duration: 200 – 600 ns; 50 – 250 μs
Max. laser power: 120 W; 300 W
Water jet pressure: 50 – 500 bar
Water jet diameter 30 – 150 μm
Water quality: filtered to 0.2 μm, de-ionized
Water jet length: 50 mm

Water jet speed up to 300 m/s (at 500 bars)

References

[1] Synova S.A.:Damage Free Cutting of Stents with SYNOVA Laser-Microjet - Application note No 111 by Synova S.A.

[2] Synova S.A.:Damage Free Cutting of Medical Devices with SYNOVA Laser-Microjet - Application note No 115 by Synova S.A.

[3] J.M. Wilkinson, Micro- and Nanotechnology Fabrication Process For Metals, Medical Device Technology, June 2004 p. 21-23

[4] Igor Malinowski, Delivery System for Laser Cataract Surgery and Method Thereof, US patent application and Ph.D. work at California Institute of Technology, USA

[5] Merdan, Kenneth M., Vertical stent cutting process and system, European Patent EP 1534462

[6] Ophardt Heiner, Water Jet Guided Laser Disinfection, World Patent publication WO2007000039, Canadian Patent CA25 10967

[7] Bernold Richerzhagen, Roy Housh, John Manley, New Hybrid Material Process: The Water Jet Guided Laser, SCI 2004

[8] Daniel Colladon, On the reflections of a ray of light inside a parabolic liquid stream, Comptes Rendus 15, pp. 800-802 (1842).

[9] Delphine Perrottet, Tuan Ahn Mai, Bernold Richerzagen, Wet Laser for Micromachining of Medical Devices, International Medical Devices Magazine Fall 2006

[10] Delphine Perrottet, Roy Housh, Bernold Richerzagen, Avoiding Material Damage with Cold Laser Cutting, MD&DI Magazine June 2005

[11] Delphine Perrottet, Frank Wagner, Roy Housh, Bernold Richerzagen, Hybrid Laser Process Cuts Medical Stents , Photonics Spectra August 2004

[12] http://www.synova.ch/pdf/microjet.pdf

[13] Drozd Z., Lasocki J., Sokolowski Z., Szwech, M., Miros A.: Research of Laser Micromachining of Silicon Wafers in Water Jet (in Polish) Final Report, Politechnika Warszawska 2001.

SELECTED PROBLEMS of MICRO INJECTION MOULDING of MICROELEMENTS

D.Biało, A.Skalski, L.Paszkowski *
* Warsaw Uniwersity of Technology, Institute of Precision and Biomedical Engineering, ul. A. Boboli 8 p. 152, Poland

Abstract

Tests results of filling micro-channels by synthetic polymers were displayed. Special test mould was applied, having ability of being heated or cooled. The material tested was polyethylene. Test results have been shown on Fig. 3 to 5. Filling tests were carried out depending on injected material temperature, injection pressure and mould temperature. The research has shown that the principal parameter is temperature of the mould, whereas temperature of the injected material and injection pressure play secondary role.

1. Introduction

Strong tendency for the development of microsystems technology (MTS) in the last decade is followed by constant growth of demand for microelements, i. e. products of sizes below 1 mm [1-2].

The microelements are both metallic and non-metallic. An important role among them play microelements made of synthetic thermoplastic materials, made in a way of moulding [3-6], which are usually of weight below 0.1 mg. Manufacturing of microelements is much more difficult than that of commonly produced macroelements. It is essential to change the injection parameters: pressure, temperatures of injected material and of the mould. Proper filling of a mould micro-chamber requires usually air evacuation at the moment preceding the material injection, due to difficulties in injective replication of smallest structural details of microelement. Injection process problems need to be a subject of detailed research. The presented paper shows test results on micro-channels filling in injection moulding of a thermoplastic synthetic material – polyethylene.

2. Research Procedure

Tests were conducted on the injection moulding machine C4/b made by ARBURG. Special injection mould was elaborated that has had a possibility of heating in a wide temperature range as well as of cooling.

Table 1. Dimensions of micro-channels.

Width	D, mm	0.05	0.15	0.2	0.3	0.4	0.5
Depth	h, mm	0.025	0.13	0.17	0.27	0.35	0.4
Cross-section	S, mm^2	0.001	0.018	0.033	0.077	0.13	0.18

Basic part of a mould is a moulding insert having shape of a plate provided with 8 micro-channels – Fig. 1. Micro-channels have shape similar to cross-section proportions ($h/D \approx 0.8$), but differ in respect of width and depth – Table 1.

Fig. 2 shows an example of an injection moulded test element. It replicates exactly the micro-channels existing in the moulding insert , also including the material supplying channel.

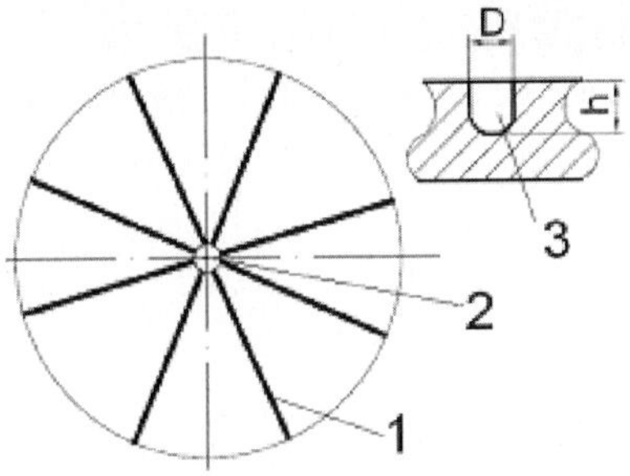

Fig. 1 Moulding insert: 1- micro-channel, 2 – cup cavity, 3 – micro-channel cross-section

Fig. 2. Example of a test element injection moulded

The mould set-up enables to simultaneously obtain information on several features of the moulding material behavior during one injection cycle. As it can be noticed, penetration of the material into individual channels, defined as the way of flow L or inflow distance L, is different in the case of each channel.

Following parameters of conducting tests were applied:

- Material temperature of 130°C to 160°C .
- Mould temperature of 20°C to 110°C.
- Injection pressure of 80 MPa to 180 MPa.
- Research material – polyethylene (PE).

3. Test Results

Tests determining following relations were carried out:

- Material inflow distance L into the micro-channel depending on injection pressure at a constant temperature value of injected polymer and mould temperature;
- Material inflow distance L into the micro-channel depending on mould temperature at a constant temperature of injected polymer and a constant injection pressure.

Test results are presented on the Fig. 3 to 5. The Fig. 3 is related to the length of polymer flow in the mould micro-channels depending on injection pressure at a constant value of polymer temperature and injection temperature.

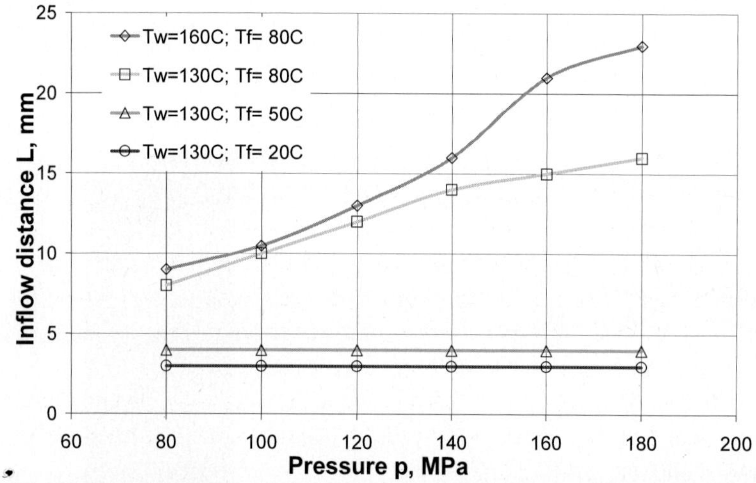

Fig. 3. Relation of micro-channels filling, depending on injection pressure p, for mould temperature T_f =20°C, 50°C and 80°C, as well as polymer temperature T_w 130°C and 160°C.

As it can be seen, impact of injection pressure is observed only in the range of higher mould temperature values and higher values of injected polymer temperature. For the values T_f = 20°C and 50°C (as well as T_w = 130°C and 160°C) pressure impact on flow distance does not exist, so that injected polymer stretches have similar length.

Result of the tests carried out related to polymer inflow distance L into the micro-channel depending on mould temperature at constant polymer temperature (130°C) and constant injection pressure (135 MPa) is shown on the Fig. 4.

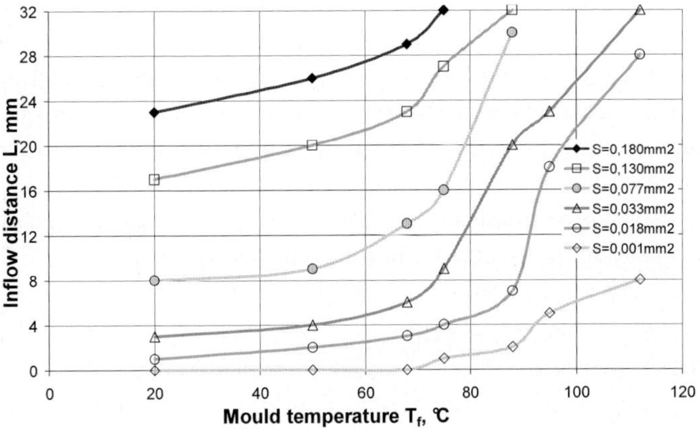

Fig. 4. Way of polymer flow in micro-channels of 0.05; 0.15; 0.2, 0.3; 0.4 and 0.5 as function of mould temperature at polymer temperature of 130°C and pressure of 135 MPa.

Basing on the measurement results it can be stated, that the mould temperature has considerable influence on the micro-channels being filled with polymer. Way of flow L grows with the temperature rising, particularly after certain border limit of temperature is exceeded. Then the way of flow becomes considerably longer. It was also noticed during the tests, that the higher is the mould temperature, the better and more tightly is the mould filled with polymer (even surface roughness irregularities are filled), which is not observed in the case of macro-channels.

It was also determined during the tests, what is the impact of channel cross-section change on thoroughness of the channel filling. The said relation is shown on the Fig. 5.

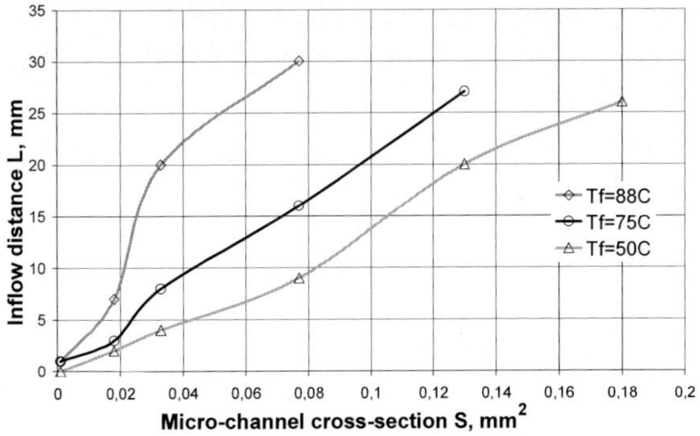

Fig. 5. Way of polymer flow L in relation to micro-channel cross-section S at different mould temperatures: 50°C, 75°C and 88°C. Injection conditions: polymer temperature of 130°C and pressure of 135 MPa.

It is visible on the Fig. 5, that the way of polymer flow grows very quickly for micro-channels of bigger cross-sections $10^{-1}mm^2$, whereas micro-channels of very small cross-sections at the range of $10^{-3}mm^2$ are almost not filled at all. It can also be seen from the Fig. 5, that filling a micro-channel of the smallest cross-section begins not sooner than at temperatures close to that of polymer contained inside the injection machine cylinder. It is worth mentioning, that a hot mould enables performing injection for producing tiny parts, but it additionally requires mould cavity to be of high precision in manufacture and finish.

4. Summary

On the basis of the test results achieved it can be concluded, that the proposed research method makes possible attaining a lot of valuable information concerning micro-moulding of elements made of thermoplastic polymers. In order to produce micro-elements it is necessary to heat injection moulds to high temperature. It is due to quick cooling-down of polymer in the mould, following small material volume when taking into account size of contact surfaces of the injected polymer and the mould. It means, that in order to obtain good polymer injection results, the mould must be thoroughly heated to 90°C and higher, depending on size of the channel cross-section. It is additionally required to cool-down the mould under pressure, which results in a longer injection cycle than the traditional one, while process efficiency is reduced.

References

[1] K.E. Oczos, Mechanik 5-6 (1999) 309 (in Polish)
[2] V. Potter et al. Proc. of 2000 Powder Metallurgy World Congress. 1652. Kyoto. Japan. Nov. 12-16. 2000
[3] C.G. Kukla et al. 6th int. Conf. Micro System Technologies 98. Potsdam. Dec. 1-2 (1998) 337.
[4] E.M. Kaer et al. Proc. of the 1st eupsen Topical Conference on Fabrication and Metrology in Nanotechnology. Copenhagen. May 28-30. 2000. vol. 1 259.
[5] R. Zauner, G. Korb, Plansee Seminar (2005) 59.
[6] V. Piotter at al. Proc. of SPIE Conf. Design, Test and Microfabrication of MEMs. Paris . France (1999) 456.

Estimation of a geometrical structure surface in the polishing process of flexible grinding tools with zone differentiation flexibility of a grinding tool

S. Makuch (a) *, W. Kacalak (b)

(a) Technical University of Koszalin, ul. Racławicka 15 – 17 ,
Koszalin, 75 – 620 , Poland

(b) Technical University of Koszalin, ul. Racławicka 15 – 17 ,
Koszalin, 75 – 620 , Poland

Abstract

In precision processes, the little of cutting layer section causes that the flexible of grinding tools exerts of meaningful influence on efficiency and results of polishing process. The flexible grinding tools with porous flexible bond creates the distinct group of grinding tools from their specific flexible properties. This properties causes that the active surface of grinding tool fitted to shape of working surface in working processes as opposed in grinding grains which are fixed in stiffness bond of grinding wheel. The flexibility of grinding grains decrease the depth of caving grinding grains on working surface and in consequence this process enables small values of roughness parameters of working surface and suitable cutting ability properties of grinding tools. The elastic properties of flexible grinding tools can be modification by zone differentiation the macrogeometry active surface of grinding tool. This modification enables using the small values of roughness parameters as well as suitable optical effects in bright polish. This article presents the results of polishing with the application of the flexible grinding tools of zonal differetation flexibility which was used for the new method of polishing process. In this method direction of lengthwise workpiece feed creates with the plane of grinding wheel a certain angle as well as was perpendicular.

1. Introduction

In precision processes, the little section of cutting layers causes that the flexible of grinding tools exerts of meaningful influence on efficiency and results of polishing process. Nowadays, the flexible grinding tools with polyurethane bond, extends the range of applications in the method of precision processes [1]. The flexible grinding tools characterized the specific flexible properties. This properties causes that the active surface of grinding wheel undergone a deformation and it enables decrease the diversification of mechanical burdens (individual pressures) as well as the depth of caving grinding grains in workpiece. The flexibility of grinding grains decreases the depth of caving grinding grains in surface workpiece and in consequence this process enables receives the small vales of roughness parameters of working surface with retain the large of remove properties material workpiece.

The elastic properties of flexible grinding tools with polyurethane bonds can be modification by zonal differentation the macrogeometry of active surface of the flexible grinding tool. This modification enables receives the advantageous exploational properties: the accuracy dimensional and shape of surface workpiece, the small vales of roughness and waviness of surface workpiece as well as decorative effects (composition of trace machining).

In typical of polishing processes the plane of grinding wheel rotation is parallel to direction of lengthwise workpiece feed. It causes that the grinding grains creates in working zone the long and irregular grooves and this operation is not good for the quality of polishing process.

2. Methodology of experimental research

The experimental research of exploational properties of the flexible grinding tools with polyurethane bonds and the zonal differentation flexibility of active surface of the grinding wheel was realized on the universal grinding machine type 4AM which produced on Poland. In the experimental research was used the disk – type grinding wheel T1A of polyurethane elasticity bond BPE and geometrical characteristic of grinding wheels: 125x20x20. The flexible grinding tools type E (elastic) and P (half elastic) was used in experimental research. The grinding grains of alundum 99A in grinding wheels has granulation 500 (the size diameter of grinding grain 13 μm).

The zone differentation flexibility of grinding tool was creates by incision the grooves along generating line of grinding wheel on half – width of grinding wheel surface. Lately on the grinding wheel surfaces was incision the helical grooves and on the grinding wheel surface was creates the cross grooves (the grooves along generating line of grinding wheel and helical grooves). The grinding wheels with continuous surface and discontinuous surface was defined the symbol C and NC. The grinding wheel with the cross grooves was defined the symbol K.

In the analysis and experimental research was used the four technological systems:

1 – The direction of lengthwise workpiece feed was parallel to axis of grinding wheel rotation and the continuous surface of grinding wheel, this technological system was defined the symbol C,

2 –The direction of lengthwise workpiece feed was parallel to axis of grinding wheel rotation and the discontinuous surface of grinding wheel, this technological system was defined the symbol NC,

3 – The flat surface rotation of grinding wheel creates with the direction of lengthwise workpiece feed the angle 45 degrees and the discontinuous surface of grinding wheel with the cross grooves, this technological system was defined the symbol K45,

4 – The flat surface rotation of grinding wheel creates with the direction of lengthwise workpiece feed the angle 90 degrees and the discontinuous surface of grinding wheel with the cross grooves, this technological system was defined the symbol K90.

The material of workpiece was steel 45. The samples before polishing process was grinding. The grinding wheel was dressing in order to improvement the accuracy of dimensional and shape grinding tool. In experimental research the peripheral speed of grinding wheel was constant (v_s = 14,5 m/s) as well as the workpiece speed was constant too (v_p =10 m/min).

After the polishing process the workpiece surface steel 45 was achieve the measurement of surface roughness parameters on the profile machine HOMMELTESTER T8000. From the estimation of the efficiency of polishing process was used the indicator of relative decrease surface roughness which compared the value of R_a parameter after polishing process to value of R_a parameter before smoothing surfaces as well as the mean value reflexivity of polishing surface which was measured in lengthwise direction and crosswise direction. This parameters as well as the frequency analysis of the profile of polish surfaces enables the comparison the new method of polishing process which was defined the symbol K90 with other methods which was defined the symbol C, NC, K45.

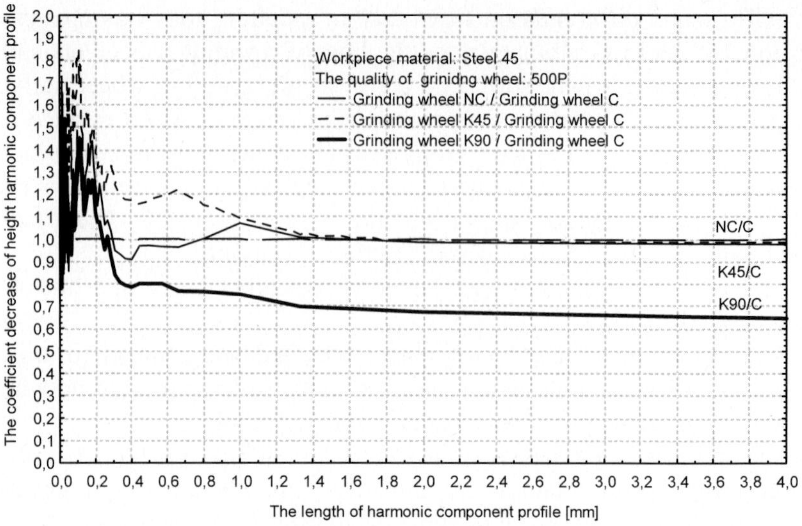

Fig. 1. The influence of technological system (NC, K45, K90) on the coefficient decrease of harmonic component of the profile surfaces steel 45 which was polishing the wheel 500P in comparison to technological system C

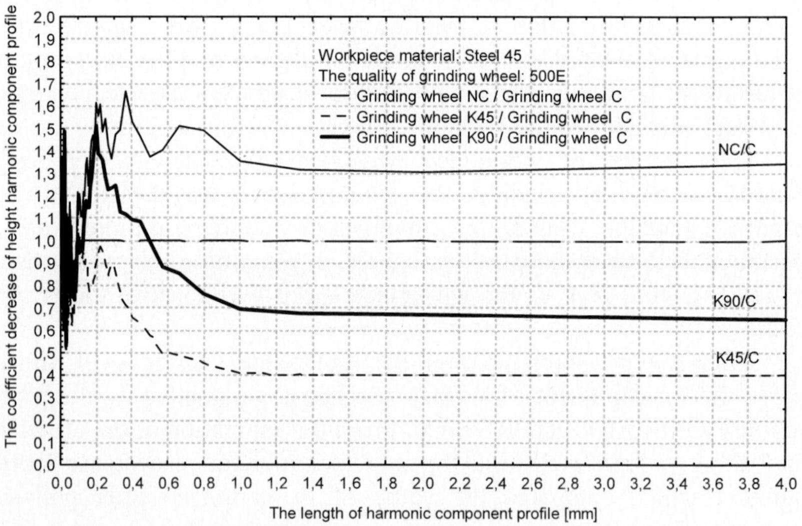

Fig. 2. The influence of technological systems (NC, K45, K90) on the coefficient decrease of harmonic component of the profile surfaces steel 45 which was polishing the grinding wheel 500E in comparison to technological system C

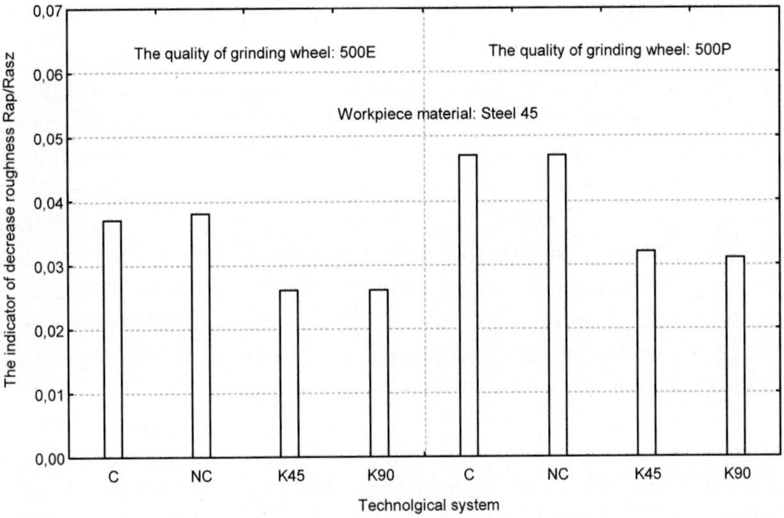

Fig. 3. The influence of technological system on the indicator of decrease roughness Ra_p/Ra_{sz} of the steel 45 surface which was polishing the grinding wheels 500E and 500P of polyurethane bond

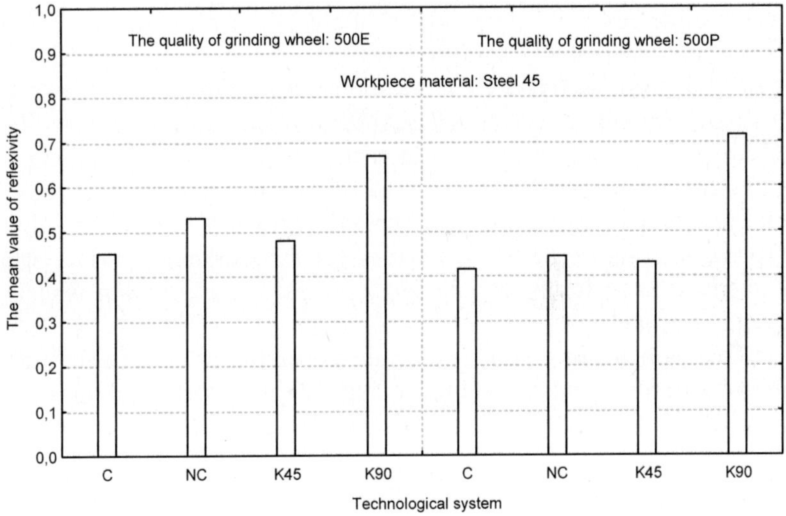

Fig. 4. The influence of technological system on the mean value of reflexivity steel 45 surface which was polishing the grinding wheels 500E and 500P of polyurethane bond

3. Conclusion

This article presents the estimation of technological effects the new method of polishing process which the flat surface rotation of grinding wheel creates with direction of lengthwise workpiece feed the angle 90°. On the frequency analysis was found that the best effectively of profile harmonic components about from 0,8 to 4 millimetres have the grinding wheel which the flat surface of rotation creates the considerable angle (in experimental research 45°) or is perpendicular to direction of lengthwise workpiece feed. The positive effects this technological system are the results of the envelope very much numbers of densely packing profiles of the grinding tool surface deflection.

The analysis of figures 1 and 2 shows that the grinding wheel 500P in technological system K90 enables decease of profile harmonic components to 65% in comparison to grinding wheel 500P in technological systems C and NC. For the grinding wheel 500E in technological system K90 the profile harmonic components decreased to 40% in comparison to grinding wheel 500E in technological systems C and NC. Moreover, on the figures 1 and 2 it notice decreases the amplitude of harmonic components about wavelength from 0,02 to 0,8 millimetres but the value of decrease is not so large as the value of harmonic components for wavelength from 0,8 to 4 millimetres.

The analysis of figures 3 and 4 shows, that setting of the flat surface rotation of grinding wheel under considerable angle decreases ability to creation the long and irregular scratch in working zone of grinding tool and enables get the workpiece surface of low roughness. When the flat surface rotation of grinding wheel creates with direction of lengthwise workpiece feed the angle 90° it enables get the workpiece surface which characterized large luster in polishing process of the grinding wheels of polyurethane bond.

This conclusions shows that in polishing process necessary is applied the new method which the flat surface rotation of grinding wheel is perpendicular to direction of lengthwise workpiece feed.

References

[1] Sung-San Ch., Yong-Kyoon R., Seung-Young L.: Curved surface finishing with flexible abrasive tool. International Journal of Machine Tools and Manufacture 42/2002, 229 – 236.

Fast Prototyping of Wireless Smart Sensor

T. Bojko *, T. Uhl

KRiM AGH-UST Cracow, al. Mickiewicza 30,
Cracow, 30-059, Poland

Abstract

Recent advances in microelectronics have brought about the possibility to integrate powerful digital electronic, radio communication and sensors in one small package or simple and fast integration ready-to-use elements like controllers, radio modems and sensors in one advanced product. The new Digital Signal Controllers (DSC) make it possible to design the small smart sensors achieving computational power equal to Digital Signal Processors (DSP) but with lower requirements concerning peripheral devices like ADC, DAC, PWM, IO pins and memory. Wireless data transfer and DSC enhanced features enables to build advanced wireless sensors for many applications.
Application of wireless sensor networks and distributed computing results in leads to new advances in measurement and monitoring applications of large scale structures like buildings and bridges. Small dimensions, low power consumption and digital signal processing functions are especially advantages in design of applications, where smart wireless sensors are distributed over the large civil engineering structures for monitoring and damage detection purposes.
In the paper fast prototyping of the advanced smart wireless sensor based on DSC is presented.

1. Introduction

Currently available advanced microcontrollers provide to its user with great potential for development of smart applications. Many domestic appliances are equipped with microcontrollers which rise up its functional properties. Modern cars are equipped with tens of microcontrollers working in engine controls, brake systems and safety circuits.

There are many other applications like communication, web services and multimedia which require fast and specialized chips like ARM based microcontrollers [1].

In many applications of microcontrollers processing of digital signals, acquisition and processing of analogue signals are required. Computational procedures are based on fixed point arithmetic and are performed with high computational effort. This is the critical point of those applications and a very important limitation in design of smart sensors, where many operations like: signal filtering, data decimation, spectral analysis, model parameter estimation, should be performed. To overcome this problem microelectronics manufactures deliver new hardware solutions like: Digital Signal Controllers (DSC), which combines powerful I/O capabilities of standard microcontrollers and computational power of Digital Signal Processors (DSP) [2].

The second advanced technique which is nowadays widely used is wireless data transfer. Advances in microelectronics and strong demand for cable reduction give rise to market of ready-to-use RF solutions based on: Bluetooth, ZigBee or 802.15.4 standards [3]. Most of commercially available radio modems are characterized by: possibilities to implement different networking strategies, long range operation, low power consumption during idle stage and simple connection with existing systems with aid of serial interfacing. Cost of radio modems is decreasing rapidly in due to vast amount of available solutions and RF components manufacturers.

Authors of the paper are currently working on implementation of smart wireless sensors networks (SWSN) [4] for monitoring and diagnostics of structures like buildings, bridges, towers, ships and aircrafts. Data from SWSN can be used for development of modal models or as an input to structural health monitoring (SHM) systems [5]. The technology of the online SHM and damage assessment are complex and sill under development. SWSN are the best solution for monitoring applications and SHM, since they make it possible to increase the number of measurement points and systems simplification due to cable reduction. The currently available wireless sensors solutions are collecting data from the object and perform data transfer to the central base station for further processing [6]. In the paper there is presented development of active wireless sensor for SHM purposes. At the development stage the commonly available software for fast prototyping and the advanced radio modem based on 802.15.4 standard were applied.

2. Wireless smart sensor architecture

The block diagram of the designed wireless smart sensor (WSS) is presented in the figure 1.

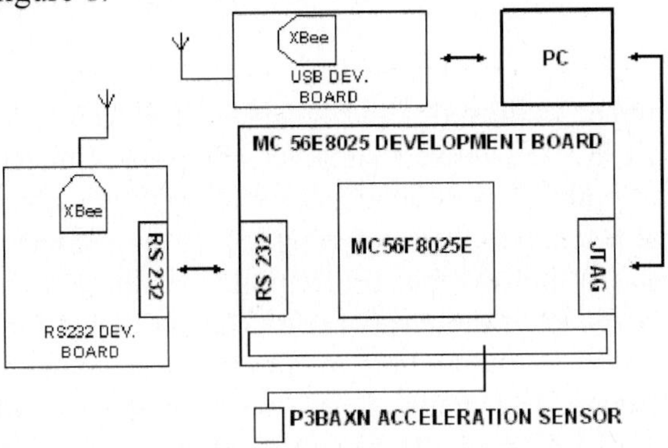

Fig. 1. Block diagram of designed wireless smart sensor

The designed WSS is built on the basis of the Freescale ultra low cost MC56F8025E DSC [7]. This new device gives microcontroller functions like: extended bit manipulations, flash programming and JTAG debugging functionalities, combined with advanced DSP processor possibilities like: fixed point extended set of arithmetic functions including 16x32 bit fractional multiplication, up to four parallel data move instructions in one cycle clock. The device has built in 12 bit – 8 channel analogue to digital converter and offers a lot of configurable hardware resources (timers, counters, PWM, serial communication). The possible applications of this DSC include: automotive applications, motor control, smart sensors and power supply units. The PB3AXN from Oceana Sensors Inc. was chosen for acceleration sensing. This piezoelectric based accelerometer is equipped with integrated voltage output amplifier. Finally, as a wireless data link to the sensor, the XBee™ module from the MaxStream was chosen [8]. This newly developed low cost wireless module works with 2.4MHz and gives great configurational capabilities allowing free configuration for 802.15.4 protocol or low power ZigBee standard. The module has low power requirements equal to 50 mA in receiving and 10μA in standby mode. The working range of the XBee™ lies between 30 and 1600 m depending on surrounding conditions and version. This module works with protocol 802.15.4 and gives of direct RS232 replacement of hard wired serial communication to wireless standard.

Based on standard DSC and XBee™ development boards the complete initial circuit of the wireless smart sensor was assembled and tested.

3. Fast software prototyping and testing

The software development was based on the CodeWarior (CW) intergraded environment, delivered with MC56E8025 DSC development board. The integrated CW IDE is well suitable for project management, code edition, code compilation and debugging with minimal effort of programmer. Additional fast software prototyping technology based on the Processor Expert (PE) [9] allows fast configuration of the DSC resources and makes it easy to develop the target code. The PE technology is based on ready-to-use hardware (bit handling, memory management, serial configuration, ADC converter operations) and software beans (advanced DSC based functions like filtering, PID controllers, and spectral analysis). The selected bean, after placement in the project, must be configured with the appropriate hardware resources. After initialization, bean deliver ready-to-use methods related to connected hardware. All the necessary code for interrupts used by the bean is automatically generated. Programmer can easily use all bean resources by the mouse click operations.
During development stage it is very important to have direct access to the processed signals and possibilities to adjust or monitor program variables. By means of the FreeMaster tool, delivered with CW, it is easy to visualize and adjust program variables. This software enables the users to monitor and adjust up to eight variables. During development of the wireless sensor this software was used with described in the previous chapter wireless XBee™ modules. The measured data was send-out using the radio module connected directly to RS232 output of the development board. The maximum data transfer achieved during tests was 57600 bps.
The first developed program is dedicated to calculations of the Discrete Fast Furrier Transformation (DFFT) from acquired acceleration data, its schematic block diagram is presented in the figure 2. This was done using internal PTO1 DSC timer and proper handling of the ADC end-of conversion interrupt. The sampled data was written to internal buffer. The size of the buffer was defined by the further applied DFFT procedure. In the next step the DC offset form the data stored in the buffer was eliminated, buffer was filtered by the Hanning window and the DFFT procedure was executed. In the beginning the 16 samples data buffer was used for the DFFT calculation. The execution time of the complete procedure was measured. The measured time was equal to 52μs, which gives great possibilities to the

real time calculations of the DFFT with higher resolutions. The limitation of the buffer length is to the size of internal RAM memory of DSC, which must handle application data and measurement data. Initial trials proved that it is possible to implement 256 data buffer, which gives 1Hz resolution sufficient for most of SHM applications. In the figure 2 the schematic block diagram of developed software is presented.

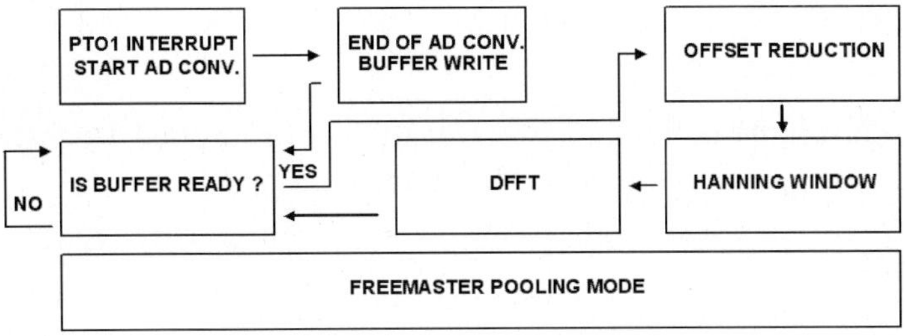

Fig. 2. Schematic block diagram of developed software

4. Conclusions

Advances in modern microelectronics and development of software tools simplify design of smart sensors. In the paper there was presented fast development procedure as well as brief description of the used hardware and software for development of the wireless smart sensor for SHM purposes. The applied software allow to fast design and test DCS procedures, which are necessary for advanced monitoring and damage detection sensors.

This research was financed from the research project 4T07B01029 (2005-2007).

References

[1] www.arm.com

[2] www.freescale.com

[3] IEEE 802.15.04 – 2006 ,Wireless MAC and PHY Specifications LR-WPANs

[4] A. Hac, “Wireless sensors network design”, John Wiley, January 2004

[5] D. Balageas, C.P. Fritzen, A. Guemes, “Structural Heath Monitoring”, ISTE Ltd Wydanie I, London, 2006

[6] Ning Xu and others, “A wireless Sensor Network For Structural Monitoring”, proc of SenSys 04, November 3-5,Baltimore 2004, USA

[7] 56F8000 16-bit DSC, MC56F8025, Freescale Rev.3, 01/2007

[8] Product Manual v1.xAx - 802.15.4 Protocol, MaxStream ,2006

[9] www.processorexpert.com

Microscopic and macroscopic modelling of polymerization shrinkage.

P. Kowalczyk

Warsaw University of Technology, Department of Applied Mechanics, Św. Andrzeja Boboli 8 Street, 02-525 Warsaw, Poland

Abstract

During polymerization of monomers into polymer chains a reduction in volume is observed. Shrinkage may cause internal stresses and strains of a final product. It makes some difficulties in technical applications of polymers, especially in dentistry and stereolithography. In the presented paper, two ways of modelling and descriptions of the polymerization shrinkage will be described: the molecular approach based on chemical analysis, and the phenomenological approach based on investigation of density changes in polymer-monomer system. Because the polymerization is time dependent process, the rate of reaction and the course of reaction are important issues. The influence of viscosity and vitrification on the rate of polymerization and finally on the polymerization shrinkage will be described.

1. Introduction

Cross-linked materials formed by the polymerization are used in a number of commercially important applications, e.g. coatings, dental restorative materials, microoptics, UV-NIL processes (UV – nanoimprint lithography), stereolithography. During polymerization a liquid monomer is converted into a solid polymer. Inseparable phenomenon of the process is a volumetric shrinkage of the polymer-monomer system. The polymerization shrinkage is especially important in dentistry (results in gaps between tooth and filling), and in stereolithography (results in shape distortion of a model). In general, shrinkage results in internal stresses in a cured system. Shrinkage can vary for different materials: about 20% for PMMA (poly-methyl-methacrylate) and 3-6% for stereolithography resins. It can be calculated from data based on chemical construction of polymer or from measured densities of polymer and monomer. Evolution of the shrinkage during curing process is an important issue. Prediction of a course of polymerization shrinkage as a function of time may give an information about evolution of internal stresses during curing process.

2. Molecular approach

Shrinkage is due to the replacement of relatively weak long distance intermolecular Van der Waals bond by strong, shorter, covalent bonds between the carbon atoms of different monomer units. It results in density changes during proceeding process from monomer to polymer.
Simple molecular approach for polymerization shrinkage for polymerizaton of MMA (methyl methacrylate) is described in the paper [1]. The percentage relative change in volume (volumetric shrinkage-strain) is given by the rule:

$$S[\%] = 22{,}5 \cdot p \cdot f \cdot \frac{V\rho_m}{M_m} \cdot 100\,, \tag{1}$$

where: p – degree of double bond conversion, f – functionality of monomer, ρ_m – density of monomer, V – system volume, M_m – molecular weight of monomer.
According to experiments, the volume change per mole of methacrylate groups (C=C) in MMA takes the value $\Delta V_{C=C} = 22{,}5$ cm^3/mol [1].
The above semi-empirical equation assumes a linear relationship between shrinkage S and the degree of conversion p. The assumption is not always correct. For most resins this relationship is linear for low degrees of conversion. For high degree of conversion (about 60%), the relationship is no longer linear due to vitrification. The polymer chains lose their ability of displacement when system vitrifies. Conversion is continued, but shrinkage remains relatively constant [2]. This phenomenon is mainly observed for highly functionalised monomers ($f \geq 3$), which can produce cross-linked structures. High functionalised monomers may provide us with high reactivity without too large shrinkage. On the other hand, a high escalation of internal stresses occurs after vitrification of a system. For low degrees of conversion (<50%), relatively low internal stresses are observed [3].

3. Phenomenological approach

Polymerization volume shrinkage can be defined as follows:

$$S[\%] \equiv \frac{|\Delta V(\Delta t)|}{V_1} \cdot 100, \tag{2}$$

where: V_1 – the initial volume of a cured system, $\Delta V(\Delta t)$ – an increase in volume during polymerization in the time period $\Delta t = t_2 - t_1$, for t_1 – the beginning of polymerization process and t_2 – the end of the process.
During polymerization the total mass of the monomer-polymer system remains constant. Change in polymer and monomer fraction in the system results in a volume decrease. Total mass of a cured resin at the moment t is:

$$m(t) = m_m(t_1) + \Delta m(\Delta t) = const, \tag{3}$$

where: $t_1 \leq t \leq t_2$.
Increment of the total mass is a sum of the mass increments of monomer and polymer. According to the mass conservation law the total mass increment is equal to zero:

$$\Delta m(\Delta t) = \Delta m_m(\Delta t) + \Delta m_p(\Delta t) = 0. \tag{4}$$

Equation (4) expressed by density of monomer ρ_m and density of polymer ρ_p is the following:

$$\rho_p \Delta V_p(\Delta t) + \rho_m \Delta V_m(\Delta t) = 0, \tag{5}$$

where: $\Delta V_p(\Delta t)$ is the increase of the polymer volume, and $\Delta V_m(\Delta t)$ – the increase of the monomer volume.
Total increase of the volume of the system is as follows:

$$\Delta V(\Delta t) = \Delta V_p(\Delta t) + \Delta V_m(\Delta t). \tag{6}$$

From equations (5) and (6) we gain the following relationship:

$$\frac{\Delta V(\Delta t)}{\Delta V_m(\Delta t)} = \frac{\rho_p - \rho_m}{\rho_p}. \tag{7}$$

The above result may be presented in the form:

$$S_t[\%] = \frac{|\Delta V(\Delta t)|}{V_1} \cdot 100 = \frac{\rho_p - \rho_m}{\rho_p} \cdot 100 \tag{8}$$

The equation (8) is a main relationship for the total polymerization shrinkage and it is widely used for experimental determination of shrinkage [2][4].

Now, let us consider polymerization shrinkage at arbitrary moment t_a of the polymerization process, Denote by $\rho(t_a)$ – a density of polymer-monomer system at the moment t_a. The total mass of the system at that time is:

$$m(t_a) = \rho(t_a)V(t_a) = \rho(t_a) \cdot (V_1 - \Delta V(\bar{\Delta} t)) = \rho_m V , \tag{9}$$

where: $\bar{\Delta} t = t_a - t_1$.

Similarly to the relation (8), one can obtain from the above, the shrinkage of the monomer-polymer system at an arbitrary time t_a:

$$S(t_a) = \frac{\Delta V(\bar{\Delta} t)}{V_1} = \frac{\rho(t_a) - \rho_m}{\rho(t_a)} = 1 - \frac{\rho_m}{\rho(t_a)} . \tag{10}$$

Equation (10) describes a temporal shrinkage of the system, which is not fully cured. Moreover, it can be used for prediction of thermal strain by modification of densities.

Other approach to determine shrinkage is to use function of shrinkage changes gained by experiments [5]. Such approach is applied for prediction shrinkage of photocured resins (Fig. 1).

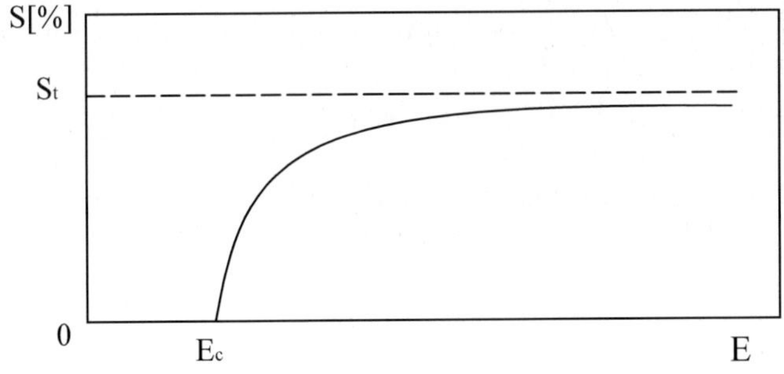

Fig. 1. Polymerization shrinkage S vs. exposure E

In that case, the shrinkage is proportional to the light exposure dosed during curing process

$$S(E) = S_t\left(1 - e^{-K_S(E-E_c)}\right), \tag{11}$$

where: S_t – total shrinkage of fully cured resin, K_S – constant value, E_c – critical exposure (exposure which start polymerization). Similar equation is given in paper [6].

4. Conclusion

Two ways of description of polymerization shrinkage are presented: microscopic (molecular) approach and macroscopic (phenomenological) one. Molecular approach describes essential principles of polymer behaviour, but it demands more information about monomers. Phenomenological approach is based on the macroscopic quantities those can be easy obtained by measurements. Moreover simple time dependent equations of the shrinkage are shown in the paper. These relations may be helpful for determination of stresses evolution. Phenomenological approach may be expanded to thermal effects, which can affect the shrinkage. Reaction of polymerization is always exothermal. Then the thermal expansion and the polymerization shrinkage result in final strain and internal stresses of fotocured resins.

References

[1] N. Silikas, A. Al-Kheraif, D. C. Watts, Biomaterials 26 (2005) 197-204
[2] UV Curable Monomer Properties: Shrinkage and Glass Transition, Startomer – application bulletin.
[3] J. W. Stansbury, M. Trujillo-Lemon, H. Lu, X. Ding, Y Lin, J. Ge, Dental Material (2005) 21, 56-67.
[4] D. A. Tilbrook, G. J. Pearson, M. Braden, P. V. Coveney, J. Polym. Sci.: Part B. Vol. 41, 528-548 (2003).
[5] P. F. Jacobs, "Rapid prototyping & manufacturing, fundamentals of stereolithography", Society of manufacturing engineers, Deaborn, MI, USA 1992.
[6] Y. M. Huang, C. P. Jiang, Int. J. Adv. Manuf. Technol. (2003) 21: 586-595.

Study of Friction on Microtextured Surfaces

G. Ionascu (a), C. Rizescu (a), L. Bogatu (a), A. Sandu (a), S. Sorohan (a), I. Cernica (b), E. Manea (b)

(a) "POLITEHNICA" University, 313, Spl. Independentei, Bucharest, 060042, Romania

(b) Institute of Microtechnology, Str. Erou Iancu Nicolae 32 B, Bucharest, 077190, Romania

Abstract

This is a part of a study concerning influence of surface material and topography on tribological behaviour. The determinations were developed using an experimental setup, which mainly consists in a safety clutch.
The clutch, basically, consists of a steel disk and a special frictional disk manufactured from a ferodo disk. The steel disk was changed with another friction material manufactured as silicon wafers. Friction coefficient determining tests for the textured silicon/ferodo disk interface were performed. There are given, also, results of modeling of the constructive elements in contact, for the motion transmission, by using the finite element method in order to evaluate the contact pressures.

1. Introduction

From the group of intermittent clutches, there can consider being also a part of them, the safety clutches, which permit the automatic disengage of the link between two shafts, when their load or speed exceeds one admitted limit.

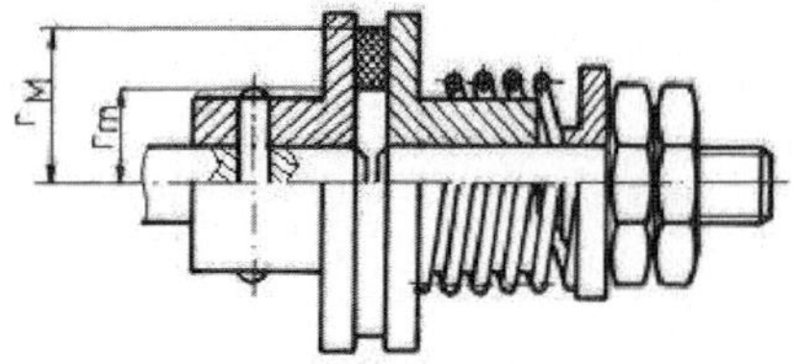

Fig. 1. The principle of a safety clutch

The most practically ones are the slipping clutches that permit the relative slipping of the surfaces making the link, at overcharge, when the transmitted torque is greater than the friction torque.
In fig. 1 is represented the usual safety clutch, which can be a slipping clutch. The clutch, basically, consists of a steel disk and a special frictional disk manufactured from a ferodo disk [1].

2. Experimental setup and results

In fig. 2 is shown the experimental setup, for which were made the notations: 1 - rotational electric motor; 2 - belt transmission; 3 - safety clutch with 2 steel disks and a friction material (ferodo); 4 - brake with blocks; 5, 5' - speedometers block; 6 - electric command panel; 7 - brake actuated with a screw; 8 - brake spring; 9 - clutch spring; 10 - brake graduated ruler; 11 - clutch graduated ruler.

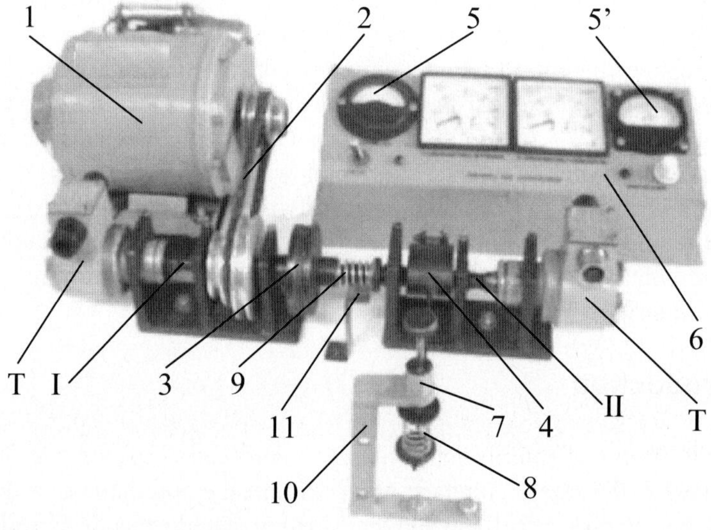

Fig. 2. The experimental setup

The driver shaft was marked with I and the driven shaft was marked with II. Both shafts are connected to the rotational transducers T. The limit situation is considered when the torque from the clutch equals the torque given by the brake crank.
The steel disk is changed each time with another friction material manufactured as silicon wafers. The wafers were stuck on steel disks having a

channelled surface by means of an uniform and thin layer (only a few micrometers) made from a mixture of colophony, bitumen and bee's wax. Each time, for each wafer, there were measured several spring tension deformations. For each determination there were computed the friction coefficients μ between ferodo and wafers [2].

The silicon was anisotropically etched to a depth varying with the time: 5 µm ($t_{etch\ 1}$ = 4 min.), 20-22 µm ($t_{etch\ 2}$ = 15 min.) and 50-80 µm ($t_{etch\ 3}$ = 40 min.).

The obtained values of the friction coefficient are the followings:

- μ = 0.08…0.27 for the silicon wafers texturized with microgrooves (sample no. 1, 2 and 3);
- μ = 0.32…0.55 for the silicon wafers texturized with quadratic openings(sample no. 4, 5 and 6).

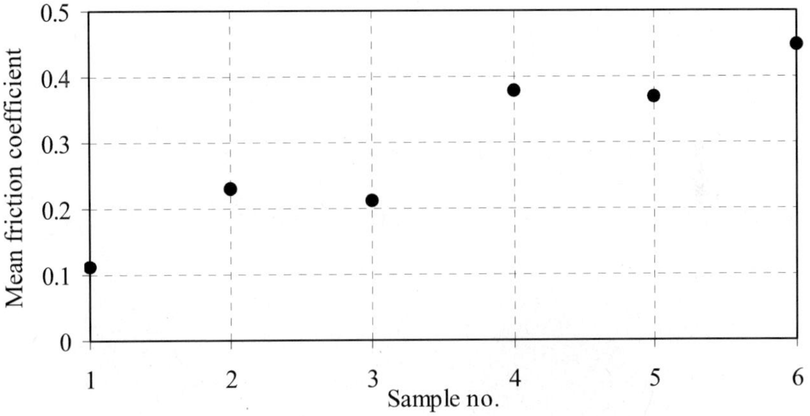

Fig. 3. Centralization and comparison of the results

The friction coefficient of all experiments on microtextured wafers is initially relatively low and then it increases and stabilizes. All the results are centralized in the diagram from the fig. 3.

3. Contact pressures computation by using the finite element method

There was considered 1/8 from the structures consisting of the ferodo ring and the silicon wafer stuck on a steel disk. The structure meshing was made in three-dimensional solid elements and contact elements were introduced on the ferodo/silicon wafer interface. The friction coefficient experimentally determined was considered on the contact of the steel disk with the wafer surface. The displacement along the symmetry axis (x) was blocked, while the ferodo was completely blocked and back charged by a pressure at the interface with the wafer:

$$p_{ax} = \frac{Q}{\pi\left(r_M^2 - r_m^2\right)} \tag{1}$$

where Q - axial force of the clutch spring.

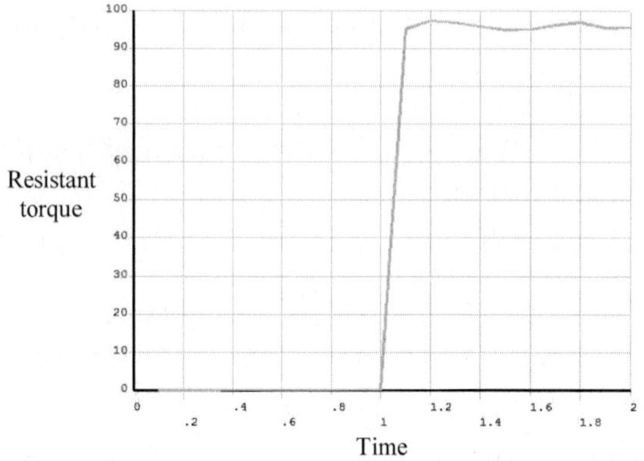

Fig. 4 The variation by time of the resistant torque for textured wafer model

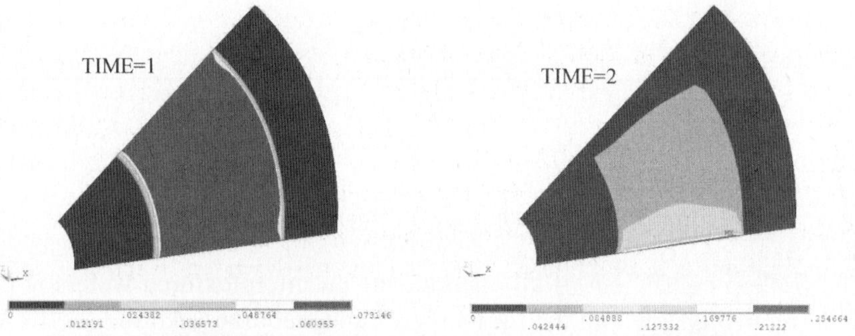

a) Non-textured wafer

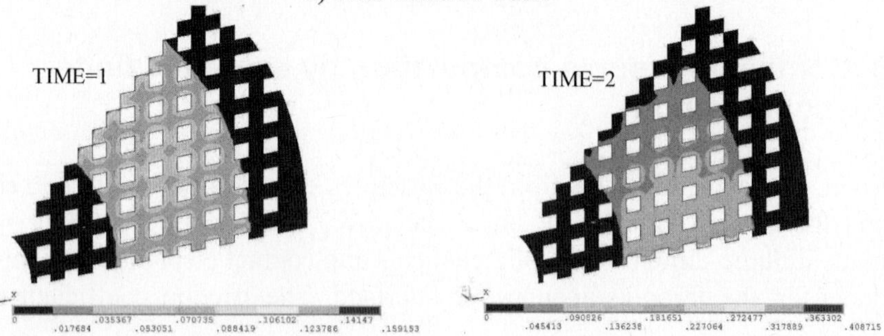

b) Textured wafer with quadratic openings (t_{etch3} – sample no. 6)

Fig. 5. The contact pressure distribution

A rotation by an angle φ = 0.5° was imposed to the steel disk with the wafer and the evolution in time of the resistant torque was watched, according to the fig. 4. For time = 0 to time = 1, charging is applied only by pressure and, beginning with time = 1, a rotation is applied too, the resistant torque reaching a maximum value for time = 2. The contact pressure distribution of the non-textured (a) and textured (b) wafer model, for time = 1 and time = 2, respectively, is shown in fig. 5.

The contact pressures have maximum values higher on the textured surface (as was expected: 0.159 N/mm^2 (time = 1) and 0.409 N/mm^2 (time = 2) for textured wafer; 0.073 N/mm^2 (time = 1) and 0.255 N/mm^2 (time = 2) for non-textured wafer), but their distribution is more uniform compared to the smooth (non-textured) surface.

The authors will continue the research regarding the contact modeling in order to study the structure behaviour and evaluate the surface wear.

4. Conclusions

The authors have already considered some improvements of the setup for friction study. The first one is referred to the control of the brake force and of the axial force in the clutch, both of them being obtained by means of two compression helical springs. The values of the forces in these elastic elements were only theoretically evaluated. In the future, an experimental determination of these forces will be made. The second improvement aims to a kinematic and dynamic study of the clutch, regarding strictly the moment of relative slip beginning between the clutch disks. This supposes the determination of the real position, speed and acceleration of the two semi-clutches. The authors will develop a new study based on the friction behavior, for different materials and geometry of microtextured surfaces.

References

[1] C. Rizescu "Mechanism and Machine Design Laboratory Textbook", POLITEHNICA University of Bucharest, 1995.

[2] G. Ionascu, C. Rizescu, L. Bogatu, D. Rizescu, I. Cernica, E. Manea, Acta Technica Napocensis, Series: Applied Mathematics and Mechanics 49, vol. III, Cluj-Napoca, Romania (2006) 765.

Design of the magnetic levitation suspension for the linear stepping motor

Krzysztof Just (a), Wojciech Tarnowski (a);

(a) Politechnika Koszalińska, ul. Śniadeckich 2;
76-100 Koszalin, Polska

1. Abstract

A design concept of an active magnetic suspension for the linear motor is presented. The mathematical model and the control system is proposed. Synthesis of PID parameters, are completed. Special laboratory set-up is made for verification the idea. Results of simulations and laboratory tests are quoted which have proved the adequacy of the mathematical model and, what more the correctness of the design idea.

2. Introduction

A magnetic levitation is one of the most effective ways to remove the friction force, which is an obstacle for ground vehicles to achieve high speeds. Magnetic suspensions are quite reliable systems, and because they are friction-free, they do not require any costly lubrication system, making them simpler and less expensive to operate [4].

3. Configuration of the active magnetic suspension

A magnetic suspension (Fig.1) includes an automatic control system, which consists of the elements: object of the control, proximity sensors with associated electronics, a controller, an electromagnets and a set of power amplifiers which supply current to the electromagnets. The electromagnetic force F_w depends on the current in the coil *(i)* and distance (x) between the suspended mass and the poles of the electromagnets.

$$F_w(i,x) = 2k_x x + 2k_i i \tag{1}$$

where: $k_x = \frac{K}{2}\frac{i_0^2}{x_0^3}$, $k_i = \frac{K}{2}\frac{i_0}{x_0^2}$ is displacement and current stiffness.

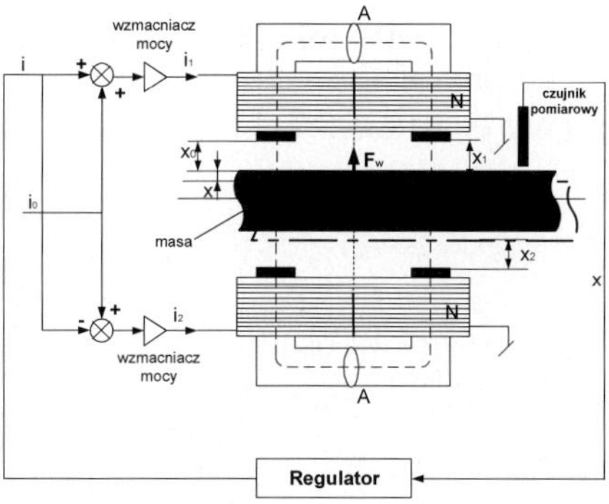

Fig.1. General design of an active suspension system

4. Mathematical model of the object

Linear equations of the mathematical model are valid if the assumptions and simplifications are acceptable: small displacements of the suspended mass, small air gaps, uniform magnetic field in the gaps, an own dampening in the electric circuit may be neglected. The equations are as follows:

$$m\ddot{x} = 2k_i i + 2k_x x + F_z \tag{2}$$

$$\frac{di}{dt} = \frac{u}{L_o + L_S} - \frac{Ri}{L_o + L_S} - \frac{k_i}{L_o + L_S}\frac{dx}{dt} \tag{3}$$

where: m – suspended mass; x – displacement; F_z – external force; R – resistance; L_o=K/2x_o – inductance in the point of the work x_o; L_s – inductance of the leakance; i – control current.

5. Control system of the magnetic suspension

The magnetic suspension is an unsteady system, so it must be connected with a negative feedback loop comprising a controller. A controller of the PID type is applied in this design. After linearization, the Laplace transform of the displacement X(s) for the system with the PID controller is:

$$X(s) = \frac{(1/m)s}{s^3 + [(2k_i k_d)/m]s^2 + [(2k_i k_p - 2k_x)/m]s + (2k_i k_I)/m} F_z(s) \tag{4}$$

The denominator of the transmittance is of the third order, so the system has three poles. Two poles being in the dominating area determine the dynamic properties [4]. Now it is possible to investigate dynamic properties of the closed system. On the Evans complex variable plane (Fig. 2), dominating poles as the functions of the natural oscillation frequency and the non-dimensional damping coefficient ζ for an equivalent second order system may be defined [3]:

$$p_{1,2} = -\omega_o \zeta \pm \sqrt{1-\zeta^2} \quad (5)$$

where: ω_o – the natural frequency of oscillation $\omega_o = \dfrac{3.2}{t_r \zeta}$, t_r – control time.

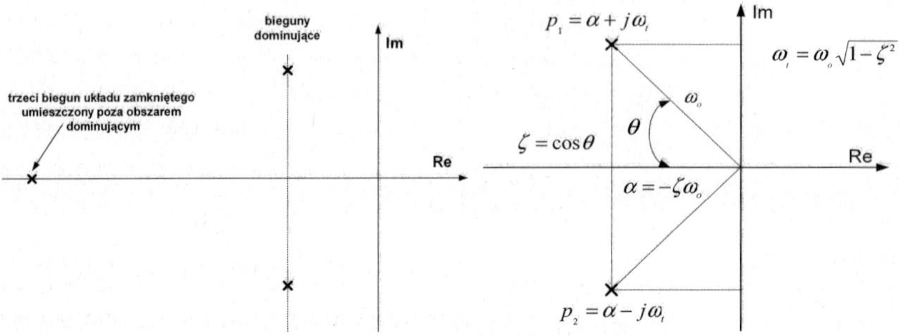

Fig.2.Poles and parameters on the complex plane

A method of moving location of poles was chosen to design the control system. PID settings are:

$$k_p = \frac{(p_1 p_2 + p_2 p_3 + p_1 p_3)m - 2k_x}{2k_i} \quad k_d = \frac{(-p_1 - p_2 - p_3)m}{2k_i} \quad k_I = \frac{-p_1 p_2 p_3 m}{2k_i} \quad (6)$$

6. Simulation model of the magnetic suspension

Simulation experiments have been performed by means of the Matlab 6.1 program pack, using Simulink Toolbox (Fig. 3).

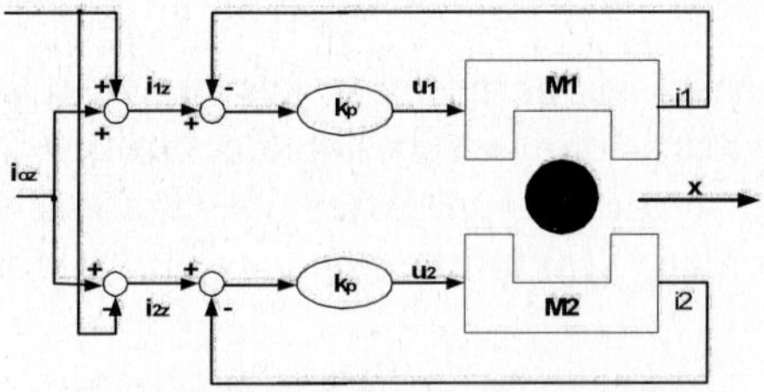

Fig.3. Simulation model of the suspension

Simulation tests: **a)** **b)**

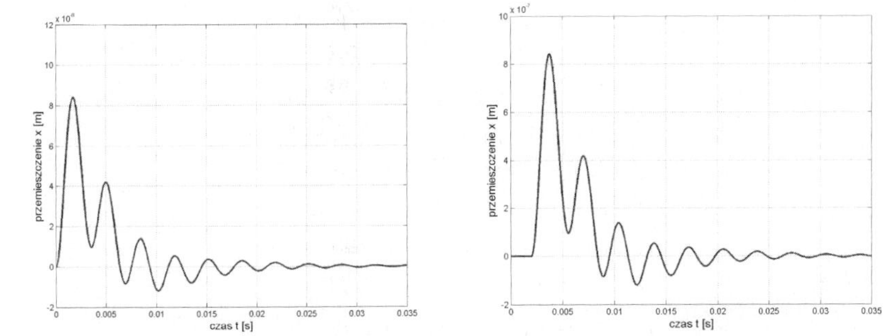

Fig.4. Displacement: a) after a step signal, b) with external force F_z=10 [N]

7. Laboratory model and experiments

The laboratory model (Fig. 5) consists of two boards of the diamagnetic material connected with the steel trusses frame.

a) the side view **b)** the front view

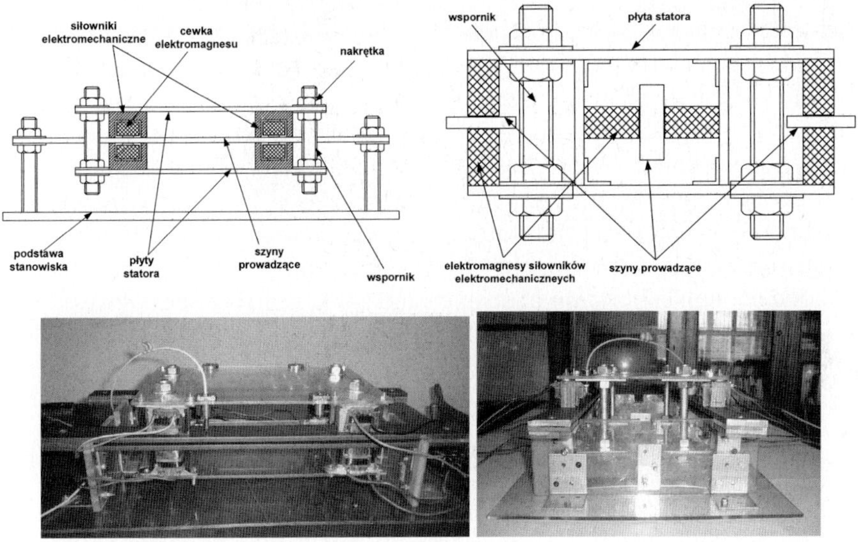

Fig.5. Schematic diagrams and the laboratory model of the levitation system

The stator is kept in the levitation state by the system of the magnetic suspension, which is built of four electromechanical suspensions, of two driving tracks and of a trusses system assuring the mechanical and magnetic symmetry. Experiment tests were carried out on the laboratory model for different values sets of the controller, what makes possible to determine the influence of the controller settings for the operation of the suspension.

Tests: **a)** k_p=1,4; k_d=0,035; k_i=0,0003, **b)** k_p=1; k_d=0,035; k_i=0,0004

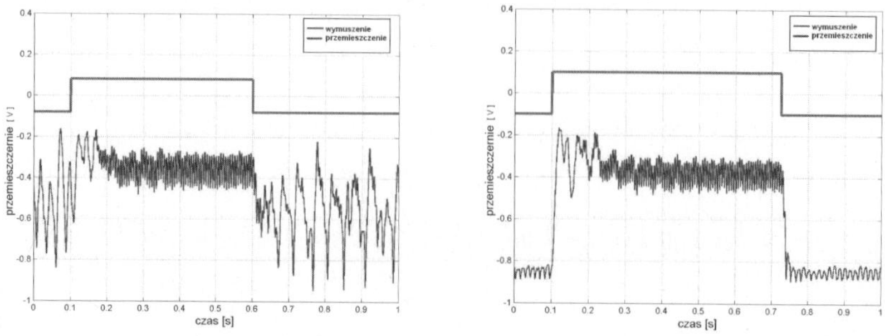

Fig.6. Displacements response of the mass for a step signal of the force

8. Conclusions

The design concept and the proposed control system has been successfully verified by simulations and on the laboratory model. It assures sufficient stability, fast enough response for external perturbation, and the steady state accuracy against the long-lasting disturbances, due to the integral action of the controller. During the tests on the laboratory model some vibrations of the object were observed, probably caused by a not symmetrical distribution of mass and not sufficient rigidity of the frame. Reasons of various oscillations and noise of signals are coupling of magnetic fields, not good enough calibration of sensors and some signals interferences.

9. References

[1] Z. Gosiewski, J. Osiecki – „*Mechatronika z przykładami*", skrypt WAT, Warszawa 2003;

[2] B. Heimann, W. Gerth, K. Popp – „*Mechatronika, komponenty, metody, przykłady*", Wydawnictwo Naukowe PWN, Warszawa 2001;

[3] Z. Gosiewski, K. Falkowski – „*Wielofunkcyjne Łożyska Magnetyczne*", Biblioteka Naukowa Instytutu Lotnictwa, Warszawa 2003;

[4] Z. Gosiewski – „*Łożyska magnetyczne dla maszyn wirnikowych*", Biblioteka Naukowa Instytutu Lotnictwa, Warszawa 1999.

Analyze of Image Quality of Ink Jet Printouts

L.Buczyński (a), B. Kabziński (a), D. Jasińska Choromańska, (a)

(a) Warsaw University of Technology , Institute of Micromechanics and Photonics, A. Boboli 8, 02-525 Warsaw, Poland

Abstract

This paper describes factors influencing quality of digital printing, how each parameter changes quality.

Introduction

In recent years there is a rapid growth of ink jet printers appliance. It is a result of widespread introduction of digital printing, development of production engineering of ink jest printing, modern inks and photo-quality papers for printing.

As a result, a role of ink jet printouts quality rating is increasing, especially objective printout quality estimation and normalization of its control methods.

Development of ink jet color printing techniques requires analysis of influence on printouts quality of new print technology, properties of newly introduced printing materials(ink, papers), construction of ink jet mechanisms, control systems of printing and color management.

Leading producers are constantly developing their printing improving systems. Analysis of ink jet printouts would allow to estimate the efficiency of introduced improvements and would show directions of further quality estimation development of printouts.

Factors influencing the quality of the ink jet printouts

The quality of color printouts is influenced by many factors which are linked between each other. The main goal is to reduce volume of ink drops. While printing in thermal ink jet printing technology color points consist of many small ink drops (each different volume). Using piezo Ink jet printing technology, volume of ink drops is also differentiated and grouping into 2 or 3 in a set to achieve better saturated images.

Reducing the volume of ink drop and simultaneously increasing its number has also negative consequences, because it increase time of input data computing and positioning a ink drop on the paper, the result: printing process lasts longer.

Another important quality parameter of ink jet printing is increasing the number of compound colors. Adding light grey (LG), light light grey (LLG), light cyan (LC), light magenta (LM) colors to basic color set CMYK, extends color gamut of printed images.

Such printed photos require appropriate color pigments and photo papers, complicates printer control process and color management system.

Quick development of supplies materials (inks and papers) is observed. Commonly used solvent and pigment inks are gradually modified and improved by producers. Except colorants, inks include additional components improving ink jet process and color stability. Moreover, key meaning for enhancing printing quality is developing form of printer input file (RIP) and color management systems. .

Researches on colorant and paper attributes, color analysis, interactions between colorant and paper have significant influence on quality printouts improvement. Great contribution to that process have international normalization procedures.

New colorants are developed such as modified pigments with additional protective layers giving uniform distribution of colorant particles, and as a result better quality of printout.

Paper with a rough surface may serve as an example because ink soaks into the paper easier. While ink thinner vaporizes, pores close so colorant stays in paper protected against light and vapor. A significant problem is mutual fitting of paper and ink to each other. Very important part in high quality printing, especially color, is preparing file for printing, creating most appropriate mixture d component color on paper (RIP), color correction and calibration between different devices - scanner, monitor, printer (CMS), management of printing process [2].

Many construction improvements is observed, especially shape of printing heads, number of ink nozzles, placement of piezoelectric converters in printing head etc.

Worth mentioning is that each factor influences not only printing quality but also other factors. Changing one of the makes necessary other to be changed. That is the reason why improvement of printing quality is complex process and is realized by producers systematically.

Examining of quality ink jet printing

There were quality studies lead using six different ink jet printers There were used printers using 4 base colors (CMYK) and 6 base color – photo printers. Test page of each control printout consisted of regions of differentiated saturation of colors: C, M, Y, K, C+M+Y, 100% saturation of R, G, B. Printouts were carried out on a plane and glossy paper. On printed test samples a line and surface parameters were measured (described in ISO/IEC 13660): darkness, blurriness, raggedness, mottle, line width, moreover there were proposed testing parameters of color printing: color difference ΔE and mean color saturation ΔS.
Raggedness is defined as width R (Fig. 1).

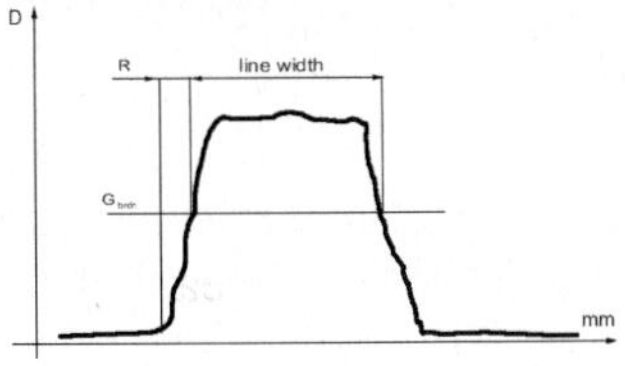

Fig.1 Definition of raggedness

Parameters described in ISO/IEC 13660 were using calculation program PRINJ TEST after previous scanning of selected area (line or field) on AGFA ARTIC SCAN in 1200 lpi.
Comparison of quality parameters of thermal ink jet printers and piezo printers did not prove definitive advantage of any printing technology. Piezo printers reproduce most accurately Cyan color, thermo printers reproduce black color- K and C+M+K. Both types have difficulties with Yellow color reproducing, especially of low saturation. Raggedness of printed test lines were most precisely reproduced by printer II (Fig 2).

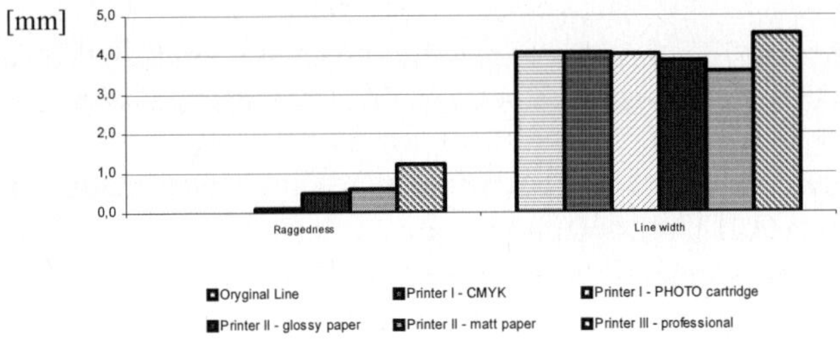

Fig. 2 Example values of raggedness for different printing conditions [3]

Influence of paper type on printing quality is big (Fig. 3). Glossy paper gives much better color reproduction. Using glossy paper results in better color reproduction.

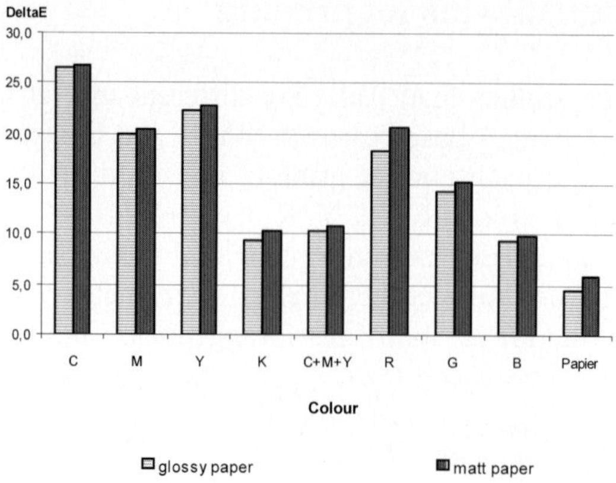

Fig. 3 . Quality of color reproduction for glossy and matt paper [3]

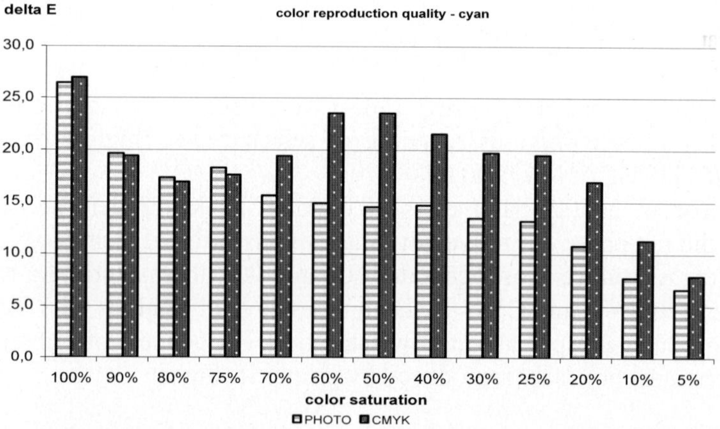

Fig. 3 Influence of number of base colors on color reproduction quality. PHOTO – 6 compound colors, CMYK – 4 compound colors[3]

Besides measurements a subjective visual research was made. Quality of standard line (ISO/IEC 13660) and color on base fields was analyzed [1]. An observation was made that even field of 100% saturation of base color consists drops of other component colors. Its number depends on printer and printing technology – thermal or piezo. Areas of smaller saturation have worse color reproduction. Results of observation agree with measurements of mean color saturation ΔS.

Conclusions

Measurement of printing quality parameters is complicated and results depend on many cross linked factors.
Pigment colorants give more saturated colors.
Higher printing quality is achieved on glossy paper.
More base colors reproduces colors better and more subtle tonal gradients.
Further researches of print quality defining influence of each parameters seems necessary.
Introduction of subjective visual observation and linking it with changes in of ISO/IEC 13660 standard seems also to be relevant.

References

[1] Standard ISO/IEC 13660 Information technology — Office equipment — Measurement of image quality attributes for hardcopy output — Binary monochrome text and graphic images" - ISO/IEC 13660:2001(E).
[2] Buczynski,.L Histogram Analyze as the Tool for Image Quality Estimation of Color Printouts *IS&T's NIP21:* Baltimore, September 18, 2005; p. 110-112.
[3] Deuszkiewicz D. Analyze of color ink jet technology as image quality estimating Diplom work in Institute of Micromechanics and Photonics Warsaw University of Technology , Warsaw , 2005. (Polish).

Development of Braille's Printers

R. Barczyk, L. Buczynski, D. Jasinska – Choromanska (a) *

(a) Institute of Micromechanics and Photonics, Warsaw University of Technology, Sw. A. Boboli 8
Warsaw, 02-525, Poland

Abstract

This paper describes course and results of works on the development of new hardware aiding blind people and use new technologies in well known hardware. The result of this work are three patents of parts of braille printers.

1. Introduction

The convex printouts (Braille texts and convex graphics) are very important in intellectual progress of blind and weak-sighted people. This printouts are very important in education, developing their perception. It allows recording the information in a way, that is understandable for every blind person. Hardware which prints convex copies is expensive, speed limited, and mostly allows only for text printing, without graphics.
There are three ways of development recently introduced technologies of convex printing.

- the quick single-line printing
- embossing with using elastic roller
- thermal copying on a microcapsule paper

There is also the newest introduced technology (nowadays in the experimental phase) of ink-jet convex printing. Economy is the reason that it is not popular.

The newest software allows for optimal control of Braille printers. Integration the natural language – Braille translation software with popular office software makes possible printing directly from text editor.

2. Present state

For the weak-sighted people besides convex printing additionally a colour plain printing is used. The good example of such appliance, is making of colour maps with embossed contours of country borders, rivers, etc. There is also used different height of embossing for different colours.

Currently available braille printers are divided in three groups:

- portable printers,
- office printers,
- high-volume printers.

Usually in each group is used traditional technology of Braille point embossing by metal stamps hitting the die.

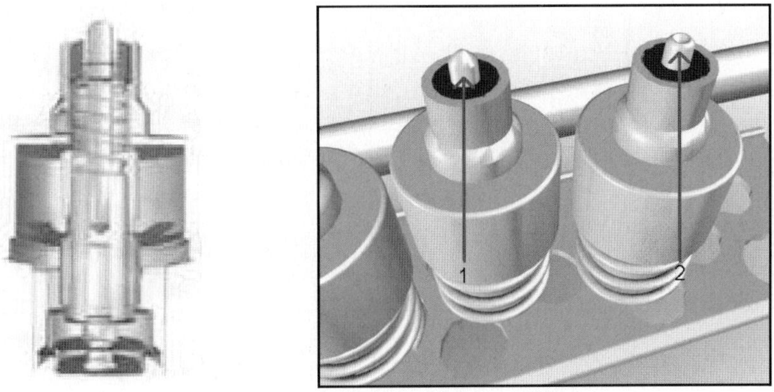

a) b)

Fig. 1. Braille points embossing scheme [1]
a) cross-section of electromagnet
b) view of electromagnets with "positive" (1) and "negative" (2) stamps

The electromagnets (Fig.1) which propel stamps need a lot of energy and are working loudly. The traditional printers achieve speed about 45 signs per second and noise level up to 76dB (for office printers) and 90dB (for high-volume printers).

Technology, that is commonly used for text and graphics printing, is copying on a microcapsule paper. The first step of such copying is preparing the plain copy on a special paper (laser-jet printing). Then, in the special device the light is emitted on the paper surface, what causes embossing in dark places.

3. Achievement of authors

Authors actively take part in research work connected with improvement and development of new braille printers.

Our research team invited single-line printing module (Fig.2). Printing one line in the same moment increases speed of printing. This solution was patented [Polish patents PL 191544 B1 and PL 191545 B1, 2005]

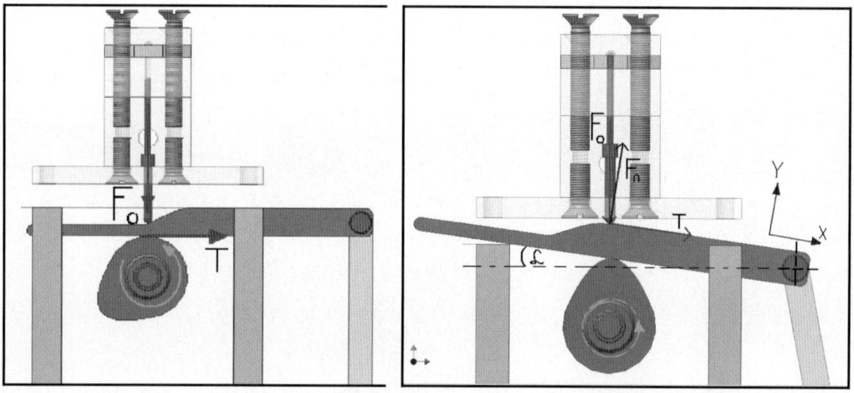

a) b)

Fig. 2. Functioning scheme of single-line printing module for high-volume printer
a) non emboss action
b) emboss action

Appliance of elliptic mirror (Fig.3) is a next step in modernization of device for copying on a microcapsule paper. A source of light is in a one focus of an elliptic mirror. The emitted rays of light are reflected and focused in line of paper transport. This solution allows to increase the efficiency of device and decrease power of lamp. As result of this an active cooling module is not necessary.

This solution was also patented Polish patent PL 191543 B1,2005.

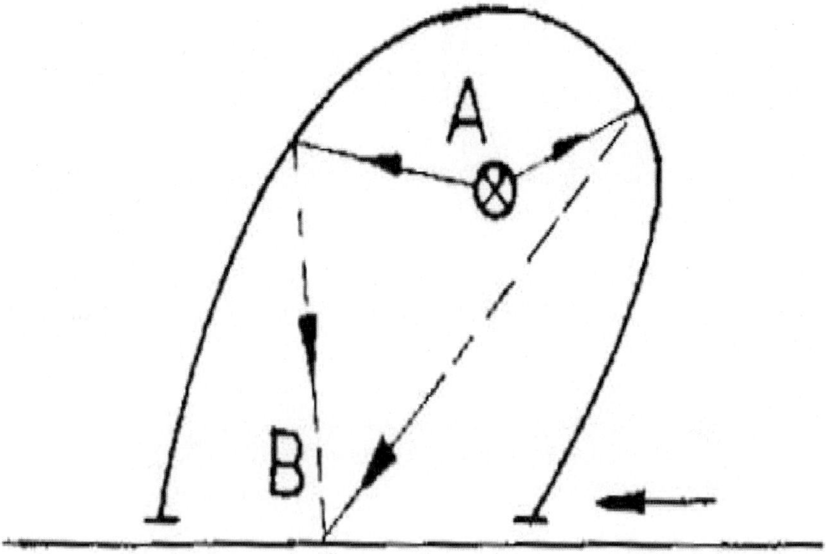

Fig. 3. The elliptic mirror used in the device (A – source of light is in one focus, and B – paper's surface is in second focus of elliptic mirror)

4. Results

Research work in the Institute of Micromechanics and Photonics is not only connected with improvement of devices, as well with quality of made convex copies. Main principle in optimization of construction of devices is quality of copy. The research [2] proves, that quality of samples differ significantly. The samples were made with different methods, devices and papers. Roughness of convex surface is factor in which the biggest changes are observed.

Example results of measurement of quality parameters are presented below (Fig.4.). The presented diagram of roughness was made for the ink-jet printing, where surface of convex line had roughness about 0.7 μm.

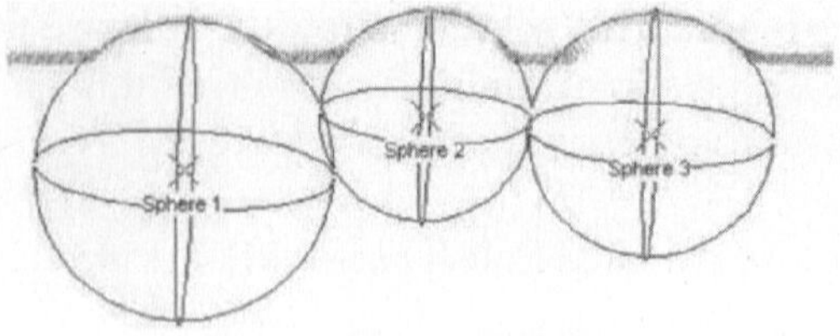

	h [mm]	d [mm]	D=2r₀ [mm]	R [µm]
A1	0,325	1,833	6,079	15
A2	0,321	1,849	5,975	17
A3	0,321	1,771	5,549	16
A4	0,256	1,501	4,930	8
A5	0,227	1,326	4,788	6
A6	0,150	0,856	2,767	6

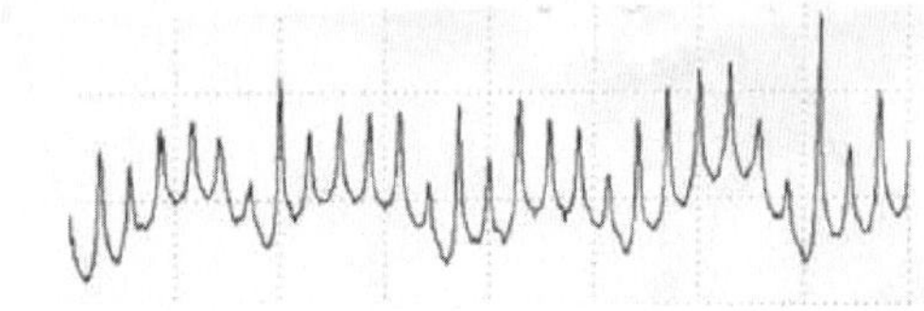

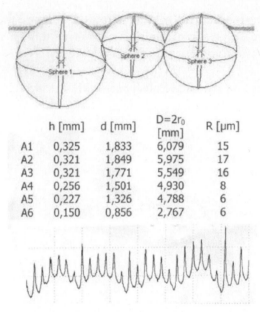

Fig. 4. Results of measuring convex samples
(h – height of point, d – diameter of point,
D – diameter of sphere, R – roughness of surface)

5. Conclusions

The quality of convex copies is very important for the blind people. Possibility of tactile reading depends on quality. The readable printouts with low quality are read reluctantly because it causes hardening of fingertips. The quality of made convex copy strongly depends on printer construction. The research regarding improvement of braille printers is helpful, so is needed.

References

[1] materials of Everest - Index
[2] R. Barczyk, „Analysis of the Quality Parameters of Convex Copies" Master Thesis, Warsaw University of Technology, 2004 (in Polish)

The influence of Ga initial boundaries on the identification of nonlinear damping characteristics of shock absorber

J. Krejsa, L. Houfek, S. Věchet

Brno University of Technology, Faculty of Mechanical Engineering, Technická 2, 616 69, Brno, Czech Republic

Abstract

The paper is focused on the simulation identification of nonlinear damping characteristics of the shock absorber using genetic algorithm (GA). 2 DOF quarter car model is used with nonlinear characteristics of damping force defined as piecewise functions. The characteristics are found using time courses of position, velocity and acceleration of both masses under kinematic excitation. The influence of initial boundaries determining the limits within the parameter values are searched is shown in the paper.

1. Introduction

The identification of shock absorber nonlinear characteristics can be provided through the genetic algorithm (GA) [1]. In this paper we use kinematically excited quarter car model with two degrees of freedom with linear stiffness of the tyre and main spring and nonlinear course of damping factor. The excitation corresponds to the overrun of defined bump. The nonlinear characteristics of shock absorber are found by GA, using the time course of kinematic variables of both masses during the overrun. As it is not practical to measure all the variables in real experiment, the reduced cost function (CF) can be introduced, using only the relative position and acceleration of suspended mass. Such reduction effects the results and the identification, however the GA is still capable of successful identification [2]. Another factor which plays the role is the setting of initial boundaries determining the limits within the parameter values are searched. One can expect that broader limits will prolong the search and reduce the precision of found results. Narrower limits set with the advantage of information about the expected course of nonlinearity represent less general approach, however, at least some idea about the expected results is usually available and faster search with more precise results can be expected.

2. Model

A quarter car model is used. Motion equations are as follows:

$$\begin{aligned} m_2\ddot{x}_2 + f_{b2}(\dot{x}_2) + f_{k2}(x_2) &= f_{k2}(x_1) + f_{b2}(\dot{x}_1) \\ m_1\ddot{x}_1 + f_{b2}(\dot{x}_1) + f_{k2}(x_1) + k_1x_1 &= f_{k2}(x_1) + f_{b2}(\dot{x}_2) + k_1Hk(t) \end{aligned} \qquad (1)$$

where m_1 is suspension mass, m_2 is suspended mass (mass of portion of the vehicle supported by given wheel), k_1 is radial stiffness of the tyre, f_{b2} is the function of damping force course, f_{k2} is the function of main spring stiffness force course and $Hk(t)$ is kinematic excitation of the model. For the simplification purposes the f_{k2} function was selected linear for the fitness function reduction experiments. Kinematic excitation was evoked by bump overrun with speed of 20 km/h.

Function of damping force f_{b2} is defined as piecewise function defined with independent slope on 6 intervals. Limit points of the intervals and slope in each interval are defined independently thus giving 12 free parameters for the function.

The total of 6 parameters $x_1, \dot{x}_1, \ddot{x}_1, x_2, \dot{x}_2, \ddot{x}_2$ are observed in each time step during the bump overrun. Those parameters courses are used by the CF to generate a measure of success/failure for given set of parameters.

3. Identification

The identification task is formulated as finding the f_{b2} parameters using the time courses of observed variables, or its combination. The CF uses simple Euclidian distance metric:

$$C = \sum_i \sqrt{(x_{1,i} - \tilde{x}_{1,i})^2 + (\dot{x}_{1,i} - \dot{\tilde{x}}_{1,i})^2 + (\ddot{x}_{1,i} - \ddot{\tilde{x}}_{1,i})^2 + (x_{2,i} - \tilde{x}_{2,i})^2 + (\dot{x}_{2,i} - \dot{\tilde{x}}_{2,i})^2 + (\ddot{x}_{2,i} - \ddot{\tilde{x}}_{2,i})^2}$$

where $X = (x_1, \dot{x}_1, \ddot{x}_1, x_2, \dot{x}_2, \ddot{x}_2)$ denotes the given observed (measured) variables values and $\tilde{X} = (\tilde{x}_1, \dot{\tilde{x}}_1, \ddot{\tilde{x}}_1, \tilde{x}_2, \dot{\tilde{x}}_2, \ddot{\tilde{x}}_2)$ denotes the values of corresponding variables calculated by the model for certain parameters of f_{b2} and i denotes $i-th$ time step of the course, with expected data acquisition frequency of 100 Hz.

Above defined CF will be further denoted as full definition. As only the acceleration $\ddot{x}_2$ and the relative position $x_1 - x_2$ are measured in the real experiment CF is reduced and calculated as:

$$C = \sum_i \sqrt{\left[\left(x_{1,i} - x_{2,i}\right) - \left(\tilde{x}_{1,i} - \tilde{x}_{2,i}\right)\right]^2 + \left(\ddot{x}_{2,i} - \ddot{\tilde{x}}_{2,i}\right)^2}$$

Classical version of genetic algorithm was used as identification method. Solution representation consists of limit intervals and corresponding slopes of f_{b2}. Population size was set to 30 individuals and initialized randomly. Two sets of initial population limits were used, forming the wide and narrow limits, corresponding to general case when expected values of the data to be found is unknown (wide) and the case when at least certain knowledge of the values is available (narrow). The limits used are shown in Tab. 1., where # denotes the number of the parameter, LIN denotes limits – intervals – narrow, LIW denotes limits – intervals – wide, LVN denotes limits – values – narrow and LVW denotes limits – values – wide.

Tab.1. Limits of randomly generated values of initial population

#	LIN	LIW	#	LVN	LVW
1	$\langle 0.1, 1.0 \rangle$	$\langle 0.01, 1.0 \rangle$	7	$\langle 100, 1000 \rangle$	$\langle 10, 1000 \rangle$
2	$\langle 0.1, 0.5 \rangle$	$\langle 0.01, 1.0 \rangle$	8	$\langle 50, 500 \rangle$	$\langle 10, 1000 \rangle$
3	$\langle 0.001, 0.1 \rangle$	$\langle 0.001, 1.0 \rangle$	9	$\langle 10, 100 \rangle$	$\langle 10, 1000 \rangle$
4	$\langle -0.1, -0.001 \rangle$	$\langle -1.0, -0.001 \rangle$	10	$\langle -100, -10 \rangle$	$\langle -1000, -10 \rangle$
5	$\langle -0.5, -0.1 \rangle$	$\langle -1.0, -0.01 \rangle$	11	$\langle -500, -50 \rangle$	$\langle -1000, -10 \rangle$
6	$\langle -1.0, -0.1 \rangle$	$\langle -1.0, -0.01 \rangle$	12	$\langle -1000, -100 \rangle$	$\langle -1000, -10 \rangle$

4. Simulation results

Matlab environment was used to create the model of the shock-absorber. GAOT toolbox extension was used as genetic algorithm implementation. Damping force f_{b2} characteristics were generated and course of observed variables was calculated and used for calculation of the fitness of individuals in GA for both full and reduced CF definitions. The search lasted 200 generations and was repeated fifty times for all four combinations groups (full and reduced cost function definition and wide and narrow initial population boundary limits).

During the search the actual fitness value and best individuals found were saved, later used for reconstruction of the search error development during the GA evolution. As each group experiment was repeated fifty times, both the means and variations of the results were obtained to give the idea on random initial population seeding influence.

Results are shown on Fig. 1 and 2. Fig. 1a shows the fitness development comparison for the wide and narrow range for the full CF definition. Similarly Fig. 1b shows the same data for reduced CF definition. As the fitness is calculated differently, the absolute values of the fitness are not directly comparable between 1a and 1b. However the shape of the curves is important rather than the absolute values of fitness. From the graphs we can see that the shapes are very similar, for both definitions wide initial range shows slightly higher values at the beginning of the search, however after 60 generations the difference is negligible.

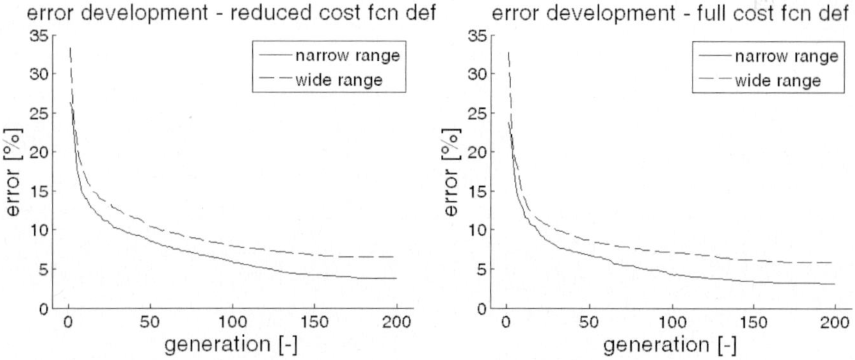

Figure 1. Fitness development comparison - means

Figure 2 shows the error development mean during the search. To get the idea of how the actual identification error evolves, the best individuals from each population were indirectly compared with the data to be identified. Indirect comparison was used due to the irregular scaling of the parameters to be found. The values found by GA were used for generating the damping force course and identification error was calculated as the percent difference between damping force found and damping force to be identified. On the contrary to the fitness curves, those data are directly comparable. We can see that in both definitions the wide range cases exhibit higher error during the search, and the difference slightly increases.

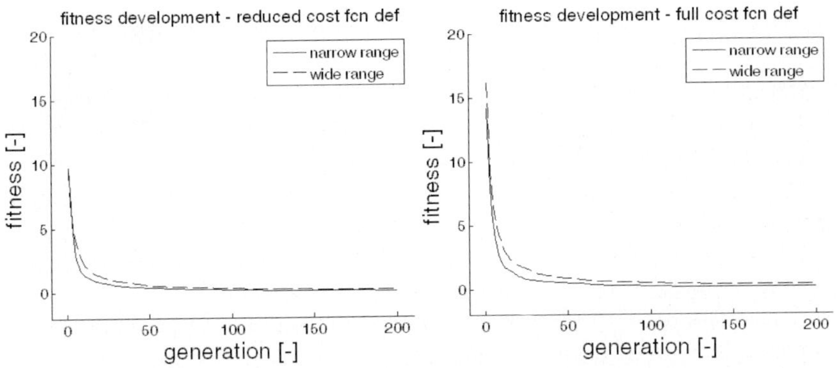

Figure 2. Error development comparison - means

Final values are about the same for both definitions, with reduced cost function being slower during the search. This is an interesting observation indicating the indirect relation between CF actual values and the resulting identification error.

5. Conclusion

The paper shows the influence of initial boundaries determining the limits within the parameter values are searched for two different definitions of cost function – full taking into account all observable variables and reduced taking into account only those practically measurable. Genetic algorithm was capable of finding correct parameters of damping force function in all the cases in reasonable 200 generations.

Acknowledgement

Published results were acquired with support of research plan MSM 0021630518 „Simulation modelling of mechatronic systems“.

References

[1] Krejsa J., Houfek, L.: Identification of Shock Absorber Nonlinear Model Characteristics, In proceedings of 10th International seminar of Applied Mechanics, Wisla, 2006. p. 99 - 102.

[2] Krejsa J., Houfek L., Věchet S.: Cost Function Definition Reduction in Shock Absorber Identification Task Through GA, In proceedings of 11th International seminar of Applied Mechanics, Wisla, 2007

Digital diagnostics of combustion process in piston engine

F. Rasch

Brno University of Technology, Institute of Automotive Engineering, Technická 2
Brno, 616 69, Czech Republic

Abstract

Modern engines are complex mechatronic systems and require periodic maintenances during whole lifecycle. In present time mobile diagnostic methods are very effective tools applicable for all combustion engines. This paper deals with an out of montage diagnostic of internal combustion engines. Principles of the diagnostic method are based on sound evaluation of a running engine. This method can control dysfunctions in mechanical and thermodynamic side of the engine

1. Introduction

At present diagnostics mechatronic systems are necessary fast and with very short time for preparation of diagnostic operation. Combustion engine belongs to complicated and very expanded mechatronic system. As general purpose availability method for petrol and diesel engine is to analyse emmited noise when the engine is running. The principle identity of piston combustion engines makes it possible to below describe method in all piston combustion engine with different setup and numbers of cylinders.

Faults in internal combustion engines can be classified into two groups, namely combustion and mechanical faults. Examples of these faults are misfiring, knocking, valve leakage, (intake and exhaust), fuel leakage or shorting, cylinder ring gumming, cylinder ringing, bearing wear, gear damage, worn timing belt, etc. Since all faults are related to excitation events, faults would alter the force-time profile of the excitation

associated with that moving element or event. As such, faults are expected to manifest themselves in the engine vibration and sound signatures. However, detecting small faults is limited by the signal to noise ratio, signal path attenuation factor, and the discrimination ability of the selected diagnostic technique [1].
This paper deals with a diagnostic of the thermodynamic cycle in combustion engines relating to the cylinder, which, due to a mechanical fault (fault of injector, valve leakage, etc.) is not working correctly.

2 Characterization principle of diagnostic

This diagnostics method of combustion engine is based on reading and evaluating the sound of a running engine. This signal in time range is process synchronic filtration according to the revolutions of the crank-shaft.
Concerning four-stroke engines, it is necessary to monitor mechanical and thermodynamical events during two revolutions of the crank-shaft. The other way is to monitor the signal from the cam-shaft. For reasons of speed oscillation during one revolution of the crank-shaft, and many other factors generating noise, synchronic filtration is a good method for getting and evaluating the signal.

3 Description of measuring chain

Fig. 1 shows a diagram of a measuring chain. Measuring microphone **8**, which is directed to the combustion engine, because we monitor only thermodynamic events, the microphone is directed at the cylinder head at a distance of 1.5 meter. Next neccesary signal for make synchronic filtration is the so-called synchronic signal, which is from revolutin sonde **10**. This is directed to the cam-shaft **6**. Both of these signals are evaluated in PULSE, Brüel & Kjærs platform for noise and vibration analysis [2].

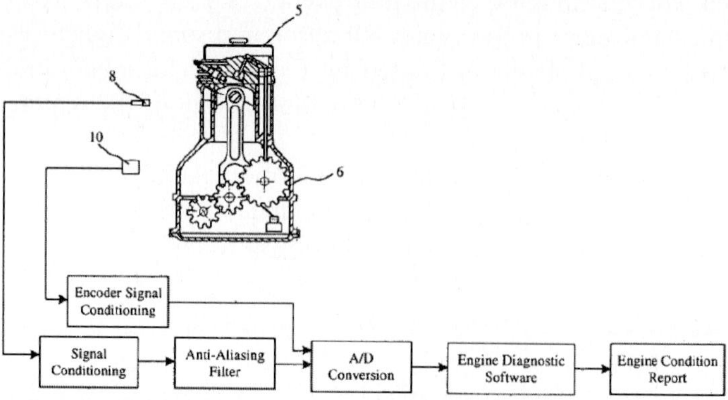

Fig. 1. Schematic diagram of an engine diagnostic system [1]

4 Measurement and evaluation signal

Measuring data is evaluatied by synchronic filtration. This method requires, along with the main signal, a monitor synchronic signal. According to synchronic signal is execute the average of the main signal (noise of combustion engine in time range).

For marking out the first cylinder in filtered signal in the time range **Fig. 3,** it is necessary to locate the synchronizing mark accurately before compression stroke of the no.1 cylinder.

Preview evaluated signal – recorded noise of a 6 cylinder engine **Fig. 2**. Every local peak **Fig. 3** matching one working stroke of one cylinder. Comparate all local peak, we can value quality combustion process.

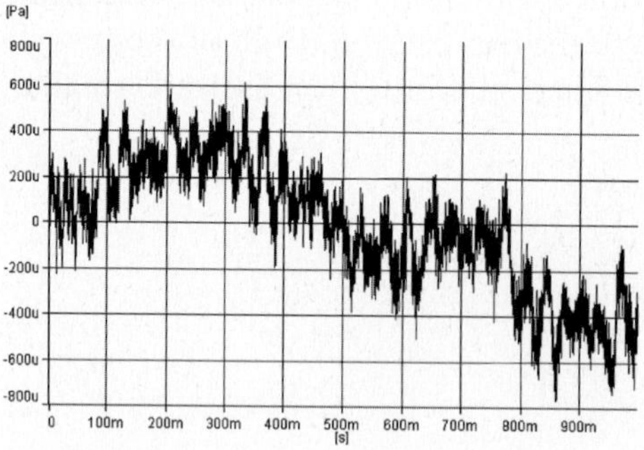

Fig. 2. Acoustics signal of 6 cylinder engine in the time range

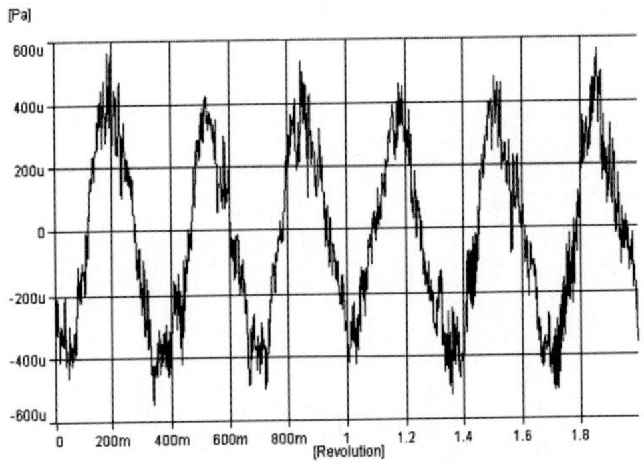

Fig. 3.Filtered signal in the time range

Experimental measuring on car Alfa Romeo 1,7 **Fig. 5,** four cylinder engine with symmetrical crank-shaft (uniform working stroke after 180°) engine rpm 980. **Fig. 6** shows evaluated acoustics signal in the time range, which is composed of 100 average wokring cycles (200 revolutions of the crank-shaft.)

Fig.5 Measuring station 4 cylinder engine Alfa Romeo

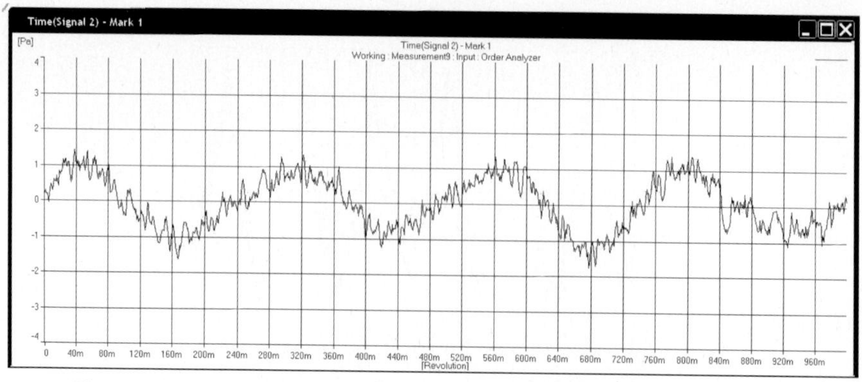

Fig. 6. Filtered acoustics signal of 4 cylinder engine in the time range

5 Conclusion

This method is suitable for finding combustion process failures in engine, because of its universal usage. Next step is diagnostics of each mechanical parts in out of montage diagnostics combustion engine. Mostly is going about diagnostics of valve operation mechanism and cranktrain.

Published results were acquired using the subsidization of the Ministry of Education, Youth and Sports of the Czech Republic, research plan MSM 0021630518 "Simulation modelling of mechatronic systems".

References

[1] Otmar A. Basir, Vibro-acoustic engine diagnostic system, Patent Application Publication, US 2004/0260454A1

[2] PULSE – Getting Started, B&K Lecture Note, BR 1505-12

Superplasticity properties of magnesium alloys

M. Greger, R. Kocich
VSB – Technical university Ostrava, 17. listopadu 15, 708 33 Ostrava-Poruba, Czech Republic

Abstract

Ratio of exploitation of magnesium based materials very rapidly increases at present. This is given not by its service properties, but also by its very low mass and also certain possibility of its use as replacement of Al based materials. Production of final products made of Mg alloys is, however, accompanied by many factors, which must be mastered for its successful implementation into practice.

2. Experimental methods

Forming of Mg-Al-Zn alloys mentioned above (namely AZ91 after T4) was realised by conventional way, i.e. by rolling. There were, nevertheless, used tow different ways of rolling in order to enable determination of differences of different approach at deformations as such. These rolled products were in the next stage subjected to the technology of Equal Channel Angular Pressing (ECAP) [1-3]. Materials processed in this manner were subjected to a hot tensile test for determination of the obtained mechanical values.

Another SPD method that was used was the ARB technology, which was applied on alloys AZ91+T4, AZ61+T4. Table 1 gives their chemical composition.

Table 1. Chemical composition of AZ61

Alloy	Chemical composition %						
	Al	Zn	Mn	Si	Cu	Fe	Mg
AZ91	8.95	0.76	0.21	0.04	0.002	0.008	Bal.
AZ61	5,92	0,49	0,15	0,04	0,003	0,007	Bal.

2.1 Conventional rolling and ECAP

Materials made of the alloy AZ 91 (Fig.1) and AZ 91+T4 (Fig. 2), which were first rolled by:

a) single pass
b) 3 passes with intermediate heating to rolling temperature, Fig. 3.

and then pressed, were subjected to hot tensile test in order to determinate a possibility of super-plastic behaviour. Equal channel angular pressing was made in two stages.

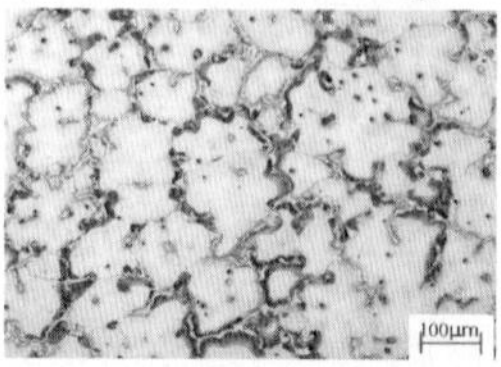

Fig.1 AZ91 alloy without T4

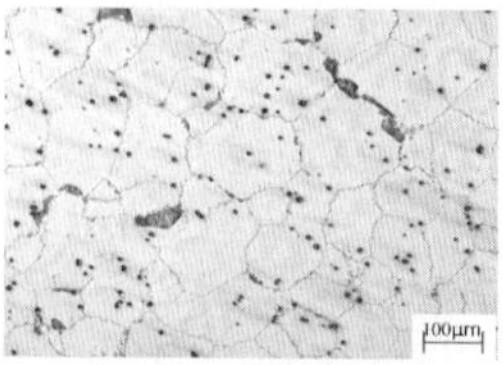

Fig.2 AZ91 alloy after T4

The first stage consisted of 4 passes at the temperature 250°C. It was followed by the second stage consisting of 1 pass ECAP at the temperature 180°C (Fig. 4).

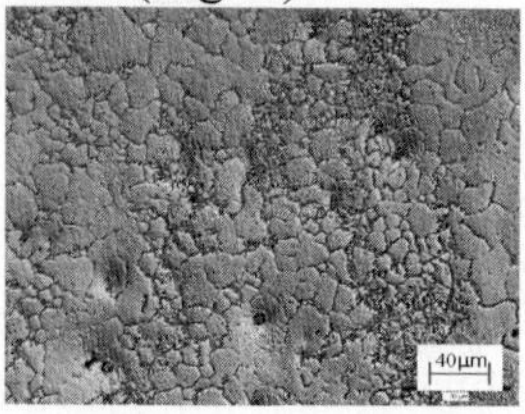

Fig.3 AZ91+T4 after rolling (3rd pass)

Fig. 4 AZ91 + T4 after rolling and ECAP

The samples were similarly as in the previous cases re-heated to the chosen forming temperature in a muffle furnace with connected inert atmosphere Ar_2. After obtaining of the required temperature and a 5-minute dwell at this temperature the material was charged into thermally insulated matrix with resistance heating, the temperature of which was identical to that of the chosen forming temperature.

2.2 Hot Tensile test

Temperature used at the tensile test was 250 °C and strain rate was $\dot{e} = 2x10^{-4}$. The samples obtained after processing by ECAP technology were adjusted to the required shape and then subjected to the tensile test, during which the set temperature was controlled by PID regulator, which used a thermo-couple situated directly on the tested sample.

Material rolled first by single pass (I, II, III) and then pressed, achieved elongation of approx. 200 %, while materials first rolled by several passes and then pressed, achieved elongation of up to 413 %. Before the tensile test microstructures of both groups did not differ significantly from each other.

Table 2 gives obtained values of elongation in individual samples after hot tensile test, where there are apparent the differences mentioned above between various methods of rolling applied prior to application of the ECAP technology, which has important influence on final plastic properties of obtained materials. An increase of plasticity with growing applied deformation can be observed at rolling by both methods, i.e. at rolling by single pass and rolling by several passes. In the latter case the obtained ductility was higher, which was probably caused by more homogenous structure obtained byre-crystallisation processes, which at this type of rolling could have developed more than at single-pass rolling.

Table 2. Values of strength and elongation of AZ 91 alloy after ECAP

Marking of sample	AZ 91 + T4	
	Elongation [%]	UTS [MPa]
I	294	15
II	286	19
III	-	-
K1	418	28
K2	384	32
K4	358	58.7

Sample taken from the alloy AZ 91 elongated at the temperature of 250°C under constantly applied strain around the value of 15 MPa to rupture.

2.3 ARB

For experimental verification of the ARB process there were produced two strips from the alloys AZ91+T4, AZ61+T4, which served as initial material. Initial dimensions of each strip were the following: thickness 4 mm, width 50 mm and length 200 mm. Experiment was made at the temperature of 380°C. The heat distortion temperature for this technology was chosen also with respect to the results of previous experiment, at which gradual samples were rolled. The samples were rolled at the first pass by deformation of 62.5% in direction of height. In all other passes by 50% height deformation. Strain rate varied in the interval from 16.83 to 17.78 s^{-1}.

At several places marked by an arrow there are visible traces of original boundaries of individual rolled layers, which have mostly disappeared. This was observed both for the alloy AZ91 and the alloy AZ61. Number of

visible places of original boundaries decreased with increasing number of accomplished cycles.
As it is demonstrated by the enclosed photos, there can be seen evident traces of crystallisation (Fig. 5), which refined the structure already after 3 cycles almost 20x, if we take into consideration the original structure with average size of 120 μm (Fig. 1).

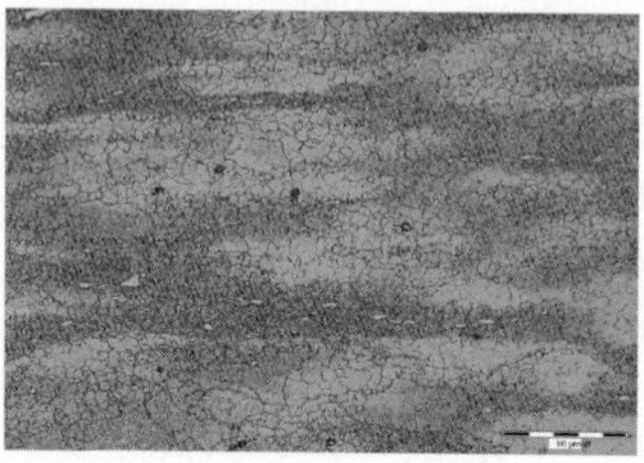

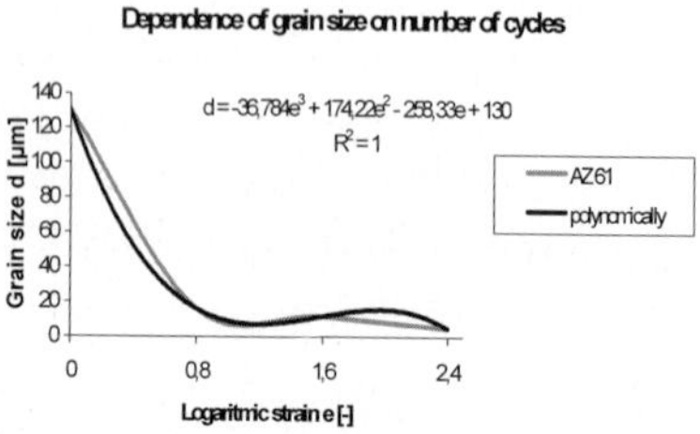

Fig.5 Final microstructure of of AZ61 after 3rd pass at ARB process

Micro-structure of rolled materials indicates formation of new grains inside the original grains, elongated in direction of rolling. Central parts of the rolled product are represented by fine-grain structure more than surface parts. The original boundaries disappeared at many places and new grains began to form at their place. High efficiency of this process is demonstrated also in the Fig. 6, which shows growth of strength of the alloy AZ91 in dependence on number of realised cycles in relation to the original non-deformed state. The values of strength increased more than 2.5 times after five accomplished cycles [4].

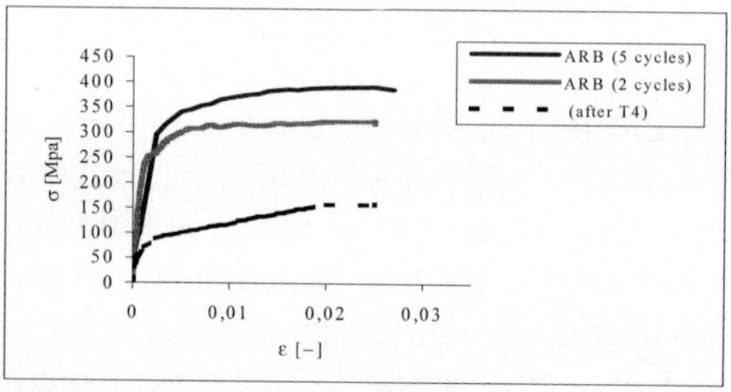

Fig.6 Mechanical properties of AZ91 at the temperature 360°C

Interposed deformation at the ARB process sufficed already after the 3rd cycle for decreasing of the grain size from the original size down under 10 μm in both types of alloys. Comparison of obtained strength in individual types of alloys after application of various forming technologies. It is evident, that the best method for obtaining the highest values of strength is the ARB process, however, this is achieved at the expense of plastic properties. Contrary to that the ECAP technology is an optimum compromise.

3. Conclusion

It is evident from micro-structures and mechanical tests that at high temperatures big elongation and lower strength are achieved after ECAP in comparison with conventional methods of forming, which is caused probably by the following factors :

1) There occurred disintegration of original precipitates to small particles, which facilitated movement of dislocations (e.g. by transversal slip), resulting in recovery of microstructure.
2) Comparatively small grain size, which enables slip deformation mechanism at the grain boundaries.

It means that during plastic deformation realised by the ECAP technology there occurred disintegration of staminate precipitates. There is also obvious occurrence of precipitates in the form of formations, the size of which exceeded 10 μm, but only in materials that were rolled by single pass. In materials rolled by several passes the distribution of precipitates is comparatively homogenous, with decreasing magnitude of deformation there is visible a growing proportion of longer staminate formations, which did not disintegrate into these smaller particles, which indicates also influence of magnitude of previous deformation at rolling. It was therefore proved that the used ARB technology is a perspective tool for obtaining of highly fine-grain structures in Mg-Al alloys. It contributes at the same time to homogenisation of micro-structure and to substantial limitation of negative consequences of dendritic segregation on mechanical properties of these alloys.

Acknowledgements

The works were realised under support of the Czech Ministry of Education project VS MSM 619 891 0013 and project GAČR no. 106/04/1346

References

[1] L.Čížek, M.Greger, L.A.Dobrzanski, R. Kocich, I. Juřička, L. Pawlica, T.Taňski, Mechanical properties of magnesium alloy AZ 91 at elevated temperatures. Journal of
[2] M Greger, et al. Structure development and cracks creation during extrusion of aluminum alloy 6082 by ECAP method. *In Degradacia.* Žilina 2005, pp. 152-156
[3] M.Greger, L.Čížek, S.Rusz, I. Schindler, *Aluminium '03*, Alusuisse Děčín 2003, p. 288.
[4] I. J. Beyerlein, R. A. Lebensohn, C. N. Tome, *Ultrafine Grained Materials II*. TMS, Seattle, 2002, p. 585

Technological Process Identification in Non-Continuous Materials

J. Malášek

Brno University of Technology, Faculty of Mechanical Engineering,
Technická 2896/2, 616 69 Brno, Czech Republic,
Phone: +420 541 142 428 Phone/Fax: +420 541 142 425
e-mail: malasek@fme.vutbr.cz

Abstract

The common reality at processing with deformation of non-continuous materials is a zone of deformation, as a cubic formation determined by a system of shear curves and streamlines. No continuum-physical equations and characteristics can be used for a mathematical-physical description of these deformation processes. The zone of deformation of non-continuous materials can be identified by border conditions state of stress (tactile transducers and strain gauge sensors) and by image identifications of shear curves and streamlines. These identifications respect the relevant discontinuity of reshaped areas at the technological processes.

1. Introduction

When processing the non-continuous materials (powdery materials, dispersions, suspensions, liquids with high viscosity) the materials are being deformed by mechanical effects - mixing, compacting, transport, storage. As a result of these deformation processes, the formed stress state determines the stress of machine parts (mixer-blades, compact-machine jaws, sides of bunkers) being in contact with the deformed material. Identifications of boundary conditions of state of stress by tactile transducers and strain gauge sensors together with image identifications of deformed materials are very important information about the relevant processes. The main problems are many variable physical properties of non-continuous materials and complicated mathematical descriptions.

2. State of stress determination – theoretical possibility

Instead of traditional physical variables the important examined entity can be an image of the reshaped volumes of the non-continuous materials with its mathematical processing together with the boundary stress state conditions of at least in a section of the image. [1]
Stress state relations at a selected point of shear curve are displayed in an osculating plain of a shear curve in Fig. 1 and displayed in the respective Mohr´s plane in Fig. 2.

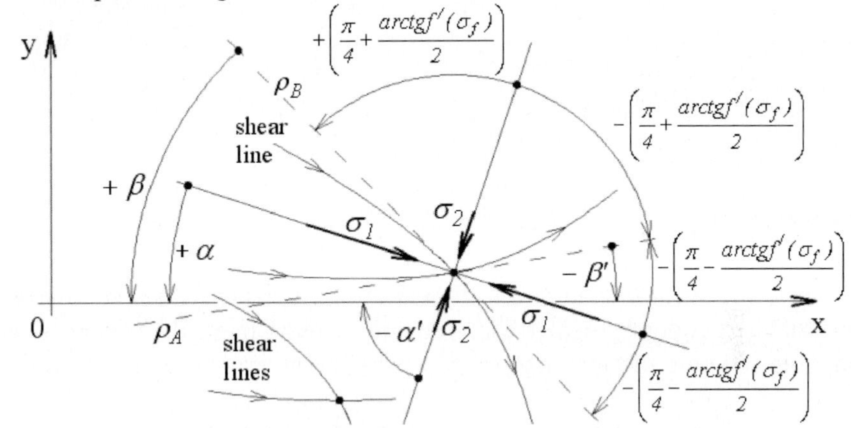

Fig. 1. Osculating plane of selected point of shear curve

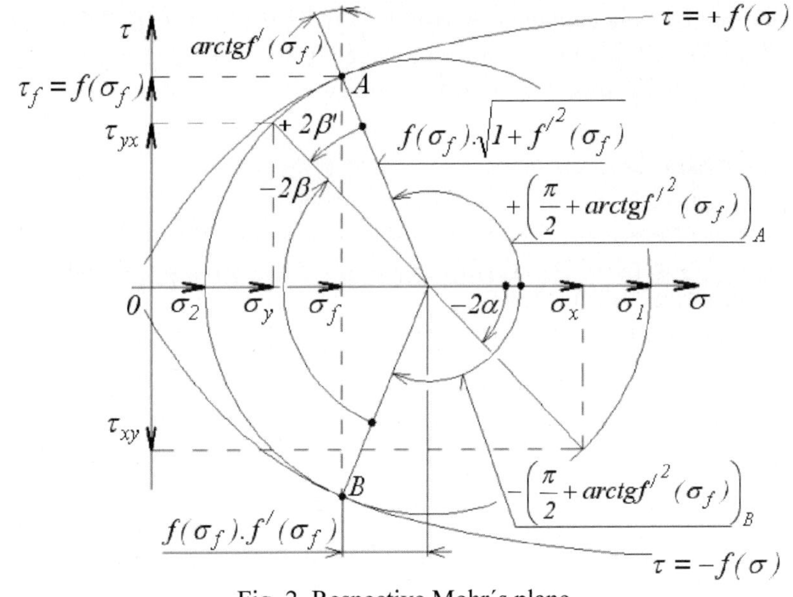

Fig. 2. Respective Mohr´s plane

The state of stress distribution can be described for example by Cauchy´s differential equilibrium equations - (1),(2) together with the mathematical description of analytical relations - (3),(4),(5) between the osculating plane of shear curve and of the respective Mohr´s plane.[2]

$$\frac{\partial \sigma_x}{\partial x} + \frac{\partial \tau_{yx}}{\partial y} = 0 \tag{1}$$

$$\frac{\partial \tau_{xy}}{\partial x} + \frac{\partial \sigma_y}{\partial y} = \left[\frac{\Delta \sigma_y}{\Delta y} \right] \tag{2}$$

$$\sigma_x = \sigma_f + f(\sigma_f).f'(\sigma_f) + f(\sigma_f).\sqrt{1+f'^2(\sigma_f)}.sin[2\beta + arctgf'(\sigma_f)] \tag{3}$$

$$\sigma_y = \sigma_f + f(\sigma_f).f'(\sigma_f) - f(\sigma_f).\sqrt{1+f'^2(\sigma_f)}.sin[2\beta + arctgf'(\sigma_f)] \tag{4}$$

$$\tau_{xy} = \tau_{yx} = f(\sigma_f).\sqrt{1+f'^2(\sigma_f)}.cos[2\beta + arctgf'(\sigma_f)] \tag{5}$$

Equations (3),(4),(5) shall be substituted in equations (1),(2). It is possible and difficult enough to solve these equations by numerical methods. It is possible to solve these equations by parametric interpolated spline on the basis of more measured values. The values $\left[\frac{\Delta \sigma_y}{\Delta y} \right]$ are determined after measurements and calculations using equations (3),(4),(5). If y-axis in Fig. 1 is identical with the line of acceleration of gravity g (or of the resultant acceleration), it is possible validity of the next equation (6) for continuous materials only:

$$\rho.g = \left[\frac{\Delta \sigma_y}{\Delta y} \right] \tag{6}$$

3. State of stress determination and measurement

The special transducers consist of these parts – miniaturized pressure sensors in matrix arrangement and a special strain–gauge bridge. Distribution of normal stress is measured by the matrix tactile sensors on the measuring surface in contact with processed materials. [3] The total normal force together with the total shear forces in two axes are measured by the special strain-gauge bridge. The appropriate software is involved. Identification of deformation consists of digital interface – camera and the appropriate software.

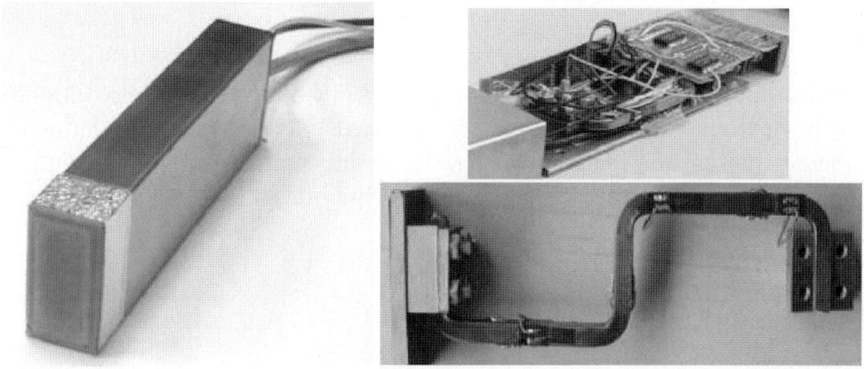

Fig. 3. Design of the transducer

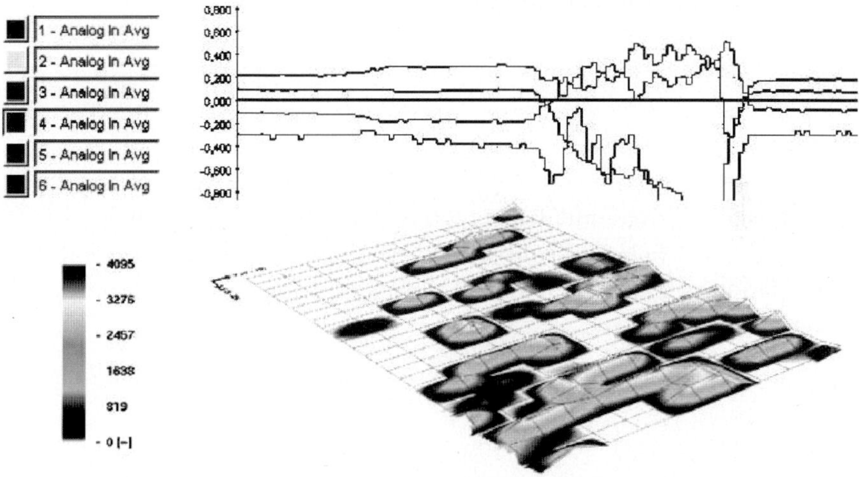

Fig. 4. Evaluation of state of stress boundary conditions

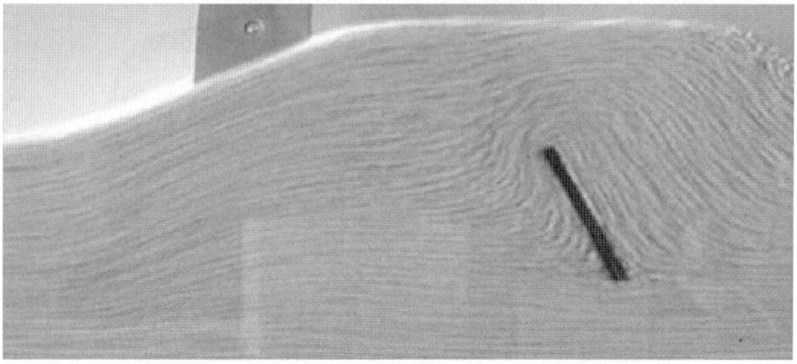

Fig. 5. Identification of shear curves and streamlines

4. Conclusion

Discontinuity of boundary conditions of state of stress and discontinuity with flexions and torsions of shear curves define the non-continuous characteristics of processed materials. Mathematical modeling of these processes is complicated and usually involve - describe the typical process only.

5. Acknowledgement

Published results were acquired using the subsidization of the Ministry of Education, Youth and Sports of the Czech Republic, research plan MSM 0021630518 "Simulation modeling of mechatronic systems".

References

[1] J. Malášek, Mísení a kompaktování partikulárních látek, (2004), ISBN 80-214-2603-9
[2] J. Malášek, Diserta□ní práce. (2003), Brno, ISSN 1213-4198.
[3] J. Volf, S. Papežová, J. Vl□ek, S. Holý, Measuring system for determination of static and dynamic pressure interaction between man and enviroment, EAN 2004.

Problems in Derivation of Abrasive Tools Cutting Properties with Use of Computer Vision

A. Bernat *, W. Kacalak **

* TU of Koszalin, Mech. Faculty of Engineering, Fine Mechanics Div., Raclawicka street 15-17, Koszalin, 75-620, Poland

** TU of Koszalin, Mech. Faculty of Engineering, Fine Mechanics Div., Raclawicka street 15-17, Koszalin, 75-620, Poland

Abstract

Nowadays, fully automated and flexible systems are more and more frequently used in grinding of advanced materials, such as for example ceramics. However, mainly due to elements dimensions, and moreover, due to their extremely high brittleness and hardness (as for instance ground and finally lapped tiny ceramic gaskets, used in high-pressured hydraulic circuits), the influence of unknown input elements must be minimized. Among these factors are those, which are closely correlated to cutting properties of grinding wheel (GW) active surface, used in the machining. Therefore, there is substantial need for such methods of estimation of cutting properties of GW, and for monitoring of tool wearing, as to enable to introduce necessary adjustment of the machining process parameters, simultaneously without altering of the initial geometry of the elements in the whole machining system. In this paper some innovative method for *in-situ* data colleting and processing has been proposed, based on computer vision techniques.

1. Introduction

Used in the past, standard 2D/3D profilometric measurement methods are mostly tedious in handling, biased with time- and labour-consuming proceedings, thus lowering the productivity. What is more, they usually need of temporary realized dismantling of GW out from grinding machine, unavoidable leading to altering in the initial geometrical orientations of GW, accordingly to the ground surface of small ceramic elements. Consequently, the ground elements might be cracked. Regarding output data set of the 2D/3D profilometric measurements, one comes to conclusion, that though that data are of high measurements accuracy, simultaneously they are redundant and irrelevant in their contents, accordingly to aimed task of estimation of cutting properties of GW.

Resuming and taking all the arguments presented above, in this paper some alternative approach to the problem considered has been presented, based on computer vision methods, used in *in-situ* data collection and processing, in main tasks of reliable, fast and effective estimation of cutting properties of GW, within short time (of few minutes) of grinding machine shutdown.

2. Methodology

In application of computer vision methods, a modified PS method [4] had been previously introduced. Surface of abrasive tools are characterized by locally-depended reflectance properties, and moreover are of complex densely spaced topographic features, such as grains summits of steep slopes, randomly spaced and occurring cutting edges, ravines and hinges. Moreover, reflectance borne (depended) properties are characterized by complex co-occurrence of both desired (diffuse or another words matte reflections) and undesired phenomena. Among undesired phenomena, there are occurrence of specular reflections of locally dominant character, self-shadowing (attached shadows) and self-masking (cast shadows).Therefore it was decided, that the monoscopic and multi-2D-image-based approach would be adapted, in presence of the mentioned phenomena, to face hard initial conditions of data acquisition, regarding surfaces of abrasive tools visually inspected.

For this aim, both classical and adapted PS methods, at lest theoretically, allow for disjoint (i.e. separately) extraction of reflectance borne (i.e. of albedo map) features in form of reflectance coefficients, and topographic borne features in form of 3D surface reconstructed.

However, the adapted PS method, previously introduced [4], allows for pixel-wise classification and filtering of data of 2D images intensities, at any (x, y) locations on the images, stacked column-wise, excluding those areas, which are related to undesired phenomena of locally dominant specular reflections and shadowings.

Thus, considering data individually, for each of the pixels points (i.e. pixel-wise), a variable number of 2D images intensities, stacked column-wise, due to initial step of data classification and filtering, will be further processed. Consequently, the whole process of 3D surface reconstruction will be based on exclusively matte (i.e. diffuse) reflections.

As to commonly assumed conformity of diffuse reflections phenomena with basic Lambert's reflectance law, it is said, that its application to real surfaces, even of metallic or glossy reflectance characters, is quite reasonable [2-3]. For the process of determination of reflectance properties with use of basic linear algebra (a) or SVD decomposition (b) [7], it is implicitly assumed, that Lambert's reflectance law is valid. Stage of reflectance determination is crucial in proper and valid further data processing, which consequently leads to accurate 3D reconstruction process.

$$\rho(x,y) = \left\| \left(\left[L^T\right]_{3xn} \cdot \left[L\right]_{nx3} \right)^{-1} \cdot \left[L^T\right]_{3xn} \cdot \left[I\right]_{nx1} \right\|,\ at \left\|N\right\| = 1 \tag{1a}$$

$$\rho(x,y) = \left\| \left[L^+\right]_{3xn} \cdot \left[I\right]_{nx1} \right\|,\ at \left\|N\right\| = 1 \tag{1b}$$

In relations (1a) and (1b) original [*L*] matrix is a matrix of incidence light sources directions taken row-wise (each row of each of the light sources). Moreover [*I*] is column of 2D images intensities for considered currently pixel point, at any (x, y) image location, [*N*] is vector normal to the surface regarded, assumed as normalized in stage of ρ(x,y) determination (reflectance coefficient). In 1st equation some kind of pseudo-inversion of [*L*] matrix, implicitly assumed as rectangular, has been applied, while in 2nd equation a pseudo-inversion of [*L*] matrix, based on Singular Value Decomposition (SVD), has been applied, thus giving in the result pseudo-inverted [L^+] matrix.

Accordingly to (1a) and (1b), a stage of *N* vector determination, (giving up complex optimization techniques used in previous works [4-5]), is of the following form, respectively:

$$\frac{[N]_{3x1}}{\sqrt{1+p^2+q^2}} = \frac{([L^T]_{3xn} \cdot [L]_{nx3})^{-1} \cdot [L^T]_{3xn} \cdot [I]_{nx1}}{\rho(x,y)}, \tag{2a}$$

$$for\ i \in [1..n],\ \ with\ i \notin \{spec, self-msk, self-shdw\}$$

$$\frac{[N]_{3x1}}{\sqrt{1+p^2+q^2}} = \frac{[L^+]_{3xn} \cdot [I]_{nx1}}{\rho(x,y)},\ for\ i \in [1..n], \tag{2b}$$

$$and\ moreover: with\ i \notin \{spec, self-msk, self-shdw\},$$

In the above equations (2a) and (2b), an *i* is current index of the light source within use set of light sources,, which is being activated, and additionally, it does not provoke occurrence of one of the undesired phenomena, such as specularites, self-masking, or self-shadowing, respectively.
Resuming consideration in this section, not taking into consideration basic Lambert's reflectance law as valid, forces the need (in cases of important deviations from this law for diffused real surfaces) of introducing more evolved methods of 3D surface reconstruction, in context of *a priori* known Bidirectional Reflectance Distribution Function (BRDF) or with simultaneously derivated BRDF. However, these aspects are rather out of scope for this paper, and should be considered elsewhere.

3. Auxiliary problems and algorithm implementations

For data acquisitions, well initially tested, and previously already presented [6], some light sets of directional incidence light will be here used, in currently related works. The geometrical assumptions for this, due to too much concise contents, and the correlated topic considerations, are presented elsewhere [5]. However, some important solutions, related to performing of auxiliary conditions and settings for 2D image data acquisition process, will be here considered in brief.
With careful analysis of photometric equations, authors came to conclusion, that incidence light directions, can be known *in advance* only partially, giving, to some degree, softening in restrictions, accordingly to light sources settings. Introducing, some important additional definitions, one can actually simultaneously perform two task. First task is of derivation of unknown *in advance* elevation angle of the light sources, while mutual azimuthal orientations for each of the light sources, within fixed light set geometry are known *a priori*. Second main task is of 3D surface reconstruction process, with already acquired and *fully known* incidence light sources directions.
Taking into account *a set of critical points*, one can assume, that there exists some highly correlated set of *N* vectors, normal to the surface regarded, at occurrencies of locally maximal intensities, within 2D image, (accordingly to *a reference light source*), which on the whole in their directions are in compliance with direction of incidence light, consequently indicating and fully determining direction of actually used light sources.
During trials and experiments, initially carried out, it occurred, that conventionally used in the past, the definition of critical points, actually must be reshaped, accordingly to the needs, of data interpretation, on inhomogeneous real surfaces visually inspected.
Thus, *a set of critical points* are called *a set of real critical points*, if and only if it's a set of unique points (i.e. set of points, which are not mutually overlapping) taking as a reference , singly and subsequently activated *all* light sources within set of light sources, used in visual

inspection of the real surface, regardless of mutual similarity or dissimilarity of the incidence light directions, for all light sources.
Therefore, a reflectance borne *quasi-critical points*, will be excluded, at least theoretically, from further data processing, leaving within analysed set of points, *real critical points*, strictly correlated with topographic features of the real surface visually inspected.

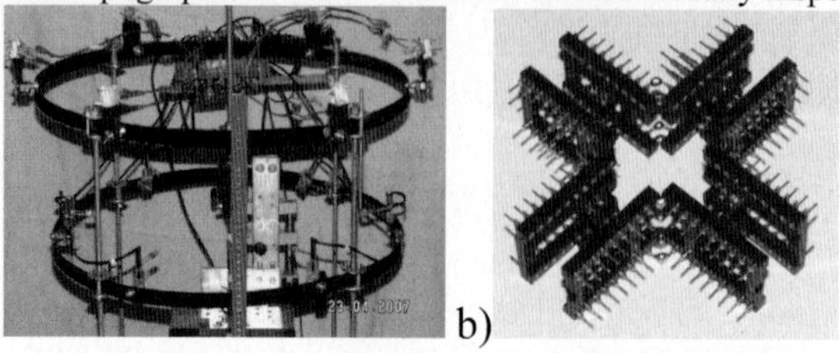

a) b)

Fig.1 A) big light sources set (halogen bulbs), B) set of SMD LED light sources,

For big light sources set with halogen bulbs, as well as, for some compact light set of SMD LED light sources, the directions known in advance are the following:

$$A:[L]_{lx3} = \begin{bmatrix} L_x & 0 & L_z \\ \cos(\frac{\pi}{3})L_x & \sin(\frac{\pi}{3})L_x & L_z \\ -\cos(\frac{\pi}{3})L_x & \sin(\frac{\pi}{3})L_x & L_z \\ -L_x & 0 & L_z \\ -\cos(\frac{\pi}{3})L_x & -\sin(\frac{\pi}{3})L_x & L_z \\ \cos(\frac{\pi}{3})L_x & -\sin(\frac{\pi}{3})L_x & L_z \end{bmatrix}_{6x3} \qquad B:[L]_{lx3} = \begin{bmatrix} L_x & 0 & L_Z \\ 0 & L_x & L_Z \\ -L_x & 0 & L_Z \\ 0 & -L_x & L_Z \end{bmatrix}_{4x3} \tag{3}$$

In equation (3), Lz implicitly represents unknown elevation angles, while azimuth angles are known and determined by distinct change from source to sources on closed ring of 60 degrees for 1st big light set, while for cross-like compact light set of 4 sources (B), change of azimuth angles from source to source is of 90 degrees.

$$f_{\min 2} = \left\| \begin{bmatrix} I_{IA} & I_{IIA} & \dots & I_{pA} \\ I_{IB} & I_{IIB} & \dots & I_{pB} \\ \dots & \dots & \dots & \dots \\ I_{Il} & I_{IIl} & \dots & I_{pl} \end{bmatrix}_{lxp} - [\rho]_{lxp} \otimes \frac{\left([L]_{lx3} \cdot \begin{bmatrix} p_A & p_A \\ q_A & \dots q_A \\ 1 & 1 \end{bmatrix}_{3xp} \right)}{\sqrt{1 + p_A^{\ 2} + q_A^{\ 2}}} \right\|, \tag{4}$$

In equation (4), $[I]$ matrix represents itself, intensities stacked column-wise, accordingly to light sources activated, and combined row-wise, accordingly to the subsequent detected and ordered critical points. Moreover there is some $[\rho]$ matrix of p mutually idempotent columns, of l reflectance coefficients, which are to be determined. Additionally, an $[N]$ matrix of p mutually idempotent (in assumption) vectors normal to the surface regarded, has been here placed, combined column-wise, side by side (i.e. vertically in p columns).

4. Output data results in derivation of illuminant direction

Trials and experiments with the method of illuminant direction determination, prior to the main 3D surface reconstruction, will be carried out on some 3D depth map of very textured surface, obtained in 3D profilometric measurements, for the abrasive tools surface samples, as well as, on sets of real 2D intensity images, taken in some acquisition systems, with advanced zooming facilities. Thus, the whole acquisition system, with digital camera, allows for taking of 2D images, starting from 10 millimeters, giving in results dimensional correspondence of one pixel on the 2D images, of a few tenths of micrometers.
Firstly, some metallic surface, of recurrent rhomboidal shapes on it, has been measured with 3D optical profilometer (wit use of Taylor Hobson 150) and within selected patch of 2x2[mm]).The resulted depth map has been exported in form of color depth map. Next, in Matlab environment, some indexing of data has been carried out, in order to extract depth information. Tthe obtained 3D map has been used in careful rendering process, giving in the results, some set of 2D images at various elevation angles, and with azimuth angles for each of 6 virtual light sources, accordingly to contents of *A* matrix from equation (3). The strictly diffuse character of modeled reflectance properties and in addition homogenous reflectance coefficient, have been used in initial experiments.

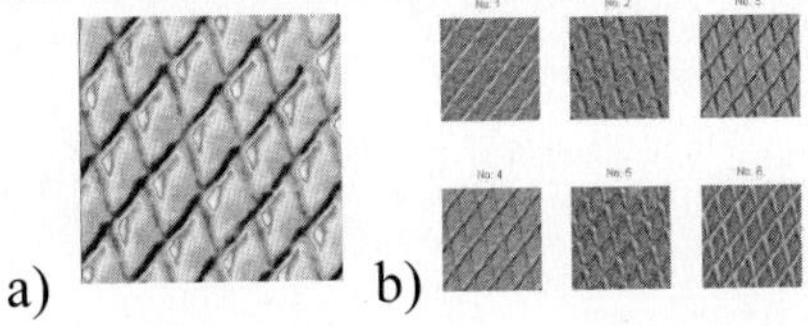

Fig.2 A) 3D depth map b) Six 2D images in virtual rendering, elevation: 30 deg.,

It occurs, that extended definition of set of real critical points, gives important improvement in derivation of unknown elevation, accordingly to originally defined set of critical points, only in cases of low values (from 20 to 40 degrees).

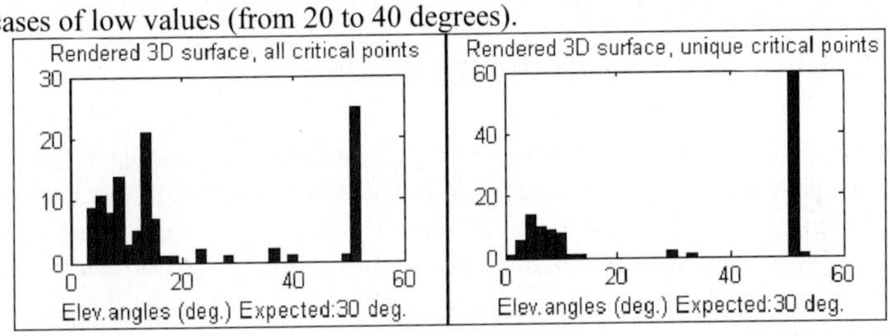

Fig.3 Histogram of elevations, left: all critical points, right: unique cps.

On fig.3 in the middle of 2nd histogram there are valid counts of elevation angle in direction derivation process. Secondly, set of real 2D intensity images, taken for Al_20_3 sample,, as a 1st sample of abrasive tool (sig.99A120MV8), next, Black SiC, as a 2nd sample (sig. 37C120JVK8), and finally green SiC sample, cut from lapping stone, as a 3rd sample (sig. 99C120N), have been all taken into consideration. Data acquisition has been realized with 1st light set (on fig. 1A), at 40 deg of elevation (following histograms), and 46 deg (examples on fig.4).

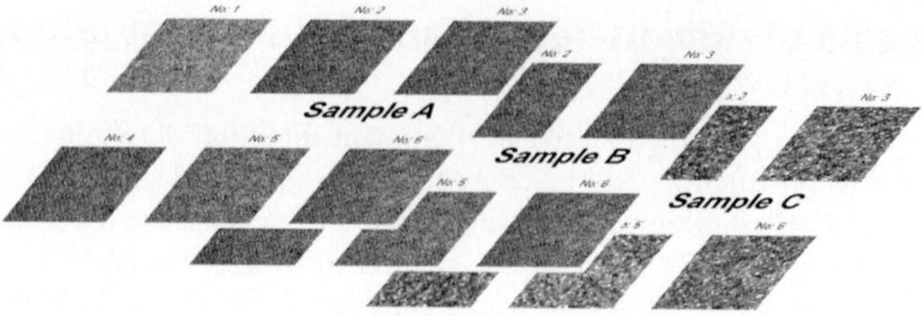

Fig.4 2D images sets for *A*, *B*, *C* samples respectively, at 46 deg..

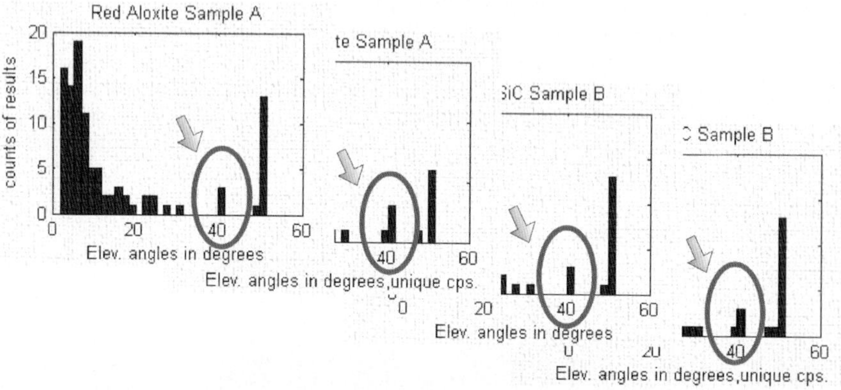

Fig.5 Histograms with marked valid counts of elevation for A &B surface samples, stacked in cascade, from top to bottom, odd histograms: all critical points, even histograms: only unique critical points

5. Concluding remarks of the methods

The intention of the authors was to basically present both some robust solutions applied prior to 2D image data acquisition stage, and algorithms used in further data processing steps. The aim of the paper was to initiate the discussion, introducing realization of the some auxiliary proceedings, such as determination of partially unknown illuminant directions, prior to main task of 3D reconstruction and acquisition of properties of the real surface of abrasive tools.

References

[1] Woodham R. J.: Photometric methods for determining orientation from multiple images, Optical Engineering, 1980, 19 (1)

[2] Pernkopf F., O'Leary P.: Image acquisition techniques for automatic visual inspection of metallic surfaces, NDT & E International 2003 (36), pp.609-617

[3] Smith M. L., Stamp R. J.: Automated inspection of texture ceramic tiles, Comp. in Ind., 2000 (43), pp.73-82

[4] Bernat A., Kacalak W.: Surface reconstruction for modeled grain geometry, based on 2D intensity images, part one/part two, XXIV Ogólnokraj. Konf.: Polioptymalizacja i Komputerowe Wspomaganie Projektowania, Mielno 2006, pp.88-95, pp.96-103

[5] Bernat A., Kacalak W.: A Method for Visual Inspection of Abrasive Tool Cutting Surface In Possible Integration of Grinding System with On-line Tool's Monitoring (Part One/Part Two), TPP'06, 19-20.X.2006, Poznan, pp.40-51, pp.52-63

[6] Kacalak W, Bernat A.: Deriving unknown illuminants parameters based on contents of 2D images of cutting surface of abrasive tools, Doktoranci Dla Gospodarki, Sarbinowo 2006, Zeszyty Naukowe. Wydz. Mech. Nr. 39 PKos., ISSN 1640-4572, p.21-26

[7] Kincaid D., Ward Cheney: Numerical Analysis. Mathematics of Scientific Computing, Ed. 3rd, The University of Texas at Austin

Mechatronic stand for gas aerostatic bearing measurement

P. Steinbauer (a), J. Kozánek (b), Z. Neusser (a), Z. Šika (a), V. Bauma (a)

(a) CTU in Prague, Faculty of Mechanical Engineering, Karlovo nám. 13, Praha 2, 121 35, Czech Republic

(b) Institute of Thermomechanics, Academy of Sciences of the Czech Republic, Dolejškova 5,
Praha 8, 182 00, Czech Republic

Abstract

Current mechatronic and mechanical devices are designed to maximize performance. They are thus driven near to stability limits. Aerostatic gas bearings are often used to support high speed rotors with higher loads. Their stability is thus one of major issues. To determine the rotor stability limits, precise gas film stiffness and damping values are necessary to be determined. There are already algorithms for calculation of aerostatic journal bearings stiffness and damping available, but up to now they were not experimentally verified and experimental results published.

1. Introduction

Gas bearings of similar size to oil lubricated bearings carry smaller loads and smaller clearances. Due to the lower viscosity of the medium the lubricant shear stresses are lower and hence the operational speed can be very high without excessive power being needed and less heat generated. The bearing can operate in environments of high temperatures, as this does not affect the lubricant properties.

There are generally two types of fluid bearings, dynamic (fluid is sucked into the bearing by shaft rotation forming lubricating wedge around the shaft) and aerostatic (fluid is pumped under the pressure into space of the bearing). Aerostatic journal bearing is subject of this paper.

Aerostatic bearings can achieve high precision of operation and low noise. In specific cases (e.g. polluted environment, required higher bearing stiffness) they are used to support high speed rotors. Successful high speed application examples are dental drills with maximum speed up to 750.000 rpm or grinding spindles (100.000 rpm).
The most important issues at high speed rotor design is their stability. The stability limit must be determined with sufficient precision before putting the machine into operation, because rotor instability leads immediately to heavy failure of the machine.
So fluid bearings are known and studied for many years ([1]). Up to now, there is lack of published experimental data about fluid layer characteristics available to designers. Thus experimental stand for fluid bearings measurement was build. It will serve to measure data which will be used for aerostatic bearing model coefficients identification.
Reliable experimental measurement procedure requires solid experimental stand, whose dynamic properties will not interfere with measured phenomena. Planned stand properties must be experimentally verified.

2. Model for the bearing identification

The considered universal dynamic model ([2]) is outlined on the Fig. 1. The stand for measurement is basically formed by stiff shaft (M_1) supported by two rolling ball bearings. The aerostatic bearing head (M_2) is placed between them and can float within clearances between aerostatic bearing and the shaft. The motion of the bearing head with respect to the shaft is measured.

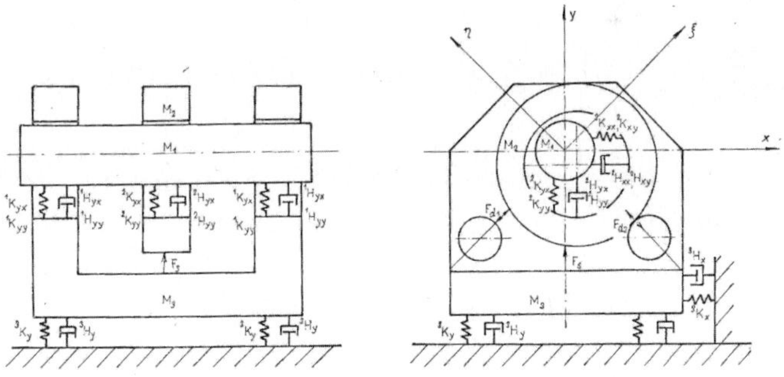

Fig. 1. Universal dynamic model of the test stand

Considered bearing dynamic properties are stiffness and damping in the form of second order matrices. They are calculated from the system re-

sponse data to harmonic excitation by force acting in two different directions relatively to static load.
The model considers flexibility of supporting sliding bearings and test stand foundation. Basic equations of motion are as follows:

$$[M_2][\ddot{x}_2]+[{}^2Z][\chi_r]=[f_k],\ k=1,2,\ \text{for test bearing} \qquad (1)$$

where

$$[M_2]=\begin{bmatrix} M_2, & 0 \\ 0, & M_2 \end{bmatrix},\ [\chi_2]=\begin{bmatrix} x_2 \\ y_2 \end{bmatrix},\ [\chi_r]=\begin{bmatrix} x_r \\ y_r \end{bmatrix}=\begin{bmatrix} x_2-x_1 \\ y_2-y_1 \end{bmatrix},$$

$$[{}^2Z]=\begin{bmatrix} {}^2Z_{xx}, & {}^2Z_{xy} \\ {}^2Z_{yxx}, & {}^2Z_{yy} \end{bmatrix} \text{ ... test bearing complex stiffness,}$$

$${}^2Z_{jk}=K_{jk}+i\Omega B_{jk},\ [f_1]=\frac{\sqrt{2}}{2}\begin{bmatrix} F_{d1} \\ F_{d1} \end{bmatrix}e^{i\Omega t},\ [f_2]=\frac{\sqrt{2}}{2}\begin{bmatrix} -F_{d2} \\ F_{d2} \end{bmatrix}e^{i\Omega t},$$

$$[M_1][\ddot{x}_1]+2[{}^1Z]([\chi_1]-[\chi_3])-[{}^2Z][\chi_r]=0,\ \text{for the shaft,} \qquad (2)$$

where

$$[M_1]=\begin{bmatrix} M_1, & 0 \\ 0, & M_1 \end{bmatrix}, \qquad [\chi_2]=[\chi_2]-[\chi_r],$$

$$[\chi_1]-[\chi_3]=\begin{bmatrix} x_1-x_3 \\ y_1-y_3 \end{bmatrix}=[\chi_2]-[\chi_r]-[\chi_3],$$

$$[{}^1Z]=\begin{bmatrix} {}^1Z_{xx}, & {}^1Z_{xy} \\ {}^1Z_{yxx}, & {}^1Z_{yy} \end{bmatrix} \text{ supporting bearing complex stiffness,}$$

$$[M_3][\ddot{\chi}_3]-2[{}^1Z]([\chi_1]-[\chi_3])-[{}^2Z][\chi_3]=-[f_k],\ \text{k=1,2,} \qquad (3)$$

for the frame, where

$$[M_3]=\begin{bmatrix} M_3, & 0 \\ 0, & M_3 \end{bmatrix},\ [\chi_3]=\begin{bmatrix} x_3 \\ y_3 \end{bmatrix},\ [{}^3Z]=\begin{bmatrix} {}^3Z_x, & 0 \\ 0, & {}^3Z_y \end{bmatrix}.$$

${}^3Z$ ... matrix of the frame support complex stiffness.

Assuming the solution in the form

$$[\chi_j]=[\bar{\chi}_j]e^{i\Omega t},\ \text{j=1,2,3,}\ [\ddot{\chi}_j]=-\Omega^2[\bar{\chi}_j]e^{i\Omega t} \qquad (4)$$

after substitution into (1), (2) and (3) we obtain

$$[\bar{Z}][\bar{\chi}] = [\bar{f}_k], \quad k=1,2 \qquad (5)$$

where

$$[\bar{Z}] = \begin{bmatrix} [^2Z] & -\Omega^2[M_2] & 0 \\ \Omega^2[M_1] - 2[^1Z] - [^2Z] & -\Omega^2[M_1] + 2[^1Z] & -2[^1Z] \\ 2[^1Z] & -2[^1Z] & -\Omega^2[M_3] + 2[^1Z] + [^3Z] \end{bmatrix}$$

,

$$[\bar{\chi}] = \begin{bmatrix} [\chi_r] \\ [\chi_2] \\ [\chi_3] \end{bmatrix}, \quad [\bar{f}_1] = \frac{\sqrt{2}}{2} F_{d1} \begin{bmatrix} 1 \\ 1 \\ 0 \\ 0 \\ -1 \\ -1 \end{bmatrix}, \quad [\bar{f}_2] = \frac{\sqrt{2}}{2} F_{d2} \begin{bmatrix} 1 \\ 1 \\ 0 \\ 0 \\ -1 \\ -1 \end{bmatrix}.$$

To determine the test bearing stiffness and damping coefficients it is sufficient to solve equation (1) for vectors of complex amplitudes $[\chi_2]$, $[\chi_r]$ measured at two different directions of excitation force F_{d1}, F_{d2}.
The dynamic model can be further simplified, because stiffness of the frame and supporting rolling bearings will be much higher than expected stiffness of aerostatic journal bearings. The movement of the foundation and of the shaft in support bearings will be very small in comparison to with the excursions of test bearing relative to shaft and could be therefore probably neglected. This assumption must be however confirmed by measurement.

2. Experimental stand design

Aerostatic bearings test stand is built on the top of Rotor Kit, experimental workplace for rotor dynamics investigation and measurement (Fig. 2).
The test shaft is supported by two rolling bearings, which are inserted into bearing bodies fastened to the frame. The test head with aerostatic bearing is located between rolling bearings. The rolling bearing outer diameter is smaller than test bearing diameter, so that the change of the bearing would be as simple as possible. Piezo-electric actuator for excitation of the bear-

ing by harmonic or arbitrary force is connected to the test head by means of butt hinge (joint) to ensure axial loading of piezoelement.

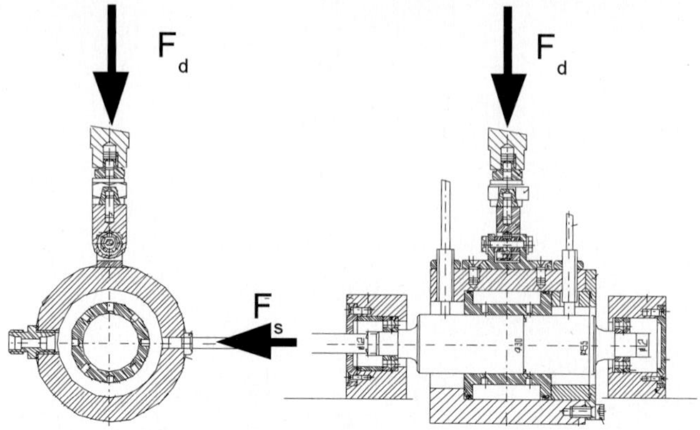

Fig. 2 Stand structure for identification of aerostatic bearings dynamic properties

Gas bearing can be loaded in two radial directions by static, harmonic or stochastic force by means of piezo-actuators, which is measured by force sensor. Redundant measurement of gas film thickness on several points of the gas bearing is provided. Assumptions in mathematical model are checked by measurement of acceleration on several points of the stand. Measurements are synchronized with rotor phase.

4. Conclusions

Aerostatic bearings identification approach was proposed. Experimental mechatronic stand based on Rotor Kit Nevada was designed, manufactured and tested. Further, its dynamic properties were evaluated to ensure they do not affect measured bearings.

References

[1] J. Glienicke. Feder- und Dämfungskonstanten von Gleitlagern für Turbomaschinen und deren Einfluss auf das Schwingungsverhalten eines einfachen Rotors, Dissertation, Technischen Hochschule Karlsruhe, 1966

[2] J. Šimek, Dynamic model of test stand for journal bearings 90 mm. Methodology of measurement and evaluation, Research rep. SVÚSS, 1982

[3] R. Tiwari, A. V. Lees, M. I. Friswell. Identification of Dynamic Bearing Parameters: A Review, The Shock and Vibration Digest, March 2004

Compression strength of injection moulded dielectromagnets

L. Paszkowski, W. Wiśniewski

Institute of Precision and Biomedical Engineering, Warsaw University of Technology, ul. Św. A. Boboli 8, 02-525 Warsaw, Poland

Abstract

The paper deals with the problems of material and processing parameters influence on axial compression values in the samples made from composites designated for producing dielectromagnets. Shapes were manufactured by injection moulding. The composite matrix constituted high- or low impact resistant polystyrene. As filler hard magnetic powders from Nd-Fe-B alloy were applied, having flake- or sphere-shaped grains. Tests were carried out on samples made from composites of various filler compactness values and grain sizes. Shapes were performed with application of varying composite injection pressure values and temperatures of injected plastic compound.
Samples made as described above were subject of axial compression test for laying down their strength value.
The experiment results have shown that the most substantial influence on the deformation sequence and on the samples compression strength has the kind of composite matrix.

1. Introduction

For performing of magnets alloys are used which are hard and bristle materials. These features pose considerable difficulties in the manufacturing process of moulding the magnetic samples.
Injected moulded dielectromagnets are made from composites, where the magnetic material is hard magnetic powder uniformly distributed in the matrix of thermoplast. Magnetic parameters of the dielectromagnets of this

kind are determined by the quantity of magnetic powder contained in the sample volume. This relationship is directly proportional. However, following the manufacturing method applied, it is necessary to obtain a compound that is injection-mouldable. Excess quantity of the powder filler used causes that the composite loses the above mentioned quality. Hence there is a limit quantity of magnetic powder, which can be added to the polymer without affecting the composite ability of being injection-mouldable.

Dielectromagnets are not structural components of which high mechanical strength is demanded. They are often subject to action of different forces, so it is necessary to know they response to these factors [1, 2].

2. Composition of Dielectromagnets

Material of excellent magnetic properties is Nd-Fe-B alloy [3]. In the research presented, the magnetic powders supplied by the Magnequentch International were applied, and namely of both; flake- (MQP-B) and spherical- (MQP-B) shaped grains [4].

Composite matrix was constituted by polystyrene supplied by the DWORY Co. [5]. For the experiments high impact polystyrene Owispol 825 and low impact strength one Owispol SX 25 were used.

3. Preparation of Granulated Products

Composite granulated material for injection moulding was prepared by a solvent method with use of toluene, which is a proper solvent for polystyrene [6].

Granulated material for the tests was prepared from powder of flake- and spherical grains, using the composite filling volumes as follows:

- For powder of flake-shaped grains: V_p = 40; 44; 48 and 54 vol. % (82,7; 84,9; 86,9 and 89,4 weight %),
- For powder of spherical grains V_p = 40; 48; 54; 60 and 64 vol. %, (82,5; 86,7; 89,2; 91,4 and 92,6 weight %).

Highest values of magnetic powder composite filling constitute filling limit values for both powder kinds. Exceeding the limit values results in the granulated material becoming nonplastic and injection-unmouldable. When filling is kept below 40 %, it results in dielactromagnets having too low magnetic values to justify their production being reasonable.

4. Preparation of Samples and Test Methodology

For conducting tests shapes were prepared in the form of short cylinders of 10 mm diameter and 4 mm height. Injection moulding was performed applying varying process parameters, in accordance with the manufacturers' recommendations. Temperatures of heating the granulated material in the injection machine cylinder amounted to 160, 200 and 240°C respectively. Injection pressure was set at 80, 110 and 140 MPa respectively.
Compression strength was determined on a modernized tester INSTRON model 1115. Velocity of the tester cross-bar travel was constant in the entire measurement process and amounted to 1 mm/min.

5. Testing of Compression Strength

Tests carried out enabled to determine influence of composites structure and their processing parameters on the samples behavior upon axial compression forces acting on them.
Fig. 1 shows co-relation between magnitudes of sample deformation and compression forces. Samples were prepared from composites having matrix composed of high- or low impact resistant polystyrene. The filling magnetic material constituted both kinds of commercially available powders. Constant magnitude of composite filling, amounting to 54%, was applied.
Analysis of the test results allows for conclusion that the nature of relation changes is close one to the other. In the beginning phase of compression force growth, deformation of samples takes place. As soon as certain specific value of compression force was reached, all samples were becoming destroyed. In the first instance tiny vertical crack lines were appearing on circumference of the shapes compressed. With growing compression stress cracks had been growing bigger and bigger, until small fragments of the samples were chipped away, leaving only the cylinder core intact. The tests allowed concluding that magnitude of compressive stresses which were destructing the samples differed depending on the composite analysis. Shapes which were subject to the soonest destruction were made from composites with matrix of low impact-resistant polystyrene, filled with powder of flake-shaped grains. Highest, in this case, deformation of samples which didn't cause their destruction, haven't exceeded 5 %. On the other hand, the highest vulnerability to deformations not causing destruction yet was demonstrated by shapes from composites of matrix made of high impact-resistant polystyrene containing spherical kind of powder. In

the latter case deformation even exceeding 20 % was possible, without cracks appearing.

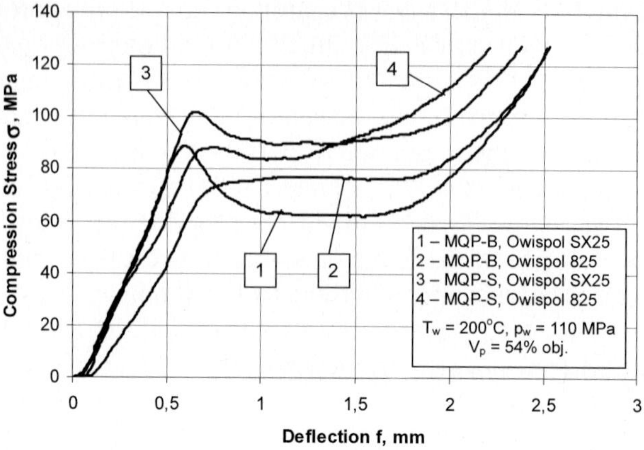

Fig. 1. Relation between deformation magnitude and compression stress in the shapes made from composites of polystyrene matrix Owispol SX 25 and Owispol 825, containing flake-shaped (MQP-B) and spherical (MQP-S) kinds of powder.

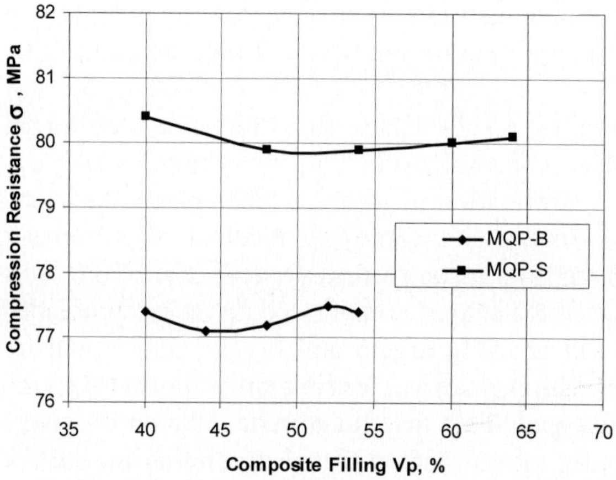

Fig. 2. Influence of filling degree of the composite with magnetic powder of flake- and spherical kind on

On the Fig. 2 influence of filling degree of the composite with magnetic powder is shown on the resulting compression resistance of the samples. As it can be seen, influence of composite filling magnitude on compression resistance of the samples is not considerable for both kinds of powder.

What is important, it is the shape of filling particles. Higher values of compression resistance were reached in the case of shapes made from composites containing spherical kind of the powder.
As it can be seen, the former case, pressure resistance of the samples is primarily determined by the shape of filler particles.

6. Summary

The tests carried out allow for drawing a conclusion that major influence on the shapes compression resistance has the composite structure, whereas its manufacturing process parameters are of lesser importance. Distinctly higher compression resistance values were attained for shapes produced on the basis of spherical powder. At the same time, samples made of composites with their matrix based on low-impact resistant polystyrene are subject of quicker destruction at less deformation, although destructive stress reaches higher values when compared with shapes with matrix from high-impact resistant polystyrene. Shape of the filler metal particles exerts also influence on easiness of shapes to become destroyed. Higher deformation values which do not entail destruction of samples, can be applied in the case of spherical kind of the powder.

References

[1] B. Ślusarek, D. Biało, J. Gromek, T. Kulesza, Journal of Magnetics. 4, 2, (1999) 52
[2] Research of powder phase Nd-Fe-B and polymer matrix influence on magnetic and mechanical properties of composite components; grant of the Warsaw Technical University (Poland), Dean of Faculty of Mechatronics, No. 503G-1142/0013-003. (in Polish)
[3] M. Leonowicz „Modern Hard Magnetic Materials. Selected Problems", Publishing House of the Warsaw Technical University, Warsaw, 1996. (in Polish)
[4] Catalogue of Magnequench International, Indiana, USA
[5] Catalogue of DWORY S.A. Co., Oświęcim (Poland) (in Polish)
[6] I. Gronowska, W. Kałuży, L. Paszowski „Injected Magnetic Components, Research on Properties by Means of a Scanning Acoustic Microscope", XIth Seminar on Plastics in Machine Construction, Kraków (Poland), 2006, "Mechanik" June 2006, 215. (in Polish)

Over-crossing test to evaluation of shock absorber condition

I. Mazůrek (a) *, F. Pražák (b), M. Klapka (c)

(a, b, c) Faculty of Mechanical Engineering, Brno University of Technology, Technická 2, Brno, 616 69, Czech Republic

Abstract

On Faculty of Mechanical Engineering, Brno University of Technology it was state to life a new methodic for check quality of suspension wheel damping of means of transport so-called over-crossing test. The target group cars are categories, for those are unsuitable existing known routes. This paper is describing principle of methodology incumbent on evaluation behavior of suspension wheel after crossing of defined obstacle. This method was conquest experimental testing on real, experimental and virtual vehicle. This method is suitable for on-car test of shocks absorber on motorcycle, delivery and truck car.

1. Excitation methods and selection of diagnostic variable

The project called "Technical diagnostic of wheel suspension, promoted by Czech Science Foundation, was successfully finished in last year. A new method of on-car diagnostic of suspension of heavy vehicles was introduced as one of the outputs. From the angle of both costs and testing method the optimum solution is when the excitation of free oscillation of the followed up masses sets in due to crossing over the defined ramp at a chosen speed. The selection of the diagnostic value and the diagnostic model for analysis of the criteria of damping quality is very important. With well-damped axles, the movement of the sprung mass is rather insignificant and the analysis would be encumbered with considerable error. As to the movement of the non-sprung mass, even the measurement of the relative deviation from the sprung mass was tested, however, this procedure was abandoned because of the higher complexity of the system on-vehicle installation.

1.1 Over crossing test

The selection of diagnostic model came from the real estimation, that, because of huge diversity of vehicles, it is not possible to identify or estimate the parameters of model during the test. For analysis only the evaluation of recorded trajectory of tail of unsprung mass comes into question. Therefore, an extremely simplified linear model of suspension in so-called "resonant" configuration has been chosen, which assumes, that in the area of natural oscillation of unsprung mass the movement of sprung mass is minor and this is replaced by fixed imbedded car spring (Fig.1.).

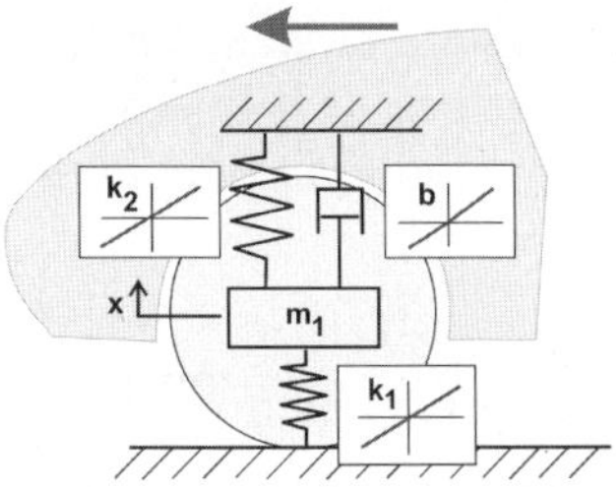

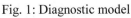

Fig. 1: Diagnostic model

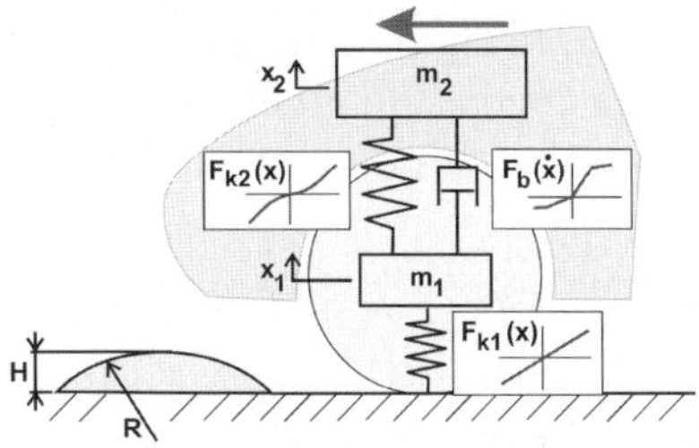

Fig. 2: Simulation model of over crossing test

On unsprung mass of suspension wheel is fixed only sensor of vertical acceleration. The communication of the accelerometer with the measuring computers is wireless [4, 5]. On the test track, the driver sets the vehicle in movement at a speed of 5 to 10 km/h and crosses over the defined ramp laid on the carriageway. The algorithm for evaluating the damping characteristic must be as simple as possible, in order to be, applicable even to 8-bit microprocessors. In principle the aim is the evolution of exponential curve of tail curve and estimation of natural oscillation of oscillating subsystem. During the analysis of monitoring movement of the wheel the first step is a determination of the beginning of free tail of the system. Motion equation of model at tail has simple notation:

$$m_1\ddot{x} + b\dot{x} + (k_1 + k_2)x = 0 \text{ or: } \ddot{x} + 2b_r\omega_0\dot{x} + \omega_0^2 x = 0, \qquad (1)$$

where k_1 and k_2 are tyre stiffness and vehicle spring, b is damper force per velocity 1m/s, $\omega_0 = ((k_1+k_2)/m_1)^{1/2}$ is natural radian frequency and $b_r = b/(2m_1\omega_0)$ is ratio damping of system. To identify the parameters of this model is not big problem.

Fig. 3 shows the individual phased stages of motion analysis of unsprung mass m_1. Thin curve is the logging of acceleration of tail mass. The natural radian frequency ω_0 is estimated from time of the first two periods free tail.

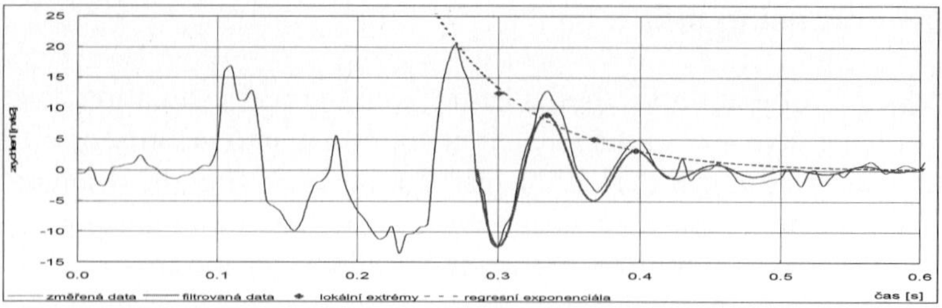

Fig. 3: Individual phases of process during the diagnosis of suspension damping

Local extremes from the first two periods are identified from the tail curve of the filtered signal. The envelope exponential curves are estimated by the following regression analysis of the curve local extremes. If it is possible to note the equation of envelope curve of particular solution in time in the form:

$$x = Ce^{-b_r \omega_0 t}, \qquad (2)$$

it is then simple to estimate a value of ratio damping b_r, which is an aim criterion of damping quality. However, the linearization of the whole problem requires the test to be used as comparative, i.e. only for comparison of vehicles of the same type under identical testing conditions. The assessed frequency of the suspension is important just for judging the identity of test conditions. It reveals quickly even a small deviation of pressure in the tyres or another change in the adjustment of the chassis.

2. Simulation testing of the method

For assessing the influence of a wide spectrum of parameters on the test result, we have used the interaction of the simulation model and the diagnostic model. In contrast to high simplified diagnostics model was model for simulation of suspension behavior drawn as optimum between complications of mathematical interpretation and faith response. The request was nonlinear interpretation of sprung and damping joins of system (Fig. 2) [6]. In this model mass m_1 is reduced unsprung mass of suspension, m_2 is ratio of sprung mass respective on wheel. Function $F_b(x')$ describes force action of shock absorber, $F_{k1}(x)$ force action of tyre and F_{k2} (x) force action of vehicle spring. The dynamic behavior of this kinematical excited system is described by next equations:

$$m_2(\ddot{x}_2 + g) + F_b(\dot{x}_2 - \dot{x}_1) + F_{k2}(x_2 - x_1) = 0$$
$$m_1(\ddot{x}_1 + g) + F_b(\dot{x}_1 - \dot{x}_2) + F_{k1}(x_1 - h(t)) + F_{k2}(x_1 - x_2) = 0 \qquad (3)$$

The usage of the damper with non-linear characteristics is the result of compromise between quality handling properties and sufficient durability of wheel suspension. Then it is not possible to model this characteristic with trivial parameter b. The parameters of stiffness and damping are therefore expressed as parametric sub-models (characteristics) in equations of simulation model (3). We decided to calculate the non-linear response of the system to the drive pulse by direct numerical integration of the equations of the motion. Hundreds of „measurements" were made on the virtual level with various chassis parameters, under various test conditions, etc. A simple procedure was chosen at appraisal of the departure of the test conditions. Then the monitored parameter was modified in large ranges; the new echoes were generated and the new values b_{ri} were evaluated. Relative error Err_{rel}, caused by the change of parameter, is described by next equation:

$$Err_{rel} = 100(b_{ri}-b_{r0})/b_{r0} \tag{4}$$

The results give an idea of the influence of the breach of the test condition on the resulting diagnoses. The following three diagrams show the influence of the non-observance of the basic settings of the tested suspension on the error of diagnose in repeated measurement (weight of suspension Fig. 4a, stiffness of suspension Fig. 4b and quality of crossing Fig. 4c). On the x-axis is the relative deviation of the given parameter from the basic setting and the relative error of the diagnosis Err_{rel} is on the y-axis.

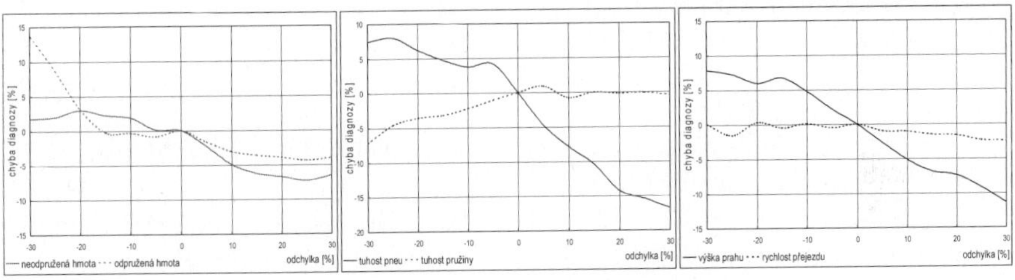

Fig. 4: Illustration of the influence of the relative deviation of the suspension parameters on the error in the diagnose in %

It is evident that the systematic error of the measurement can be kept in the value to 10%. In addition, the deviation of the sprung mass or of the tyre stiffness is unambiguously indicated by the changed proper frequency. The application of the methodology is exclusively defined as comparative for vehicles of the same technical configuration. Also the useful load should be, if possible, always the same (load, taken up quantity of fuel, etc.). A merit seems to be that the method is not too sensitive to the over crossing speed.

3. Conclusion

In the framework of the project was developed the method of the so called over crossing test for testing the technical condition of the wheel suspension. First off had to be selected the degree of simplification of the mechanical and mathematical models of the tester – tested vehicle system. Then had to be defined the measured variable and the suitable type of sensor. The following step consisted in wireless transmission and analysis of the obtained data and evaluation of the criterion selected as a measure of quality of the tested suspension damping. In the sense of detailed analysis of the shock absorber technical condition, this on-car diagnostic method is intended merely as an initial phase of conclusive diagnostics of a removed shock absorber. Even so, the economic benefit of this method, the sorting out of all cases of limiting technical conditions, is evident.

This work was developed with the support of the grant project GA ČR No.: 101/03/0304

References:

[1] EUSAMA – Recommendations for a performance test specification of an “on-car” vehicle suspension testing system – TS-02-76.

[2] Mazůrek, I.: Bezdemontážní diagnostika automobilových závěsů kol, inaugural dissertation, Brno University of Technology – Faculty of Mechanical Engineering, Brno 2000

[3] Novák, J.: Bezdrátový akcelerometr, project Science Fund Brno University of Technology – Faculty of Mechanical Engineering BD 1353029, Brno 2006

[4] Kopecký, T., Krupa, M.: Sensor Universal Wireless Unit and Acceleration Measurement. Proceedings of the International Interdisciplinary

[5] Mazůrek, I., Dočkal, A., Pražák, F.: Diagnostic model of a shock absorber. In: Engineering Mechanics, 2005, vol. 12, no. A1, p. 71-76. ISSN 1210-2717.

Laboratory Verification of the Active Vibration Isolation of the Driver Seat

L. Kupka, B. Janeček, J. Šklíba

Technical University of Liberec, Hálkova 6,
Liberec, 461 17, Czech Republic

Abstract

In the paper the introduction studies and first results of the active vibration isolation of the driver seat are presented. The actuator under examination is the air spring. The laboratory results of designed active vibration isolation system are very promising. Results of the use of the active and the passive vibration isolation systems are compared.

1. Introduction

We present the nonlinear mathematical model with concentrated parameters of the driver seat with an air spring. The linearization of this model is main idea of state space linear controller design. The active vibration isolation is based on feedback principle.

2. Model and theory

Simple mechanical scheme of the considered laboratory driver seat is shown in Fig. 1. Hydraulic damper is not used.

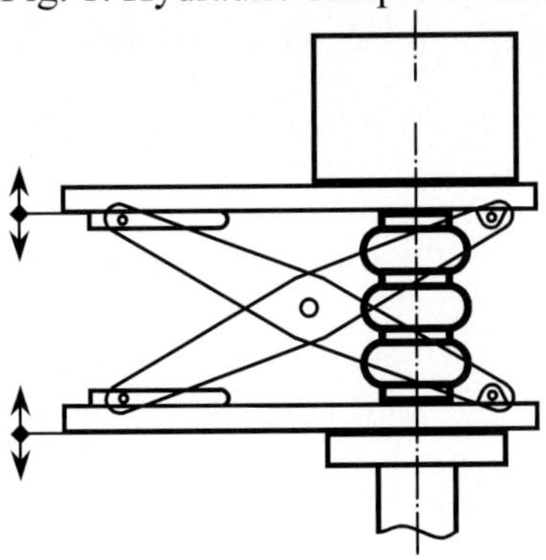

Fig. 1. Scheme of vibration isolation system with scissor mechanism

Equation of dynamic forces equilibrium on the system is

$$\frac{\mathrm{d}^2 z_2}{\mathrm{d}t^2} = \frac{1}{M}(S_{\mathrm{ef}}(p_2 - p_{\mathrm{a}})) - g - \frac{k_{\mathrm{d}}}{M}\left(\frac{\mathrm{d}z_2}{\mathrm{d}t} - \frac{\mathrm{d}z_1}{\mathrm{d}t}\right), \tag{1}$$

where M is a driver reduced mass, p_2 the absolute pressure inside the spring, p_a the absolute atmosphere pressure, g the gravity acceleration constant, k_d the coefficient of viscous friction, $S_{ef} = h_1(z_2 - z_1)$ the effective area of the air spring and h_1 the function of distance $z_2 - z_1$.

Air mass flow filling the air spring

$$Q_m = u_1 k_{\mathrm{v1}} \sqrt{p_1 (p_1 - p_2)} \ , \qquad u_1 \geq 0 \ , \tag{2}$$

where u_1 is the voltage input of electro-magnetic valve (controller output), k_{v1} the coefficient, p_1 the absolute high air pressure inside the accumulator. Air mass flow leaving the air spring into the atmosphere

$$Q_m = u_1 k_{\mathrm{v2}} \sqrt{p_2 (p_2 - p_a)} \ , \qquad u_1 < 0 \ . \tag{3}$$

The time derivative of pressure p_2 inside the air spring

$$\frac{\mathrm{d}p_2}{\mathrm{d}t} = \kappa p_2 \left(\frac{Q_m}{m} - \frac{1}{V}\frac{\mathrm{d}V}{\mathrm{d}t}\right), \tag{4}$$

$$\frac{\mathrm{d}V}{\mathrm{d}t} = \frac{\mathrm{d}V}{\mathrm{d}(z_2 - z_1)} \frac{\mathrm{d}(z_2 - z_1)}{\mathrm{d}t},$$

$$V = h_3(z_2 - z_1), \quad \frac{\mathrm{d}V}{\mathrm{d}(z_2 - z_1)} = h_4(z_2 - z_1),$$

where κ is an adiabatic air constant, m the air mass inside the spring, V is the spring's inside volume, h_3 and h_4 are the functions of distance $z_2 - z_1$.

$$\frac{\mathrm{d}m}{\mathrm{d}t} = Q_m \ . \tag{5}$$

It is possible to use inside the controller the function, which makes linearization of nonlinear air mass flow (2), (3).

In the Fig. 2 are the used variables renamed. With renamed variables the equations (5), (4) are

$$\frac{dx_1}{dt} = u_1 \ , \tag{6a}$$

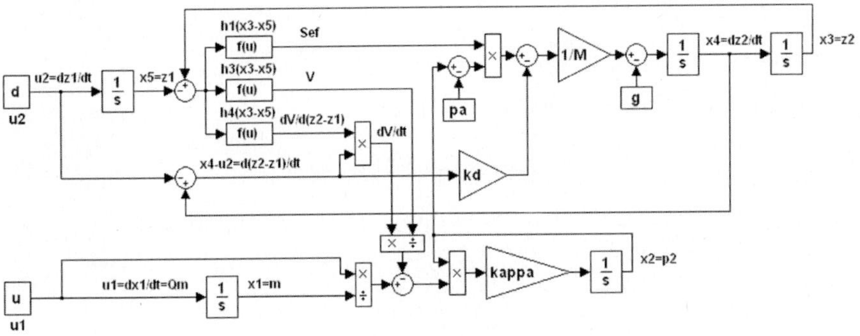

Fig. 2. Nonlinear simulation model (linearization of air mass flow is considered)

$$\frac{dx_2}{dt} = \kappa x_2 \left(\frac{u_1}{x_1} - \frac{(x_4 - u_2) h_4 (x_3 - x_5)}{h_3 (x_3 - x_5)} \right) . \tag{6b}$$

Next equation arises from Fig. 2

$$\frac{dx_3}{dt} = x_4 \ . \tag{6c}$$

Equation (1) with renamed variables is

$$\frac{dx_4}{dt} = \frac{1}{M} [x_2 \, h_1 (x_3 - x_5) - p_a \, h_1 (x_3 - x_5) - k_d (x_4 - u_2)] - g \ . \tag{6d}$$

Last equation of nonlinear model is

$$\frac{dx_5}{dt} = u_2 \ . \tag{6e}$$

In equations (6) are x_i, $i = 1, \ldots, 5$, state variables, u_1 is controller output, u_2 is disturbance, $u_2 = dz_1 / dt$. The discussed five equations are state equations of the system. The vector form of them is

$$\dot{\mathbf{x}} = \mathbf{f}(\mathbf{x}, \mathbf{u}) \ . \tag{7}$$

The state equations (7) can be linearized about the operating point ($\mathbf{x}_0$, $\mathbf{u}_0$). The linearization of ith state equation is

$$\dot{x}_i = f_i(\mathbf{x}_0, \mathbf{u}_0) + \left(\sum_{j=1}^{r} \frac{\partial f_i}{\partial x_j} \right)_{\substack{\mathbf{x}=\mathbf{x}_0 \\ \mathbf{u}=\mathbf{u}_0}} (x_j - x_{j0}) + \left(\sum_{k=1}^{s} \frac{\partial f_i}{\partial x_k} \right)_{\substack{\mathbf{x}=\mathbf{x}_0 \\ \mathbf{u}=\mathbf{u}_0}} (u_k - u_{k0}) \,. \quad (8)$$

Let we designate $\widetilde{x}_j = x_j - x_{j0}$, $\dot{\widetilde{x}}_j = \dot{x}_j$ and $\widetilde{u}_k = u_k - u_{k0}$.
The linearized state equations are

$$\dot{\widetilde{\mathbf{x}}} = \mathbf{A}\widetilde{\mathbf{x}} + \mathbf{B}\widetilde{\mathbf{u}} + \mathbf{f}(\mathbf{x}_0, \mathbf{u}_0) \quad (9)$$

and the linearized state equations of nonlinear equations (6) are

$$\dot{\widetilde{x}}_1 = \widetilde{u}_1 + f_1(\mathbf{x}_0, \mathbf{u}_0) \,, \quad (10a)$$

$$\begin{aligned} \dot{\widetilde{x}}_2 = \kappa \Bigg\{ & -x_{20} \frac{u_{10}}{x_{10}^2} \widetilde{x}_1 + \left[\frac{u_{10}}{x_{10}} - \frac{w_2\, h_4(w_1)}{h_3(w_1)} \right] \widetilde{x}_2 + \\ & + x_{20} \left[\frac{w_2\, h_4^2(w_1)}{h_3^2(w_1)} - \frac{w_2\, h_5(w_1)}{h_3(w_1)} \right] \widetilde{x}_3 - x_{20} \frac{h_4(w_1)}{h_3(w_1)} \widetilde{x}_4 - \\ & - x_{20} \left[\frac{w_2\, h_4^2(w_1)}{h_3^2(w_1)} - \frac{w_2\, h_5(w_1)}{h_3(w_1)} \right] \widetilde{x}_5 \Bigg\} + \\ & + \kappa x_{20} \left[\frac{1}{x_{10}} \widetilde{u}_1 + \frac{h_4(w_1)}{h_3(w_1)} \widetilde{u}_2 \right] + f_2(\mathbf{x}_0, \mathbf{u}_0) \end{aligned} \,, \quad (10b)$$

$$\dot{\widetilde{x}}_3 = \widetilde{x}_4 + f_3(\mathbf{x}_0, \mathbf{u}_0) \,, \quad (10c)$$

$$\begin{aligned} \dot{\widetilde{x}}_4 = & \frac{1}{M} [h_1(w_1)\widetilde{x}_2 + (x_{20} - p_\mathrm{a}) h_2(w_1) \widetilde{x}_3 - k_\mathrm{d}\, \widetilde{x}_4 - \\ & - (x_{20} - p_\mathrm{a}) h_2(w_1) \widetilde{x}_5] + \frac{k_\mathrm{d}}{M} \widetilde{u}_2 + f_4(\mathbf{x}_0, \mathbf{u}_0) \end{aligned} \,, \quad (10d)$$

$$\dot{\widetilde{x}}_5 = \widetilde{u}_2 + f_5(\mathbf{x}_0, \mathbf{u}_0) \,, \quad (10e)$$

where $w_1 = x_{30} - x_{50}$, $w_2 = x_{40} - u_{20}$,

$$h_2(w_1) = \frac{\mathrm{d}h_1(w_1)}{\mathrm{d}w_1} \,, \quad h_4(w_1) = \frac{\mathrm{d}h_3(w_1)}{\mathrm{d}w_1} \,, \quad h_5(w_1) = \frac{\mathrm{d}h_4(w_1)}{\mathrm{d}w_1} \,.$$

The linearized state space equations (10) in equilibrium state $\dot{\mathbf{x}} = \mathrm{f}(\mathbf{x}_0, \mathbf{u}_0) = \mathbf{0}$ were used for linear state space controller design. The modification of this controller was used for control of laboratory driver seat. The results of laboratory verification with disturbances measured on truck TATRA 815, during the drive on off-road track, are in Fig. 3.

For comparison are in Fig. 4 presented the measurements with industry produced driver seat with passive vibration isolation system. The used disturbances are in both figures the same.

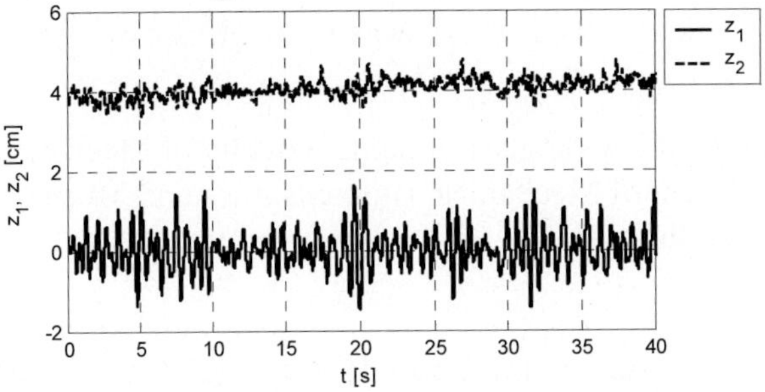

Fig. 3. Laboratory measurement with use of active vibration isolation system

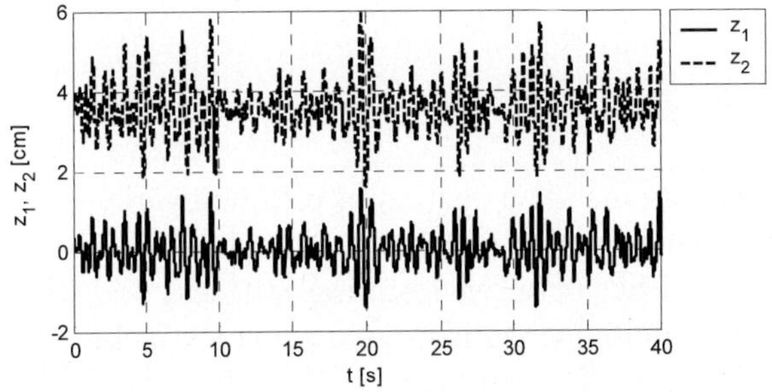

Fig. 4. Laboratory measurement with industry produced driver seat

3. Conclusion

The different penalty functions for controller design and structures of the estimators, which are the parts of state space controller, are tested at present. The linearization of system state space equations will be used for nonlinear state space controller design in future.

References

[1] L. Kupka, B. Janeček: Aktivní řízení sedačky řidiče. [Research report no. 1453/2006/10.] CEZ: MSM 4674788501. Liberec: TU, 2006.

[2] I. J. Nagrath, M. Gopal: Control System Engineering. Second edition. New Delhi: John Wiley & Sons, 1982. ISBN 0-471-09814-0.

Variants of Mechatronic Vibration Suppression of Machine Tools

M. Valasek, Z. Sika, J. Sveda, M. Necas B. (a), J. Bohm (b)

(a) Czech Technical University in Prague, Faculty of Mechanical Engineering, Department of Mechanics, Biomechanics and Mechatronics, Karlovo nam. 13, Praha 2, 121 35, Czech Republic

(b) Academy of Sciences of the Czech Republic, Institute of Information Theory and Automation,
Pod Vodarenskou vezi 4, Praha 8, 182 08, Czech Republic

Abstract

This paper deals with the investigation of different variants of mechatronic vibration suppression of machine tools. The structures of machine tools suffer from the conflict between resulting stiffness and dynamics of the machine tool. The consequence is limited accuracy and/or limited productivity of manufacturing. This problem can be solved by mechatronic modification of the machine tool instead of usual pure parametric optimization. There are several variants of such mechatronic modification for active vibration suppression of machine tools. They are the vibroabsorption by adding the auxiliary mass of vibration absorber, the vibrocompensation by adding a new parallel force connection of vibrating point to the frame, the control damping by adding a new link with damping force inside the construction and the mechatronic stiffness by adding a parallel structure to the existing one with collocated force connections between them. These variants were investigated on several examples of machine tools.

1. Introduction

The structures of machine tools suffer from the conflict between resulting stiffness and dynamics of the machine tool. The consequence is limited accuracy and/or limited productivity of manufacturing. The sufficient stiffness requires increase of used material that results into increase of machine tool mass. The increased mass of machine tool leads to the decrease of dynamics and machine tool productivity. The result of compromise is vibration of machine tools and decreased accuracy. This problem can be solved by mechatronic modification of the machine tool instead of usual pure parametric optimization. There are several variants of such mechatronic modification for active vibration suppression of machine tools. The paper deals with an investigation of different variants of mechatronic vibration suppression of machine tools. These variants were investigated for several examples of machine tools.

2. Approaches towards active vibration suppression

The mechatronic variants of active vibration suppression can be divided into traditional ones (also general purpose ones) and non-traditional ones (specific for machine tools). The traditional variants (e.g.[1-2]) are the vibroabsorption by adding the auxiliary mass of vibration absorber, the vibrocompensation by adding a new parallel force connection of vibrating point to the frame and the control damping by adding a new link with damping force inside the construction. The non-traditional ones are the active mounting of machine tool feed drives with the connection of the drive with the frame by another drive [3] and the mechatronic stiffness [4] by adding a parallel structure to the existing one with collocated force connections between them.

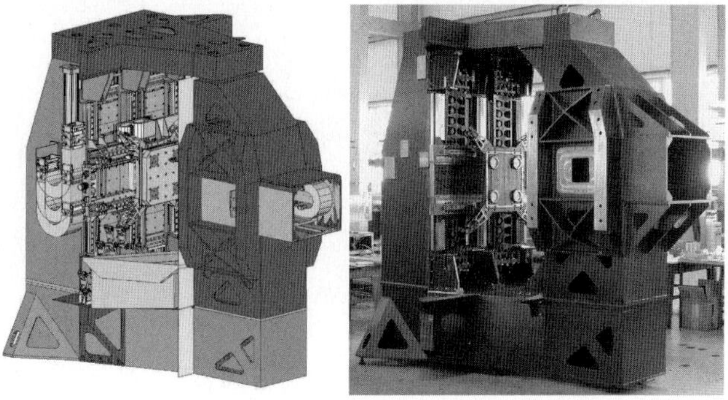

Fig. 1. Experimental milling centre LM-2

These mechatronic variants were investigated on the example of a new experimental milling centre LM-2 in the Research Center of Manufacturing Technology of CTU in Prague (Fig. 1). This machine has 3-highly dynamical axes equipped with linear motors and they excite the machine tool frame more than desired.

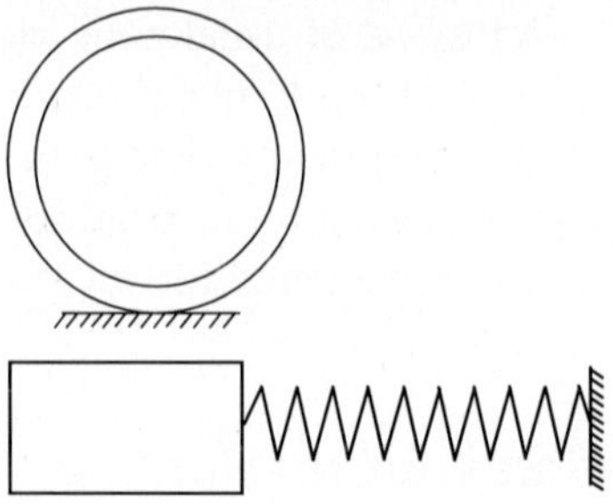

Fig. 2. Original machine tool

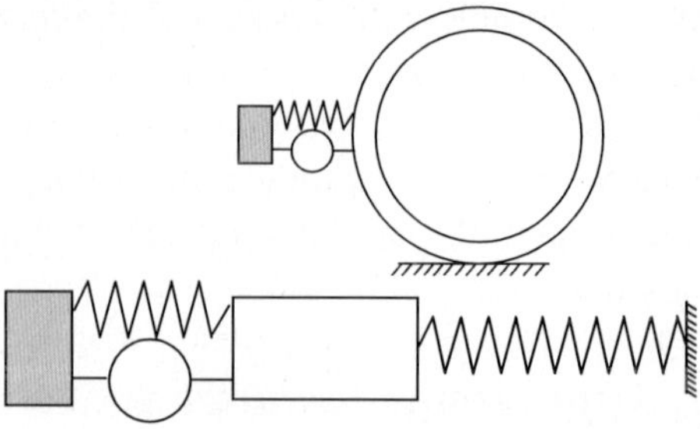

Fig. 3. Controlled dynamic absorber

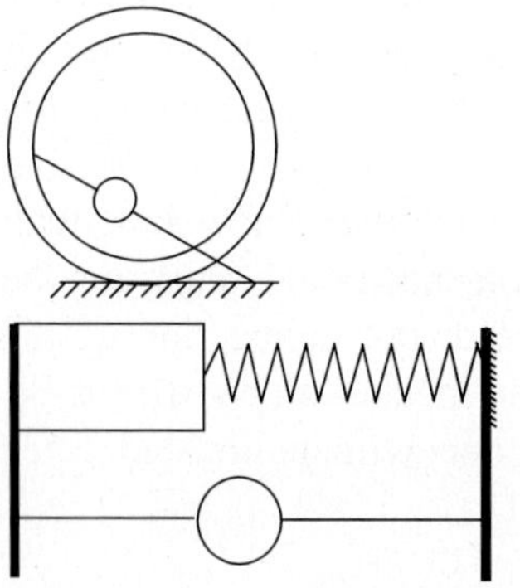

Fig. 4. Controlled vibrocompensation

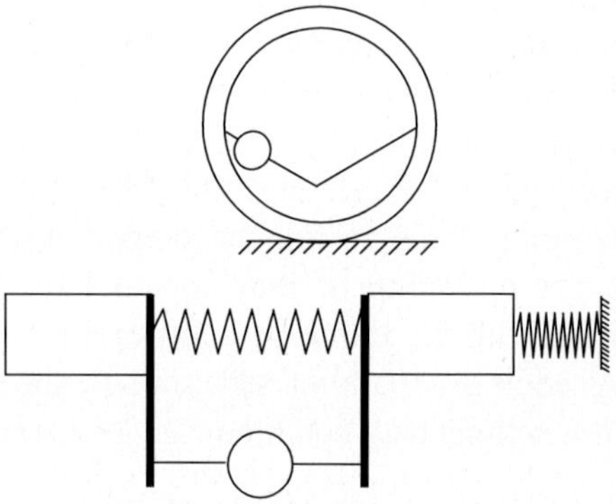

Fig. 5. Controlled active damping

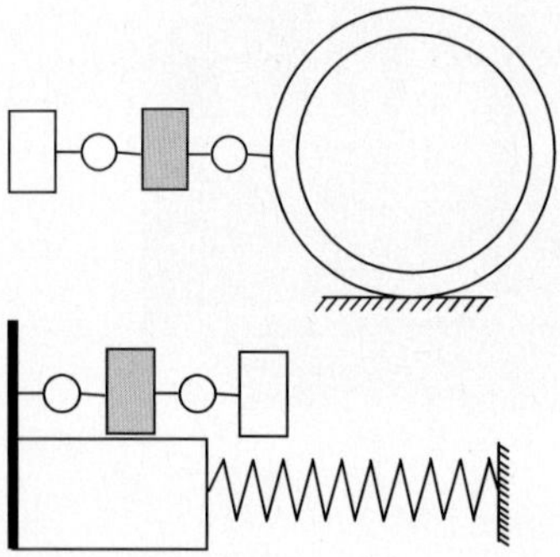

Fig. 6. Active drive mounting

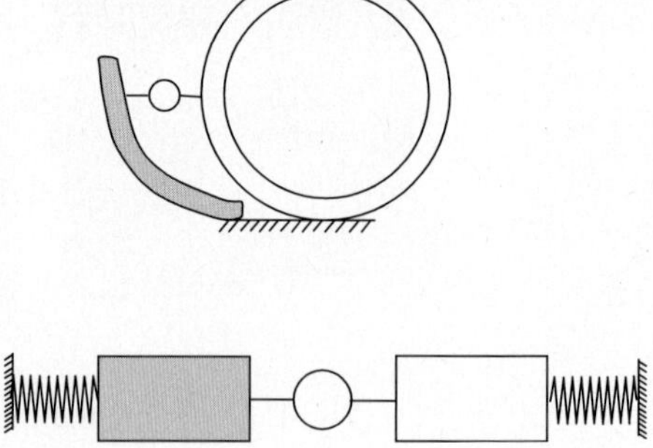

Fig. 7. Mechatronic stiffness

The possible approaches towards active vibration suppression can be applied to the machine tool LM-2 as follows. The schematic structure of the original machine is in Fig. 2. On each of the following figures there is always the schematic structure of the machine tool and its equivalent mechanical model used for the control synthesis. The dynamic absorber with additional mass is in Fig. 3. The vibrocompensation with additional direct force link to the frame is in Fig. 4. The controlled active damping with additional force link inside the structure is in Fig. 5. The new solution by active drive mounting by additional actuator is in Fig. 6 and the new solution by mechatronic stiffness, where the auxiliary structure provides flexible support for exerting additional force, is in Fig. 7.

3. Simulation results

The control methods of particular proposed variants of mechatronic solution for controlled vibration suppression have been synthetized and simulated. The results of frequency response are in Fig. 8-11 for variants in Fig. 3-6.

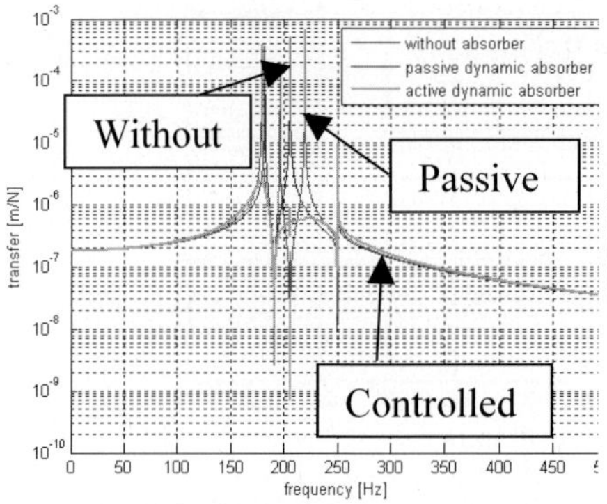

Fig. 8 Controlled dynamic absorber from Fig. 3

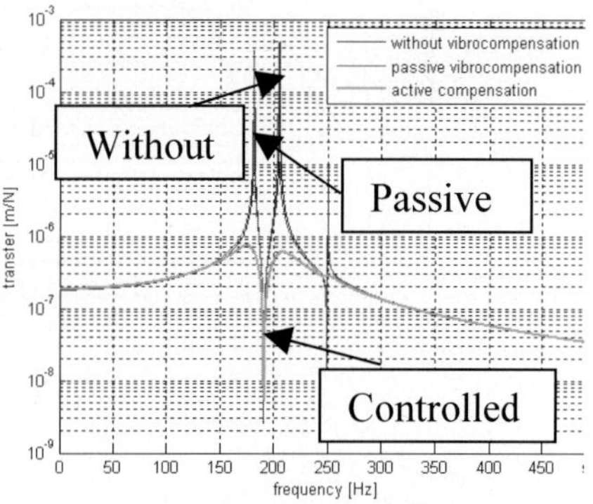

Fig. 9 Controlled vibrocompensation from Fig. 4

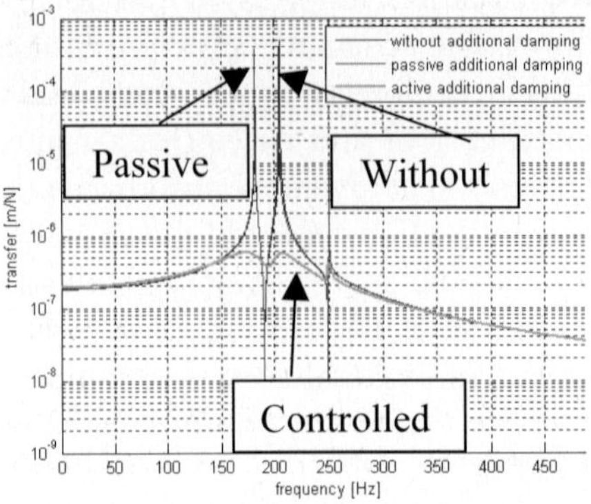

Fig. 10 Controlled active damping from Fig. 5

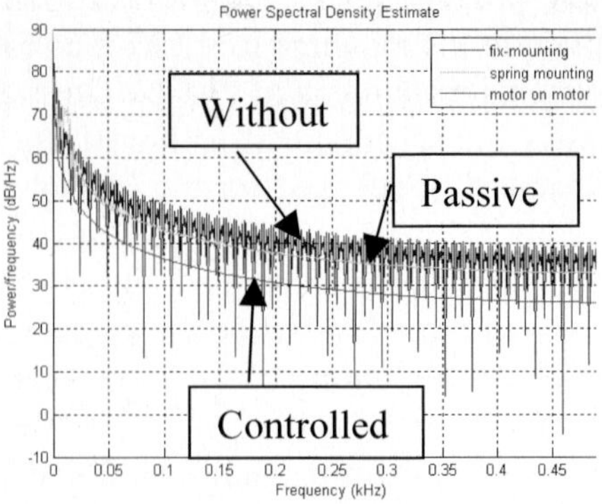

Fig. 11 Active drive mounting from Fig. 6

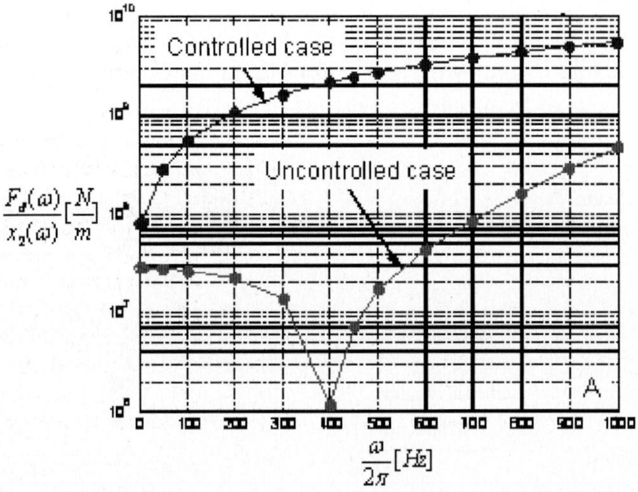

Fig. 12 Dynamic stiffness of mechatronic stiffness solution from Fig. 7

The mechatronic stiffness solution from Fig. 7 is characterized by frequency dependence of dynamic stiffness in Fig. 12. The comparison is done between cases without modification, with passive modification of the structure and with controlled modifications. All results have demonstrated the large potential of mechatronic solutions of controlled vibration suppression.

4. Conclusions

The various variants of mechatronic solutions for vibration suppression of machine tools have been described. They demonstrate the large potential of these controlled approaches. Nevertheless, the basis of all mechatronic solutions is a suitable modification of the original structure of machine tool in order to efficiently exert the additional controlled force.

References

[1] Z. Sika: Actice and Semiactive Vibration Suppression of Machines, Habilitation Thesis, Czech Technical University in Prague, Prague, 2004 (in Czech)
[2] J. Kejval, Z. Sika, M. Valasek: Active Vibration Suppression of a Machine, In: Proc. of Interactions and Feedbacks 2000, Prague 2000, pp. 75-80
[3] J. Sveda, Z. Sika, M. Valasek: Active Mounting of Machine Tool Feed Drives, In: Proc. of WAM 2007, Prague 2007, pp. 1-6
[4] M. Valasek: Method and Device for Change of Stiffness of Mechanical Constructions, PV2006-123, Patent pending

Flexible Rotor with the System of Automatic Compensation of Dynamic Forces

T.Majewski (a) *, R. Sokołowska (b)**

(a) Universidad de las Americas-Puebla, CP 72820, Tel (52)(22)229 26 731, tadeusz.majewski@udlap.mx , Mexico

(b) Politechnika Warszawska, 02-525 Warszawa, ul.A.Boboli 8, Tel (22)234-8447, roza@mech.pw.edu.pl, Poland

Abstract

The paper presents dynamic analysis of a rotor with elastic shaft and the dynamic force that generates its vibration. To balance the rotor, free elements (balls or rollers) are placed in one or two drums. The balls can compensate the rotor's unbalance or increase it depends on the parameters of the system. The balls and the rotor are in different planes and it is not obvious if the system can be balanced. The vibrational forces that act on the balls push them to new positions in which the balls can compensate the rotor unbalance, entirely or partially. Computer simulation shows what part of the rotor's unbalance can be compensated by the balls and what the final positions the balls occupy.

1. Introduction

E. L.Thearle proposed a method of automatic balancing of the rotors [1]. In earlier author's papers [3-5] and other publications [6-8] the rotor was taken as rigid one. Depends on its lengths the balls were placed in one or two planes. For the balls in one plane they should be very close to the rotor unbalance and therefore this method is affective for the short rotor. For longer rotor the balls can be placed on its end. In many situations the rotor cannot be taken as a rigid one. When the deformations of the shaft are too large then they change the behavior of the rotor and the balls. The dynamic forces generated by the rotor unbalance and the balls are in different planes. It is not clear in what way they will be transformed between these planes and in what way they effect on the behavior of the balls. The deformation of the shaft plays greatly impacts the behavior of the balls. The relations that define the relations between forces in two planes and there deformations should be given. The rotor on the elastic shaft and a drum with two balls is shown in Fig.1. The rotor mass center is at C which is in the distance *e* from the axis of rotation. The distance between the rotor and the drum is L_2 and later during the analysis of the system its influence on the possibility of the system balancing will be verified. The force generating by the rotor unbalance is in the plane *E* and the centrifugal force of the balls are in the plane *D*.

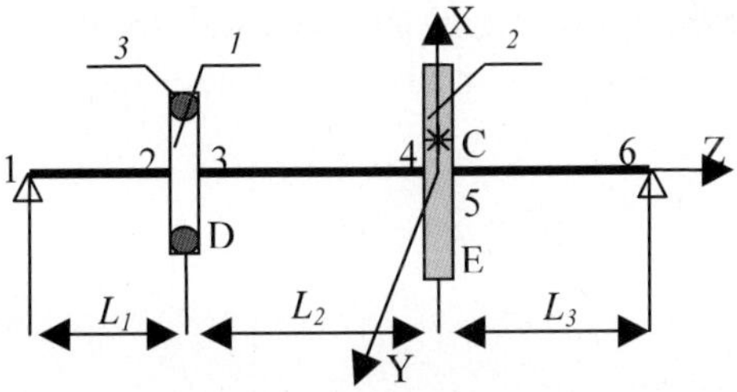

Fig.1. Rotor and two balls
1– disk, 2 – drum, 3 - balls

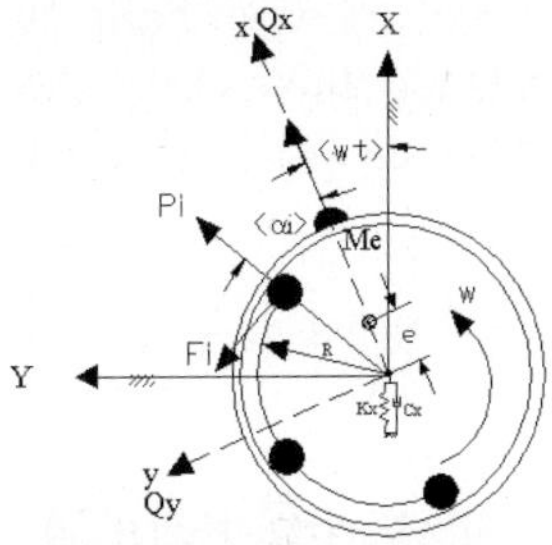

Fig.2. Position of the ball with respect to the rotor

The position of *ith* ball in the drum is defined by an angle α_i that is measured with respect to the position of the rotor center C – Fig.2. The displacement of the rotor is defined by the linear x_4, y_4 and angular Φ_4, Θ_4 coordinates. The vibration of the drum are described by x_3, y_3, Φ_3, Θ_3 . The relation between the displacements x_3, θ_3, x_4, θ_4 and the forces in the plane XZ is defined by the relation (1) - Fig.3.

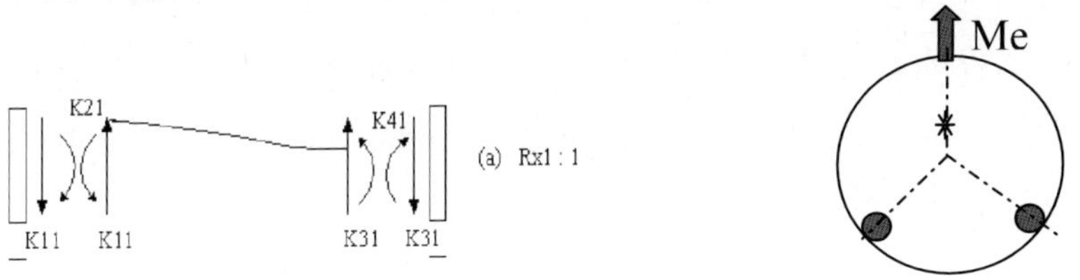

$$\begin{Bmatrix} F_{x3} \\ M_{y3} \\ F_{x4} \\ M_{y4} \end{Bmatrix} = -\frac{2EI}{L_3^3}\begin{bmatrix} 6 & 3L_3 & -6 & 3L_3 \\ 3L_3 & 2L_3^2 & -3L_3 & L_3^2 \\ -6 & -3L_3 & 6 & -3L_3 \\ 3L_3 & L_3^2 & -3L_3 & 2L_3^2 \end{bmatrix}\begin{Bmatrix} x_3 \\ \theta_3 \\ x_4 \\ \theta_4 \end{Bmatrix}. \tag{1}$$

Deformation of the shaft in the plane YZ is defined by a similar matrix with some change with sign of the matrix elements. For elastic elements 1-2 and 5-6 the relations are similar with another length of the shaft. At the points 1 and 6 the moments are zero $M_1=M_6=0$.
If there are two balls in the drum and they really compensate the rotor unbalance then they should occupied the positions opposed to the rotor unbalanced [4]– Fig.4. The theoretical final positions of the balls are defined by (2)

$$\alpha_{1t} = \arccos(-\frac{Me}{2mR}), \qquad \alpha_{1t} = 2\pi - \alpha_{1t}. \tag{2}$$

2. Mathematical Model

The disk has four degrees of freedom. The positions of the disk are defined by $x_4, y_4, \phi_4, \theta_4$ with respect to the fixed coordinates system XYZ. If the rotor is equipped with two balls then there are two degrees more with coordinates α_1, α_2. The equations of motion can be obtained from Lagrange'a equation. The forces acting on the rotor and the balls are presented in Fig.2. The equation of the motion of the disk are defined by

$$M\ddot{x}_4 + c_x\dot{x}_4 + k_x x_4 + c_{x\theta}\dot{\theta}_4 + k_{x\theta}\theta_4 = Q_x + k_{11x}(P_{1x} + P_{2x}), \tag{3}$$

$$I_x\ddot{\theta}_4 + c_\theta\dot{\theta}_4 + k_\theta\theta_4 - J\omega\dot{\phi}_4 + c_{x\theta}\dot{x}_4 + k_{x\theta}x_4 = k_{21x}(P_{1x} + P_{2x}), \tag{4}$$

$$M\ddot{y}_4 + c_y\dot{y}_4 + k_y y_4 + c_{y\phi}\dot{\phi}_4 + k_{y\phi}\phi_4 = Q_y + k_{11y}(P_{1y} + P_{2y}), \tag{5}$$

$$I_y\ddot{\phi}_4 + c_\phi\dot{\phi}_4 + k_\phi\phi_4 + J\omega\dot{\theta}_4 + c_{y\phi}\dot{y}_4 + k_{y\phi}y_4 = k_{21y}(P_{1y} + P_{2y}). \tag{6}$$

Q_x, Q_y - components of the forces from the static unbalance in the plane X-Z and Y-Z. P_x, P_y - components of the force generated by the ball,

$k_{1x}, k_{1y}, k_{21x}, k_{22y}$ - coefficients of the influence of the balls on the disk behavior.

If the balls are in the plane of the disk then the coefficients k_{1x}, k_{1y} are equal to one and the coefficients k_{21x}, k_{22y} are zero. The first coefficients decreases and the second one increases when the distance L_2 between the disk and the drum increases. The motion of the balls with respect to the rotor are governed by the following equations

$$mR\ddot{\alpha}_1 = m[\ddot{x}_3 \sin(\omega t + \alpha_i) - \ddot{y}_3 \cos(\omega t + \alpha_i)] - n_1 R\dot{\alpha}_1, \quad (7)$$

$$mR\ddot{\alpha}_2 = m[\ddot{x}_3 \sin(\omega t + \alpha_2) - \ddot{y}_3 \cos(\omega t + \alpha_2)] - n_2 R\dot{\alpha}_2. \quad (8)$$

It is seen from (9, 10) that the motion of the balls depends on the inertial forces generated by the rotor vibration $x_3(t)$, $y_3(t)$. The vibration force for one ball

$$\overline{F}_i = mL(\ddot{x}_3 \sin(\omega t + \alpha_i) - \ddot{y}_3 \cos(\omega t + \alpha_i)), \quad (9)$$

and this force define the motion of the ball and its equilibrium position. The displacement of the drum depends on the displacement of the disk and the forces produced by the balls. The relations between the displacement of the disk and the drum, with the balls inside it, are defined by

$$x_3 = b_1 x_4 + c_1 \theta_4 + d_1 P_x, \qquad y_3 = b_2 y_4 + c_2 \phi_4 + d_2 P_y, \quad (10)$$

where b, c, d are the coefficients that present the relations between the displacements of the rotor and the drum. The symbol 1 is for the plane XZ and 2 for the planeYZ. The coefficients k_{11x}, k_{11y}, k_{21x}, k_{21y}, and b_1, c_1, d_1, b_2, c_2, d_2 are calculated from the relation (1). From (10) the acceleration $\ddot{x}_3, \ddot{y}_3$ can be calculated as a function of the vibration of the disk and then the vibrational force F_i. The final positions of the balls depend on these forces and at the positions of equilibrium these forces are equal to zero. From the previous author works it is know that the motion depends on the average force

$$F_i = \frac{1}{T}\int_0^T \overline{F}_i(t) \cdot dt. \quad (11)$$

At the final position of the balls these forces are zero

$$F_1(\alpha_{1f}, \alpha_{2f}) = 0, \qquad F_2(\alpha_{1f}, \alpha_{2f}) = 0. \quad (12)$$

3. Results of Simulation

The analysis was done for different parameters of the system. Some of the results are presented in the diagrams Fig.5 and 6. It can be seen that the balls move to a new positions and the vibrations of the disk decreases in time. It means that the system goes to the balanced state. The Fig. 5 presents the vibration of the disk in the plane XZ when the balls are inside the disk and the disk is in the middle of the shaft ($L_1=L_3=0.55$m and $L_2=0$). There is no angular vibration of the disk because all dynamic forces are in the same plane. Other parameters; mass of the rotor M=35 kg, anular velocity ω= 100 rad/s, R=0.15 m, Me= 2.25 kgcm, ET= 1650 Nm^2.

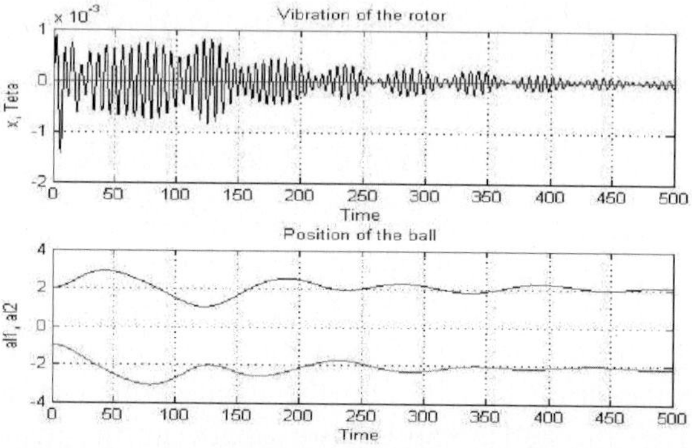

Fig.5. Vibration of the disk and behavior of the balls in time when $L_1=L_3$ and $L_2=0$

When the disk and the balls are in the same plane ($L_2=0$) then the balls compensate the rotor unbalance in 100%. The linear vibration vanishes as a result of balancing of the system. The diagram in Fig. 6, presents the vibration of the disk when the drum with the balls is close to the disk ($L_1=300$ mm, L_2 =100 mm). The balls go to the positions of equilibrium that are very close to the theoretical one. The system is not completely balanced because there are small vibrations and dynamic reactions of the bearing.

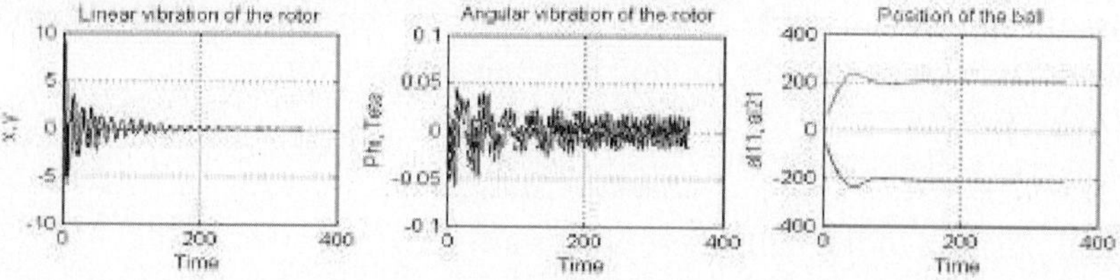

Fig.6. Linear and angular vibration of the rotor and the positions of the balls in time

If the distance between the rotor and the drum increases then the residual unbalance also increases. The balls try to compensate the static unbalance of the disk but at the same time the disk and the balls generate a dynamic unbalance and therefore the diagram present much greater angular vibration. When there is only one drum and $L_2 \neq 0$ then the balls cannot compensate the rotor unbalance in 100%. The rotor can be equipped with two drums, each containing two balls. The balls in two different planes can produce a force that can compensate the disk unbalance and also a moment which can decreases the dynamic unbalance.

4. Conclusions

When some of the coefficients of the influence in eqs. 3-8 take a magnitude zero or one then the system can be balanced in 100%. But for any position of the drum with respect to disk the system cannot be completely balanced. The computer simulation presents in what way the balls change their positions and in what way the rotor's vibrations vanish. The examples given in this paper were obtained for the rotor speed greater than its natural frequency.

References

[1] Ernest L.Thearle 1934 *United States Patent Office No. 1 967 163*. Means for Dynamically Machine Tools

[2] Majewski Tadeusz, Synchronous Elimination of Vibrations in the plane. *Journal of Sound and Vibration*. No. 232-2, 2000. Part 1: Analysis of Ocurrence of Synchronous Movements, pp.555-572. Part. 2: Method Efficiency and Stability, pp. 573-586.

[3] T. Majewski, *Synchronous Elimination of Vibrations in the Plane. Method Efficiency and its Stability*. Journal of Sound and Vibration, No. 232-2, 2000, pp.573-586

[4] Majewski T. *Position error occurrence in self balancers used on rigid rotors of rotating machinery*. Mechanism and Machine Theory, v. 23, No 1, 1988, pp71-78

[5] Majewski T. *Synchronous vibration eliminator for an object having one degree of freedom*. Journal of Sound and Vibration, 112(3), 1987

[6].- C. Rajalingham and S. Rakheja 1998 *Journal of Sound and Vibration* 217, 453-466. Whirl suppression in hand-held power tool rotors using guided rolling balancers.

[7].- J. Chung and D. S. Ro 1999 *Journal of Sound and Vibration* 228, 1035-1056. Dynamic analysis of an automatic dynamic balancer for rotating mechanisms.

[8].- C. H. Hwang and J. Chung 1999 JSME *International Journal* 42, 265-272. Dynamic analysis of an automatic ball balancer with double races.

Properties of High Porosity Structures Made of Metal Fibers

D. Biało, L. Paszkowski, W. Wiśniewski, Z. Sokołowski

Institute of Precision and Biomedical Engineering, Warsaw University of Technology, ul. Sw. A. Boboli 8, 02-525 Warsaw, Poland

Abstract

Subject of the paper is manufacturing technique of porous structures made of stainless steel fibers. Preparatory operations on fibers of various diameters and lengths, compacting and sintering the structures were discussed. Samples 30 mm in diameter and 4 mm high were investigated.
Filters permeability was evaluated on the basis of so called viscosity type permeability coefficient α. Influence of permeability as well as that of diameter and length of fibers contained in the samples, on coefficient α was determined.

1. Introduction

Sintered materials of high porosity are applied in technology widely and for various applications. As an example one can mention [1] applications in manufacture of machines and measuring equipment, in aircraft-, chemical, foodstuff, pharmaceutical and nuclear energy industries, in metallurgy, etc. An important group of the a. m. semi-products is constituted by filtration materials for purifying liquids and gases [2]. Metallic filters have a number of beneficial properties compared with filters made of organic materials (like paper, textile, plastic), or inorganic ones (ceramics, glass and mineral fibers). Their principal advantage is a possibility to attain a wide range of porosity and permeability, while maintaining relatively good strength values.
Basic stuff for fabrication of sintered filtration materials are powders and metal fibers [3 - 6].

When applying powders, it is possible to reach maximum porosity as much as 45%. Use of metal fibers enables reaching maximum porosity value up to 90%.

The presented paper pertains to manufacture of compacted components made from acid resistant steel fibres and to investigate their permeability. Fibres applied were of differentiated diameter and length. The permeability coefficient α was applied for evaluation of permeability.

2. Preparation of the Samples

The initial stock for preparing fibers was an stainless steel wire 0H18N9 in softened state (R_m=750 MPa) of diameter as follows: 0.08, 0.2 and 0.32 mm. The wire was cut into predetermined pieces, 4, 8 and 12 mm long. Cutting was done on a special device of own design [7].

The precut wire pieces were used for forming investigation samples of 30 mm diameter and 4 mm high, which were made by means of die compacting on a hydraulic press. Compacting pressure between 12.5 and 700 MPa was applied that enabled to achieve widely differentiated density range (2.3 to 6.7 Mg/m^3).

Fibers were characterized by good compactibility, particularly the thinnest ones, i. e. those of 0.08 mm diameter. At pressure as low as 12.5 MPa, compacts obtained were of structural integrity and free from chippings.

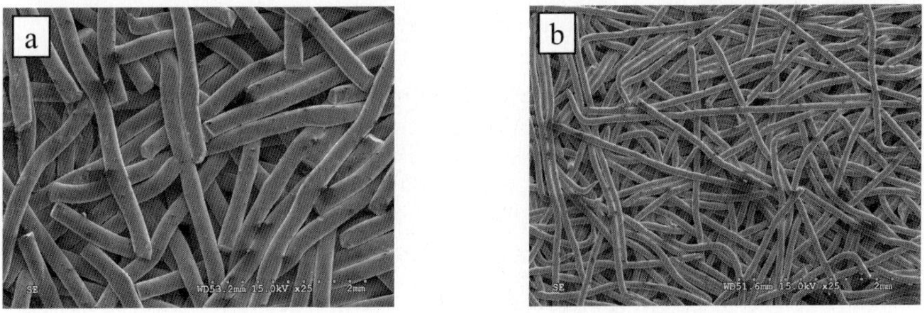

Fig. 1. SEM wives of the samples compacted from fiber:
a) Φ 0,20x8 mm at pressure of 500 MPa,
b) Φ 0,08x8 mm at pressure of 100 MPa

Surface images of samples made from fibers were shown on the Fig. 1. It can be seen that the fibers are tangled and undergo deformation when being compacted, particularly on the crossing spots. Pores between the fibers are of relatively big sizes compared with those in samples compacted from powders and sintered. It must be mentioned, that a few small, strange particles seen on fibers surfaces constitute remainders of impurities originating from air, left after permeability tests.

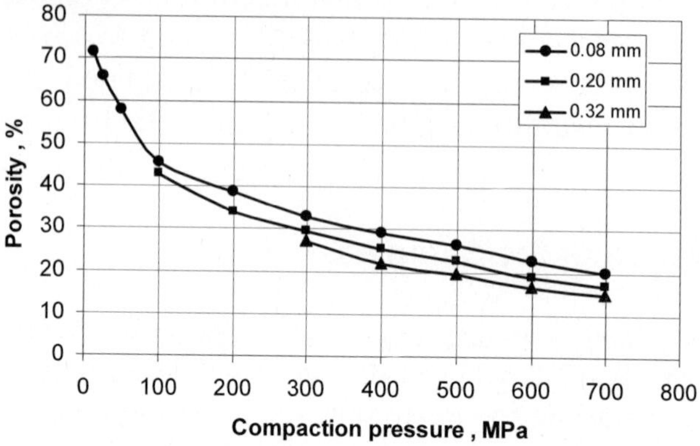

Fig. 2. Porosity of the samples made of the fiber 0.08, 0.20 and 0.32 mm in diameter and constant length of 12 mm as a function of compaction pressure

On Fig. 2 relation of porosity to compaction pressure is shown, for the samples made of fibers of constant length l = 12 mm.
The highest curve pertains to the samples made of fibers of 0.8 mm in diameter. It can be seen that attaining the highest porosity values, exceeding 70% is possible for the lowest compacting pressure, i.e. 12.5 MPa.
In the case of higher fiber diameters, a 4 to 7 % reduction of samples porosity took place at the determined compaction pressure value.

3. Investigation of the Samples Permeability

Permeability of the samples prepared was investigated in a way described in PN-92/H-04945 [8] with application of air. Core of the investigation lays in carrying out a series of measurements on volumetric rate of flow and air pressure drop while penetrating a sample under conditions of non-laminar flow. Values of viscosity type (α) and inertial (β) permeability coefficients were also determined in the course of the investigation.
On the Fig. 3 to 6 selected results of viscosity type permeability coefficients α are shown as a function of the samples porosity.
Porosity of the samples has an essential influence on the coefficient α value. As expected, the coefficient value grows with increase of porosity. Fig. 3 pertains to the samples made from fibers of 0.2 mm diameter and differentiated length of 4, 8 and 12 mm.

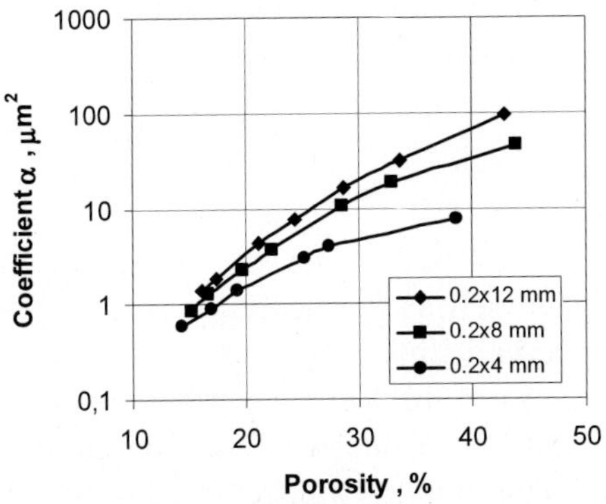

Fig. 3. Permeability coefficient α for the samples made of fiber with 0.2 mm in diameter and different length as a function of porosity

The lowest permeability is shown by samples made of the shortest fibers, for which relatively highest compacting density was attained. As much as the fiber length increases, the coefficient α takes higher values.

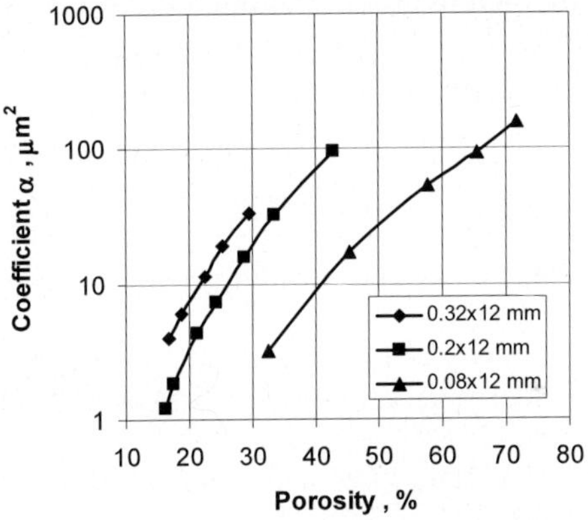

Fig. 4. Influence of fiber diameter with constant length of 12 mm on permeability coefficient α of samples.

Similar dependence was attained for samples made from fibers of the smallest diameter [7]. In this case influence of fiber length on the samples permeability is much lower than for fibers of bigger diameter.
Much higher influence than that of fibers length on the coefficient α has their diameter, what is substantiated by the data shown on the Fig. 4.
At determined fibers length (12 mm) permeability is growing considerably with increase of fiber diameter. At comparable porosity in the samples

made of thicker fibers bigger pores are formed, hence better conditions for air flow are attained.

4. Summary

The investigations carried out in respect of performing porous samples from metallic fibers as well as permeability tests of the samples allow to formulate following conclusions:

1. Preparation of samples from metallic fibers is more difficult than performing porous samples from powders. The thinner and longer are the fibers, the easier is to obtain compact structures free from chippings.
2. Permeability of samples from metallic fibers determined by so called viscosity type of permeability coefficient α depends first of all on the compacting pressure applied and, ultimately, from the samples porosity achieved.

The actual value of coefficient α is also influenced by fiber sizes, i. e. its diameter and length. With growing fiber length the coefficient attains higher values. Changing the fibers diameter results in higher permeability changes than that of their length. At comparable porosity in the samples made of thicker fibers bigger pores are formed, which results in better conditions for air flow being attained.

References

[1] W. Schatt, K.P. Wieters, Powder Metallurgy Processing and Materials. EPMA (1997).
[2] S. Borowik, Filters for Work Fluids, Warsaw, 1985 (in Polish)
[3] G. Hoffman, D. Kapoor, The Int. Journal of PM and PT, vol. 12, No 4, (1976) 281.
[4] W. Cegielski, M. Czepelak, Ores and Nonferrous Metals, MP, R44, No 1, (1999) 33. (in Polish)
[5] R. De Bruyne, Advances in Powder Metallurgy and Particulate Materials vol. 5, Part 16, (1996) 99.
[6] D. Bialo, Z. Ludyński, R. Bala, Int. Conf. MECHATRONICS 2000, Sept. 2000, Warsaw, Poland, vol. 2, pp. 304-306. (in Polish)
[7] L. Paszowski et al, Ores and Nonferrous Metals, No 2 (2005) 87. (in Polish)
[8] Powder Metallurgy. Determination of the viscous Permeability Coefficient. PN-92/H-04945. (in Polish)

Fast prototyping approach in developing low air consumption pneumatic system

K. Janiszowski, M. Kuczyński

Warsaw University of Technology Institute of Automatic Control and Robotics,
ul. Św. A. Boboli 8, Warsaw, 02-525, Poland

Abstract

In the paper consecutive steps of fast prototyping with pneumatic positioning drive were outlined. The model of asymmetrical pneumatic cylinder, fed with compressed air buffer was presented. This model was determined using an IDCAD software package, developed in IAiR of WTU. The fast prototyping was carried out using PExSim (Process Explorer and Simulation) software package developed recently in IAiR of WTU too. Results received during the tests and simulation were compared and presented.

1. Introduction

Fast prototyping is the methodology of carrying out research and development activities being more and more widely applied in designing of modern mechatronics systems. This idea assumes eliminating the real object from the process of working out the control policy and replacing it by its model. Simulation of the behaviour of the object and controller in normal operation and extreme conditions reflects the phenomena that usually can be observed only in a laboratory or real industrial conditions after long preparation of a proper stand and measurement equipment.

Positioning pneumatic drives, controlled using proportional valves are quite well examined. The usage of cheap, two-position valves (alternatively working) with PWM-like control technique in low-cost control systems has led to significant consumption of the compressed air [1]. A concidered drive, based on the bang-bang principle control, is to achieve fast positioning with moderate final position accuracy, significantly reduced positioning time and consumption of the compressed air. The idea of time-optimal control of a pneumatic drive relies on usage of fast switching valves, that are controlled directly and final position is reached with acceptable overshoot.

2. The model of the pneumatic actuator

Having a complex set of information about the pneuamtic drive allows to formulate equations, which describe the relations in pneumatic cylinder [2], [3]. As far as geometrical parameters of the actuator, masses of the moving elements, supply conditions can be easily measured and recognised as constant, the measurements of the friction force parameters are difficult to carry out and are dependent on stoppages, conditions of lubrication, etc. To sum up, in formulated equations there is a number of unknown coefficients which determination is problematic.

Another approach to developing the model of the pneumatic actuator is a statistical identification, based on measured input-output data. The main advantage is the fact, that it is not necessary to declare many technical parameters of the pneumatic drive. Relying on discrete time data, recorded with a sampling time interval, it is also possible to estimate the parameters of the linearised model which are velocity gain, eigenfrequency, damping factor and to write down the linear transfer function between the piston velocity and control input [2].

The investigated system consisted of pneumatic actuator FESTO DSNU-25-400 PPV-A, four fast switching valves FESTO MHE4-MS1H-3/2G-QS8 and 10 liter air reservoir. The measurements of the piston position, supply pressure and pressures in cylinder chambers were realised using respectively linear encoder and piezorezistive pressure sensors. The control of the identification experiment and data acquisition were carried out using PC computer equipped with dSPACE 1102 board.

The following conditions of the identification experiment were assumed: experiment carried out without feedback loop, sampling time interval 1ms, binary signals u_1, u_3 control inlet valves of the respectively left and right chabmer and binary signals u_2, u_4 control outlet valves of the respectively left and right chabmer, between control signals there is a relationship $u_1 = \overline{u}_2, u_3 = \overline{u}_4$, groups of valves of the left and right chamber excitited by two pseudo random binary signals (PRBS) with constant amplitude, based on the 4-bit register, and generation periods 10ms and 16ms.

Linear ARMA (Auto Regression Moving Average) model which takes into account the control signal u and n previous values of its output was searched. Its difference equation can be written as

$$\hat{y}(k)=\sum_{i=0}^{n} b_i u(k-i-d)-\sum_{i=0}^{n} a_i y(k-i) \quad (1)$$

or in case of m input signals as

$$\hat{y}(k)=\sum_{j=0}^{m}\sum_{i=0}^{n} b_{ji} u_j (k-i-d_j)-\sum_{i=0}^{n} a_i y(k-i) \quad (2)$$

where: $\hat{y}(k)$ model output, $y(k)$ object output, $u_j(k)$ input signals, a_i, b_{ji} model coefficients, d_j time delays.

Fig. 1a shows the structure of the model being identified firstly.

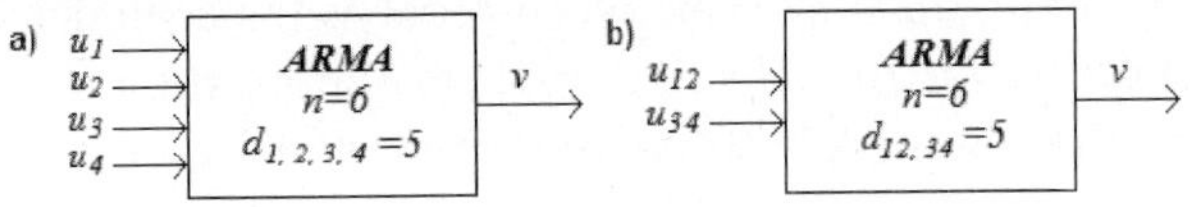

Fig. 1. Structures of identified models

In order to simplify the structure of the model, two modified input signals u_{12}, u_{34} were used. They were determined on the basis of input signals gains of the model presented in Fig 1a and defined as follows

$$u_{12}=\begin{cases}\dfrac{k_1}{k_1+|k_2|}: u_1=1\wedge u_2=0\\ 0: u_1=0\wedge u_2=0\\ \dfrac{k_2}{k_1+|k_2|}: u_1=0\wedge u_2=1\end{cases} \qquad u_{34}=\begin{cases}\dfrac{|k_3|}{|k_3|+k_4}: u_3=1\wedge u_4=0\\ 0: u_3=0\wedge u_4=0\\ \dfrac{-k_4}{|k_3|+k_4}: u_3=0\wedge u_4=1\end{cases} \quad (3)$$

where k_i is the gain of the input signal u_i. Fig. 1b shows the structure of the model with two modified input signals. This approach allows to reduce the number of inputs, takes into account the fact that the gain of each input signal is different and preserves the quality of the model. The model was verified by parallel simulation [5] of the model output $\hat{y}(k)$. The results of the verification of the model were presented in Fig. 2.

Although during the identification experiments 20% drop of the supply pressure were observed, introduction of additional input signal as supply pressure caused only 0,3% reduction of the model error.

3 Simulation software

The usage of simulation software in combination with data recorded during identification experiment allows to tune the parameters of the function blocks which were used to model the pneumatic system. This approach can be particulary useful while determining unknown parameters of the friction, nominal pressure drops, air flows, etc.

The function blocks model of the pneumatic system was prepared using PExSim simulation software package. Masses of the moving elements, initial conditions and all the parameters of used function blocks were known except the parameters of the Stribeck friction model, which were constant friction α_0, linear friction coefficient α_1, expotential friction coefficient α_2, friction values for zero velocity while accelerating FRC1 and braking FRC2. In order to identify them the *Quality index* block defined as an absolute difference between recorded and modeled piston velocity was added to the model structure. The optimization task was carried out using PExSim Optimizer software. Fig. 3 shows an exemplary plot of recorded and modeled velocity.

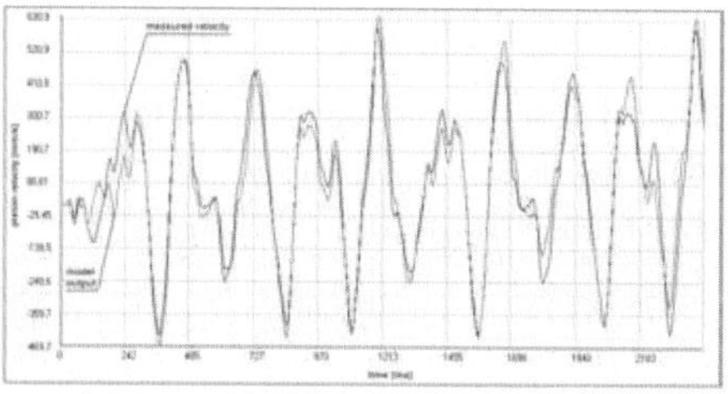

Fig. 2. Results of the verification

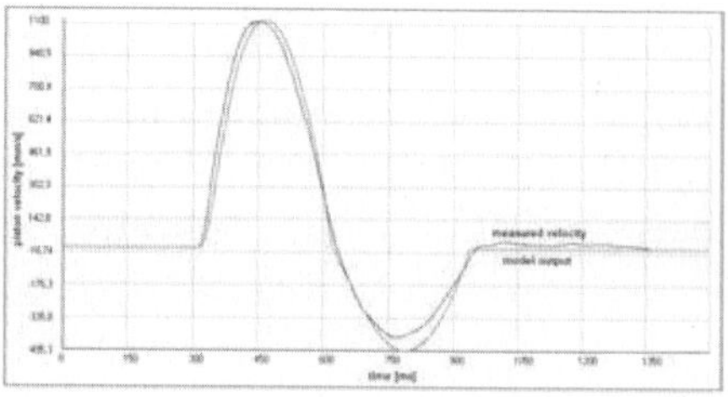

Fig. 3. Plot of modeled velocity

4 Simulation and fast prototyping

Identified model of the pneumatic drive was implemented in PExSim simulation software package as a C++ script. In order to verify the idea of fast prototyping, preliminary, simplified control policy was developed. The displacement was divided into acceleration and braking phases. During acceleration one chamber is fed and other is evacuated. While braking phase previously evacuated chamber is fed also. It was assumed that during braking phase of the movement the kinetic energy E_K of the mass is converted into the work of the friction force W_F and work to compress gas W_C from volume corresponding to braking phase beginning position to volume corresponding to set position. It was also assumed that air is being compressed from atmospheric pressure to supply pressure during adiabatic process. This policy was implemented as a condition $W_C + W_F \geq E_K$, written in C++ and checked every sampling period during simulation after the start ot the movement. Fig. 4 shows exemplary results obtaind during the simulation.

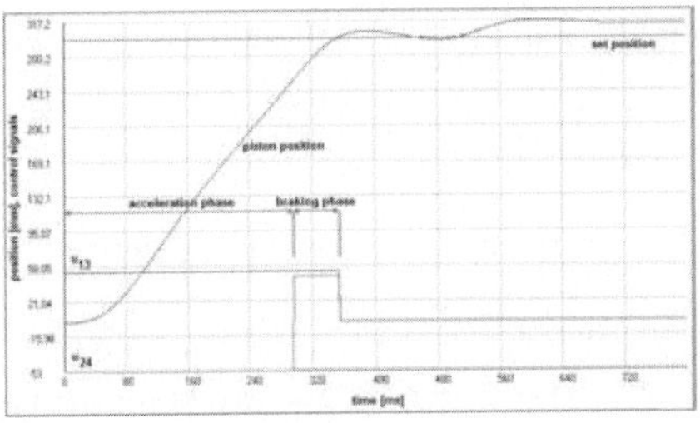

Fig. 4. Simulated course of positioning

The initial position during simulation tests was 150mm. Displacements from 250 to 350 were simulated. The average positioning error was 5%.

5 Summary

The usfulness of any simulation software is limited by the degree of complexity of the utilized equations. As far as they can cope with modelling traditional pneumatic servosystems, where cylinder chambers are fed and evacuated alternatingly, the results of simulations performed for independent control of each chamber, just like during the identification experiment, are not satisfactory.

Presented in the paper control policy was simplified as much as possible and was only a contribution to presentation the idea of fast prototyping approach in developing low air consumption pneumatic systems.

Obtained results lead to the conclusion that the usage of statistically identified dynamic models assures the fastest and most adequate way of simulating pneumatic drives. Moreover such models can be used to develop and test control laws. This approach allows to implement, test and modify the control algorithm much easier, faster and cost efficient.

References

[1] Wiślicki K.: *Implementation of Pneumatic Servo-System based on Switching Valves.*, IaiR/FESTO AG & Co., Wrszawa 1998.

[2] Janiszowski K.: *Identyfikacja modeli parametrycznych w przykładach*, Akademicka Oficyna Wydawnicza EXIT, Warszawa, 2002

[3] Chudzik Z., Janiszowski K., Olszewski M.: *Modelowanie obiektów sterowania na przykładzie opisu siłownika pneumatycznego*, PAK nr 10, str. 231-236, 1994.

Chip card for communicating with the telephone line using DTMF tones.

Igor Malinowski
Department of Mechatronics
Warsaw University of Technology, Warsaw, Poland, EU

igorma@interia.eu

Abstract

To communicate with the telephone line **chip-cards (IC cards)** require terminals or other special apparatus, equipped with a card reader for retrieving the information on a card.
Also in the case of disabled or sick people problem exists of dialing of a phone number in distress. To solve these problems a special chip-card was designed, equipped with an electro-acoustical transducer and a DTMF (Dual Tone Multi Frequency) tone generator, for acoustic communication with a telephone line through the microphone of a regular telephone apparatus

Introduction

Problem:
To communicate with the telephone line **chip-cards (IC cards)** require terminals or other special apparatus, equipped with a card reader for retrieving the information on a card.
Also in the case of disabled or sick people problem exists of dialing of a phone number in distress.

Solution:
Design of a chip-card equipped with an electro-acoustical transducer and a DTMF (Dual Tone Multi Frequency) tone generator, for acoustic commu-

nication with a telephone line through the microphone of a regular telephone apparatus.

DTMF (Dual Tone Multi Frequency) tones:
Is a system of tone pairs of pre-determined frequency combinations, when re-played in pairs are received by the telephone line as numbers or symbols to be dialed. DTMF tones correspond to so called „tone dialing" of a telephone number, (as opposed to older - „pulse dialing" method).

Keypad The DTMF keypad [2] is laid out in a 4×4 matrix, with each row representing a *low* frequency, and each column representing a *high* frequency (Fig. 1) . Pressing a single key such as '1' will send a sinusoidal tone of the two frequencies 697 and 1209 hertz (Hz) (Fig. 2). The original keypads had levers inside, so each button activated two contacts. The multiple tones are the reason for calling the system multifrequency. These tones are then decoded by the switching center to determine which key was pressed.

DTMF keypad frequencies

	1209 Hz	**1336 Hz**	**1477 Hz**	**1633 Hz**
697 Hz	1	2	3	A
770 Hz	4	5	6	B
852 Hz	7	8	9	C
941 Hz	*	0	#	D

Figure 1. DTMF tones keypad symbols and frequencies

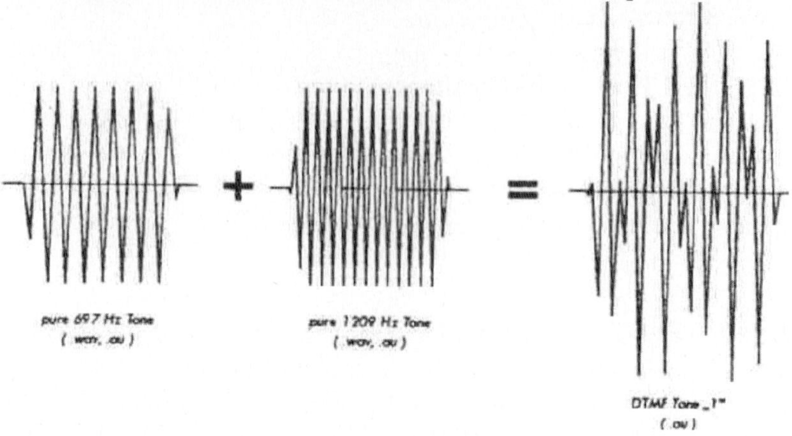

Figure 2. Combination of DTMF tones for digit 1

Requirements There are certain requirements [5] on the receiver for performing several checks on the incoming signal before accepting the incoming signal as a DTMF digit:

1. Energy from a low-group frequency and a high-group frequency must be detected

2. Energy from all other low-group and all other high-group frequencies must be absent or less than -55dBm.

3. The energy from the single low-group and single high-group frequency must persist for at least 40msec*.

4. There must have been an inter-digit interval of at least 40msec* in which there is no energy detected at any of the DTMF frequencies. The minimum duty cycle (tone interval and inter-digit interval) is 85msec*.

5. The receiver should receive the DTMF digits with a signal strength of at least -25 dBm and no more than 0 dBm.

6. The energy strength of the high-group frequency must be -8 dB to +4 dB relative to the energy strength of the low-group frequency as measured at the receiver. This uneven transmission level is known as the "twist", and some receiving equipment may not correctly receive signals where the "twist" is not implemented cor-

rectly. Nearly all modern DTMF decoders receive DTMF digits correctly despite twist errors.

7. The receiver must correctly detect and decode DTMF despite the presence of dial-tone, including the extreme case of dial-tone being sent by the central office at 0 dBm (which may occur in extremely long loops). Above 600Hz, any other signals detected by the receiver must be at least -6 dB below the low-group frequency signal strength for correct digit detection.

* The values shown are those stated by AT&T in Compatibility Bulletin 105 [3]. For compatibility with ANSI T1.401-1988 [4], the minimum inter-digit interval shall be 45msec, the minimum pulse duration shall be 50msec, and the minimum duty cycle for ANSI-compliance shall be 100msec.

The proposed chip-card is an „active chip-card" – it has:

A) own source (or coupling) of energy, (for instance „paper thin " lithium battery made by Panasonic- type CS1634 or CS2329 thickness 0,5 mm),

B) own telephone number(s) memory chip, with a DTMF tone generator (such as KS5820 made by Samsung Electronics),

C) thin electro-acoustic transducer; electrodynamic (such as for instance MSD 791701 manufactured by TDK), or a piezoelectric made of material such as piezoelectric plastic PVDF (trade name Kynar, made by Atochem),

D) a switch means for re-playing, to the microphone of the telephone apparatus, the sequence of DTMF tones programmed in the card – at the demand of the user.

Application:

Electronic business cards, self-dialing of the telephone number, or access cards for telephone and telecommunication services, allowing simple, effortless and quick access to certain telephone numbers. Also phone dialing cards for people who need to dial a number in distress.

US patent 4,995,077 was granted to the author of this article for the „Chip-card for communicating with the telephone line by means of DTMF tones". The invention of the above chip-card was awarded Bronze Award at 6th International Exhibition of Inventions in Gdańsk, 2005, and Gold

Medal at Brussels Eureka 2006, 55 World Exhibition of Inventions, Brussels, Belgium.

References

[1] I. Malinowski, „Chip-card for communicating with the telephone line by means of DTMF tones" US Patent 4,995,077
[2] http://en.wikipedia.org/wiki/DTMF
[3] AT&T Compatibility Bulletin No. 105, Issue #1, August 8, 1975
[4] ANSI T1.401-1988, Interface between Carriers and Customer Installations - Analog Voicegrade Switched Access Lines Using Loop-Start and Ground-Start Signaling
[5] http://nemesis.lonestar.org/reference/telecom/signaling/dtmf.html

CFD Tools in Stirling Engine Virtual Design

V. Pistek (a) *, P. Novotny (b)

(a) Brno University of Technology, Technicka 2,
Brno, 616 69, Czech Republic

(b) Brno University of Technology, Technicka 2,
Brno, 616 69, Czech Republic

Abstract

A successful realization of Stirling engines is conditioned by its correct conceptual design and optimal constructional and technological mode of all parts. Initial information should provide computation of real cycles of the engines. Present calculation models of thermodynamic cycles of the external heat supply engines, e.g. Stirling or Ericsson engine, arise from ideal theoretical cycles which are known from basics of thermodynamics that are, for this purpose, modified by various methods [1, 2, 3, 4, 5]. High-level CFD (Computational Fluid Dynamics) models, arising from the description of real processes which run in external heat supply engine are used for virtual prototype of Stirling engine.

1. Introduction

Requirements of computational modeling of different physical phenomena rise in the present time. Dynamics of Stirling engine parts and dynamics of fluid processes in the respective characteristic areas (or volumes) of external heat supply engines are specific which is given by the fact that the course of observed values (force, temperature, pressure, heat transfer, etc.) is periodical.

Modern computational models deliver relative accurate results but only if correct inputs are included. This represents a fundamental drawback of modern computational methods. The correct inputs can be greatly obtained from measurements and therefore the measurements are continuously a fundamental part of the Stirling engine development.

2. Computational Models of Stirling Engine

The development of a computational models starts with CAD model of the Stirling engine (Figure 1). Stirling engine geometry is set to an initial cranktrain position. A volume of a working medium is created by a subtraction of a Stirling engine CAD model from a properly chosen volume (Figure 1). A geometrical symmetry of an inner working medium volume can be used. Consequently a high quality hexa mesh is generated.

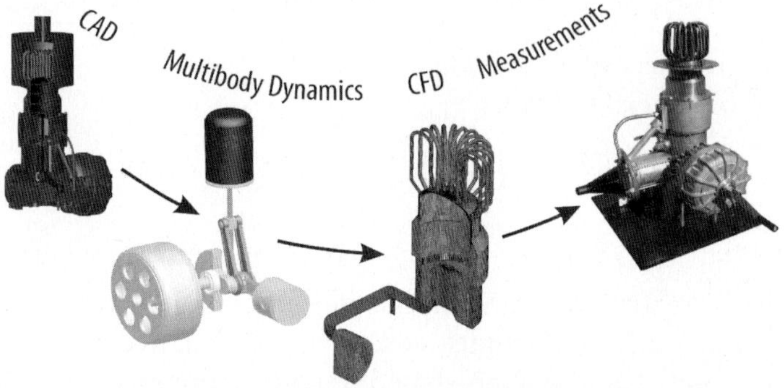

Fig. 1. Stirling engine development cycle using CAD, MBS and CFD tools and measurements

2.1. Multibody Dynamics

Multi-body systems (MBS) can be applied as effective tools for solving Stirling engine dynamics. Multi-body systems enable solving different dynamic issues of complex systems combining rigid and flexible bodies. In the case of Stirling engine mechanisms, they can be used to find the optimum alternative for balancing the driving mechanism [6]. When the mechanism moves, inertial forces of different moving parts take effect. These cause vibrations and must be "captured" by means of the machine seating. A virtual mechanism prototype has been created in the multi-body system and the Stirling engine driving mechanism has been optimised to produce low vibrations.

2.2. CFD

The application of known computational models derived for stationary states is therefore not possible. In the recent years computational fluid

dynamics (CFD) has been enormously developed and therefore is applied to the development of Stirling engine thermodynamic cycle.
The aim of the project is to develop calculation models of real thermodynamic cycles of external heat supply engines, which will enable us to calculate the thermodynamic cycle parameters necessary for structural design of engines with higher order accuracy precision. In the second phase, a virtual prototype of external heat supply engine will be made as a complex calculation model. This complex calculation model enables to find optimal parameters of the engine, for example, design and materials of a regenerator, proper design of a combustor modulus or Stirling engine working medium.

3 Stirling Engine Thermodynamic Cycle Results

The computational model of a Stirling engine thermodynamic cycle can be used in many ways. Generally, the first question can be what sort of a working medium should be used. The CFD model can give relatively precise answer.

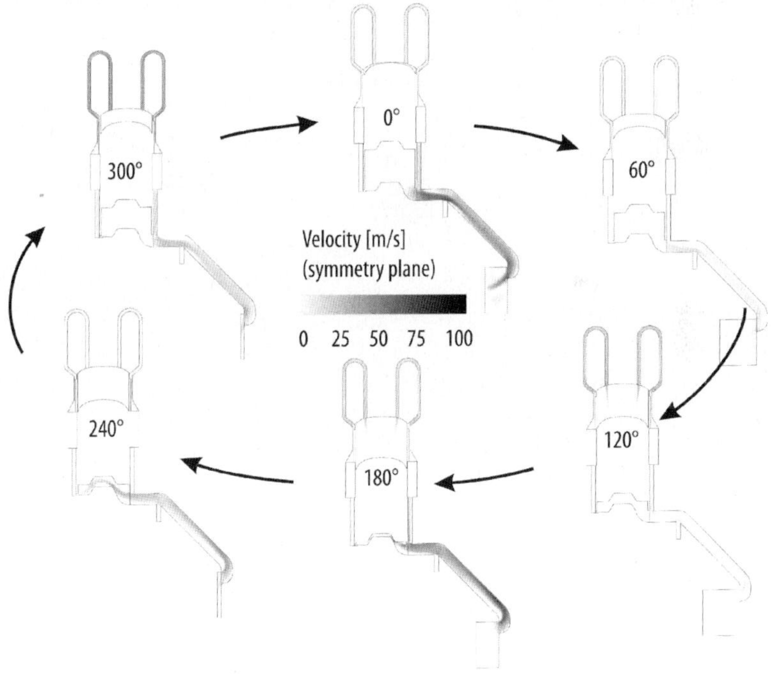

Fig. 2. Computed velocity distributions on symmetry plane vs. crank angle (air)

A fundamental request for a Stirling engine construction is a fluent flow without any pressure losses. Figure 2 presents computed velocity distribution on a symmetry plane of the engine vs. a crank angle if air is used as a working medium. Initial static pressure in an engine volume is set to value of 1 MPa.

The complex computational model enables to solve the ◘-type Stirling engine combustor modulus in detail. A heat source distribution is optimized to uniform heating of the combustor modulus. Various types of design and materials of a regenerator are also discussed to ensure maximal thermal efficiency of the Stirling engine.

A computed temperature distribution on a symmetry plane of the engine vs. a crank angle is shown in Figure 3.

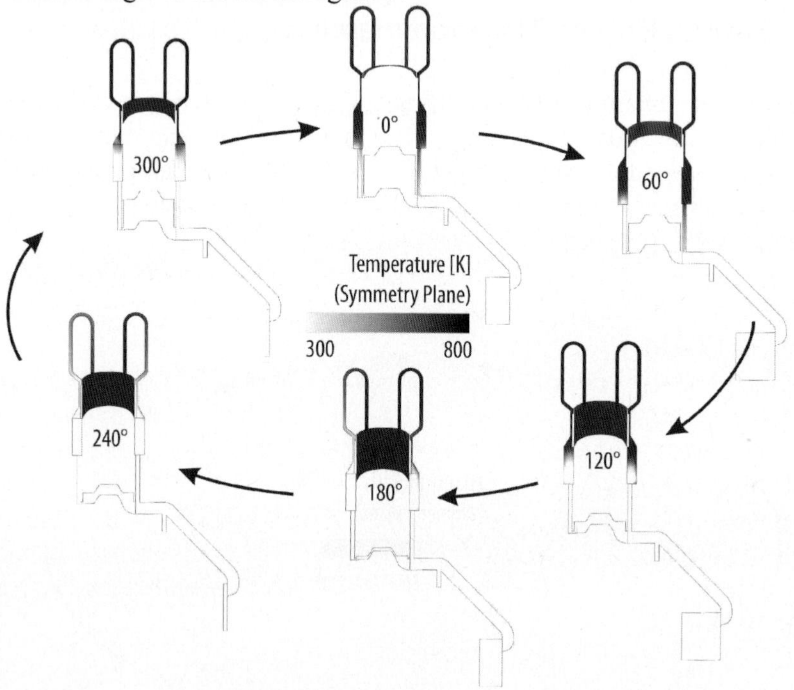

Fig. 3. Computed temperature distributions on symmetry plane vs. crank angle (air)

4. Conclusion

Computational models of thermodynamic cycles of the Stirling engine are being created as higher-level computational models based on CFD models of physical processes occurring in real units, using only the minimum simplifying assumptions. New computational models will be created after the necessary number of technical experiments is made. They will speed up the development of Stirling engines with better technical and economic parameters.

Acknowledgement

Published results were acquired using the subsidization of the Ministry of Education, Youth and Sports of the Czech Republic, research plan MSM 0021630518 "Simulation modelling of mechatronic systems".

References

[1] Schmid, G. “Theorie der Lehmann´schen calorischen Maschine”, ZVDI, XV, 1871, 99-111

[2] Finkelstein, T. “Generalized thermodynamic analysis of Stirling engines”, Paper 118B, Proceedings of the Winter Annual Meeting, Society of Automotive Engineers, Detroit, Michigan, USA, 1960

[3] Finkelstein, T. “Computer analysis of Stirling engines”. Adv. in Cryogenic Engineering, 20, pp: 269-282, Plenum Press, New York and London, 1975

[4] Organ, A.J. “Thermodynamics and Gas Dynamics of the Stirling Cycle Machine”. Cambridge University Press, ISBN 0-521-41363-X

[5] Woschni, G. “Verbrennungsmotoren“. Technische Universität München, 1999

[6] Pistek, V., Kaplan, Z., Novotny, P “Micro Combined Heat and Power Plant Based on the Stirling Engine”. MECCA - Journal of Middle European Costruction and Design of Cars, Vol.2005, No.4, pp.8-16, ISSN 1214-0821

Analysis of viscous-elastic model in vibratory processing

R. Sokołowska (a), T.Majewski (b)

(a) Politechnika Warszawska, 02-525 Warszawa, ul.A.Boboli 8, Tel (22)234-8447, roza@mech.pw.edu.pl, Poland

(b) Universidad de las Americas-Puebla, CP 72820, Tel (52)(22)229 26731, tadeusz.majewski@udlap.mx, Mexico

Abstract

The paper presents a model of the technological process in the vibro-energy machines. The abrasive medium has viscous-elastic properties. As a result of container's vibration the medium with finishing elements translate with oscillation and it results in machining the elements that are in the container. The forces between the medium and the elements are result of friction. The mathematical model is defined and the properties of the system are determined through a numerical simulation. Trajectory of the elements, their velocities and the forces between them are determined as a function of the container's vibrations.

1. Introduction

Efficiency of mass production process of small elements in vibratory machines depends on the forces acting on the element's surface. The motion of loose abrasive medium with the finishing elements inside it is a result of vibration of the container and the properties of abrasive medium (stiffness and friction). The elements move with vibrations and there are also impacts between the elements. Technological liquid in the container helps the process of finishing and control the properties of abrasive medium. Vibratory finishing is used for deburring, rounding, cleaning, and brightening of small elements in mass production.

2. Modelling the vibratory machining

The motion of the medium filling the container and the machining elements inside it comes from the container's vibration. The lower part wall of the container has a circular shape. The vibration of the container is generated by an unbalanced rotor. The amplitude and frequency of container's vibration can be controlled by changing the unbalance and the speed of the rotor. The trajectory of a point B on the container's wall is an ellipse. The vibrations of the centre point O of the container are harmonic

$$x_o = A_x \sin \Omega t, \quad y_o = A_y \sin(\Omega t + \psi_1), \varphi_o = A_o \sin(\Omega t + \psi_2) \qquad (1)$$

where A, Ω, ψ are the amplitude, frequency, and shift angle of the components of vibration [2].
The tangential and normal components of vibration of the point B of the wall – in the natural coordinates XBY-Fig.1

$$x(t) = x_o \cos\alpha + y_o \sin\alpha + \varphi_o R, \quad y(t) = -x_o \sin\alpha + y_o \cos\alpha \qquad (2)$$

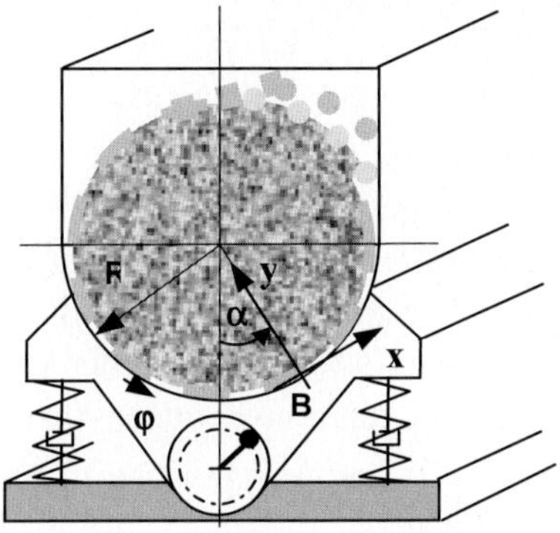

Fig.1. Model of the vibratory machine

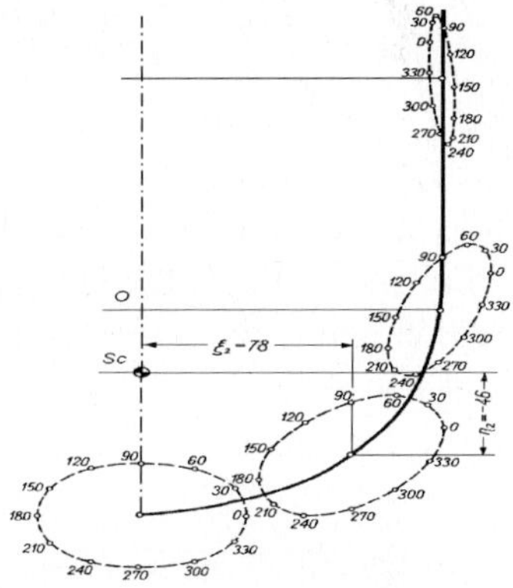

Fig.2. Trajectories of the select points on the container's wall.

The processing elements are taken as a rigid objects and the abrasive medium as a viscous-elastic material – Fig.3. The motion of the elements is defined in coordinate frame XBY fixed to the container and therefore the inertia forces J_x, J_y have to be introduced. The components of vibration x(t), y(t) are harmonic so the inertial forces are determined by the following relations;

$$J_x = m\Omega^2 x(t) \qquad J_y = m\Omega^2 y(t) \qquad (3)$$

The first layer of elements contacts with the rough wall. The friction between the wall and the elements forces them to move. The friction between the first layer and the next one pushes the last one and so on with the next layers.

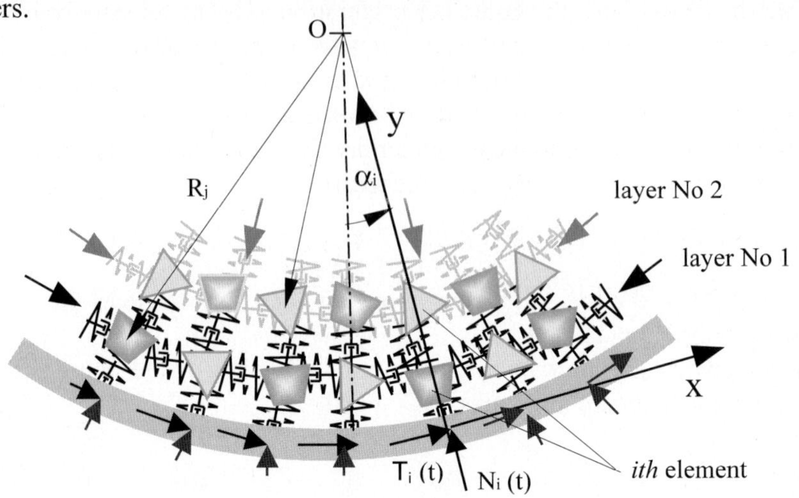

Fig. 3. Interaction between the elements

The layers moves with different velocity and in this way the elements are finished. The motion of one element depends on the forces acting on it; inertial forces J_x, J_y, normal reaction N, friction T, gravity mg, viscous F, and elastic forces S between the elements and abrasive medium. They should be projected on tangential and normal direction. The forces that act on the elements are shown in Fig.4. Position of the elements in the container is defined by the angle α_i or the coordinate z_i. The relation between them is $\alpha_i = (z_j + x_i)/R_j$.

The motion of the element is defined by the following equations

$$m_i \ddot{x}_i = P_{xi} = J_x + T_1 - T_2 + S_{1x} - S_{x2} + F_{x1} - F_{x2} - F_{x3} - m_i g x_i \,, \tag{4}$$

$$m_i \ddot{y}_i = P_{yi} = J_{yi} + N_{i1} - N_{i2} + F_{y1i} - F_{y2i} + F_{y3i} - z_j m_i g \cos \alpha_i \tag{5}$$

where S_{1i}, S_{2i}, F_{1i}, F_{2i} are the elastic and damping interactions in x and y directions from the adjacent elements, N_1, F_3 are the reactions of the wall on the *ith* element, T_2, N_2, F_{y3} are the reactions from the next layer, z_j is the number of layers.

Each element has two degrees of freedom. So the number of degrees of freedom of the system depends on the number of elements in the layer and the number of layers.

3. Results of computer simulation

Calculations were executed using software MATLAB 6.5. The results of computer simulations are shown in Figures 4, 5.

The following parameters were taken for the calculation: the radius of the first and second layers container $R_1 = 0.5\ m$, $R_2 = 0.45\ m$ the difference between layers *0.05 m*, the mass of the element $m_i = 0.1\ kg$, the coefficient of friction $\mu = 0.2$, the amplitude of vibrations $Ax = Ay = 0{,}5 \div 1.5\ mm$, the frequency of vibrations $\omega = 100$ rad/s, the stiffness of medium $kx = ky = 300\ N/m$, coefficient of damping $cx = cy = 10\ kg/s$.

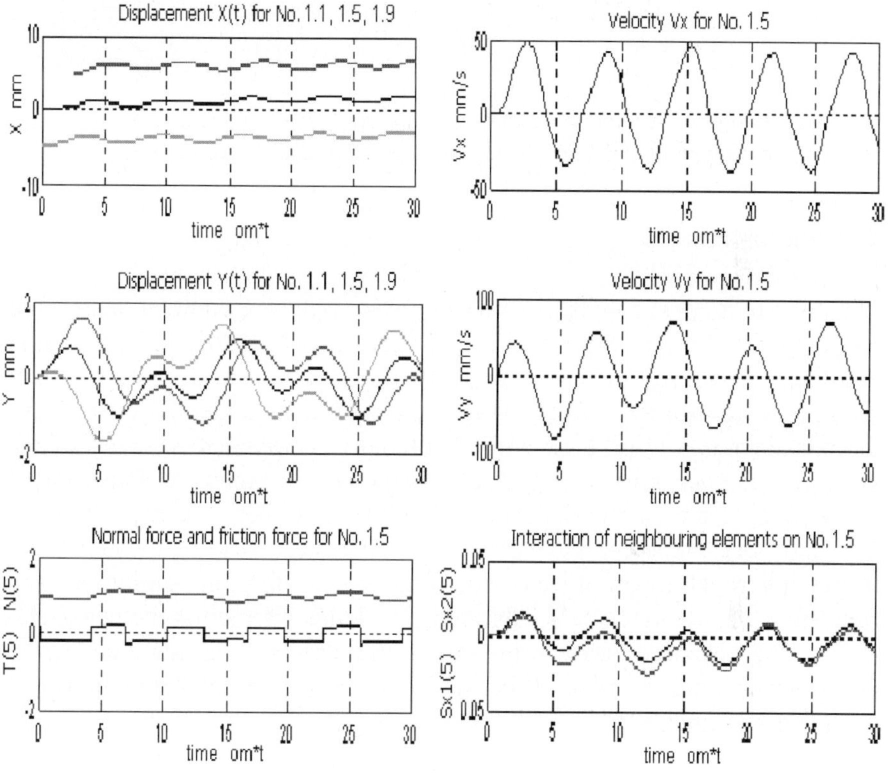

Fig.4. Behaviour of the elements in the first layer (z_i=1) when Ax=Ay=0.5 mm

The diagrams in Fig.4 present displacements *x* of three elements in the first layer with respect to the initial position, normal reaction of the wall, friction force and the interaction between the adjacent elements. Each element has two components of displacement *x(t)* and *y(t)*. The normal displacement is periodical with variable amplitude. In the direction tangential to containers wall the element moves with variable velocity. It means that there is an interaction between the elements what gives the finishing of the elements. An average velocity depends on the parameters of the container's vibration.

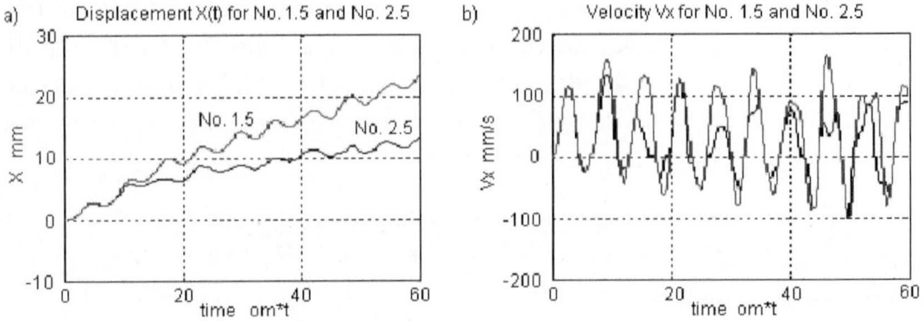

Fig.5. Behavior of the two elements in adjacent layers; a) displacements x(t) b) velocities vx(t) for the elements in the first and second layer.

4. Conclusions

The discrete model of vibratory finishing was applied. Computer analysis of the motion and interactions between the elements was done. The model is non-linear as a result of interaction between the wall and the elements that are in contact with it. The motion of the elements in tangential direction to the container is a translation with oscillation. The average velocity of the layer increases with the amplitude of vibration of the container. Direction of the motion of the elements depends on shift angle between the vibrations $x_o(t)$ and $y_o(t)$. Using computer simulation allows establishing the behaviour of the individual layers of processing medium for different amplitudes and frequency of the machine's vibrations, and the interaction between elements. The distances between the layers and also between the elements in one layer change. It has an effect on the density of abrasive medium and parameters of finishing.

References

[1] J.B.Hignett., J.Coffield : *Automated high energy mass finishing,* Soc. Manufacturing Eng. Tech. Paper No. 693 p. 1-17, 1983.

[2] R. Sokołowska: *The simulation analysis of the abrasive media motion in the vibro-energy round bowl machine*, Proceedings of the International 7th Conference on Dynamical Systems - Theory and Applications, Łódź, 2003.

Improvement of performance of precision drive systems by means of additional feedback loop employed

J. Wierciak

Warsaw University of Technology, Faculty of Mechatronics,
ul. Św. A. Boboli 8,
02-525 Warszawa, Poland

Abstract

Precision drive systems are expected to fulfil still higher and higher performance requirements regarding their speed of operation, accuracy etc. This is being achieved on various ways e.g. by changing construction of mechanical parts or modifying electronic circuits. The other approach is to modify control algorithms using additional data from the system, which was previously not considered. In the paper there are examples of such solutions presented. One of them, electrical linear actuator controlled using signal of loading force, has reached an experimental stage. It demonstrated its ability to operate efficiently under changing load with synchronous motion of driving stepping motor kept. Results of those experiments are added.

1. Introduction

There are numerous solutions developed to make traditional electrical drives more efficient. One of them is to measure current temperature of selected parts of a drive to protect it from overheating. Another idea uses signal of loading torque to stop a spindle motor before a drill is broken due to its wear during technological operation. In case of a drive with stepping motor there is a risk of loosing its synchronicity when load exceeds the value defined by characteristic for a given stepping rate. An idea to prevent such events is presented below.

2. New idea of control of stepping linear actuator

Users of linear actuators usually expect them to position driven objects with a given accuracy, sometimes – to develop high forces, and almost always – to move with high speed. When mechanism converting rotational to linear movement is powered by stepping motor (Fig. 1) then increase of pusher velocity v can be obtained by increasing stepping rate f.

$$v = P\frac{s}{2\pi}f \tag{1}$$

where: P – pitch of screw gear, s – nominal step of motor.

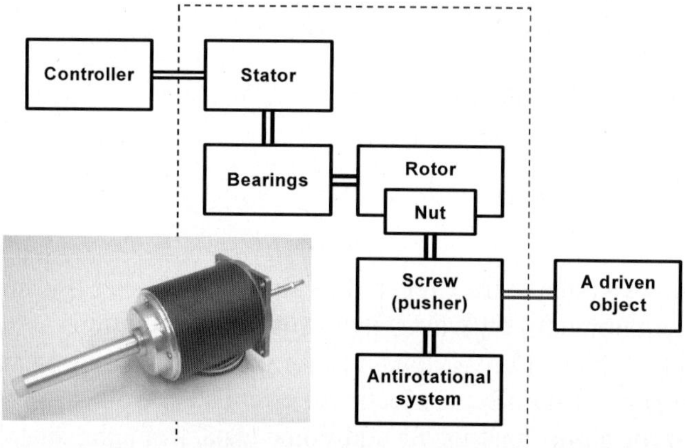

Fig. 1. Linear actuator with screw gear driven directly by motor's rotor

Depending upon mode of operation of stepping motor adopted for the drive the maximal stepping rate is limited either by its pull-in or pull-out characteristic (Fig. 2).

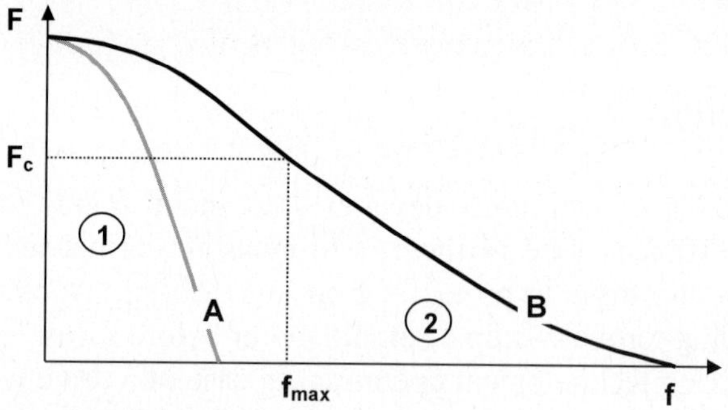

Fig. 2. Mechanical characteristics of stepping actuator: A – pull-in characteristic, B – pull-out characteristic; f – stepping rate, F – load force

When loading force acting on pusher varies, stepping rate is set at the level assuring synchronism of motion for the highest expected force. It means that for smaller loads the actuator operates at speeds lower from those possible to be achieved. In order to eliminate such restrictions it is proposed to continually adjust stepping rate to current load force using characteristic of motor. Thus during the positioning cycle control algorithm shall repeat the following functions (Fig.3):

- acquisition of instantaneous force value F_i,
- determination of pull-out frequency f_{gi} for this force,
- adjustment of stepping rate to the admissible level f_{maxi} with the constant, previously fixed acceleration a_f.

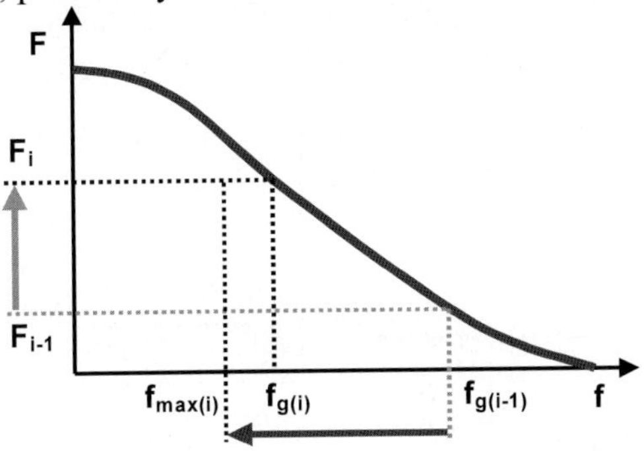

Fig. 3. The new idea of control

Realization of the specified tasks by the system requires:

a) mechanical characteristics of the actuator to be recorded in the control unit memory,

b) admissible acceleration of the actuator to be fixed,

c) constant measurement of load force.

In Fig. 4 block diagram of modified actuator with force sensor located on the pusher is presented.

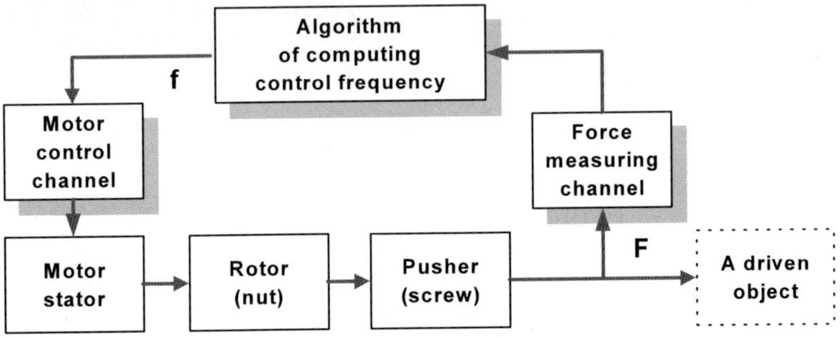

Fig. 4. Block diagram of modified actuator F – measured force, f – stepping rate

In order to verify the above idea simulation experiments were performed. Mathematical model of modified actuator was developed. The so called "idealised" model of stepping motor as well as classical relations for screw gear were used in the model [1]. Numerical data of actuator built with *FA 34* [3] hybrid stepping motor equipped *with M8 x 1,25* thread in movement converting mechanism were applied in the simulation model. A set of experiments was performed to determine pull-in characteristic of the actuator, which was subsequently approximated with polynomial function of the 4th order. Force measurement channel was modelled as the 1st order inertial element. Stepping rate f was computed as a subtraction of pull-in frequency f_{max} determined from approximating function and two correcting components: Δf_{max} for approximation and Δf_g as a safety margin.

$$f = f_g - \Delta f_g = f_{max} - \Delta f_{max} - \Delta f_g \, . \quad (2)$$

Responses of actuator to stimuli input as the step of force were computed (Fig. 5).

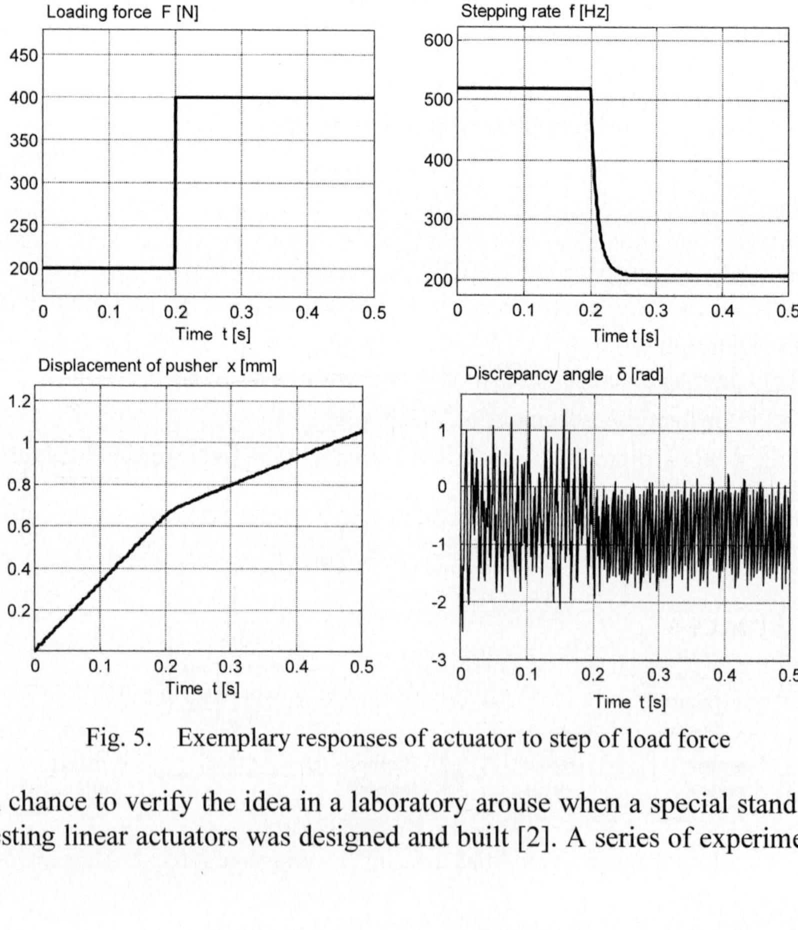

Fig. 5. Exemplary responses of actuator to step of load force

A chance to verify the idea in a laboratory arouse when a special stand for testing linear actuators was designed and built [2]. A series of experiments were carried out with approximately linear increase and decrease of load force (Fig. 6 a). In these conditions proper operation of the actuator was obtained (Fig. 6 b).

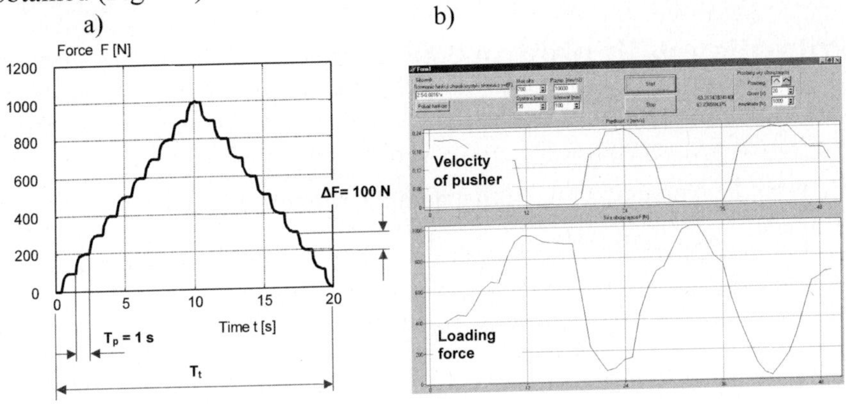

Fig. 6. Input signal of load force during tests (a) and right responses of the actuator (b) [2]

3. Summary and conclusions

The new idea of control of stepping linear actuator was confirmed by both simulation and laboratory experiments. The following conclusions can be formulated upon the results of tests.

1. Experiments proved that under certain conditions the actuator driven by a stepping motor with load force feedback loop can operate with possibly high stepping rates having synchronicity of motion guaranteed.
2. The modified actuator has a valuable property – ability to stop under overload, wait for better conditions and next start to continue synchronous motion.
3. Applying additional feedback loops to well known, reliable drive systems can be a good way for improving their performance.

References

[1] W. Oleksiuk at all “Konstrukcja przyrządów i urządzeń precyzyjnych” WNT Warszawa, 1996
[2] J. Wierciak, J. Lisicki, 5. Polish-German Mechatronic Workshop, 16-17.06.2005, Serock, Poland, 114
[3] MIKROMA S.A. „Silniki skokowe”. Katalog

"Manipulation of single-electrons in Si nanodevices -Interplay with photons and ions-

M. Tabe, R. Nuryadi, Z. A. Burhanudin, D. Moraru, K. Yokoi and H. Ikeda

Research Institute of Electronics, Shizuoka University, 3-5-1 Johoku, Hamamatsu 432-8011, Japan

Abstract

Recently, we are entering a new stage of electronics, in which time-controlled transport of individual electrons can be achieved by using nanodevices, so-called single-electron tunneling devices. Also, it is recognized that single-electron transport is highly sensitive to ultimately small environmental charges such as a photogenerated electron and a doped ion, leading to a new paradigm in electronic devices working with a few elemental particles, i.e., electrons, phonons and ions.

1. Introduction

Since almost two decades ago, single-electron-tunneling (SET) devices have been intensively studied[1]. The SET devices basically consist of ultimately small capacitors with the order of 10^{-19}Farad, in which even only one electron stored results in a huge potential difference of ~1 volt between the parallel electrodes of the capacitor. This is simply derived from the equation

$$\delta V = \delta q / C. \quad (1)$$

When δq is corresponding to an elemental charge, 1.6×10^{-19}Coulomb, the order of 10^{-19} for δq and C is cancelled out. Such a small capacitance can be attained by nm-scale fabrication, because the capacitor area size is proportional to the capacitance value. Under this condition, if the capacitor is thin enough to allow electrons to tunnel through, electron transport is dominated not by tunnel resistance but by Coulomb charging energy. This mechanism, so-called Coulomb blockade mechanism, is completely different from the conventional devices. When a quantum dot inserted between tunnel capacitors are biased by a gate, a stable number of electrons in the dot is controlled and electron transport can take place only when the stable number of electrons in the dot has double values, i.e., when (n)-electrons and (n+1) electrons have the same charging energy. The SET transistor, which is the most popular SET device, is based on this mechanism, and the dot potential is controlled by the gate.

In the ordinary SET transistors, however, timing of electron tunneling is not controlled. More than fifteen years ago, single-electron turnstile [2] and single-electron pump [3] devices, consisting of precisely designed multiple-capacitors, were proposed as those that can achieve time-controlled tunneling of individual electrons by means of ac-gate voltages. Each cycle of the ac-gate conveys exactly one electron from the source to the drain, leading to the resultant current of the circuit, $I=ef$. Most recently, we have discovered that even random multiple-tunnel capacitors have a capability to transfer electrons one by one [4, 5].

In this work, we present this result of single-electron transfer in the random system, as well as other important results on interplay of the SET devices with photons and dopant ions. These results, we believe, will open up new and wide possibilities for electronics applications.

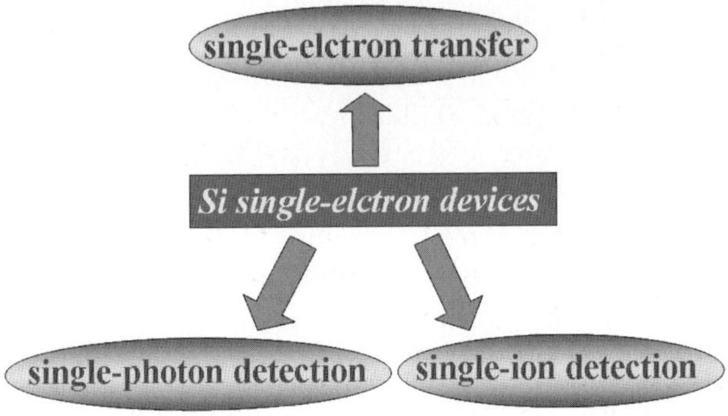

Fig.1 Recent exciting research topics based on Si single-electron-tunneling devices; single-electron transfer, single-photon detection and single-ion detection.

2. Manipulation of single-electrons

Recently, we have found by collaboration with Ono (NTT) that a P-doped Si-nanowire transistor can transfer electrons one by one by means of ac-gate bias [5]. This is quite surprising because multiple-tunnel-capacitors (or -junctions) are naturally formed by randomly distributed P-ions and it was not evidenced from the conventional theory that such random junctions have a capability of single-electron turnstile operation. We have analyzed these phenomena by theoretical simulations and found that most of non-homogeneous capacitance arrays statistically lead to the successful turnstile operation with unexpectedly high probabilities. Figure 2 shows a schematic view of the P-doped Si nanowire field effect transistor (FET). Figure 3 shows measured dc and ac characteristics of SET current vs gate voltages. Plateaus at I=ef is indicating that each ac-cycle of gate bias conveyed one electron.

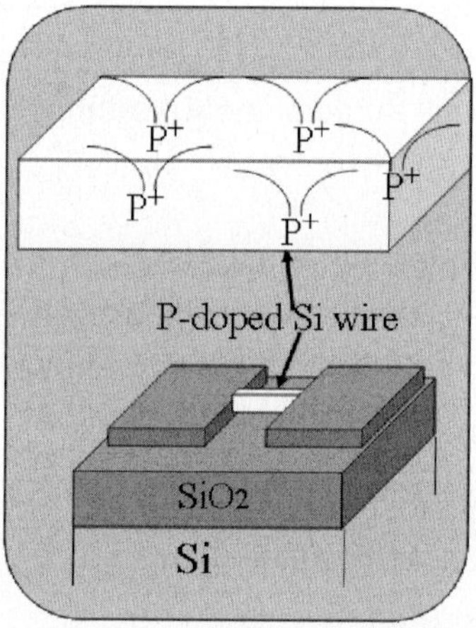

Fig. 2 A schematic view of a Si nanowire field effect transistor, which works as a SET multiple-junctions device. Phosphorous ions work as quantum dots and generate naturally formed multiple-tunnel junctions. A top metallic gate is covering the channel entirely (not shown for clarity).

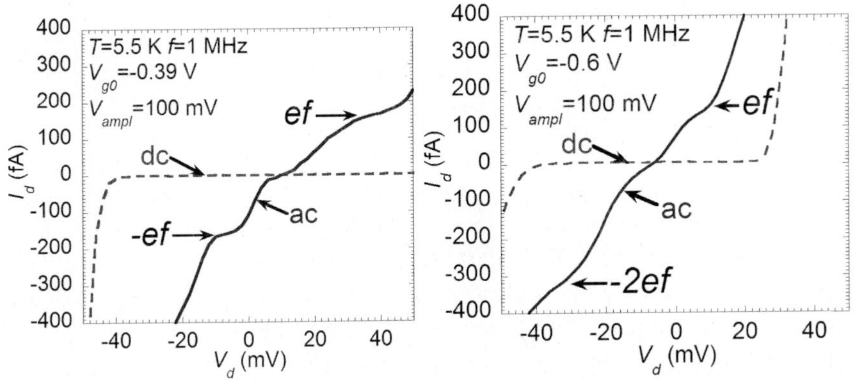

Fig. 3 (a) and (b) I_d-V_d characteristics measured at T=5.5 K under dc operation (dashed curves) and ac operation (solid curves) for different gate voltage offsets. For ac operation, frequency was set at f=1 MHz. Current plateaus appear aligned around ±*ef* (±0.16 pA) levels indicated by the horizontal lines as guides for the eyes.

Figure 4 shows our two-dimensional (2D) multiple-dots (multiple-tunnel junctions) FET, working as an SET device [6]. We have demonstrated [7, 8] that this device can detect a single-photon through random-telegraph-signal in the SET characteristics, i.e., each current revel switching in the random-telegraph-signal is ascribed to a photo-generated charge effect. (See Fig. 5)

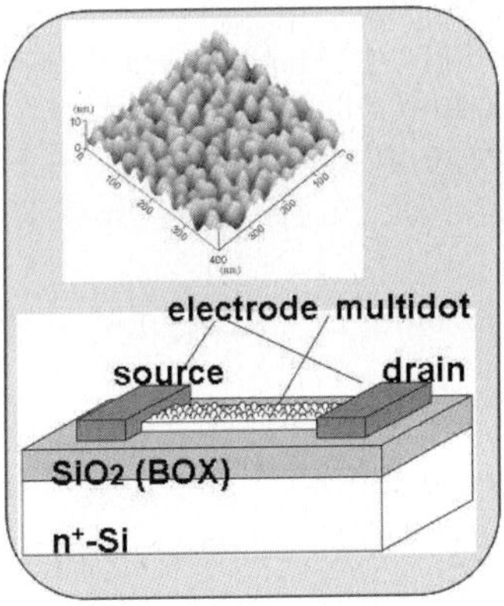

Fig. 4 A schematic view of our multiple-dots FET. The channel part consists of randomly distributed Si dots, as indicated by the AFM image.

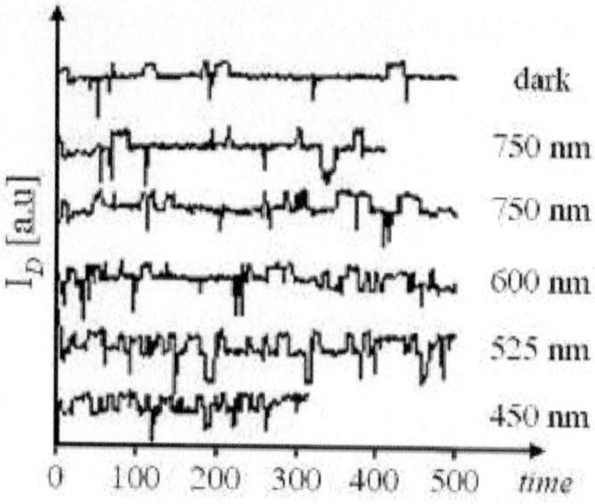

Fig. 5 Random-telegraph-signal observed for the dark and light-illuminated conditions. The frequency of the RTS increases with decreasing wavelength. This current switching in RTS is ascribed to individual photon absorption.

In order to improve the quantum efficiency, we have tried to detect photons absorbed in the underlying Si substrate by changing the substrate structure from n^+-Si to p on p^+ layered one. As a preliminary result, we succeeded in detecting individual boron ions by monitoring the single-hole-tunneling current [9].

References

[1] D. V. Averin and K. K. Likharev, *Single charge tunneling*, edited by H. Grabert and M Devoret (Plenum, New York, 1992).

[2] L. J. Geerligs, V. F. Anderegg, P. A. M. Holweg, J. E. Mooij, H. Pothier, D. Esteve, C. Urbina, and M. H. Devoret, Phys. Rev. Lett. 64, 2691 (1990).

[3] H. Pothier, P. Lafarge, C. Urbina, D. Esteve, and M. H. Devoret, Europhys. Lett. **17**, 249 (1992).

[4] H. Ikeda and M. Tabe, J. Appl. Phys. **99**, 073705 (2006).

[5] D.Moraru, Y.Ono, H.Inokawa, and M.Tabe, unpublished.

[6] R.Nuryadi, H.Ikeda, Y.Ishikawa, and M.Tabe, IEEE Trans. Nanotechnol. **2**, 231 (2003).

[7] R. Nuryadi, H. Ikeda, Y. Ishikawa, and M. Tabe, Appl. Phys. Lett. **86**, 133106 (2005).

[8] R. Nuryadi, Y. Ishikawa, and M. Tabe, Phys. Rev. B **73**, 045310 (2005).

[9] Z. A. Burhanudin, R.Nuryadi, and M. Tabe, unpublished.

Calibration of normal force in atomic force microscope

M. Ekwińska, G. Ekwiński , Z. Rymuza

Warsaw University of Technology,
Institute of Micromechanics and Photonics,
Św. A. Boboli 8,
Warsaw, 02-525, Poland.

Abstract

Investigation with the use of the atomic force microscope enables to estimate material properties in micro/nanometer scale. During such investigation load applied by the sensor (in this case cantilever) on the investigated surface is a crucial parameter. Under that circumstances there were elaborated several methods of normal force calibration. In this work a new calibration method with calibration gratings is proposed as well as advantages of this method are discussed. On the basis of this method also method for stiffness measurements of MEMS structures was proposed.

1. Introduction

Atomic force microscope (AFM) is one of the most commonly used devices for investigation of the tribological properties of the material in micro and nanoscale. This information is essential especially when construction of the MEMS (Micro Electro Mechanical Systems) is taken into account. Unfortunately load applied during tests performed on the AFM (such as: force – distance – curve measurement, wear and friction tests) is given in arbitrary units. In order to change qualitative information about applied load into quantitative information a calibration of the normal force has to be done.

2. Theoretical approach

In order to determine real values of the load applied during test performed on the AFM the calibration of the normal force has to be done. There are many different calibration methods which are described elsewhere [1 – 11]. In some of those methods only geometrical parameters of cantilever are needed. In other methods the way of laser beam is analyzed and out of these information the stiffness of cantilever is established. Using already established stiffness of cantilever, normal force applied to the system is established.

In all cases the biggest problem is connected with high inaccuracy of the method and with considering machine stiffness. Under these circumstances there still was a need to make a new approach to the normal force calibration in the AFM. In this paper a new easy- to – operate approach to the problem of normal force calibration was proposed and a new method of normal force was created. The method is called Black Box Method. In this method a whole measuring path of normal force is treated as a black box to which a known parameter is introduced. Then a reaction of the AFM on the introduced parameter is being observed. In other words the idea of the calibration is to cause the change of the normal force signal in arbitrary units by introducing to the system a known parameter (force). The introduced parameter is known value of force, which is applied at the very end of the cantilever's tip. This causes displacement of the cantilever's surface from which laser beam is reflecting. The change of the reflection angle causes the change of the normal force signal in arbitrary units [a.u.].

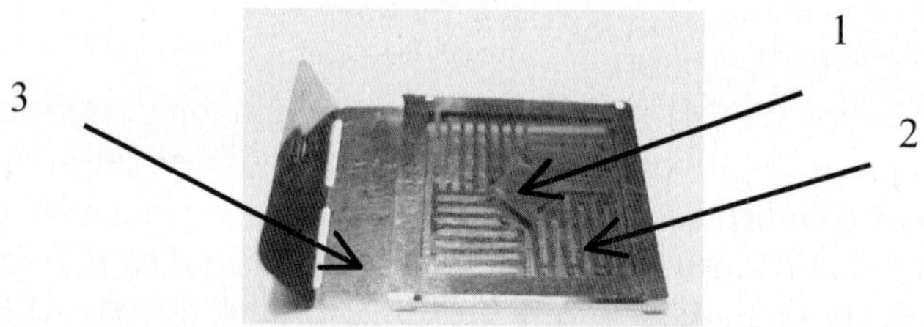

Fig.1: System for calibration of normal force in AFM, 1 – area where AFM's tip stands during calibration, 2 – plane which is elastically deformed during calibration, 3 – holder of device.

In order to calibrate normal force, a system with elastic element was elaborated [12, 13] (Fig.1.) There are three main parts of the elastic element: surface on which cantilever's tip is located (1), flat surface with known stiffness (2), surface to which AFM table is mounted (3). During calibration of the AFM with this calibration device it is placed on the AFM table and cantilever's tip is approached to the surface (1). After obtaining contact a force distance curve can be performed. During the experiment the surface (1) is pushed by the cantilever what causes the deformation of the surface (2). The deformation is registered. Bending of the cantilever as well as bending of the calibration device can be established. Then the normal force can be counted out of the deformation of the calibration device and stiffness of the surface (2). The ratio of the normal force estimated after calibration and the normal force in arbitrary units is the factor between arbitrary units and real units.

On the basis of this calibrating method a method for investigation of stiffness on MEMS structures, such as beams and bridges was created. The test is performed in the same scheme as the calibration procedure. The only differences are: the use of a cantilever without tip (in order not to damage investigated structures) and the fact that cantilever has earlier established stiffness. During this investigation the tip is approached to the investigated structure. After reaching the structure's surface the tip is pushed in order to achieve elastic deformation of the investigated structure. Then the tip is withdrawn. The result of the measurement is a force distance curve with additional bending on the approaching part..

3. Experimental details

The investigations were divided into two sections. First section was devoted for checking a normal force calibration method. The second one was usage of a new measuring technique for investigation of stiffness of MEMS structures.

Both investigations were carried out under laboratory conditions: temperature 22 ± 0.5 °C, humidity 40 ± 2 %, atmospheric pressure, air atmosphere. In first step the calibration gratings were elaborated (Fig.2). Then a calibration device for calibration of these gratings was built. Using this calibration device the stiffness of the calibrating gratings was established.

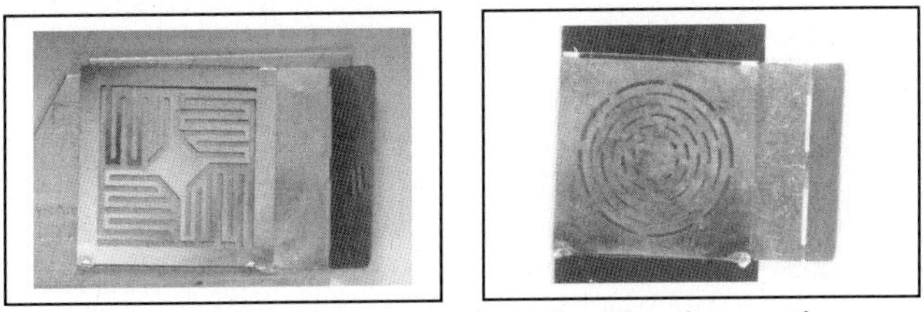

Fig.2. Different geometries of calibration gratings

After that set of cantilevers, which parameters are presented in Table 1 was calibrated.
In the second step already calibrated cantilevers were used for the calibration of stiffness of MEMS structures (microbridges and microcantilevers).. During these measurements MikroMasch cantilever NSC12 type tip B was used. It's parameters are presented in Table.1.

Table.1. Information about investigated AFM cantilevers according to producer ; "+" cantilever with tip, "-" cantilever without tip

AFM cantilever type	NSC12	CSC37	CSC37	NSC12	NSC12
Tip	-	-	-	+	+
Cantilever	E	B	B	F	B
Length [μm]	350 ± 5	350 ± 5	350± 5	250 ± 5	90 ± 5
Width [μm]	35 ± 3	35 ± 3	35 ± 3	35 ± 3	35 ± 3
Thickness [μm]	2 ± 0.3	2 ± 0.3	2 ± 0.3	2 ± 0.3	2 ± 0.3
Typical stiffness [N/m]	0.3	0.3	0.3	0.65	14.0
Minimum stiffness [N/m]	0.1	0.1	0.1	0.35	6.5
Maximum stiffness [N/m]	0.4	0.4	0.4	1.2	27.5

4. Results and conclusions

The stiffness of calibration gratings was established using Black Box Method. The results of the calibration of MicroMasch cantilevers are presented in Table 2.

Table 2. Comparison of established stiffness for investigated cantilevers; k – typical stiffness given by producer, Δk – interval between the biggest and the smallest value of the stiffness (given by manufacturer), k_E –stiffness established using calibration gratings, Δk_E – inaccuracy of the estimation of stiffness using calibration gratings.

Denotation of cantilever	k [N/m]	Δk [N/m]	k_E [N/m]	Δk_E [N/m]
NSC12 cantilever E	0.3	0.1 – 0.4	0.33	±0.03
CSC37 cantilever B	0.3	0.1 – 0.4	0.34	±0.03
CSC37 cantilever B	0.3	0.1 – 0.4	0.15	±0.02
NSC12 cantilever F	0.65	0.35 – 1.2	0.42	±0.04
NSC12 cantilever B	14.0	6,5 – 27.5	13.4	±1.35

Stiffness of MEMS structures established using calibration gratings is presented in Table 3.

Table 3. Comparison of established stiffness for investigated MEMS structures; k –stiffness established, Δk – inaccuracy of the estimation of stiffness ; C – cantilever like structure, materials out of which structures were made: first batch of devices are surface micromachined from 1.0μm thick cold-sputtered aluminium; polyimide film is used as a sacrificial layer; this gives an airgap of approximately 1.5-2μm, bottom metallisation layer is 0.5μm thick aluminium/1%silicon, covered by a 100nm thick layer of PECVD silicon oxide.

Sample denotation	k [N/m]	Δk [N/m]	Width [nm]	Length [nm]
CI01	0.032	0.005	30	100
CI02	0.042	0.006	30	200
CI03	0.126	0.019	10	100
CI04	0.489	0.075	30	100
CI05	0.037	0.006	30	200
CI06	0.110	0.017	10	100

Results of the investigations proved that presented method is easy to operate. Investigations were held on two different AFM microscopes and nearly the same results were achieved. The method enables also to estimate stiffness of cantilever, which is more precise than information given by manufacturer of the cantilevers. In this case also stiffness measurement of the same cantilever were done on two different AFM microscope and results of the establishment were close to each other. Under these circumstances it can be said that proposed method of calibration of normal force is correct.

This method is also good for establishing stiffness of other MEMS structures (e.g. .microbridges , microcantilevers). Especially when it is hard to establish stiffness because of structure is multiplayer one and the thickness of it is not known precisely.

References

[1] T. R. Albrecht, S. Akamine, T. E. Carver, C. F. Quate, J. Vac. Sci. Technol. A 8 3386

[2] J. M. Neumeister, W. A. Ducker, Rev. Sci. Instrum., 65, (1994), 2527

[3] J. D. Holbery, V. L. Eden, J. Micromech. Microeng. 10 (2000), 85 - 92

[4] J. P. Cleveland, S. Manne, D. Bocek, P.K. Hansama, Rev. Sci. Instrum,

64 (2), (1993), 403 - 405
[5] J. E. Sader, I. L. Larson, P. Mulvaney, L. R. White, Rev. Sci. Instrum. 66 (7), (1955), 3789 - 3798
[6] J. L. Hutter, J. Bechhoefer, Rev. Sci. Instrum. 64 (7), (1993), 1868 - 1873
[7] E. L. Florin, V. T. Moy, H. E. Gaub, Science 264, (1994), 415
[8] M. Radmacher, J. P. Cleveland, P. K. Hansma, Scanning 17, (1995), 117
[9] R. W. Stark, T. Drobek, W. M. Heckl, Ultramicroscopy 86, (2001), 207
[10] Ch. T. Gibson, G. S. Watson, S. Myhra, Nanotechnology 7, (1996), 259 – 262
[11] N. A. Burnham, X. Chen, C. S. Hodges, G. A. Matei, E. J. Thoreson, C. J. Roberts, M. C. Davies, S. J. B. Tendler, Nanotechnology 14, (2003), 1 - 6
[12] M. Ekwińska, Z. Rymuza, Tribologia, No5/2006, (2006), 17 - 27
[13] M. Ekwińska, Z. Rymuza, International Tribology Conference AUSTRIB 2006, 3-6 December Brisbane Australia., Proceedings on CD, Brisbane 2006

Advanced Algorithm for Measuring Tilt with MEMS Accelerometers

S. Łuczak

Institute of Micromechanics and Photonics, WUT, ul. A. Boboli 8, Warsaw, 02-525, Poland

Abstract

An advanced algorithm for tilt measurements to be realized by means of standard MEMS accelerometers is presented. It ensures to determine the pitch and the roll over 360° with accuracy of ca. 0.2°, and regards such problems as: calibration of MEMS accelerometers, checking correctness of their indications, increasing the accuracy of determining the tilt.

1. Introduction

The problem of determining tilt by means of a miniature sensor built of accelerometers belonging to Micro Electromechanical Systems (MEMS), characterized by miniature dimensions, satisfactory metrological parameters, and low cost, has been presented in detail in [1], while a way of increasing accuracy of the related measurements has been described in [2]. As far as mechatronics is concerned, the most typical applications of the considered sensor are control systems of mobile microrobots [3].

2. Calculating the tilt

An arbitrary tilt angle φ can be defined as two component angles α_1 and β_1 [1] (called pitch and roll), determined according to [2]:

$$\alpha_1 = arctan \frac{g_x}{\sqrt{g_y{}^2 + g_z{}^2}} \tag{1}$$

$$\beta_1 = arctan\frac{g_y}{\sqrt{g_x^{\;2} + g_z^{\;2}}} \tag{2}$$

where: g – gravitational acceleration; g_x, g_y, g_z – component accelerations.

3. Calibration of the tilt sensor

In order to apply a MEMS accelerometer it is often required to have it calibrated beforehand, as in the case of sensors presented e.g. in [4].
While building a dual-axis tilt sensor one must use either one tri-axial or multi-axial (such as e.g. in [5]) accelerometer, two two-axial, or three uni-axial accelerometers. Each accelerometer must be calibrated, preferably while embedded into the structure of the tilt sensor. Then, it is possible to determine two essential parameters of the analog output signal generated by the calibrated accelerometer: the offset and the gain (amplitude). Using the simplest way of calibrating tilt sensors it is possible to obtain their operational characteristics represented by the following formulas [6,7]:

$$U_x = a_x + b_x \sin\alpha_1 \tag{3}$$

$$U_y = a_y + b_y \sin\beta_1 \tag{4}$$

$$U_z = a_z + b_y \cos\alpha_1 = a_z + b_z \cos\beta_1 \tag{5}$$

where: $U_{x...z}$ – voltage signal related to axis $x...z$; $a_{x...z}$ and $b_{x...z}$ – offset and amplitude of the respective signal $U_{x...z}$.
The calibration process makes it also possible to evaluate uncertainties of determining the tilt angles by the way of defining appropriate prediction intervals assigned to the variables $U_{x...z}$ in equations (3)–(5) [6], whose maximal values $\Delta U_{x...z}$ are used in the considered algorithm.

4. Algorithm for determining tilt angles

Values of the parameters $a_{x...z}$ and $b_{x...z}$ obtained in the calibration process are indispensable during a standard operation of the tilt sensor. In order to determine the tilt angles it is advantageous to use equations derived from:

$$\frac{g_{x...z}}{g} = \frac{U_{x...z} - a_{x...z}}{b_{x...z}} = m_{x...z} \tag{6}$$

After a readout of the output voltages $U_{x...z}$ of the sensor it is possible to verify correctness of its indications, i.e. to check whether it is affected by any external constant acceleration. If that is the case, it is impossible to determine the tilt properly [1], as the gravitational acceleration geometrically sums up with the mentioned acceleration (variable accelerations can be eliminated by appropriate filtering the output voltage of the accelerometer). Then, the geometric sum of the Cartesian component accelerations has a value other than $1g$. When the gravitational acceleration affects the sensor exclusively, the following idealized relation is true:

$$\sqrt{g_x^2 + g_y^2 + g_z^2} = g \tag{7}$$

However, it is highly probable that the occurrence of random errors will result in a situation where the formula (7) is not satisfied [8]. So, one should take into account the mentioned above errors $\Delta U_{x...z}$ determined while calibrating the sensor. Under a rational assumption that their values will be approximately the same for each sensitive axis, and equal to ΔU, the correct indications of the sensor can be found within the following interval (assuming a statistical character of the regarded errors):

$$1 - \Delta U < \sqrt{m_x^2 + m_y^2 + m_z^2} < 1 + \Delta U \tag{8}$$

If the above inequality is not satisfied, some constant external acceleration affects the sensor, and thus its indications are false. However, it should be noted that in a contrary case, it is not obvious that the sensor operates under the necessary quasi-static conditions, for there is an infinite number of acceleration vectors (having the sense opposite with respect to the gravity vector), whose geometric sum with the gravity vector is described by (8). For instance, an acceleration with the absolute value of $2g$, acting upwards vertically, yields a resultant acceleration, indicated by the tilt sensor, that satisfies the inequality (8), yet has the sense opposite with respect to the gravity vector (the respective indication error reaches then the possibly highest value of 180°).

If one has no additional knowledge about the accelerations affecting the sensor, or about the real position of the sensor with respect to gravity, it is impossible, in a general case, to state whether its indications are correct. However, if we are sure that no constant accelerations act, the inequality (8) may be disregarded.

Yet, there is one more case to be considered. Values of the parameters $m_{x...z}$ determined according to formulas resulting from (6) should be contained within the range of sine function, i.e. $\langle -1; 1 \rangle$. However, some random errors may cause them to slightly exceed this interval, so in the further steps one should use new variables $n_{x...z}$ defined by relations derived from:

$$\begin{Bmatrix} m_{x...z} > 1 \Rightarrow n_{x...z} = 1 \\ m_{x...z} \leq 1 \Rightarrow n_{x...z} = m_{x...z} \end{Bmatrix} \tag{9}$$

Having verified correctness of the sensor indications and eliminated the case defined by (9), it is possible to determine values of the tilt angles on the basis of formulas derived form the initial equations (1) and (2):

$$\alpha_2 = arctan\frac{n_x}{\sqrt{n_y^2 + n_z^2}} \tag{10}$$

$$\beta_2 = arctan\frac{n_y}{\sqrt{n_x^2 + n_z^2}} \tag{11}$$

While using such formulas (arc tangent type), the related accuracy is constant at any value of the determined tilt angle, what means that measurements of the tilt are realized by a constant value of the respective sensitivity [2].

Since the range of the employed arc functions is ⟨-90°; 90°⟩, and the measuring range of the sensor equals ⟨-180°; 180°⟩, the last step is to find values of the measured tilt angles over the full angular range by the way of checking the sign of the component acceleration g_z, represented by the variable m_z. Both for the pitch and the roll, three cases are possible:

$$\begin{Bmatrix} m_z > 0 \Rightarrow \alpha_3 = \alpha_2 \\ m_z < 0, \alpha_2 > 0 \Rightarrow \alpha_3 = 180° - \alpha_2 \\ m_z < 0, \alpha_2 < 0 \Rightarrow \alpha_3 = -180° - \alpha_2 \end{Bmatrix} \tag{12}$$

$$\begin{Bmatrix} m_z > 0 \Rightarrow \beta_3 = \beta_2 \\ m_z < 0, \beta_2 > 0 \Rightarrow \beta_3 = 180° - \beta_2 \\ m_z < 0, \beta_2 < 0 \Rightarrow \beta_3 = -180° - \beta_2 \end{Bmatrix} \tag{13}$$

5. Summary

An algorithm for determining the tilt on the basis of measuring Cartesian components of the gravitational acceleration has been presented. Since it employs an arc tangent relation between the components and the tilt, a constant value of the related sensitivity is ensured over the whole measurement range [2]. The algorithm can be realized in the following steps:

1. Input of the calibration data: a_x, a_y, a_z, b_x, b_y, b_z, ΔU.
2. Readout of the indications of the sensor: U_x, U_y, U_z.
3. Determination of the variables $m_{x..z}$ in accordance with (6).
4. Verification of the condition (8).
5. Determining the variables $n_{x..z}$ in accordance with (9).
6. Initial calculation of the pitch and the roll according to (10) – (11).
7. Determination of the pitch and the roll according to (12) – (13).

Experimental studies, performed on a tilt sensor built of two dual-axis accelerometers ADXL 202E by Analog Devices, have proven correctness of the algorithm and yielded a relatively high accuracy of the sensor, as far as MEMS devices are concerned. Assuming that it may be evaluated by the maximal value of the difference between the value of the roll angle applied by means of the used test station and the value determined with respect to the average of respective indications of the sensor, the accuracy was found out to be of 0.18° (0.05 % as referred to the measuring range).

References

[1] S. Łuczak, W. Oleksiuk, M. Bodnicki, IEEE Sensors J., Vol. 6, No. 6 (2006) 1669.
[2] “Tilt Sensing with Kionix MEMS Accelerometers” Kionix, Ithaca, NY, 2005.
[3] S. Fatikow, U. Rembold “Microsystem Technology and Microrobotics” Springer-Verlag, Berlin Heidelberg, 1997.
[4] “Sensors & Sensory Systems Catalog” Crossbow, San Jose, CA, 2006.
[5] S. Bütefisch, A. Schoft, S. Büttgenbach, J. Microelectromech. Syst., Vol. 9, No. 4 (2000) 551.
[6] S. Łuczak, Elektronika 8-9 (2004) 215.
[7] “Low g Accelerometer Non-Linearity Measurement” MEMSIC Inc., North Andover, MA, 2003.
[8] J. C. Lötters, J. Schipper, P. H. Veltink, W. Olthuis, P. Bergveld, Sensors & Actuators A 68 (1998) 221.

Theoretical and Constructive Aspects Regarding Small Dimension Parts Manufacturing by Stereophotolithography

L. Bogatu, D. Besnea, N. Alexandrescu, G. Ionascu, D. Bacescu, H. Panaitopol

"POLITEHNICA" University, 313, Spl. Independentei, Bucharest, 060042, Romania

Abstract

StereoPhotoLithography (SPL) by laser allows three-dimensional objects obtaining.

The paper presents some aspects related to a device consisting of a stage performing displacements on x and y directions, controlled with PC, by means of a controller and a motion interface. This device is designed to be a subassembly of a μSPL installation.

There are given different geometric patterns, as preliminary test, generated by using a program in LabView to command the servomotors that act the x-y stage. There are given, also, the results of the authors as regards the 2 ½ D configured parts by using the standard photoresist technology, starting from the idea to combine μSPL with thick resist UV lithography.

1. Introduction

Stereophotolithography utilizes the principle of successive superposition of plane-profiled layers, obtained by selective polymerization of a synthetic resin under the action of a laser emitting ultraviolet or ultraviolet-near-visible radiations.

Applying of this technology in the microfabrication field requires a noteworthy improvement of the spatial resolution. In the same time, this approach proves itself very attractive, presenting a number of advantages over other micromanufacturing processes [1, 2]:

- direct obtaining of three-dimension objects;
- possibility to obtain complex shaped objects in a versatile manner, without using of masks, starting from the numerical coordinates of the objects, generated by a compatible CAD system;
- obtaining of micro-moulds for forming by micro-galvanizing (if a conducting support/substrate is used).

2. Description of the experimental setup and obtained results

For the time being, the research team has performed the designing, simulation and execution of a device – stage with displacements on x and y directions, controlled by PC, by means of a controller and a versatile interface for the motion controlling. The device is designed to be a subassembly of a μSPL installation [3], being placed on the supporting plate of a UV laser. In fig. 1 and 2, there are presented the kinematic scheme and the 3D model, in Solid Works, of this device. The experimental setup building is shown in fig. 3.

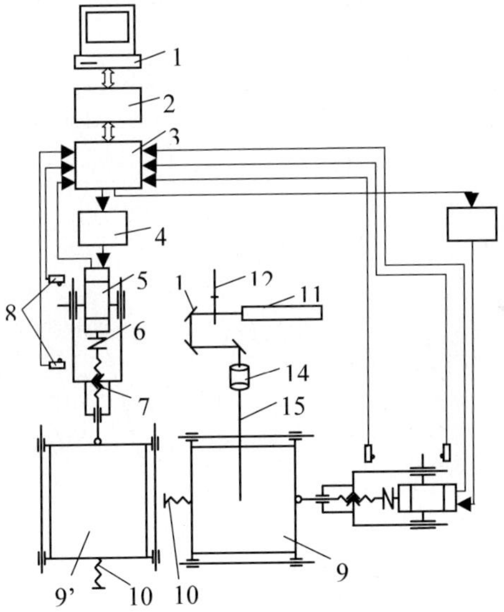

Fig. 1. Kinematic scheme of the x-y table:1 - PC; 2 - controller; 3 - motion control interface; 4 - d.c. servo-amplifier; 5 - servomotor with incorporated planetary gearing and encoder; 6 - coupler; 7 - screw and nut mechanism; 8 - course limiters; 9, 9' - x, y tables; 10, 10' - springs; 11 - laser; 12 - beam splitter; 13 - mirrors; 14 - objective; 15 - laser beam

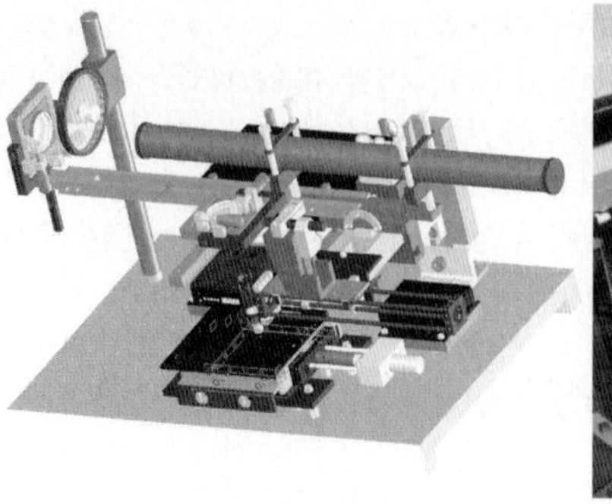

Fig. 2. 3D model, in Solid Works, of the x-y stage

Fig. 3. Measurement of the position errors on y direction

A laser emitting in the visible field will be used for the UV laser lining-up, as well as for the resin thickness control.

The maximum course on x-y directions is 25 mm, with a displacement increment (the resolution, R) of 74 nm, as results from the relationship given below:

$$R = \frac{p_s}{2\pi} \cdot \frac{1}{i_g} \cdot \Delta\varphi \qquad (1)$$

where: p_s - the pitch of micrometer screw (0.5 mm), i_g – the transmission ratio of planetary gearing (3375/64), $\Delta\varphi$ - the encoder increment ($2\pi/128$ rad).

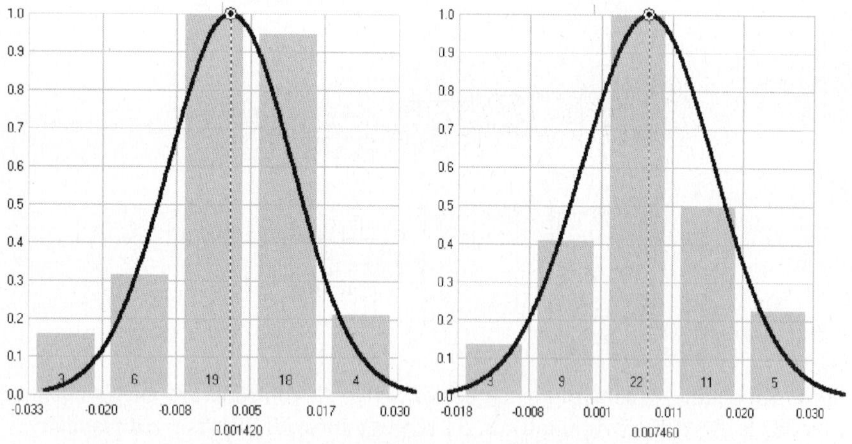

a. x direction b. y direction

Fig. 4. Normal (Gaussian) distributions for 50 measurements, representing the displacement errors on x and y directions

By initiating of commands of the stage in different positions on x-y directions with return in a reference position and, then, by displacing the two (x-y) tables successively, with a same number of steps, the positioning error and the repeatability error were determined. Thus, the influence of the errors of elements from the kinematic chains of motion transmission was established. An inductive transducer connected by an electronic interface to PC was used for measurements. More measurements (50 determinations) were performed and the Gauss distribution curves were drawn, fig. 4. Based on the experimental results (table 1), a positioning error of ± 5 μm and a repeatability error of 10 μm can be estimated.

Table 1: Experimental results

Measurement direction	μ	σ	$\mu - 3\sigma$	$\mu + 3\sigma$
x	0.001420	0.010540	-0.030199	0.033039
y	0.007460	0.008925	-0.019315	0.034235

In fig. 5 there are given different geometric patterns generated by using a program in LabView to command the servomotors that act the x-y stage.

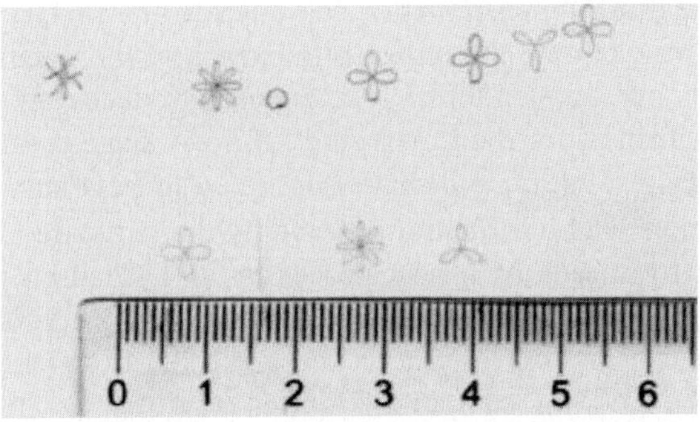

Fig. 5. Patterns generated and drawn based on the program in LabView

In the following fig. 6 and 7 are shown examples of 2 ½ D parts obtained by using standard photolithography (UV) processes for configuration. The covered technological stages are: base preparing, photoresist layer deposition (solid photoresist applied by lamination, for the parts shown in fig. 6 and, respectively, liquid photoresist deposited by spinning for the structures presented in fig. 7), UV-rays exposure, selective galvanic deposition (in the gaps of the photoresist mask), photoresist removing and releasing of the parts from the deposition support.

Fig. 6. Armature core disk and relay lamella: permalloy (thick. 0.13 mm); copper (thick. 0.16 mm), nickel (thick. 0.04 mm)

Fig. 7. Copper mechanical microstructures galvanically grown (thick. 27 μm) on a silicon wafer (SEM photography, x43)

3. Conclusions

In conventional microfabrication techniques, multiple deposition, etching and lithographic steps are required in order to produce miniature parts or microstructures. These techniques are applicable for two-dimensional processing, the third dimension (perpendicular to the surface) being fixed at one value, for example the uniform thickness of the deposited (grown) layer, as is in our applications. This paper introduces an experimental setup, which consists, for the time being, of a x-y stage controlled by PC, with nanometer resolution, and that is designed to be a subassembly of a μSPL installation. The information offered by the structure 3D model will be transposed by means of a post processor in a standard programming language that commands the displacement on the x-y-z coordinate axes by a controller and a motion interface. These are the research development directions of the working team, in the future.

References

[1] H. Yu, B. Li, X. Zhang, Sensors and Actuators A 125 (2006) 553.
[2] G. Ionascu "Technologies of Microtechnics for MEMS" (in Romanian), Cartea Universitara Publishing House, Bucharest, 2004.
[3] L. Bogatu, D. Besnea, N. Alexandrescu, G. Ionascu, D. Bacescu, H. Panaitopol, Acta Technica Napocensis, Series: Applied Mathematics and Mechanics 49, vol. III, Cluj-Napoca, Romania (2006) 727.

Comparative Studies of Advantages of Integrated Monolithic versus Hybrid Microsystems

M. Pustan, Z. Rymuza

Warsaw University of Technology
Institute of Micromechanics and Photonics,
ul. Sw.A.Boboli 8, Warsaw, 02-525, Poland

Abstract

This paper shows a comparative study of differences between monolithic microsystems and hybrid microsystems. The manufacturing technologies, the performance, and financial aspects are the main criteria which were taken into consideration in this study. The establishment of the manufacturing cost of monolithic versus hybrid microsystems was carried out, respectively. In this way collaborations with over 50 manufacturing companies had been performed for estimation of manufacturing cost of monolithic microsystems versus hybrid microsystems.

1. Introduction

A microsystem is defined as an intelligent miniaturized system comprising sensing, processing and/or actuating functions. These would normally combine two or more of the following: electrical, mechanical, optical, chemical, biological magnetic or other properties integrated onto a single chip or a multichip hybrid.

The microsystem can carry out four basic functions: (a) perception of the environment with a sensor; (b) signal processing, data analysis and decision-making, with a microelectronic circuit; (c) reaction upon environmental input according to data received, with an actuator; (d) communication with the outside world, with signal receivers or generators.

Microsystems meet the growing demand of the market for systems that are increasingly reliable, multifunctional, miniaturized, cheap, possibly self-managed and/ or programmable. As previously indicated, two main requirements account for the evolution towards system miniaturization:

- Manufacturing at very low unit cost for mass application;
- Reducing the size of devices for applications aimed at very narrow spaces or requiring minimal weight.

2. Monolithic and hybrid microsystem

The two constructional technologies of microengineering are *microelectronics* and *micromachining*. Microelectronics, producing electronic circuitry on silicon chips, is a very well development technology. Complementary metal-oxide-semiconductor (CMOS) technology is a major class of electronic circuit. Micromachining is the name for the techniques used to produce the structures and moving parts of microengineered devices.

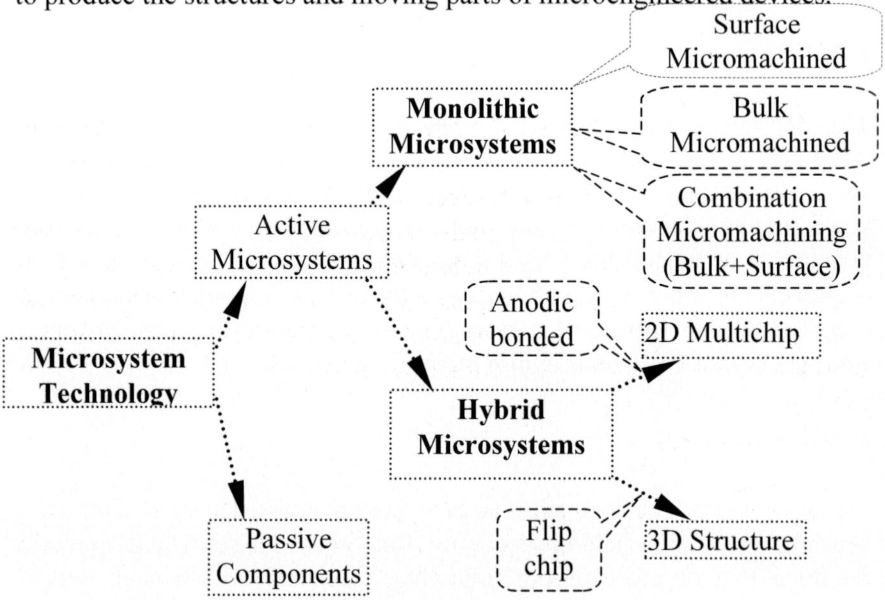

Fig. 1. Classification of micromachining for monolithic and hybrid microsystems

The various microsystems technologies (MST) can be classified as shown in the Figure 1. The diversity of microsystem technologies is a result of the wide range of materials that can be used and the number of different forming or machining techniques. Materials used are silicon, quartz, ceramics, metals, plastics, glass, piezoelectric layers, etc.

The *passive microcomponents* cannot realize signal transformation, information processing or system control. This category includes filters, resistors, capacitors, inductors, transformers and diodes. The *active microcomponents* can detect, process, transform and evaluate external signals, can make decision based on the obtained information and finally can convert the decision into corresponding actuator commands.

In many cases, mechanical structures are combined with active electronic circuitry. There are two major techniques employed, *monolithic* and *hybrid.* Microsystems may be constructed from parts produced using different technologies on different substrates, connected together, i.e. a hybrid microsystem (Fig.2a). Alternatively, all components of a system could be constructed on single substrate, i.e. a monolithic microsystem (Fig.2b).

Fig.2. Hybrid approach (a) versus monolithic integration (b) of MEMS and CMOS

The decision to merge CMOS and Microelectromechanical Systems (MEMS) devices to realize a given product is mainly driven by performance and cost. On the performance side, co-fabrication of MEMS structures with drive/sense capabilities with control electronics is advantageous to reduce parasitic, device power consumption, noise levels as well as packaging complexities, yielding to improved system performance [1-3]. With MEMS and electronic circuits on separate chips (Fig.3) the parasitic capacitance and resistance of interconnects, bond pads, and bond wires can attenuate the signal and contribute significant noise.

Fig. 3. Accelerometer showing control IC on the left and sensing cell on the right (Freescale Semiconductor, Inc.)

On the economic side, an improvement in system performance of the integrated MEMS device would result in an increase in device yield and density, which ultimately translates into a reduction of the chip's cost. Moreover, eliminating wire bonds to interconnect MEMS and integrated circuits (IC) could potentially result in reduced packaging complexities which will eventually lead to more reliable systems, and to lower manufacturing cost. Modular integration will allow the separation of development and optimization of electronics and MEMS processes. There are three main integration strategies (Fig.4): "Pre-CMOS", "Post-CMOS", and the "Interleaved approach".

MEMS process insertion before the electronics process	**MEMS process insertion after the electronics process**	**MEMS process insertion during the electronics process**
(a)	(b)	(c)

Fig. 4. Schematic description of the MEMS – CMOS monolithic integration (a) Pre-CMOS; (b) Post-CMOS; (c) Interleaved approach

The first integration approach is the "Pre-CMOS" scheme (Fig.4a) that was first demonstrated by *Sandia National Laboratory* through their IMEMS foundry process [4]. The second integration approach is the "Post-CMOS" scheme (Fig.4b) which was successfully demonstrated by *Texas Instruments Inc* through the Digital Micro- Mirror Device (DMD), which uses an electrostatically controlled mirror to modulate light digitally, thus producing a stable high quality image on a screen [5]. The third integration approach is the interleaved approach (Fig.4c). This approach has been successfully demonstrated by *Analog Devices Inc* in their 50G accelerometer (ADLX 50) technology which was the first commercially proven MEMS-CMOS integrated process [6].

3. Manufacturing cost of monolithic and hybrid microsystems

A complete cost analysis of monolithic versus hybrid microsystem had been made together with over 50 (received answers from 90 companies to which the question was sent) manufacturing MEMS companies from Europe, USA and Asia. The problem which was discussed is as follow: *which of monolithic or hybrid MEMS integration solutions are more expensive*. Of course, the answers were different for each of company and some companies could not give us a definite answer. Figure 5 shows a graphically repartition of the answers which were gave by MEMS companies. Most of them estimated that the monolithic integration is a cheaper solution for MEMS.

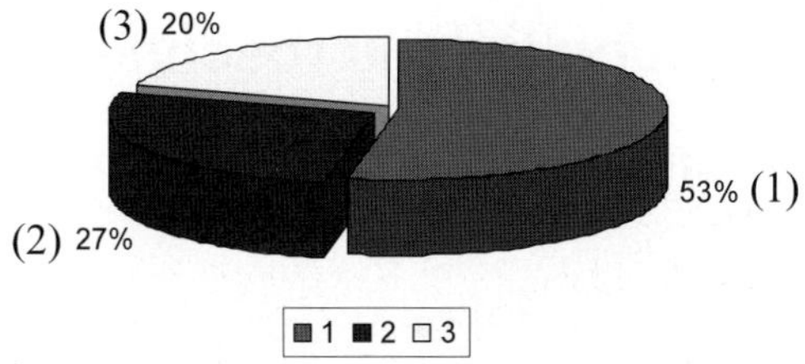

Fig. 5. Distribution of company answers on MEMS manufacturing cost (1)Hybrid; (2) Monolithic; (3) Depend by manufacturing volume.

4. Conclusion

In general, monolithic systems are only considered were the production volumes are expected to be very high (several 10's or 100's of millions of parts per annum). Main conclusions of monolithic versus hybrid microsystem are: the hybrid MEMS solution represents a bigger unit cost that an alternative, monolithic solution; the assembly and packaging cost is higher compared to the monolithic approach; for monolithic approach the MEMS above CMOS (Post-CMOS) is more cost competitive than MEMS below CMOS (Pre-CMOS); the package of MEMS is very expensive - packaging currently represents more than 80 percent of the cost of some systems and is often the leading cause of system failure.

Acknowledgement

The work was carried out in the frame of the European Project ASSEMIC MRTN-CT-2003-504826.

References

[1] M.A.N. Eyoum, "Modularly Integrated MEMS Technology", Dissertation, Dept of Elec. Eng. and Computer Sciences, UC Berkeley Fall (2006).
[2] H. Xie, L. Erdmann, X. Zhu, K. Gabriel, G. Fedder, Solid-State Sensor and Actuator Workshop, Hilton Head -SC (2000) 77.
[3] K.A. Shaw, N. C. MacDonald, Proc. 9th Int. Workshop on MEMS, San Diego CA (1996).
[4] J.H. Smith, S. Montague, J.J. Sniegowski, J.R. Murray, R.P. Manginell, P. J. McWhorter, SPIE (1996) 306.
[5] P. F. Van Kessel, L. J. Hornbeck, R. E. Meier, M. R. Douglass, Proc. of IEEE (1998)1687.
[6] T.A.Core, W.K.Tsang, S.J.Sherman, Solid State Tech. (1993) 39.

New thermally actuated microscnner – design, analysis and simulations

A. Zarzycki (a) *, W. L. Gambin (b)

(a) (b) Warsaw University of Technology,
Department of Applied Mechanics
ul. Sw. Andrzeja Boboli 8
02-525 Warszawa, Poland

Abstract

A design of a new thermally actuated microscanner is proposed. The device is capable of two-dimensional scans for optical raster imaging. It consists of a micromirror and four thermal actuators. A special location of the mirror with respect to the cantilever beams assures the high precision of scanning action. The distance of the centre of the mirror from the light source is the same during the whole scanning process. It allows projecting an image with fewer distortions. The rest position of the mirror and resonant frequency for raster scanning action are investigated.

1. Introduction

Scanning micromirrors are used in devices for imaging, bar-code reading, laser surgery, laser machining, etc. Modern MEMS and MOEMS technologies [1] enable to produce the microscanners smaller and smaller. One can classified microscanners with respect to its actuation principle. The most common are: electrostatic, piezoelectric, electromagnetic and thermally activated devices.

Micromirrors activated thermally, used in the proposed project, have some interesting properties. Thermal actuators, formed as thermo-bimorph beams, provide large scan angle, nearly linear deflection versus power relationship and moderate power consumption. Moreover, fabrication of such microscanners is based on very cheap technology. Thermal actuators do not operate with as high frequency as electrostatic or piezoelectric ones, but high enough for some optical applications.

2. Micromirror design

Proposed microscanner is composed of a round micromirror and four thermal-bimorph beams, Fig. 1 [2]. The beams play a role of actuators deflecting the mirror in two orthogonal planes. One pair of parallel actuators (moving with a high speed – for raster scanning) is fixed directly to the mirror, and to a rigid movable frame. Second pair of the bimorph beams (moving with a low speed – for frame scanning), are situated perpendicularly to the first one and joins the frame with a stationary silicon substrate. The actuators are composed of two layers with different coefficients of thermal expansions (CTE) and a thin insulator layer between them, Fig. 2c.

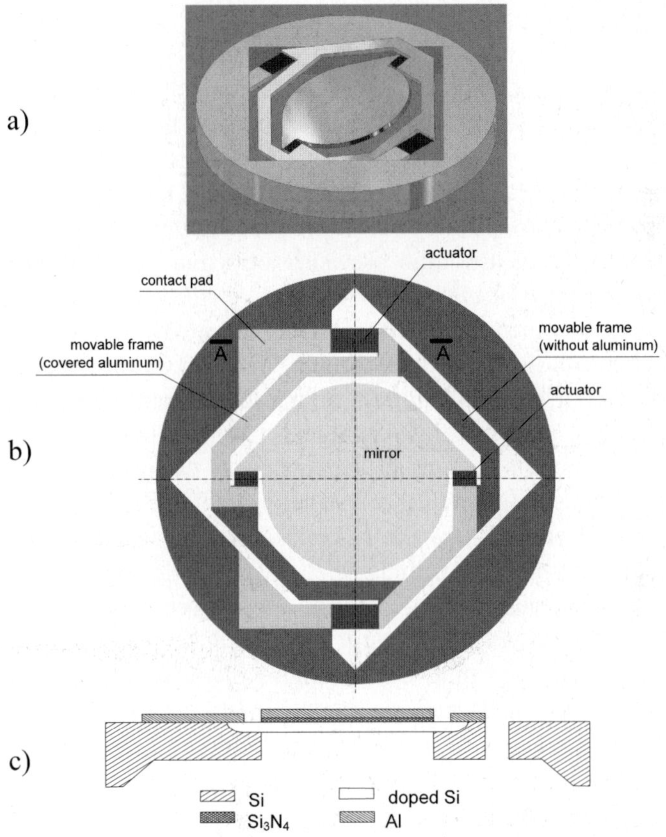

Figure 1: a) General view (mirror diameter – 400 μm). b) Planar projection. c) Cross-section A–A (frame and scanning actuators). Materials and layers thickness: Si (frame and mirror) – 15 μm; doped Si (passive layer) – 1 μm; silicon nitride (insulator) – 0.15μm; Al (active layer) – 0.7 μm

The bottom layer (the *passive* one) has lower CTE, whereas the top layer (the *active* one) has higher CTE. In our project, the bottom layer has properties of an electric heater resistor connected with an electric supply. Due to mismatch between the CTE of the materials the bimorphs after the forming process curl out-of-plane and make the whole device a 3D structure. When electric current is passed through the heater resistor, the temperature of the actuators increases and the structure, initially deflected out-of-plane, deflects downward to the substrate plane. Actuators directly connected to the mirror are short and move together with a high, resonant frequency, whilst the two others are longer and move together with low, non-resonant frequency. It enables the laser beam reflected by the mirror draws a 2D raster image on a screen. The raster scanning system creates a light beam that produces a single bright pixel. The last one is scanned in two dimensions to create an image. The light beam moves with a high speed along a horizontal line, and next, with a lover speed, come back to the beginning of the second line situated below the first line. The movement of the light beam along the horizontal line (with a high speed) is called the raster scanning, whereas the movement of the light beam in vertical direction (with a low speed) is called the frame scanning.

In the proposed design the high precision of scanning action is achieved due to a *special position of the mirror centre with respect to its rotation axes*. Namely, the mirror centre lies on the line going through the centres of two attached cantilevers, Figure 1b. The above position assures, that the distance of the mirror from the light source is the same during the whole scanning process. Additional advantage is the fact that the inertial moments of movable parts, as well as, the influence of air damping are minimized. It allows achieving a higher frequency for the frame scanning and a higher resonant frequency for the raster scanning. Higher frequencies result more accurate projecting image because of greater image refreshing. On the other hand, more precise motion of the mirror causes less image distortion. The kinematical behaviour of the scanner was presented in [2] in details. It was proved that for the angles of mirror rotations from the interval $0 \le \varphi \le 45^0$, the distance of the mirror centre from the light source may be taken as equal to *L/2* with sufficient accuracy.

3. Micromirror behaviour

The initial flexure of actuators appears during the *forming process*. When flat bimorph beams are cooled from the temperature 300^0C to the room temperature 20^0C, its take a shape of arc with the bend angle φ_0. The rest position of the mirror is described by the initial bend angles of actuators. During the *exploitation process*, when a driving voltage is applied to the device, the free ends of the actuators oscillates, bending back from its initial positions determined by the angles φ_0 to smaller angles $\varphi = \varphi_0 - \Delta\varphi$. Using the procedure described in [3], one can find quantities $\Delta\varphi$.

ΔT [°C]	50	100	150	200	250
$\Delta\varphi$ [°]	1.070	2.010	3.060	4.110	5.160

Table 1: Changes of bending angle for the raster scanning actuators versus changes of temperature

Presented in Table 1 values of mechanical angles $\Delta\varphi$ create the optical angles $\Delta\alpha$, those are twice of $\Delta\varphi$ and are large enough to use the device for image display.

4. Thermal analysis

The highest frequency exciting the non-resonance vibrations of the scanner is called thermal cut-off frequency. In order to state it, we assumed simple 1D model of clamped beam with heat transport from free end to the fixed one, Fig.2. Next, the thermal cut-off frequency of cantilever was obtained from the heat transport equation [3]. For the case of short actuators, we have obtain: f_{cut_off} = 3250 Hz.

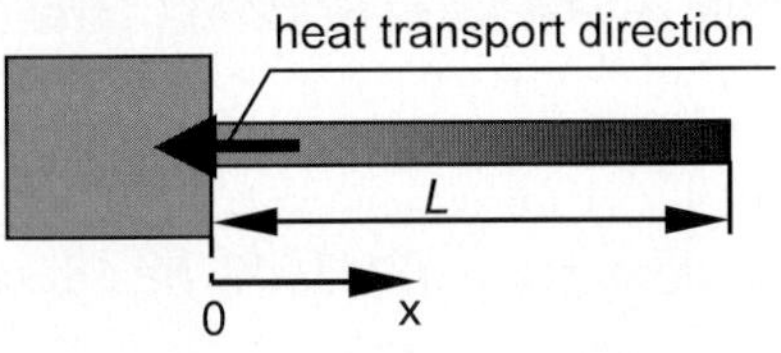

Fig. 2. Considered model.

5. Resonance frequency calculations

To find the fundamental resonance frequency analytically, we consider model of the system *shorter actuators + micromirror* shown at Fig.3. When the mirror oscillates with rotating motion around its centre C_0, the force F and the moment M act at the end of the cantilever. Because The point C_0 is practically motionless [2], one can assume that the force F is equal to zero. For the considered system the fundamental natural frequency took the value: $\omega_0 = 9266$ rad/s was obtained from. It correspond to the frequency of short actuators $f = 1475$ Hz.

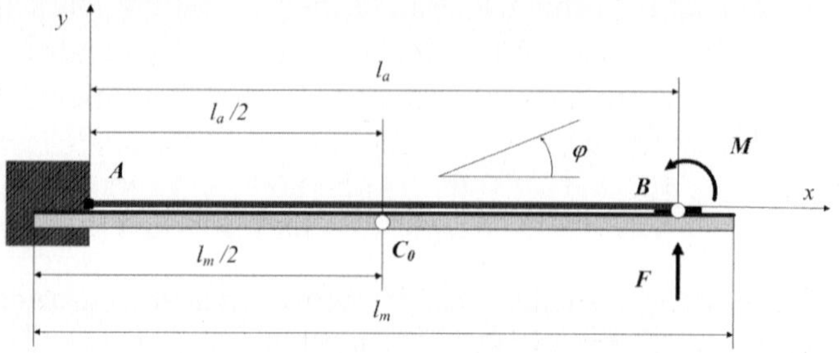

Figure 3: Cantilever loaded by mirror attached at point *B*

To obtain more realistic estimation, it was necessary to perform numerical simulations using 3D model with deformable mirror element. The obtained five succeeding resonance modes with associated frequencies are: I mode: 1052 Hz, II mode: 13688 Hz, III mode: 31056 Hz.

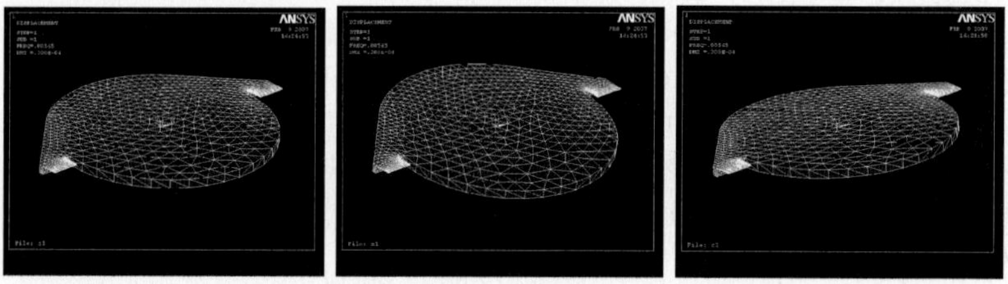

Figure 5: First mode of the resonance frequency of the mirror

Notice that the obtained frequency $f_0 = 1052\ Hz$, described by I mode, is large enough for scanning process.

References

[1] M. Madou "Fundamentals of Microfabrication", CRC Press, Boca Radon, Florida, USA.
[2] W.L. Gambin, A. Zarzycki "Optimal design of a new thermally actuated microscanner of high precision", 1-st Int. Conf. Multidisciplinary Design Optimisation and Application, Besançon, 17-20 April 2007
[3] W.L. Gambin, A. Zarzycki "Thermal and dynamical analysis of a new thermally actuated microscanner", 1-st Int. Conf. Multidisciplinary Design Optimisation and Application, Besançon, 17-20 April 2007

Influence study of thermal effects on MEMS cantilever behavior

K. Krupa (a) *, M. Józwik (a), A. Andrei (b), Ł. Nieradko (b), C. Gorecki (b), L. Hirsinger (b), P. Delobelle (b)

(a) Institute of Micromechanics and Photonics, Warsaw University of Technlogy, 8 Sw. A. Boboli St.,
Warsaw, 02-525, Poland

(b) Institute FEMTO-ST, Université de Franche-Comté, 16 Route de Gray, Besançon, 25030, France

Abstract

Aluminum nitride (AlN) films have piezoelectric properties that are already used for acoustic wave propagation in miniature high frequency bypass filters in wireless communication [1,2]. This is a promising material also for MEMS applications and sensors using surface acoustic waves, what have already been proposed. For actuation purposes, even if PZT films are frequently used for their better piezoelectric properties, AlN material still represents an alternative that have to be explored [3].
The subject of the study is the investigation of the high quality cantilevers with AlN layer operating as reliable actuation elements in more complex MEMS systems. It makes necessary to study the mechanical and fatigue behaviors of such components allowing to understand and analyze their failure mechanisms. This paper focuses on device characterization and determination of thermal effect on the cantilever parameter evolution as one of the reliability study aspects. In this process the precise measurements play a key role therefore the interferometric platform has been proposed as an appropriate method of testing.

1. Introduction

During the last tens years a large progress in microelectronics fabrication technology has taken place. Nowadays the MEMS/MOEMS systems find a lot of applications in many fields of industry like in telecommunication or automotive market. Further development of microsystems taking into account their advantages will bring for the MEMS/MOEMS technology wide range of potential applications also in biomedical, pharmaceutical and military domain. Therefore, it is crucial to be able to determine the reliability of silicon devices defined as the probability that these structures will perform a certain function within a set of pre-defined specifications for a given period in a required time. Manufacturers have to guarantee the correct operation of their products for a certain time. Therefore, it is essential to know the reliability of these devices since it has a fundamental influence on the competitiveness, on the acceptance and on the commercialization of this new technology. Nevertheless, the researchers are still more interested in developing the new fabrication technology then in studying the reliability and the failure mechanisms of the new structures. Till now there is no sufficient knowledge concerning the behavior of the material in micro scale what makes necessary to study 'the micro world' allowing to determine the microdevice reliability. Such investigation requires to develop the special methodology and the accurate measurement system playing a key role in the reliability study. In this paper we focus on the accelerated aging through thermal cycling and its influence on the micromechanical and the physical parameters of the AlN driven cantilevers applying interferometry as a measurement method.

2. Measurement set-up

2.1 Device presentation

From the reliability study point of view it is essentially to investigate the microelements operating as the actuators since they are the most often run a risk of damaging because of the friction or the material fatigue. Therefore, for the object under test it was chosen the silicon microcantilever driven by AlN piezoelectric layer placed between two metal electrodes (the bottom electrode from Al and the top one from CrNi). Aluminum nitride is a material which represents the alternative for PZT but till now is rarely used in the MEMS/MOEMS systems. The studied elements have been fabricated in the Institute FEMTO-ST Université de Franche-Comté in Besançon in France. They have had a width of 50μm, a length of 200 ÷ 800μm and an AlN thickness of 400mn. Detailed description of the tested devices and their process fabrication has been presented in the literature [4].

2.2 Measurement system

For the characterization of such fragile elements in a micro scale it is essential to chose the appropriate measurement system allowing to investigate the microdevices in the non-destructive way with nanometric accuracy of out-of-plane displacement. It was chosen a non-contact optical technique applied in a multifunctional interferometric platform dedicated for the MEMS/MOEMS structures [5]. Combining the capabilities of different interferometric methods (e.g. conventional interferometry, time-average interferometry and stroboscopic interferometry) the interferometric system performs the measurement of an out-of-plane displacement in both static and dynamic regimes what allows to find such parameters like an initial shape, a resonance frequency and an amplitude distribution in the vibration modes, being the most demanding features for determining the mechanical proprieties of microelements . The architecture of the measurement set-up presented in Fig.1 is based on the Twyman-Green interferometer integrated with optical microscope. It has been used a temporal phase shift algorithm (TPS) with 5 interferograms (Fig.2) acquired and visualized by 768x576 CCD array, coupled with a frame grabber. A basic sensitivity of this system is λ/2 per fringe (345 nm) and an accuracy after AFPA ± 20 nm.

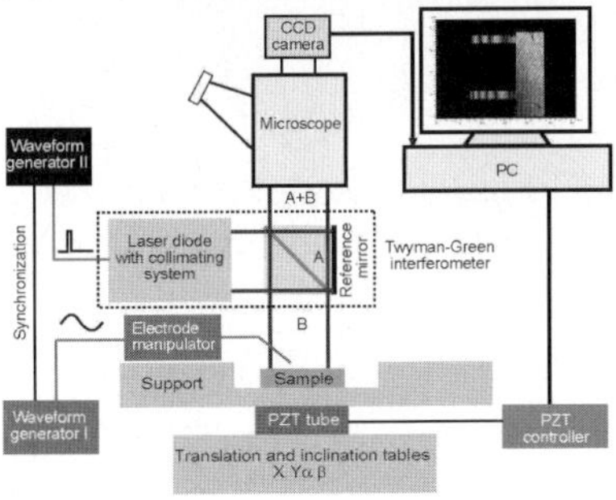

Fig.1. Scheme of the interferometric platform

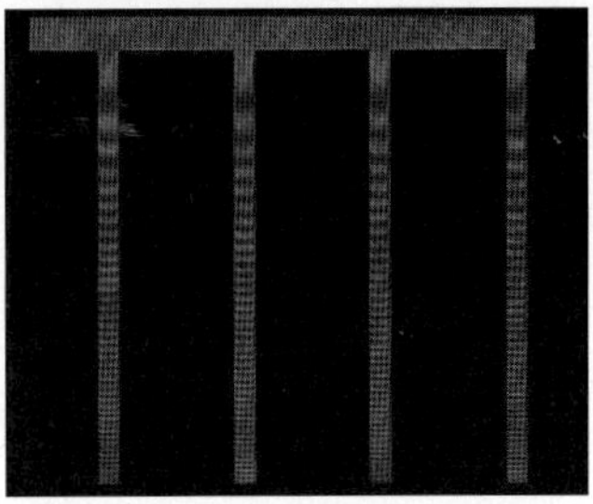

Fig.2. Interferogram from the measurement of the 700μm long cantilever's initial deflection

3. Results

The initially straight Si cantilevers after a deposition of metal electrodes and piezoelectric layer on them show bending caused by introducing an internal stress into thin films (Fig.3). The evaluation of the initial deflection from performed measurements can be used to estimate the magnitude of stresses and certain material properties and micromechanical features like an AlN piezoelectric coefficient d_{31}. Therefore, after the initial characterization of AlN microelements they have been thermally exposed for 87 hours at 130°C in the thermal chamber under vacuum, in order to avoid possible deterioration of the samples (metal oxidation). Then the characterization of the devices was carried out for the second time and the changes of their behavior were studied. The results of measurement for different length of cantilevers (200, 500, 700, and 800 µm) are presented in Table 1. In all cases of tested cantilevers the decreasing of initial deflection values has been observed. For proving the repeatability of this phenomena the exemplary results for cantilevers of 200µm length have been presented in Fig.4.

The vibration amplitudes of cantilevers were estimated by use of the interferometric platform with stroboscopic technique. The structures were brought to the vibration by applying an excitation sinusoidal signal with an amplitude of 130V and an offset of 140V. The results of this measurement for different length of cantilevers have been presented in Table 2.

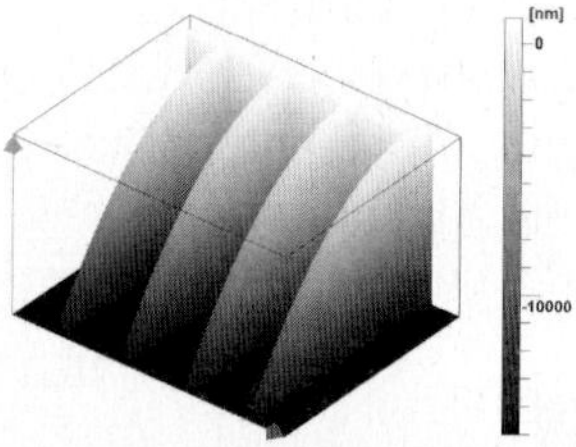

Fig.3 3D view of the initial deflection of 700□m long cantilevers

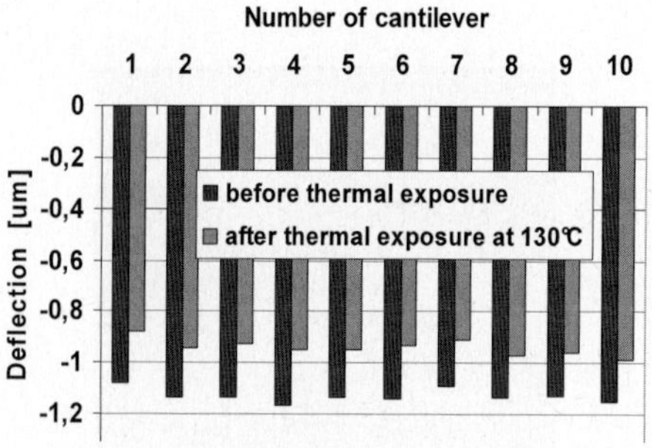

Fig.4 Results of deflection measurement for 200µm long cantilevers

Length [µm]	Initial deflection [□m] before	after T=130°
200	-1.13	-0.94
500	-3.66	-3.49
700	-13.48	-13.01
800	-11.93	-11.65

Tab.1 Initial deflection measurement before and after thermal exposure

Length [μm]	Initial deflection [μm]	Resonance frequency [kHz]	Amplitude of vibration [μm]
200	-0.94	563.56	0.34
500	-3.49	94.8	33.15
700	-13.01	41.48	19.45
800	-11.65	31.52	15.21

Tab.2 Results of vibration measurement

4. Conclusions

The applicability of the proposed measurement interferometric platform for the microelements characterisation, for the determination of operational behaviour of the MEMS/MOEMS systems and for the reliability study has been proved. Results of the thermal ageing investigation presented in this paper seem to show that temperature decreases the initial deflection of cantilevers releasing their internal stresses. In the future we will focus on ensuring more stable measurement conditions by using the thermal control module based on Peltier device for improving accuracy of measurement. Presented studies are the beginning of developing methodology which will bring in the future the evaluation of the AlN cantilevers lifetime depending on the environmental conditions what allow to improve their reliability by feedback to the design and to the fabrication process of these microelements.

Acknowledgments

This research was supported by European Network of Excellence in Microoptics (NEMO) and grant no N505 004 31/0670 of the Polish Ministry of Science and Higher Education.

References

[1] M. Clement, L. Vergara, J. Sangrador, E. Iborra, and A. Sanz-Hervas, Ultrasonics 42 (2004) 403-407.

[2] Ju-Hyung Kim, Si-Hyung Lee, Jin-Ho Ahn and Jeon-Kook Lee, Journal of Ceramic Processing Research, 3 (1) (2002) 25-28.

[3] M. A. Dubois, P. Muralt, Appl. Phys. Lett. 74 (20) (1999) 3032-3034.

[4] A. Andrei, K. Krupa, M. Jozwik, L. Nieradko, C. Gorecki, L. Hirsinger and P.Delobelle, Proc. of SPIE, vol. 6188 (2006).

[5] C. Gorecki, M. Jozwik, and L. Salbut. Journal of Microlithography, Microfabrication, and Microsystems, 4(4), (2005)

Comparison of mechanical properties of thin films of SiN_x deposited on silicon

M. Ekwińska , K. Wielgo , Z. Rymuza

Warsaw University of Technology,
Institute of Micromechanics and Photonics,
Św. A. Boboli 8,
Warsaw, 02-525, Poland.

Abstract

Nanoindentation studies of SiN_x layers deposited on silicon substrate with PECVD, PECVD LF and PECVD LF + HF methods were performed. During investigation Young`s modulus, hardness and energy needed to make plastic deformation of the material were established.

1. Introduction

When construction of the MEMS (Micro Electro Mechanical Systems) is taken into account the information about mechanical properties of the material is crucial for a designer. In micro and nanoscale this is a nanoindentation test, which is commonly used to investigate these features [1-6]. What is very important about this test is the fact that all information form the singular investigation of the material might be achieved within a minute. Because of that this test might be used in manufacturing of the MEMS as a check test. In this paper the comparison of results achieved during nanoindentation test on the SiN_x films deposited using different modifications of PECVD (tutaj trzeba wyjasnic te skroty: Plasma Enhanced Chemical Vapor Deposition, LF – low frequency, HF- high frequency, method of film deposition on the silicon substrate will be presented.

2. Experimental details

The investigations were carried out in clean room under laboratory conditions: temperature 22 ± 0.5 °C, humidity 40 ± 2 %, atmospheric pressure, air atmosphere.
The nanoindentation test was carried out with the use of nanoindenter TriboScope made by Hysitron Inc. (USA). For the measurements diamond Berkovich indenter (Fig.4, rysunki musza byc numerowane kolejno.) was used with the tip radius 50 nm. Measurements were made for the established depth of indentation: 50 nm.
During a test sharp tip (called later on indenter) is pushed with applied load into the sample surface. That causes plastic deformation of the material. As a result of the investigation the deformation mark (indent) on the sample surface can be observed and a force displacement curve is achieved.
From the unloading part of this curve (Fig.1.) mechanical properties of the material such as Young modulus and hardness of investigated material are established by Oliver and Pharr formula[6].

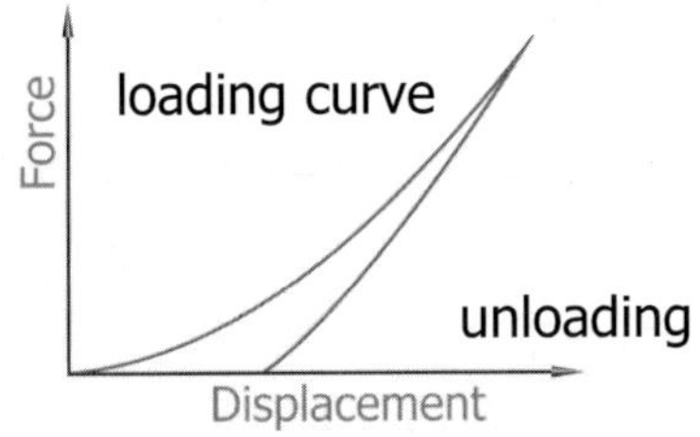

Fig. 1. Typical force –distance curves achieved during nanoindentation test.

From the same curve an information about energy needed to deform material plastically might be also achieved.
The energy E_{ii} introduced to the system in order to perform a test (Fig.2.b.) is the area under loading curve. The energy E_{io} which was refund by the system during the unloading process, in other words the elastic recovery of the material, is the area underneath unloading curve (Fig.2.c.). In order to estimate the energy E_{id} needed to plastically deform material on nano-scale during nanoindentation process (Fig.2.d.) the formula (1) was used, [6].
In this paper the comparison of results achieved during nanoindentation test on SiN_x films deposited on the silicon substrate will be presented.

$$E_{id} = E_{ii} - E_{io} \quad (1)$$

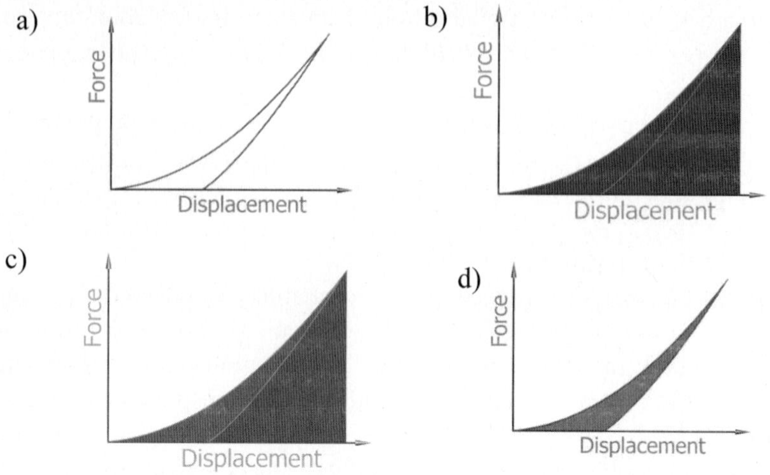

Fig. 2. The estimation of energy needed to realize a plastic deformation of the material, a) nanoindentation curve, b) energy introduced to the system, c) energy which was given back by the system during withdrowing process, d) energy needed to make a plastic deformation of the material

The samples under investigation were samples with SiN_x layer deposited on silicon substrate by PECVD method. Information about thickness of the films is presented in Table 1.

Table.1. Information about investigated samples

Sample denotation	Material	Thickness [nm]	Method of deposition
RRB_2B	SiN_x	230	PECVD LF
RRB_2C	SiN_x	660	PECVD LF
RRB_3B	SiN_x	230	PECVD LF + HF
RRB_3C	SiN_x	660	PECVD

3. Results and discussion

The results of the studies are presented in Tables 2 and 3 and in Fig. 4.

Table 2. Comparison of mechanical properties for all investigated samples, E – reduced Young modulus, ΔE – inaccuracy of estimation of reduced Young`s modulus, H – hardness of material, ΔH – inaccuracy of estimation of hardness of material, h– depth of indentation, Δh – inaccuracy of estimation of depth of indentation.

Sample denotation	Method of deposition	E_r [GPa]	ΔE_r [GPa]	H [GPa]	ΔH [GPa]	h [nm]	Δh [nm]
RRB_2B	PECVD LF	102.33	3.29	13.08	1.00	48.7	0.4
RRB_2C	PECVD LF	91.46	6.10	11.39	1.01	49.4	0.9
RRB_3B	PECVD LF + HF	56,62	3.4	8,5	0.7	47.7	0.5
RRB_3C	PECVD	87,76	5.2	9,8	0.8	47.1	0.7

Table 3. Energy needed to make plastic deformation during nanoindentation test, E_w – energy introduced into system in order to make indentation process, ΔE_w – inaccuracy of estimation of energy which was introduced to system, E – energy needed to make plastic deformation in material, ΔE – inaccuracy of estimation of energy needed to make plastic deformation in material, F_p – normal force during indentation process, ΔF_p – inaccuracy of estimation of normal force during indentation process.

Sample denotation	Method of deposition	E_w[μJ]	ΔE_w [μJ]	E [μJ]	ΔE [μJ]	F_p [μN]	ΔF_p
RRB_2B	PECVD LF	13.67	2.13	3.39	0.51	690.4	13.8
RRB_2C	PECVD LF	13.09	1.94	3.84	0.58	640.1	16.0
RRB_3B	PECVD LF + HF	8.12	1.12	2.25	0.35	381.0	11.4
RRB_3C	PECVD	10.40	1.56	2.99	0.44	533.9	11.2

SiN_x film deposited using simple PECVD method has similar mechanical properties such as reduced Young modulus and hardness only slightly smaller than for film with the same thickness but deposited using PECVD LF method (Table.2., Fig.4.). Because of that less energy is needed to deform layer made with PECVD (Table.3). In this case however the difference is significant.

As far as comparison of the properties of the films produced by PECVD LF and PECVD LF +HF methods is concerned differences in achieved mechanical parameters are more significant. The films with the same thickness deposited using PECVD LF + HF method have smaller hardness

and Young modulus than these which were produced using PECVD LF method (Table.2., Table.3.). Also the energy needed to deform plastically film deposited using PECVD LF + HF method is smaller.
What is important the energy needed to deform material plastically is nearly the same for the films with different thicknesses but all deposited by PECVD LF method

4. Conclusions

Even though mechanical properties for films deposited by PECVD method are just slightly poorer than for the films deposited using PECVD LF big difference can be seen in energy which is needed to deform material plastically. As far as PECVD LF and PECVD LF + HF films are concerned Young`s modulus, hardness and energy needed to deform material plastically are smaller for films deposited using PECVD LF + HF method. For the same method of layer creation thickness of the layer do not affect results strongly. Achieved results showed that there is a need to investigate materials with different thicknesses of the film and with layer made from different materials. The method for establishing energy needed to deform material plastically may also give qualitative information about elastic properties of the material (energy of elastic recovery).

Acknowledgement

The work was supported by Patent No.507255 project.

References

[1] B. Bhushan (ed), Nanotribology and Nanomechanics, Springer Verlag, Berlin 2005
[2] B. Bhushan, Introduction to Tribology, J.Wiley, New York 2002
[[4] Triboscope Nanomechanical Test Instrument, Operating Manual, Hysitron Incorporated, 1997
[6] A. C. Fischer – Cripps, Nanoindentation, 2nd edition, Springer Verlag, New York Berlin Heidelberg, 2004

Micro- and nanoscale testing of tribomechanical properties of surfaces

S.A. Chizhik (a), Z. Rymuza (b), V.V. Chikunov (a), T.A. Kuznetsova (a), D. Jarzabek (b)

(a) A. V. Luikov Heat and Mass Transfer Institute, National Academy of Sciences of Belarus,
15 P. Brovki St., Minsk, 220072, Republic of Belarus

(b) Warsaw Technical University, Poland, 8 A. Boboli St. Warsaw, 02-525, Poland

Abstract

The new techniques of micro- and nanotribometry, oscillation and rotation tribometry are discussed. These techniques are used to simulate real process in MEMS devices. The complete description of the methods and test device are included. The results of studies of the combinations of materials identical to those used in MEMS are discussed.

1. Introduction

One of the problems in designing micro- and nanomachines is the analysis of micromechanical and frictional properties of surfaces [1]. Difficulties in a development of methods of the analysis are stipulated by necessity of realization of movement on micro- and nanoscale, as well as of application and monitoring of small forces by the decrease in the probe influence scale to nanoscales.
The most common principle of realization of micro- and nanomovement is scanning by means of piezoelectric motors. Such type of movement is the basic in Scanning Probe Microscopy (SPM). Based on the given principle, sliding of the probe over a sample surface is achived in lateral force microscopy (LFM) [2]. The LFM techniques can be referred to the field of nanotribometry. The very small influence area (up to several nanometers) is caused by SPM probe with nanoscale radius of its tip.

The basic problems of LFM use for MEMS frictional properties characterization are slow speed of movement of the probe (which do not correspond to the actual one) and nanolocalization of the probe influence on the sample, which enables one to estimate the process of friction only within the limits of individual asperity.
For more adequate modeling of the MEMS surfaces sliding the new approaches in study of friction are presented: oscillation and rotation tribometry.

2. Methods of micro- and nanotribometry

Scanning lateral force tribometry consists in measurement of cantilever torsion during sliding of a tip at direct and inverse sliding over a sample surface. On the basis of these data, the frictional force is calculated as $F_{fr}=0.5k_t\,(dX1+dX2)$, where k_t is the spring constant of the cantilever subjected to torsion. Fig. 1 a gives the results of frictional force examination for polymeric coatings at sample heating which are obtained by the method of lateral force tribometry. Besides, the scanning lateral force tribometry can be used for modeling of the abrasion process (Fig. 1b).

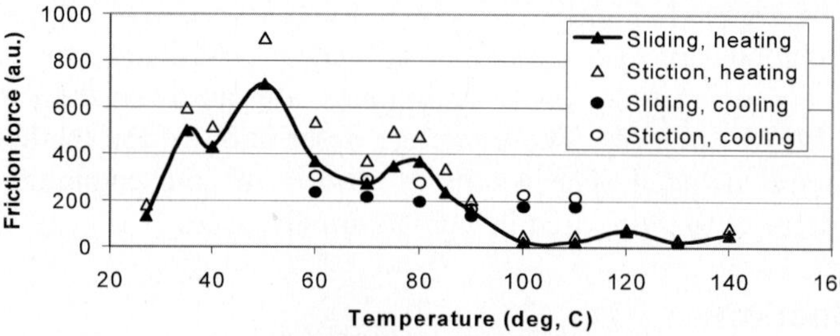

(a)

(b)

Fig. 1. Scanning lateral force tribometry: the scheme of measurement data interpretation (a); an example of measurement of the friction force on heating of a polymeric coating (b); an example of examination of abrasion of a hard surface by a diamond probe, when the probe load is increased for friction areas in direction A-B-C-D (c).

Oscillation microtribometry consists in oscillations at the resonance frequency of the probe paralell to the sample surface (Fig. 2a). The energy dissipation as a result of the friction leads to the change in the detected dynamic characteristics of a system. The frictional force can be calculated as $F_{fr} = \pi k\ (Ao\text{-}A)\ /\ (4Q)$ [3], where k is the spring constant of the oscillating cantilever, Q is the parameter of a quality-factor, A_o is the initial oscillation amplitude, A is the working oscillation amplitude. The results of using this method for characterization of different engineering surfaces are given in Fig. 2 b.

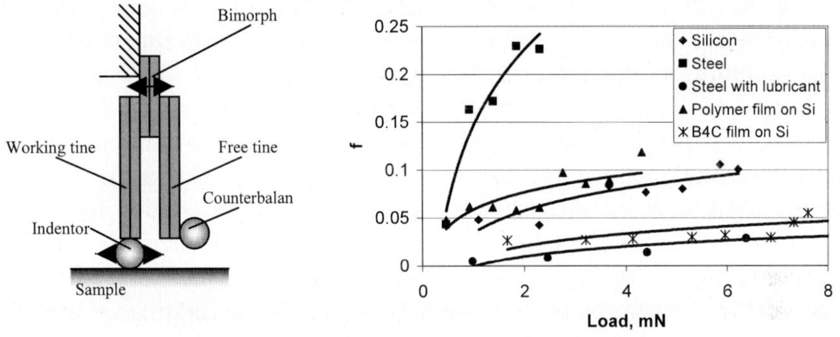

Fig. 2. Oscillation microtribometry: the scheme of measurement sensor (a); an example of measurement of the friction coefficient for engineering surfaces (b).

Experimental studies of the dependences of adhesion and friction forces were carried out by nanooscillations of the friction surface. It is shown that control of oscillation frequency and amplitude enables us to diminish severalfold the forces mentioned (Fig. 3).

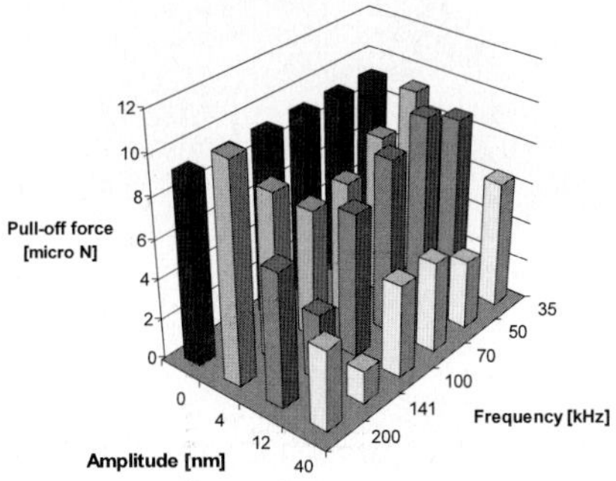

Fig. 3. Diagram of the dependence of Pull-off (adhesion) force between the stell-ball indentor and PMMA photoresist on the amplitude and frequency of the sample nanooscillations

Rotation tribometry consists in rotational movement of a micro- or nanoindentor. The movement radius can be adjusted and decreased down to tens of nanometers. Here, the area of the contact area–friction area overlapping on the sample approaches 100%. The analysis of the friction area is carried out by the SPM method, The method enables one to study the phenomena of local change in a material as a result of tribochemical reactions on the contact spots.

Friction of the silicon tip with a radius of 40 nm over the silicon surface was studied, when the latter is protected by a monomolecular polymeric layer of thickness of 2 nm. A 100 cycles of tip sliding were performed with an external load, whose value changes to several micronewtons. The rotation radius R of the tip relative to the chosen point on the sample was changed from several micrometers to 20 nm. The analysis of the friction zone by the SPM method showed substantial differences in the changes in the friction zone depending on the rotation radius of the indentor. For a greater rotation radius, the abrasive wear of protective coating to a depth equal to the coating thickness is revealed. For small rotation radius, beginning from 50 nm, where the contact and friction areas are comparable, a negative wear is seen (Fig. 4), i.e., an additional material arises on the friction area. An image of lateral forces shows high contrast (Fig. 4d) in the limits of the areas with a rotation radius of 50 nm, which is indicative of a chemical change in the material. The thickness of the layer changed is about 4 nm (Fig. 4b). These changes can be explained by the mechanochemical reaction of silicon oxidation that is due to high temperatures and shearing reactions in the friction zone.

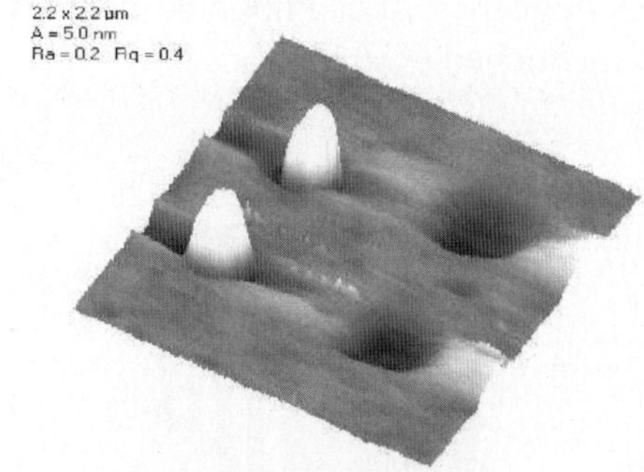

(a)

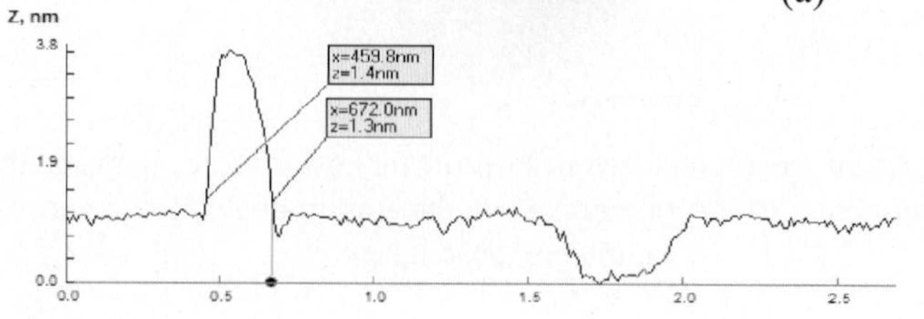

(b)

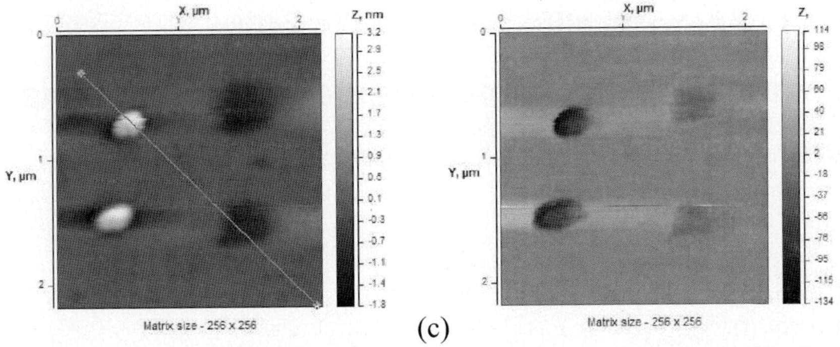

(c) (d)

Fig. 4. Rotation tribometry: 3D image of the friction zone, the left and right zones correspond to R = 200 nm and 50 nm (a); 2D image of topography (b); image of the lateral forces (c); topography profile along the trajectory shown in b) (d).

3. Conclusions

A set of nanotribometry techniques and examples of their use for studies of the MEMS surfaces is presented. It is shown that a combined application of lateral force, oscillation, and rotation tribometry techniques can characterize the friction surfaces on micro- and nanoscale most fully.

References

[1] B. Bhushan "Handbook on Micro- and Nanotribology" CRC Press, New York, 1995.

[2] E. Meyer, H. Heinzelmann, P. Grütter e.a., Thin Solid Films 181 (1989) 527.

[3] S. A. Chizhik, H.-S. Ahn, V. V. Chikunov, A. A. Suslov, Scanning Probe Microscopy (2004) 119.

Novel design of silicon microstructure for evaluating mechanical properties of thin films under quasi axial tensile conditions

D. Denkiewicz, Z. Rymuza

Institute of Micromechanics and Photonics, Faculty of Mechatronics,
Warsaw University of Technology, ul. Św. A. Boboli 8,
Warsaw, 02-525, Poland

Abstract

A new testing method to estimate mechanical properties of e.g. metallic thin films supported by a MEMS structure as a lever mechanism is developed. The MEMS support structure is chosen for coupling microspecimen with measuring macro-devices. The thin films microspecimens were designed in two shapes. The first one provides information about the Young's modulus from the tensile test, whereas the second used to the buckling test gives a value of the Kirchoff's modulus. Poisson's ratio can be estimated analytically. A FEM model was prepared to confirm the obtained analytical results.

Introduction

The analysis of existing test methods follows that values of the mechanical parameters are dependent from a particular measuring device. Moreover, there is not possibility to carry the microspecimens between different measuring devices.

The distinguishing feature of the proposed solution is possibility to standardize the measuring method; to realize the tests with the specimen it is possible to combine of the designed MEMS structure with many existing measuring systems (especially nanoindentation devices). The results of tests will provide an objective estimation of mechanical properties: Young's modulus, Kirchoff's modulus, Poisson's ratio, and fracture strain.

1. Structure configuration

A silicon chip has been designed utilizing the knowledge of MEMS devices microfabrication process (Fig.1).

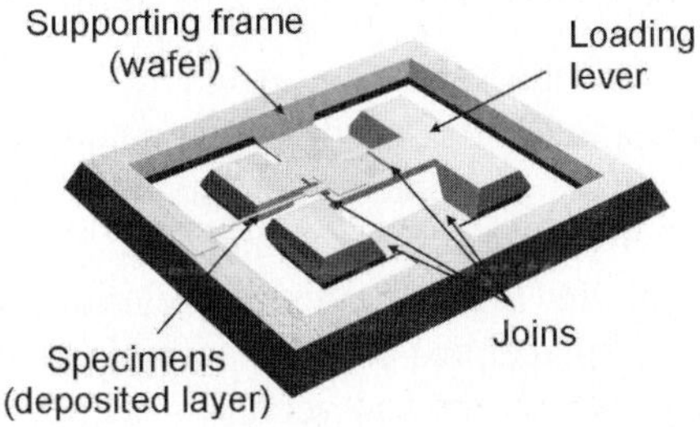

Fig. 1. Overview of chip etched in silicon substrate

The chip was designed to fulfil two fundamental functions: the silicon chip is a supporting frame preventing specimens against destroy and realizes an executive mechanism (a lever mechanism) to perform tensile test uniaxially on the specimen; the substrate with the chips is convenient to manipulate into a working area.

Typical lever mechanism described by Parszewski [1] consists of four links connected by joins (Fig.2).

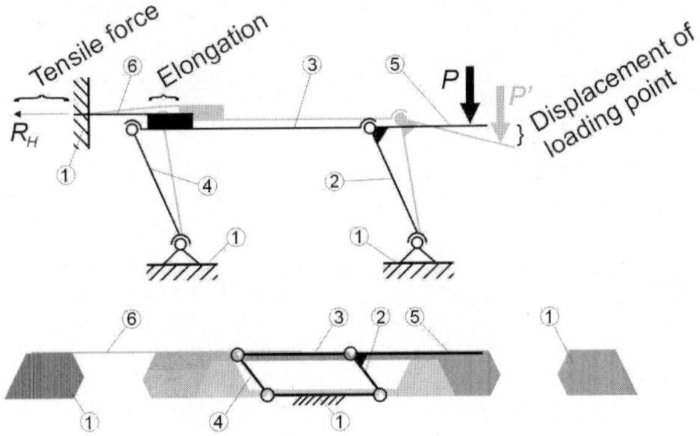

Fig. 2. Executive mechanism and cross section of silicon chip

Element 6 connected with link 3 and support 1 is the microspecimen. In this example the mechanism of driving link 2 (a winch) is extended by rigid link 5. Link 5 is a loading lever. It transfers an external vertical force P to the elements of mechanism to move them. The driving link is connected flexible to connecting link 3 and sway beam 4. Rotary movement of the driving link is changed to linear movement of the connecting link. The equal lengths of the winch and the sway beam assure a parallel displacement of the connecting link in relation to support 1. The tensile force R_H acting on the microspecimen 6 is quasi axial.

2. Testing method

The elaborated test algorithm bases on an energy balance of the mechanism. The energy accumulated in the lever mechanism is shared between joins and tensile element. The energy of the joins deformation can be measured experimentally after the lever mechanism will be released; the test element 6 will rupture. Fig. 4 represents relationship between the external force P and its displacement for two variants of scheme: first, original, when specimen 6 is present and second - without specimen. Hence, one is able to estimate participation of the spring energy accumulated in the specimen 6 as

$$E_{EB} = E_P - E_{WO} \qquad (1)$$

where: E_{EB} is the work of external force P equivalent to a spring energy of the deformed element 6, E_P - work of external force P at original configuration, E_{WO} - work of external force P at the second configuration.

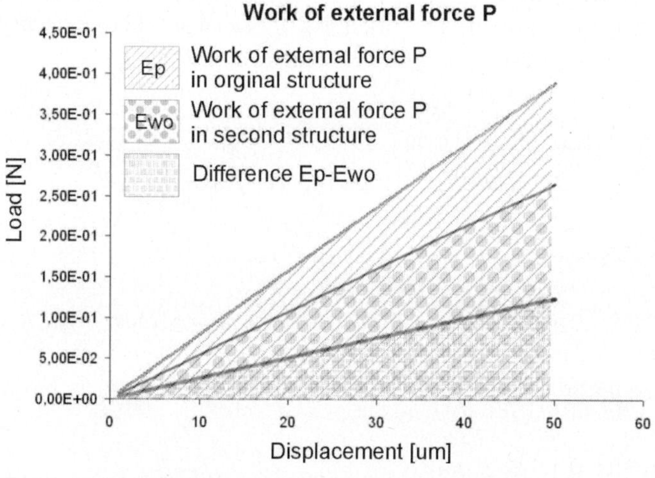

Fig. 4. Energy balance of the lever mechanism

The right side values of the equation (1) will be found experimentally. Knowledge of the articulated quadrangle mechanical characteristic gives a possibility to prepare the final graph of the accumulated energy in the test element (Fig. 5). Since, the simple calculation can be made to estimate the tensile forces R_H and the strain of the specimen. The strain is proportional to displacement of the connecting link.

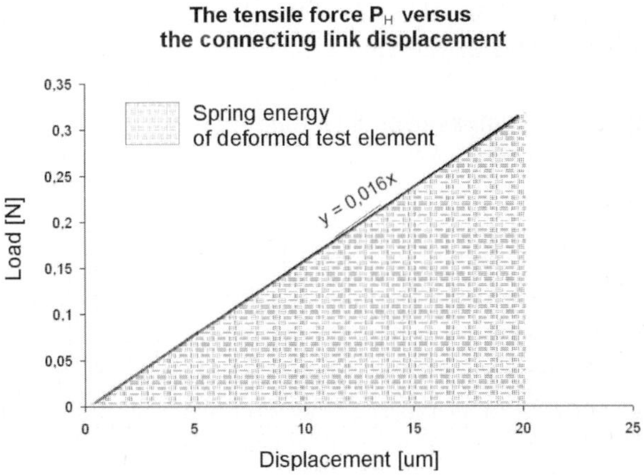

Fig. 5. Graphical representation of tensile force RH versus displacement of connecting link

The presented procedure is typical for estimation of the Young's modulus of the elongated elements. Fig. 6 shows the specimens of two different shapes that will be used to perform the tests. The first one (a) is a typical uniaxial test specimen, whereas the second (b) serves to make a buckling tests according to theorem described by Timoshenko and Gere [2].

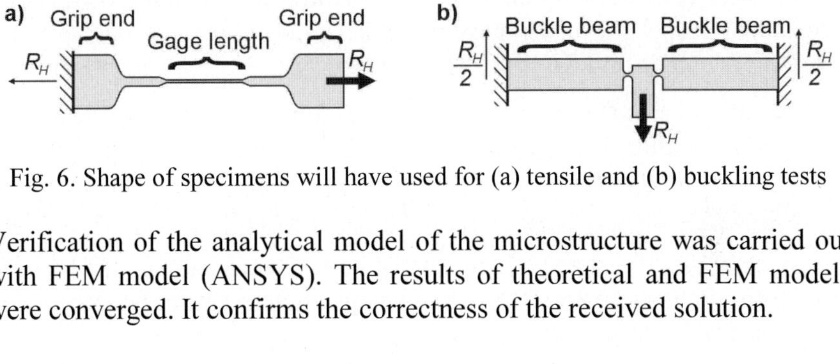

Fig. 6. Shape of specimens will have used for (a) tensile and (b) buckling tests

Verification of the analytical model of the microstructure was carried out with FEM model (ANSYS). The results of theoretical and FEM models were converged. It confirms the correctness of the received solution.

3. Fabrication process

The fabrication process has some advantages: it is possible to change kind of the evaporated metals, the evaporated metals are protected by the oxide masks against the etching processes (except HF oxide mask remover), and the specimens are released at the end of the fabrication process.

The assumed dimensions of uniaxial test specimen are: ~2 μm wide, <1 μm thick, and ~50 μm long. The silicon substrate is 400 μm thick and loading lever is 3 mm in length. A single silicon wafer can be used to prepare 180-300 structures showed in Fig. 1. They will be united in one silicon wafer. Hence, the statistic analysis will be possible. The structures can be tested using different measuring systems.

Conclusions

The first mechanical and simulation results confirmed the effectiveness of the designed structure. The structure has some advantages: there exists possibility to use the silicon wafers with chips at different measuring systems having vertical or horizontal movement of the measuring tip. The chip configuration eliminates the need for troublesome manipulation of the fragile specimen. Since, the adhesive force problem does not exist. Moreover, the specimen is not preloaded by forces before the tensile test; the test structure is a freestanding component and the microspecimen won't be preloaded.

The loading mechanism has a proportional translation of the loading force and displacement. Therefore, the uncertainty of translation depends upon the precision of the structure preparing process. If the length errors of the articulated quadrangle elements are less than 1% the mechanism transmission error is less than 3%. The loading mechanism stiffness and mechanical properties of the substrate material does not influence on values of the Young's and Kirchoff's modulus.

Owing to a difficult organisation of thin films torsion tests the Kirchoff's modulus is determined from the buckling test. Value of Poisson's ratio is out of reach in the physical experiment. It is estimated analytically.

By changing the shape of specimens the lever mechanism can be used to perform other mechanical tests like creep, fracture strain, fatigue endurance, and cracks propagation studies.

References

[1] Z. Parszewski: „Teoria mechanizmów i maszyn",WNT, Warszawa, 1974

[2] S. P. Thimoshenko, J. M. Gere: „Teoria stateczności sprężystej", PWN, Warszawa, 1963

Computer simulation of dynamic atomic force microscopy

S. O. Abetkovskaia (a), A. P. Pozdnyakov (b), S. V. Siroezkin (a), S. A. Chizhik (a)

(a) A. V. Luikov Heat and Mass Transfer Institute, National Academy of Sciences of Belarus,
15 P. Brovki St.,
Minsk, 220072, Republic of Belarus

(b) The Belarusian State University, 4 Nezavisimosti Ave.,
Minsk, 220030, Republic of Belarus

Abstract

Mathematical modelling of dynamic force spectroscopy is carried out and a character of dependences of AFM probe vibration parameters on surface properties is revealed. Simulation of scanning process in tapping mode is performed.

1. Introduction

Interest in dynamic mode of atomic force microscope (AFM) as a very promising method has not become relaxed. At the same time, AFM operators know about complexity of setting of this mode and about the dependence of quality of derivable AFM pictures on choice of the scanning parameters. The AFM probe behavior directly depends on the properties of a surface investigated with the use of the probe. Here mathematical modelling is very effective.

The work with computer models allows us to analyse and to understand the process of interaction of the probe with the sample surface. The possibility exists to give concerning the probes for researches, to justify selection of optimal conditions for scanning process and to give interpretation of phase contrast images [1]. We can use the simulation data for development of existent methods and creation of new ones for detection of material properties on micro- and nanolevels.

2. Simulation of dynamic force spectroscopy

In dynamic spectroscopy of materials, the vibration amplitude and phase shift of probe are recorded by AFM during approach of the probe to the sample surface and subsequent retraction. Changes in these parameters depend on the properties of sample material (elastic and adhesion properties).

The following mathematical model for describing interaction of the AFM probe with the sample surface is accepted for computations. Contact of the probe tip with the material sample is described by the Hertz model [2] and out-of-contact interaction implies the presence of attractive forces of the Van der Waals and short-range repulsive forces, which is represented by the Lennard–Jones potential [3]. The equation of motion of the AFM probe tip in the field of sample surface forces is written as

$$m\frac{d^2z(t)}{dt^2}+\frac{m\omega_0}{Q}\frac{dz(t)}{dt}+k(z(t)-z_{pos})=a_{\mathrm{bim}}k\sin(\omega t)+F(z(t)). \quad (1)$$

Here $z(t)$ is the vertical displacement of the tip; m is the microprobe mass; ω_0 is the intrinsic angular frequency of the probe; Q is the quality factor of the probe cantilever; k is the spring constant of the cantilever; z_{pos} is the position of the cantilever attaching point above the sample surface; a_{bim} is the oscillator amplitude; ω is the angular frequency of harmonic force; $F(z(t))$ is the force interaction of probe with the sample surface; the term $a_{\mathrm{bim}}k\sin(\omega t)$ represents the driving force.

Simulation of dynamic indentation is performed with two packages: Mathematica [4] and Simulink pack of Matlab environment. In the Fig. 1 are given the results of simulation of force spectroscopy. The following parameters of the tip-sample system were chosen for obtaining estimation dependencies: $k = 1$ N/m, $f = f_0 = \omega/(2\pi) = 100$ kHz, radius of tip curvature $R = 10$ nm, the Young modulus of tip material $E_2 = 179$ GPa, the Poisson's ratios of tip and sample materials $\nu_1 = \nu_2 = 0.3$, $a_{\mathrm{bim}} = 0.8$ nm, $Q = 100$. The Hamaker constant for modelling of dependencies for different sample materials (Fig. 1, a, c) is assumed to be equal to $0.1 \cdot 10^{-18}$ J. The Young modulus of sample material during variation of adhesion forces (Fig. 1, b, d) is equal to 0.1 GPa.

During varying sample material elastic properties (Fig. 1, a), significant influence of the material Young modulus on oscillation amplitude is observed. When the Hamaker constant was changed (Fig. 1, b, d), the amplitude curves are in close agreement in the region, where a prevalent force is one of elastic reaction of material. In the case of significant influence of adhesion forces in Fig. 1, a, b we can reveal transition between prevalence of adhesion forces and elastic reaction one. Amplitude curves in this case, as the phase shift curves, show stick-slip nature at the corresponding values z_{pos}. Sensitivity of phase shift to material properties is present. Not only quantitative changes of phase shift occur but a change in of interaction mechanism (repulsive and attractive interactions) as well.

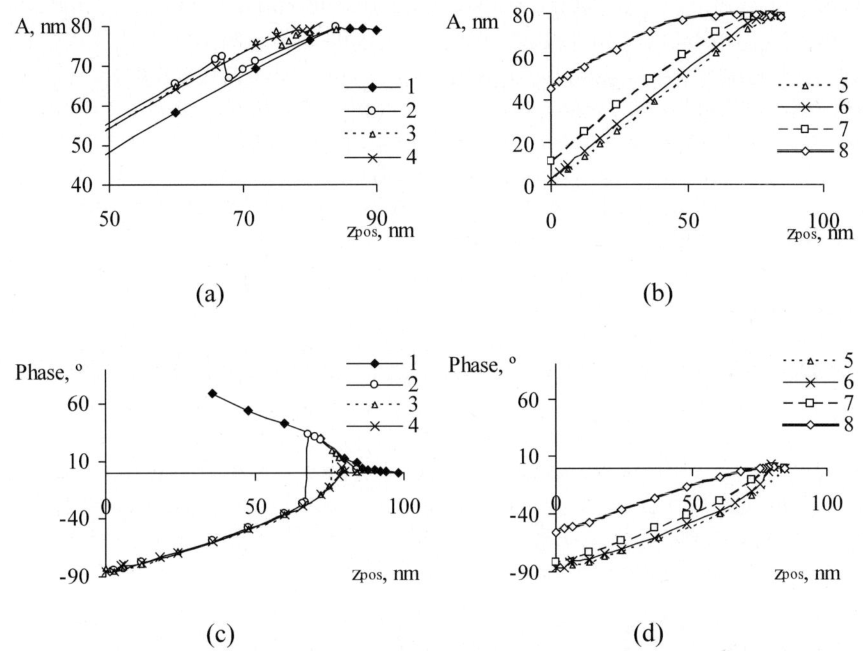

Fig. 1. Amplitude (a, b) and phase shift (c, d) of AFM probe as a functions oscillations distance between probe cantilever and sample surface: 1 – H = $10 \cdot 10^{-18}$ J; 2 – H = $3 \cdot 10^{-18}$ J; 3 – H = $1 \cdot 10^{-18}$ J; 4 – H = $0{,}1 \cdot 10^{-18}$ J; 5 – E = 1 GPa; 6 – E = 0,1 GPa; 7 – E = 0,01 GPa; 8 – E = 0,001 GPa

3. Simulation of AFM scanning process in intermittent mode

Simulation of scanning process is more complicated in comparison with modelling of indentation, since it assumes additional displacement of probe above sample surface. With horizontal shift of the sample under the probe (or the probe above the sample depending on a model of a specific microscope; this aspect in calculations is not of fundamental importance), the probe position above the surface level is changed as a result of real surface relief. Then adjustment of probe–sample distance is carried out according to information transfered from a feedback channel. In the intermittent (tapping) mode, operating amplitude of the probe oscillation as a feedback parameter is used, which is held constant by variation of tip–sample separation [5]. Analisys of scanning process further becomes complicated by the fact that feedback system responds not only to the relief changes but to the surface sample properties during movement from point to point of the sample also. One of the most significant material properties of the sample is the Young modulus of the material. At the same time, the need for a change in the probe level above the sample is caused by the fact that the probe oscillation amplitude in the tapping mode of interaction is related directly to the Young modulus of the sample material. The calculation results confirm this regularity (Fig. 2). Then, first of all, situation was modeled, when both the relief and the Young modulus of a surface are changed. Such situation is close to reality, this is the case of AFM researches of heterogeneous samples like polymeric composite materials.

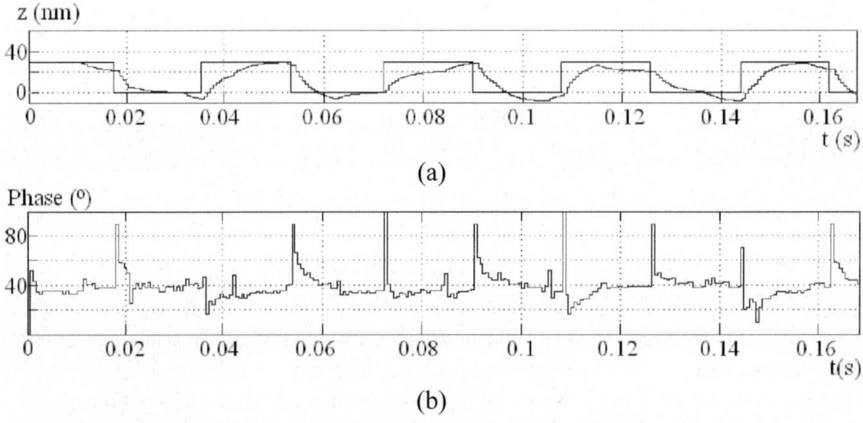

Fig. 2. Simulation of scanning process: (a) – 1 and 2 correspond to a surface relief and a profile of "scanned" topography image; (b) – profile of "scanned" phase shift image

Equation (1) is used to simulate scanning process with the package Matlab (Simulink) for describing AFM probe oscillations above investigated sample surface. The following model for probe–sample interaction is introduced: tip–sample contact takes place according to the Johnson–Kendall–Roberts theory [2]; Van der Waals and short-range repulsive

interactions are present, when a gap between the probe and the material surface [2].
Surface modelling is as follows: geometric relief shape of the surface is graded by analogy with real calibrating grate on the TGZ type. Line 1 in the Fig. 2, a is a simulated "profile" of such surface. Virtual sample properties are also chosen as varying: the Young modulus takes the values 0.1 and 20 GPa with a definite period. Velocity of displacement along the surface is 2 μm/s. Other parameters are the same as for section 2.
Line 2 in Fig. 2, a represents curve of tip–sample separation changes z_{pos}. The curve shows so far as surface topography is distorted during AFM scanning in dynamic mode at incorrect settings. In this case the feedback velocity of scanning is too high for given settings, which leads to operation at transient oscillation regime, when oscillation at a point has not managed to be stable.
Fig. 2, b demonstrates situation expected on the phase contrast image. Deviations on the phase shift curve made at points, where the Young modulus of the sample material changes but due to nonsteady oscillation regime there is a correlation of phase shift with a surface topography (line 1 in Fig. 2, b).

4. Conclusions

Sensibility of such parameters of AFM probe oscillations as the amplitude and phase shift depending on the investigated sample material is revealed. Sample topography has a distortions of in the case of incorrect setting of dynamic mode even without taking account of the influence of probe shape are shown. It is necessary to take into consideration changes of the Young modulus, when phase contrast images are interpreted.

References

[1] S. A. Chizhik, H.-S. Ahn, A. A. Suslov, A.V. Kovalev, C.-H. Kim, Physics of Low-Dimensional Structures 3/4 (2001) 39.
[2] А. И. Свириденок, С. А. Чижик, М. И. Петроковец "Механика дискретного фрикционного контакта" Навука і тэхніка, Минск, 1990.
[3] D. Sarid. "Exploring Scanning Probe Microscopy with Mathematica" John Wiley & Sons, Inc., New-York, 1997.
[4] S. O. Abetkovskaia, S. A. Chizhik, J. of Engineering Physics and Thermophysics 80/2 (2007) 173.
[5] R. García and P. Pérez, Surface Sci. Rep. 47 (2002) 197.

KFM measurements of an ultrathin SOI-FET channel surface

M. Ligowski (a, b) * , R. Nuryadi (a) , A. Ichiraku (a), M. Anwar (a), R. Jablonski (b) , M. Tabe (a)

(a) Research Institute of Electronics, Shizuoka University, 3-5-1 Johoku, Hamamatsu 432-8011, Japan

(b) Division of Sensors and Measuring Systems, Warsaw University of Technology, A. Boboli 8, Warsaw 02-525, Poland

Abstract

In this work, we investigate surface potential of the thin silicon-on-insulator field-effect-transistor (SOI-FET) by Kelvin Probe Force Microscope (KFM) at different temperatures. It will be shown that the surface potential changes in the range of temperatures from 15 K to 100 K, indicating increasing number of ionized dopants.

Introduction

As continuous miniaturization of silicon nano-devices is still in progress, there is an effort focused on single atom investigation. In particular, single dopant ionization is being a subject of interest at present [1-2]. However, it is desirable to find a direct observation method. Therefore, surface potential mapping of the device channel with the KFM in the range of low temperatures is very attractive. With this method direct dopant freeze-out observation seems to be possible.

KFM surface potential measurement

For measurements, a thin SOI-FET without the top gate was fabricated (Fig. 1). The Si substrate, which was used as a gate, was doped with boron ($1x10^{15}$ cm^{-3}), and whole top Si region (including channel) was heavily doped with phosphorus ($1x10^{18}$ cm^{-3}). Channel width was 2 µm and channel height was around 20 nm. Buried SiO_2 (BOX) layer thickness was 400 nm.

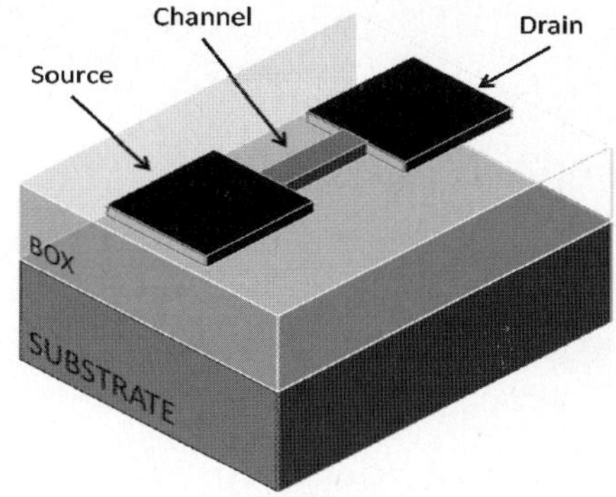

Fig. 1 Thin SOI-FET structure

The channel surface was investigated by KFM at different temperatures. Both source and drain were grounded, while back gate voltage (V_g) was changed in the range of -8 V to 8 V for each temperature. The temperature was changed from 15K until around 120K since in that range of temperatures the number of free carriers changes drastically [3].
In the resultant KFM images, we can observe a different potential distribution due to V_g changes. Furthermore potential distribution is different for different temperatures. Fig. 2 and Fig. 3 show examples of surface potential maps obtained by KFM at 117 K for V_g=-8 V and 8 V, respectively.

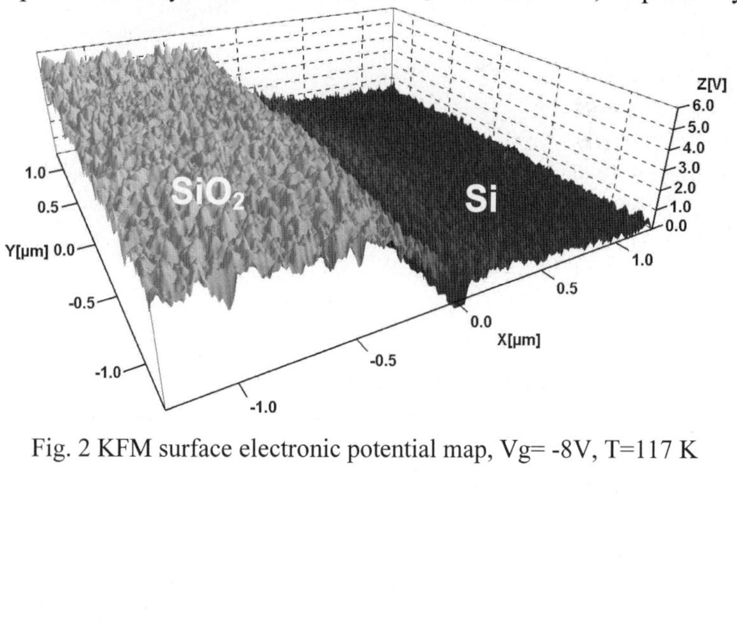

Fig. 2 KFM surface electronic potential map, Vg= -8V, T=117 K

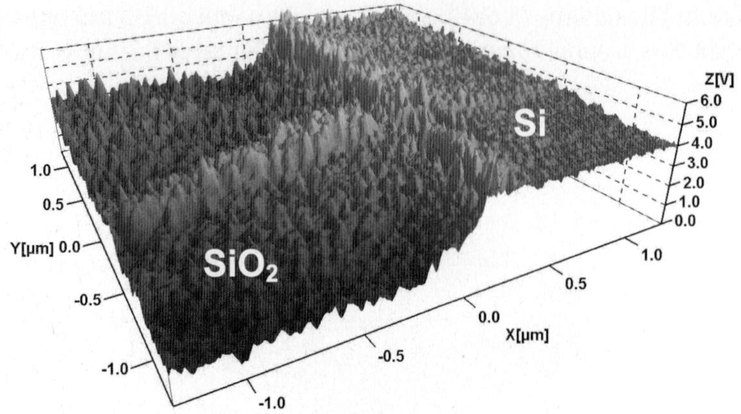

Fig. 3 KFM surface electronic potential map for Vg=+8 V, T=117 K

In Fig. 4, we can see the dependence of potential difference between BOX (SiO_2) and channel (Si) surfaces in the range of applied V_g at three different temperatures. It is noticeable that for negative V_g, with decreasing the temperature the difference in the potential between BOX and channel surface is also decreasing.

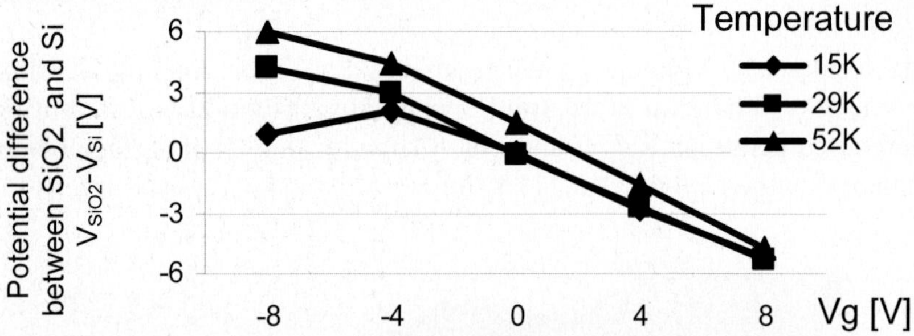

Fig. 4 Temperature dependence of potential difference between channel and BOX surface for SOI-FET with p^--Si substrate and n^+-Si channel

Possible model

The different potential distribution due to temperature changes can be ascribed to carrier freeze-out. In measured sample, both source and drain were grounded. As whole top Si layer was n^+ heavily doped, it behaved as metal and it was conductive even at low temperatures. Therefore, potential detected at the channel surface was independent of the potential in the underlying layers (BOX and substrate) in the whole range of measured tem-

peratures and remained constant at the ground level. However, potential near the channel at the SiO_2 surface was changing.
At high temperatures, we consider the p-type Si substrate to be conductive. Therefore, the potential in the substrate is almost flat and most of the voltage drop is present in the insulating BOX layer (Fig. 5).

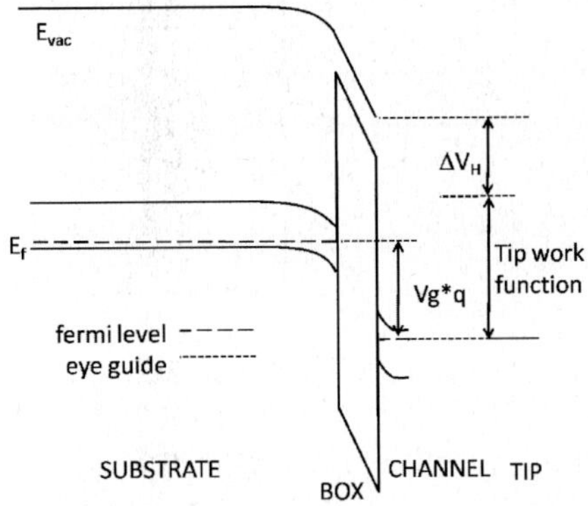

Fig. 5 Band diagram through the channel for high temperature

However, at low temperatures (Fig. 6), due to freeze-out effect, Si substrate becomes an insulator and the voltage drop is present not only in the BOX layer, but mostly in the p^- substrate.

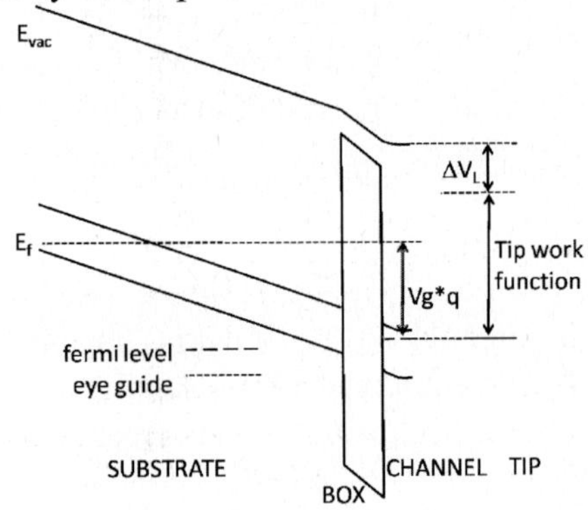

Fig. 6 Band diagram through the channel for low temperature

Consequently, at high temperatures electric field in the BOX layer is stronger than at low temperatures. Potential measured at the BOX surface (in some distance from the grounded channel) will be higher if the electric field around the channel is stronger (Fig. 7). On the contrary in low temperature, when the freeze-out effect occurs and the electric field is weaker, the potential difference between channel and the BOX surface is smaller (Fig. 4).

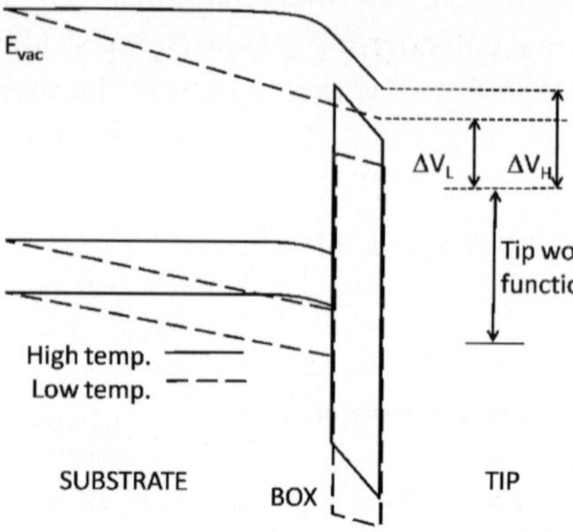

Fig. 7 Band diagram close to the channel for low and high temperature

In order to confirm our model, we have investigated another thin SOI sample with heavily doped n^+-Si substrate ($1x10^{18}cm^{-3}$) and p^--Si channel ($1x10^{15}cm^{-3}$). During measurements, substrate was grounded and channel was kept floating. For this sample, surface potential did not change when the temperature was increased from 14 K to 140 K. Comparing those two samples, we can conclude that potential difference due to temperature changes, observed in Fig. 4 is caused by freeze-out effect. This result and interpretation are consistent with the previous freeze-out effect studies. However, direct observation of freeze-out effect hasn't been reported yet. We believe this is a significant progress towards single dopant observation.

Conclusions

We have investigated potential on the SOI-FET channel surface by KFM at low temperatures. Surface potential difference has been observed due to temperature change, directly indicating dopant freeze-out release.

Acknowledgements

This work was partially supported by MEXT KAKENHI (16106006 and 18063010).

References

[1] Y. Ono et al., Appl. Phys. Lett., 90, 102106 (2007).
[2] Z.A. Burhanudin at al., SNW, Kyoto, June 2007
[3] S.M. Sze, Physics of Semiconductor Devices, Wiley 1981

The improvement of pipeline mathematical model for the purposes of leak detection

A. Bratek (a), M. Słowikowski (a), M. Turkowski (b)

(a) Industrial Research Institute for Automation and Measurements, Al. Jerozolimskie 202, 02-486 Warszawa, Poland

(b) Warsaw University of Technology, Institute of Metrology and Measurement Systems, ul. Boboli 8, 02 525 Warszawa, Poland

Abstract

The paper presents the improvements of the mathematical model of liquid flow dynamics in long distance pipelines. The model was formulated on the basis of the real pipeline system as a result of research concerning the leak detection and localization algorithms. The model takes into account the liquid pressure and velocity in the pipeline, but also the impact of the others important system elements such as pumps, valves and a receiving tank. This allows more precise detection and localization of the pipe leaks.

1. Introduction

The aim of the research was to improve the pipeline system model developed for the purposes of leakage detection and localization. The modelling of liquid flow dynamics for long distance transfer pipelines is essential for the analysing of physical phenomena that occur in the pipeline, which are extremely difficult to be detected and examined without being supported with mathematical and computational methods.

The mathematical model presented in the paper enables the simulation of various events related to liquid transportation, such as start up or shut down the pumping, switching inlet or outlet of the pipeline from one tank to another, liquid transport in steady state conditions, stand-by conditions and the situations when two or more media of diverse physical characteristics are being transported subsequently through the same pipeline.

The accurate mathematical model describing the real pipeline in all technological situations is of greatest importance for the analytical methods of the leak detection and localization. Existing models, however, rarely take into account other than the pipe elements of the system.

2. Short description of a pipeline transport system

Figure 1 presents the simplified scheme of the pipeline transport system [1]. A liquid is drawn from the supply tank T_A by the main pump 2 and the auxiliary pump 1, and pumped through a pipeline to the receiving tank T_B. The tanks are filled to different levels, with products of various physical characteristics. Remotely controlled gate valve stations are installed along a pipeline every several kilometres.

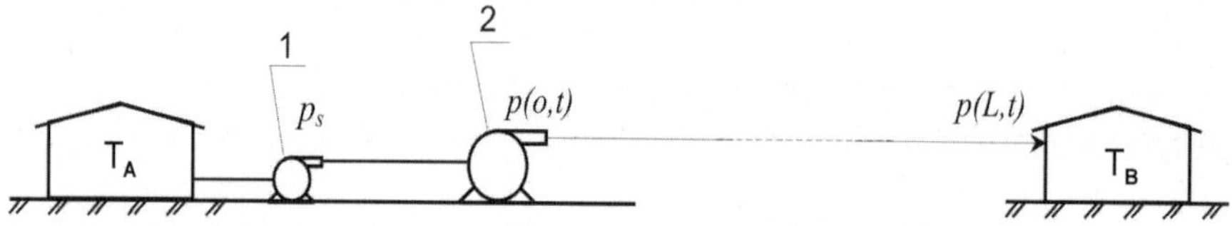

Fig. 1. Simplified scheme of a pipeline transport system

Each tank can be switched from one tank to another in the both supply A or receiving B tank set, which may result in a sudden disturbance of pressure at the pipe input or output.

3. The model of the pipeline

For the modeling purposes the pipeline was arbitrarily divided into sections at points where measuring transmitters were installed and each section between these points was divided into shorter, equal segments.
Every segment fulfils a set of partial differential equations as a result of the law of mass and momentum conservation. In the case of a leak-proof pipeline these equations, according to [4] can be written as

$$\frac{\partial w(x,t)}{\partial x} + \frac{1}{E}\frac{\partial p(x,t)}{\partial t} = 0 \tag{1}$$

$$\frac{\partial p(x,t)}{\partial x} + \rho(x)\frac{\partial w(x,t)}{\partial t} = -\rho(x) g \sin\alpha - \frac{\lambda(x)\rho(x)}{2d} w(t)\left|w(t)\right| \tag{2}$$

where E is the elasticity coefficient of a liquid-pipeline system, $\lambda(x)$ is the pipe friction factor and α - the angle of inclination of a pipeline segment. The mathematical model should however also reconstruct as accurately as possible the static and dynamic characteristics of all other elements of a pipeline system as well as interactions between these elements.

4. The model of primary pump

The static characteristics of the pump for petrol of the density 755 kg/m^3, was approximated with reference to the pump catalogue data as

$$\Delta H(q) = -2{,}6499\cdot 10^{-6} q^3 + 0{,}73238\cdot 10^{-3} q^2 - 0{,}14757 q + 340{,}95 \quad (3)$$

where $\Delta H(q) = H_T - H_S$ is the increment of a liquid column between pump suction and force sides (in m) and q is the value of the volumetric flow rate (in m^3/h).
In the pump nominal working conditions (200 m^3/h $< q <$ 300 m^3/h) the relative error of the approximation (3) is less than 0.2%. The pressure directly behind the pump is given by

$$p(0,t) = p_S(t) + \rho g \Delta H(q) \quad (4)$$

where $p_S(t)$ is the pressure at the auxiliary pump outlet, ρ is the density of the pumped medium and g is the acceleration due to the gravity.
Considering the liquid delivery, the pump can be treated as a non inertial object. The pump reaches its nominal speed in 5-7 s while the volumetric flow rate $q(0,t)$ achieves more than 80 % of its steady state value. Later variation of $q(0,t)$ is implied by the pipeline dynamics and back coupling delay in flow influence on the pressure behind the pump. Disregarding the delay reason, the transmittance obtained experimentally in form

$$G(s) = (Ts+1)^{-2} \quad (5)$$

was included into the model to relate $q(0,t)$ with $\Delta H(q)$ in the equation (3).

5. The model of receiving tank

The real pipe installation has been equipped with the outlet tank of diameter D_T = 56 m, height of 13 m with a floating roof causing small constant

overpressure above the liquid surface. The liquid flows to the tank or from the tank through manifolds installed in the bottom part of the tank.
The change of the liquid level in any tank connected to a pipeline results in a change of the pressure at either the input or the output of a pipeline, introducing in this way a disturbance to the transportation process. The change of the liquid level rate is given by

$$\frac{\mathrm{d}H_T}{\mathrm{d}t} = 3600\left(\frac{D}{D_T}\right)^2 \cdot w(x_T, t) \tag{6}$$

and the pressure variation rate at the pipeline output is given by

$$\frac{dp_T(t)}{dt} = \rho(x_T, t) g \frac{\mathrm{d}H_T}{\mathrm{d}t} \tag{7}$$

where D is the pipe inner diameter (m), x_T is the pipeline length (m) .
The liquid velocity in a pipeline, $w(t)$, is less than 1,1 m/s, then the maximum change rate of the liquid level in a tank is smaller than 0.12 m/h, so the maximum change rate of the pressure at the output is about 1000 Pa/h. However, switching the fully filled tank to the empty one causes strong disturbances, and it generates a sudden pressure jump at the output side of the system. This change can achieve even $\Delta p_T = \rho g \Delta H_T \approx 0.11$ MPa.

6. The model of gate valve

The pins of the real pipeline valves are moved by electrical drives with constant linear movement speed. The average time period of switching between one to another extreme position was about 150 s.
The valve pressure loss coefficient z is defined as the ratio of the pressure drop across the valve and the total kinetic energy of a flowing medium:

$$z = \frac{2\Delta p}{\rho w^2} \tag{8}$$

For numerical simulation the following relation was used

$$w = \sqrt{\frac{2}{z}} \cdot \sqrt{\frac{\Delta p}{\rho}} = K(x) \cdot \sqrt{\frac{\Delta p}{\rho}} \tag{9}$$

where z is the pressure loss coefficient, Δp is the pressure drop across the valve (Pa), w is the flow velocity of the liquid (m/s). $K(x)$ is the coefficient related to the valve's aperture $x=H/D$, H is the linear shift of the valve pin and D is the pipeline inner diameter. It was also assumed that $K(x) = K_Z \cdot \varphi(x)$, where $\varphi(x)$ – is the ratio of a valve's clearance surface to the cross section of the pipe given by the formula

$$\varphi(x) = \frac{1}{\pi}\arccos(1-2x) - \frac{2}{\pi}(1-2x)\sqrt{x-x^2} \tag{11}$$

The value $K_Z = 0.45$ was calculated from pressure balance along a pipe.

7. Conclusions

The inclusion of the pump, valves and tanks to the mathematical model enabled the increase of the performance of the leak detection system. It allows to detect the position of the leak with accuracy ±(100 – 200) m for steady flow and during transient process (pumping start up, switching the outlet tank, changing the medium being transported) with accuracy ±(200 – 400) m. It must be underlined that for unsteady flow the existing systems [2, 3, 4] cannot detect the leak or they generate false alarms.

Unfortunately some parameters were not measured: the density of the liquid ρ, the pressure at the primary pump inlet p_s and the pressure just before the recieving tank p_h. There is no doubt that these measurements would increase the quality of the model even more.

The paper is a result of research financed as a research project within the scope of the Multi-Year Programme PW-004 "Development of innovativeness systems of manufacturing and maintenance 2004-2008".

References

[1] *Michałowski Witold S., Trzop S.*: „Rurociągi dalekiego zasięgu", wydanie IV, Fundacja Odysseum, Warszawa, 2005

[2] Bilman, L. Isermann R.: „Leak detection methods for pipelines", Automatica, vol.23, no. 3, 1987, Pages. 381-385

[3] *H. Siebert*: "Dynamische Leckuberwachnng bei Pipelines", Erdol Ergas Kohle, 116 Jahrgang, Heft 11, November 2000

[4] *Sobczak R:* „Detekcja wycieków z rurociągów magistralnych cieczy", Nafta, Gaz, nr 2/01, pp. 97-107.

Thermodynamic Analysis of Internal Combustion Engine

M.Sc. David Svída

Brno University of Technology
Fakulty of Mechanical Engineering
Institute of Automotive Engineering
Technická 2
Brno, 616 69, Czech Republic
svida@fme.vutbr.cz

Abstract

Development in modern combustion engines presents high challenges for designers and requires complex experimental and computational methods in development phase of new power train. This paper presents compact tools for computational and experimental thermodynamic analysis of internal combustion engines. A modern single zone heat release model is used for the computational thermodynamic analyses. This new methodology is applied for modernized subject Theory of internal combustion engines at Brno University of Technology, Institute of Automotive Engineering

1. Combustion analysis

Cylinder pressure versus crank angle data offers the developer a crucial insight into the combustion phenomena occurring within internal combustion engines. The combustion analysis process is based on the simple analysis of pressure / volume data. Cylinder pressure changes varies with crank angle due to the following phenomena:

- Cylinder volume change
- Combustion
- Heat transfer to chamber walls

The figure Fig. 1 show 'log cylinder pressure versus log cylinder volume' graphs for a typical automotive SI engine.

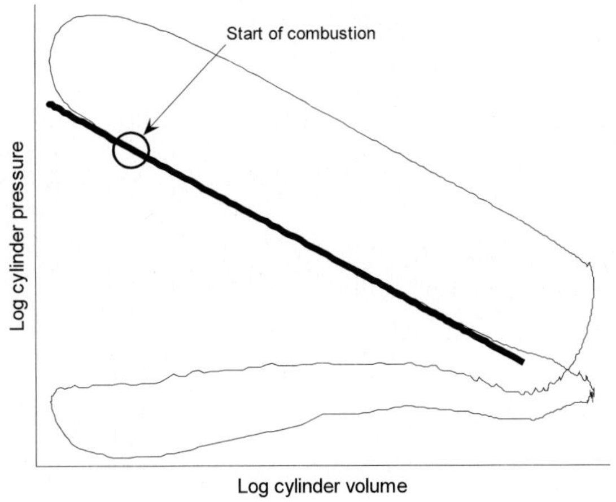

Fig. 1. Log cylinder pressure versus log cylinder volume and theoretical polytrophic pressure (thicken line)

More extensive studies show that the compression and expansion processes are well fitted by a polytrophic relation:

$$p v^{n} = \text{constant} \tag{1}$$

The exponent n for the compression and expansion processes is 1.3 (+/-0.05) for conventional fuels.

Log p – Log V plots as shown above, approximately define the start and end of combustion, but do not provide a mass fraction burned profile. One well-established technique for estimating the mass fraction burned profile from the pressure and volume data is that developed by Rassweiler and Withrow [3].

The method attempt to find out when the energy in the fuel is released. Between two samples, a crank angle interval, there is a pressure change Δp. The pressure change originates in either volume changes, Δp_v, or energy release from the fuel, Δp_c.

$$\Delta p = \Delta p_c + \Delta p_v \tag{2}$$

The pressures and volumes at the start and end of the interval, i and j, in the absence of combustion, are related by

$$p_i V_i^n = p_j V_j^n, \tag{3}$$

which gives the corresponding pressure change, due to volume change:

$$\Delta p_v = p_j - p_i = p_i \left[\left(\frac{V_i}{V_j} \right)^n - 1 \right]. \tag{4}$$

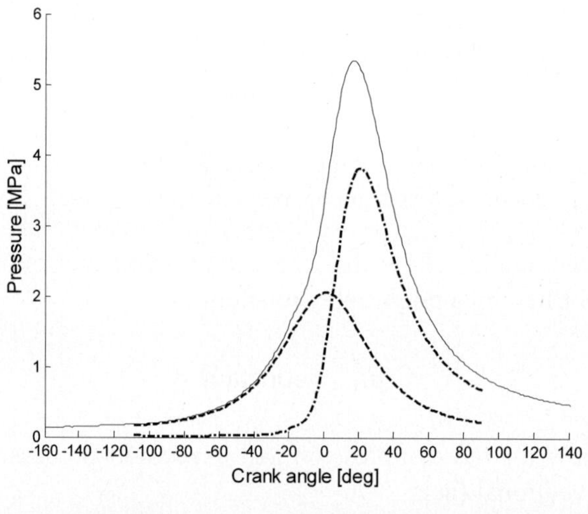

Fig. 2. A cylinder pressure, theoretical pressure (dashed line), pressure from combustion (dash dot line)

2. Combustion models

For combustion description is used a single zone heat release model [1]. This means that during combustion the heat released is used to heat the whole of the combustion space. The main implication of this assumption is that the bulk gas temperature is generally lower than the core combusted gas temperature behind the flame front. This may have an effect on detailed in-cylinder heat transfer, however since the semi-empirical heat transfer models make gross assumptions regarding heat transfer coefficient and wall temperature the effects of this assumption are small. The Viebe function define the mass fraction burned as

$$m_{frac} = 1 - e^{-A\left(\frac{\theta}{\theta_b}\right)^{M+1}}, \tag{5}$$

where

A	is	coefficient in Vibe equation (unburned fuel fraction)
M	is	coefficient in Vibe equation (burn process)
θ	is	actual burn angle (after start of combustion)
θ_b	is	total burn angle (0 – 100% burn duration).

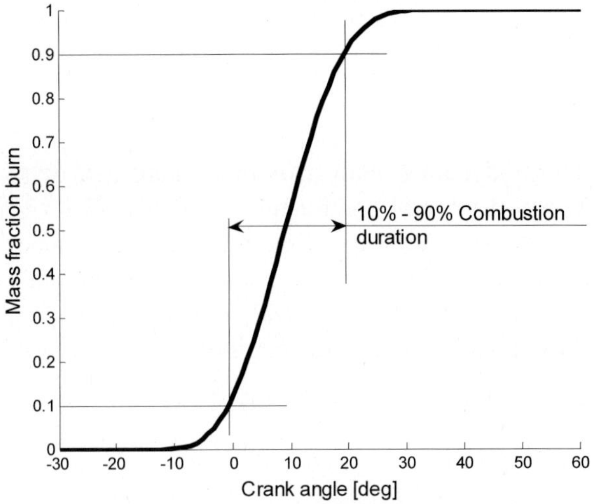

Fig. 3. Viebe function of mass fraction burned (M = 2, A = 10), 10% - 90% combustion duration

3. Measurement technique

It's necessary to take very precision measurements to indicate engine and internal combustion parameters (air/fuel ratio, cylinder pressure, water temperature, intake and exhaust temperature). For this measurements is useful to use multifunctional wide bend oxygen sensor unit and precision multichannel analyzer.

For analyzing is used Matrix analysis program where for combination engine revolutions and manifold pressures obtain actual air/fuel ratios. After configuration of fuel injectors we get input data for engine digital control unit.

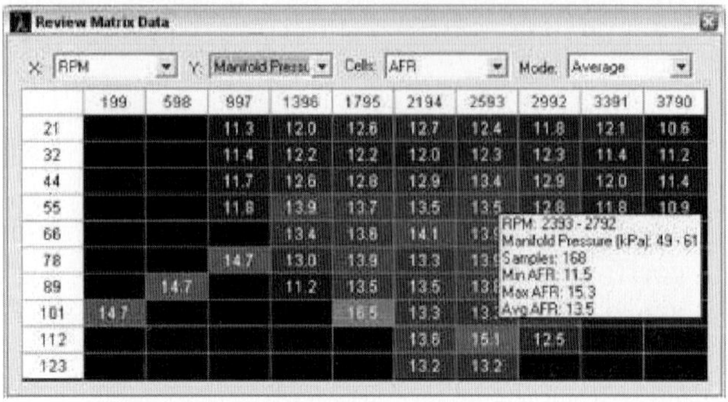

Fig. 4. Measurement data analyzer

4. Conclusion

The presented thermodynamic analysis is only small part of new modernized subject Theory of internal combustion engines at Brno University of Technology, Institute of Automotive Engineering. This subject make possible the students to obtain the knowledge of theory of internal combustion engines. Provide knowledge for practical, theoretical, experimental and scientific activity in area of combustion engines.

5. Acknowledgement

The problems mentioned above are being solved within the framework of the project No. 2029/2007 which has been given by the fund FRVŠ of the Czech Republic. The author would like to thank FRVŠ □R for the rendered assistance.

6. References

[1] V. Píšt□k, J. Št□tina "Výpo□etní metody ve stavb□ spalovacích motor□" Nakladatelství VUT v Brn□, 1991, ISBN 80-214-0368-3
[2] Inaudible Knock and Partial-Burn Detection Using In-Cylinder Ionization Signal, SAE technical paper 2003-01-3149
[3] Theory of internal combustion engines, Norfolk, Lotus Engineering 2001

Extraction and liniarization of information provided from the multi-sensorial systems

Eugenie Posdarascu, Anca Gheorghiu *

* “Hyperion” University of Bucharest, Electronics and Computers Faculty, 169 Calarasilor Way, Bucharest, Romania

Abstract

The performances of a multisensorial system are given by the fundamental and technological characteristics of the sensors, as parts of the system, no matter if these are conventional sensors, matrix sensors or intelligent sensors. This paper tries to make possible the taking over a part of the signal conditioning modules or acquisition boards functions into a temperature measuring system which collect the signals from varia types of temperature sensors with nonlinear characteristic using intelligent software techniques based on certain mathematical and physical models. The implementation of these software techniques is made using the programme package LabView of the company National Instruments.

1. Introduction

The technic-scientifical progress in microelectronic and informatics field has permitted to the industrial units to endow with modern systems, able to monitorize the function of the installations or industrial equipments, to carry out the evolution of a process by measuring in real time of the significant parameters of the process, concomitantly with the taking of functioning decisions. Sometimes certain interest parameters need to be the result of the measuring of a parameter in a lot of different points. For this reason, modern measuring and control systems are used and the multisensorial systems, as primary components of a data acquisition intelligent system from inside, have an important role.

The used soft instruments are as important as the hard components of the measuring system. The intelligent acquisition systems made nowadays by the well-known companies means the existence of a well cooperation between the hardware and software parts as a system. A lot of functions of the hardware components will be taken over by the software components when need. This situation is useful especially when an intervention during the measuring process is obligatory, in order to make some corrections (on-line). Using some software algorithms, the functions necessary to be fulfilled by the signal conditioning modules from the measuring systems way can be reduced, these elements being responsible eventually only of the filtering and adaptation of the signal level the entrance in the acquisition boards. The results of these practice is the lower cost of the measuring system reported to the obtained quality, result which can be foreseen by a rigorous design of the system.

2. Discrete temperature sensors

The temperature sensors are generally devices with non-linear transffer characteristic. Depending on the nature of the phenomenon which define a process and its carrying out conditions, various sensors are used to determine the temperature. The most known types of discret sensors are thermoresistences, thermistors, themocuples, semiconductor sensors and variuos optic and acustic sensors.

3. Mathematic modelling

The polynomial model is a satisfactory solution. Depending on precision asked by us for the proposed measuring, we can choose polinoms of 2,3 or more orders. The most used mathematic models, which describe the above-mentioned sensorial structures, are:

- model RTD: $R(T) = R_0\left(1 + AT + BT^2\right)$, with $R_0 = 100\,\Omega$, at $T_0 = 0^o C$; (1)
- model KTY: $R(T) = R_0\left(1 + A(T - T_0) + B(T - T_0)^2\right)$, with $R_0 = 1\,\mathrm{k}\Omega$, at $T_0 = 25^o C$; (2)
- model thermocouple: (TC): $U(T) = U_0 + AT + BT^2$; (3)
- model integrate (SI): $U(T) = U_0 + AT$; (4)
- model thermistor (TM): $\dfrac{1}{T} = A + B \cdot \ln R(T) + C \cdot \ln^3 R(T)$. (5)

The general model is considered as:

$$Y(T) = Y_0 + A\alpha(T) + B\beta(T) \tag{6}$$

From the above-mentioned formula it results that in measuring of a temperature it is involved o variable T (parameter to be measured) and a function Y (T) named measured parameter. The coefficients A, B and C are determined experimentally, so to find out the temperature is enough to know the Y (T) and a mathematic model.

It can be checked that the above first four models are easily according to the general model. To describe as exactly as possible a characteristic $T \rightarrow Y(T)$ it is necessary to know the parameters Y_0, A and B. The models $\alpha(T)$ and $\beta(T)$ are functions which reproduce an etalon model; the precision of the standard for our thermometers depends of characteristic of these. The three parameters are determined making three experimentals with the following results:

$$\begin{cases} Y_1 = Y_0 + A\alpha_1 + B\beta_1 \\ Y_2 = Y_0 + A\alpha_2 + B\beta_2 \\ Y_3 = Y_0 + A\alpha_3 + B\beta_3 \end{cases} \quad \text{or} \quad [Y] = [T_E] \cdot [P] \tag{7}$$

To solve this system means to determine the vector $[P]$ corresponding to the parameters, which describe the respective characteristic:

$$[P] = [T_E]^{-1} \cdot [Y] \tag{8}$$

Finally, to determine the temperature means to determine $T = Y^{-1}$. The extraction of T it could be made for the first models by extracting firstly of $\alpha(T)$ and then by $T = \alpha^{-1}$.

The relations (7) describe the dependence of a measured parameter of the models $\alpha(T)$ and $\beta(T)$ from the matrix $[T_E]$ through the parameters Y_0, A and B. This dependence can be found in the model of thermistor Steihard (5) by the physical changing of the roles: Y (T) becomes etalon function and $\alpha(T)$ and $\beta(T)$ functions depends of the measuring process. The result is the same, means the determination of the coefficients of the chosen model for calibration - relation (8).

The procedure can be extended for the calibration simultaneously of a multisensorial system discrete or integrate, composed of temperature sensors provided by the five above models. The sensors, components of the multisensorial system, can be part of the same model or can be from different models, the results of the experiences having the same form from the mathematic point of view. In this case the vectors [P] and [Y] are transformed in matrix with a dimension given by the number of sensors of the multisensorial system, the relation (7) and (8) being adapted to the new situation.

4. Hardware structure

The hardware system for calibration and linearisation is shown in the figure no.1. The calibration and linearisation procedure of some temperature sensors or a multisensorial system means to read in the same time the information taken over from these and also the information given by the etalon transducers.

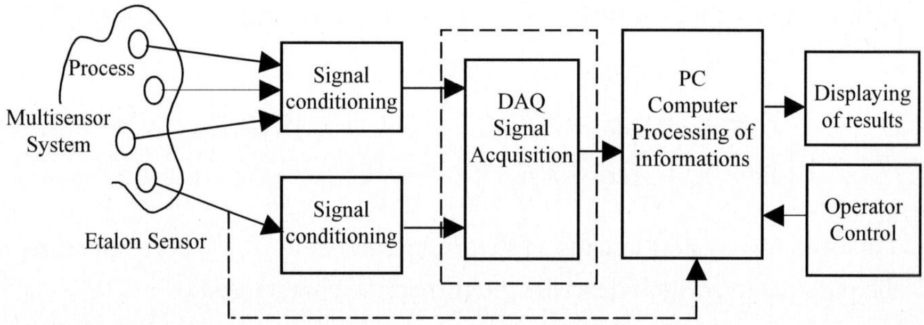

Fig. 1 Calibration and liniarisation modern system of the sensors

5. Software structure

The measuring system from figure no.1 has in his structure a computer where the Labview programme pack from National Instruments company is loaded, this meaning an intelligent measuring system, the programme written in the graphic language G of this programme ensure the implementation of the developing organigram (figure no.2). The formulas developed to calculate some parameters are computered by the programme using itself virtual instruments.

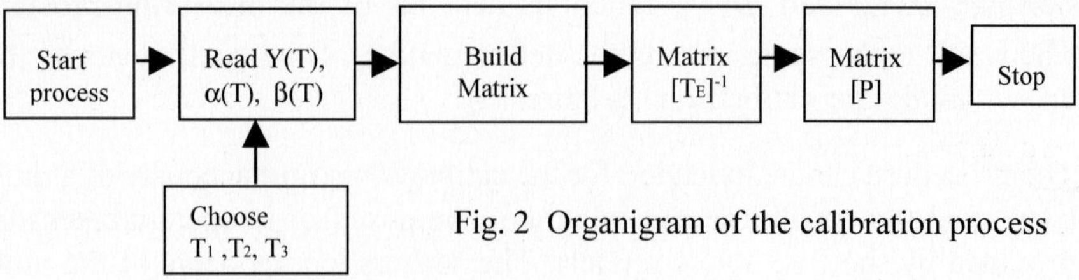

Fig. 2 Organigram of the calibration process

6. Results and conclusions

The application was checked by simulation. The heating of a room between 0 and 100 ^{0}C represents the process. The measuring of the temperature is made using a multisensorial system, which uses temperature sensors described by the shown models. Each sensor gives through its model the parameter Y (T). The input data are parameters y_0, A, B of each sensor and temperature T. Using the algorithm from figure no.2, the matrix of parameters is rebuilt. The calibration and linearisation are obtaining by extracting of the parameter T from Y(T) and for each sensor it depends on the chosen model. Un example of such linearization model is given by the relation:

$$T = \xi (1 - \frac{B}{A} \xi) \tag{9}$$

Using LabView application, in the fig. 3 is presented the soft implementation for this model.

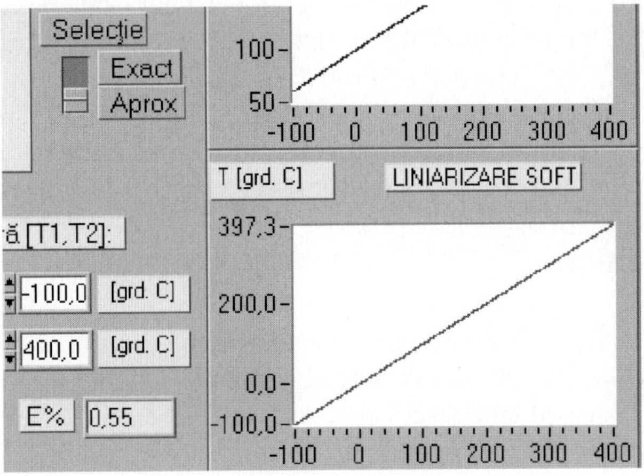

Fig. 3 Soft linearization of a sensor

References

[1] E. Posdarascu, "Contribution regarding signals acquisition from multisensorial systems", Bucharest, 2003.
[2] E. Posdarascu, "Achizitia si prelucrarea semnalelor provenite de la sisteme multisenzoriale" Matrix, Bucharest, 2007.

Contact sensor for robotic applications - Design and verification of functionality

P. Krejci

Institute of Solid Mechanics, Mechatronics and Biomechanics, Brno University of Technology,
Brno, 616 69, Czech Republic

Abstract

The information about interaction between robotic parts and surroundings is necessary for intelligent control of robot behavior. This research is connected to research of contact force vector sensor principle which was performed in last year. The paper presents working prototype of sensor and describes experimental verification of sensor functionality. The optimization of sensor body in order to improve its axial sensitivity is also discussed as well as the design of electronics driving unit of suggested sensor.

1. Introduction

This paper deals with design and verification of functionality of contact force vector sensor. The sensor and its principle suitable for various kind of task was designed and described in previous research [1]. During simulation and also by experimental verification of sensor functionality the small sensitivity of sensor in axial direction was observed. It is necessary to optimize the geometry and material of sensor in order to increase the sensitivity in critical direction.
Based on previous research [1,2], working prototype of sensor (see fig. 1) was developed. Sensor is made from aluminium alloy (AlMgSi0.5). This sensor is containing three strain gauges (*0,6/120LY13*) produces by Hottinger Baldwin Messtechnik (HBM).
For experimental verification the commercial measuring unit SPIDER 8 is used for measuring real sensor body strain. This unit is also produced by HBM Company and is designed among other things for strain gauges mea-

surement. In order to use this sensor for real application, the concept of electronic measuring and control unit is described below.

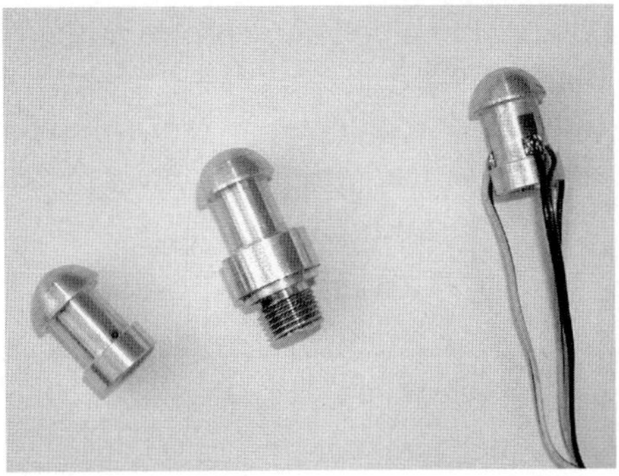

Fig. 1: Working prototype of contact force sensor

2. Experimental verification of sensor functionality

The sensor functionality was verified by experimental simulation in laboratory of Mechatronics. During experiment the loads of sensor was applied in several positions of sensor head. Load was applied by materials testing machine Zwick Z 020-TND (see fig. 2) where the real load force was measured.

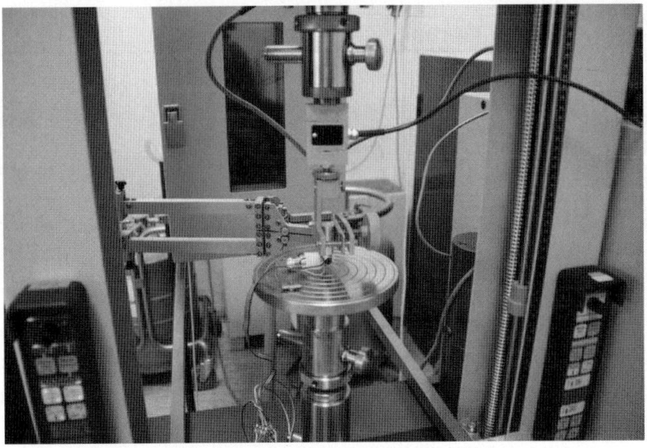

Fig. 2. Testing machine Zwick Z 020-TND

The deformation of sensor body was measured by strain gauges through HBM Spider 8 unit which is among other things designed for measuring of deformation by strain gauges. Measured deformations are transferred to information about contact force position and magnitude by neural network implemented in Matlab software.
The results of experimental verification are shown in tab. 1 for four positions of load force and shows really good accuracy of designed sensor.

Tab. 1 Results of verification

Case	Direction [mm]	**Contact force coordinates**		Accuracy [%]
		Position of force during experiment	Simulated by ANN	
1	x	0.8	0.81	98.8
	y	-2.5	-2.75	91.0
	z	20.0	20.62	96.8
2	x	-1.0	-0.97	97.0
	y	-1.9	-2.03	93.6
	z	22.0	24.1	90.9
3	x	-1.0	-1.02	98.0
	y	0.1	0.11	90.9
	z	22.0	22.91	96.0
4	x	2.0	2.07	96.6
	y	-5.0	-4.52	90.4
	z	16.0	16.38	97.7

3. Design optimization of sensor body by topological optimization using ANSYS

The ANSYS program can determine an optimum design, a design that meets all specified requirements yet demands a minimum in terms of expenses such as weight, surface area, volume, stress, cost, and other factors. An optimum design is one that is as effective as possible.
Topological optimization is a form of "shape" optimization, sometimes referred to as "layout" optimization. The purpose of topological optimization is to find the best use of material for a body such that an objective criterion takes on a maximum/minimum value subject to given constraints (such as volume reduction).
Unlike traditional optimization, topological optimization does not require explicit definition of optimization parameters (that is, independent vari-

ables to be optimized). In topological optimization, the material distribution function over a body serves as the optimization parameter. You define the structural problem (material properties, FE model, loads, etc.) and the objective function (the function to be minimized/maximized), then select the state variables (the constrained dependent variables) from among a set of predefined criteria.

The goal of topological optimization-the objective function-is to minimize/maximize the criteria selected (minimize the energy of structural compliance, maximize the fundamental natural frequency, etc.) while satisfying the constraints specified (volume reduction, etc.). This technique uses design variables (ηi) that are internal pseudo-densities assigned to each finite element.

The standard formulation of topological optimization defines the problem as minimizing the structural compliance while satisfying a constraint on the volume (V) of the structure. Minimizing the compliance is equivalent to maximizing the global structural stiffness.

FE model from previous research (see [1]) was used for sensor optimization. The sensor body was the volume which was subjected to optimization process. Volumes located under supposed locations of strain gauges are excluded from optimization process. Load force of 20N was used during optimization process and was located in axial direction of sensor As we can see from results of optimization (see fig. 3 and fig.4 for more details), the volume of sensor body was reduced. The volume reduction of sensor body is closed to 60%. Volume reduction of sensor body leads to redistribution of strain to area where strain gages are located. This means that we can measure large deformation of sensor body for same load force in comparison to non-optimized sensor.

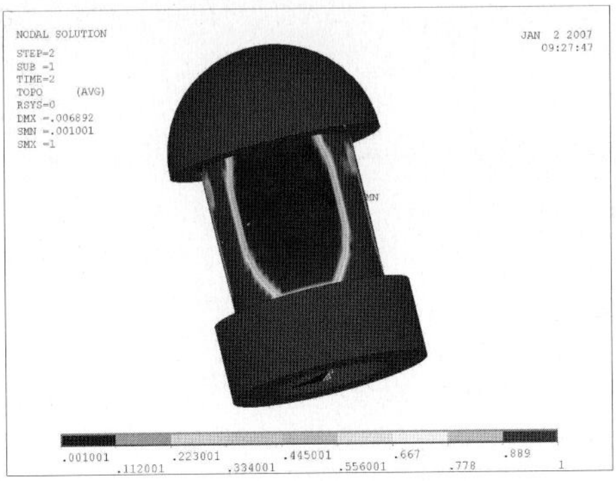

Fig. 3. Reduction of sensor body volume after 20 iterations (Topological optimization Densities)

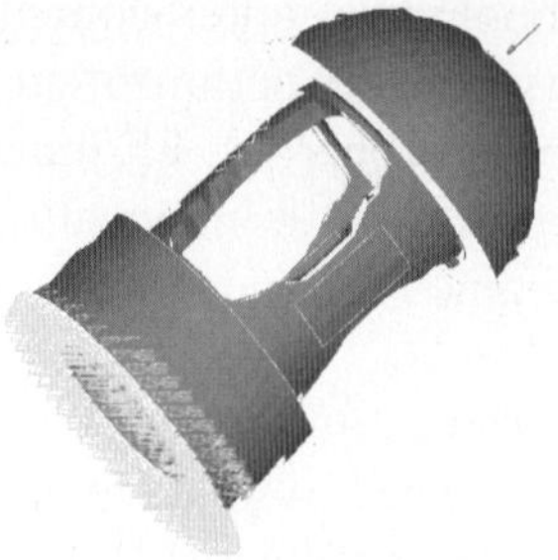

Fig. 4. Optimized shape of sensor body

3. Conclusion

Experimental verification of sensor functionality proofs sensor working principle and also shows really good accuracy of force vector identification. The problem with low sensitivity of sensor in axial direction was solved by topological optimization in ANSYS software. The reduction of 60% of sensor body volume was achieved and in this relation the sensitivity in axial direction increases.

4. Acknowledgment

Published results were acquired using the subsidization of the Ministry of Education, Youth and Sports of the Czech Republic, research plan MSM 0021630518 "Simulation modeling of mechatronic systems".

References

[1] Brezina, T.: Simulation Modeling of Mechatronics Systems I, *Brno 2006*, ISBN 80-214-3144-X

[2] Krejci, P.: Contact sensor for robotic applications, *Engineering Mechanics, Vol.12, (2005), No.A1*, pp.257-261, ISSN 1210-2717

[3] Schwarzinger *Ch., Supper L. & Winsauer H. (1992) Strain gauges as sensors for controlling the manipulative robot hand OEDIPUS: RAM vol. 8, pp.17-22.*

Two-variable pressure and temperature measuring converter based on piezoresistive sensor

H. Urzędniczok

Institute of Measurement Science, Electronics and Control
Silesian University of Technology
ul. Akademicka 10, 44-100 Gliwice, Poland

Abstract

A conception of utilizing the common existing cross-sensitivity of sensors to the additional quantities to measure both, the main and the additional one is presented. As an example a piezoresistive pressure sensor was investigated. An experimentally confirmed mathematical model of the sensor is given. An electric conditioning circuit and a method of two-variable measurand reconstruction are proposed too.

1. Introduction

The sensors are very often sensitive not only to the one (main) quantity but to some additional ones too. Many examples could be given: the gas sensors, semiconductor (piezoresistive and piezoelectric) pressure or force sensors, magnetoelastic force or torque sensors, humidity sensors. By the proper method it may be (but not always is) possible to eliminate the influence of the additional ones in satisfactory level. In most systems, on the other hand, usually is necessary to measure more than one quantity, often the same that influence the sensor. The additional sensor and measuring chain is then usually applied.

This observation lead to an idea to utilize the sensitivity of sensor to the additional quantities - in such way the sensor becomes a multivariable sensor. The proper conditioning circuit and the effective method of reconstruction of the measured values are essential, of course.

In this paper the above presented idea is discussed for the piezoresistive pressure sensor. A mathematical model of the sensor based on result of investigation is given. An electric conditioning circuit and a method of two-variable measurand reconstruction are proposed too.

2. Piezoresistive sensor

Piezoresistive sensors are designed for force or pressure measurement. They are made of some kind of semiconductor in which the resistivity depends on strain (force or pressure). The sensor resistance changes are rather small, but much higher than for metal strain gauges – that is very important advantage. Unfortunately there is a disadvantage too – the resistance of sensor is highly depended on temperature. The undesirable sensitivity for temperature is usually higher than sensitivity for strain (see Fig. 2). To eliminate (considerably reduce) influence of temperature a Wheatstone bridge is generally used to detect the changes of resistance and simultaneously to convert them to voltage signal [1]. In commercial pressure sensors based on these principle (Motorola, Fujicura, Endevco, Kulite, Vigotor) four sensors are usually used. Typical construction of such sensor is presented on Fig. 1.

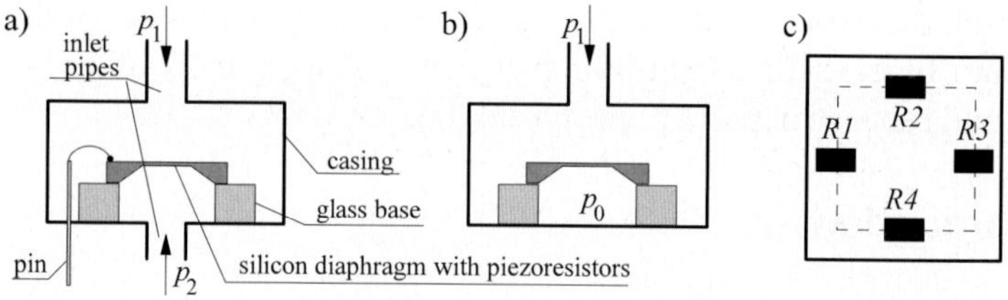

Fig. 1. Structures of silicon pressure sensors for absolute (a) and differential (b) measurement, typical layout of the four piezoresistors on the diaphragm (c). R_1 and R_3 are the sensitive ones; R_2 and R_4 are the reference ones.

3. Results of piezoresistive sensor investigations

A typical piezoresistive pressure sensor (type PS-040 Vigotor) was investigated to determine the pressure and temperature influence on the resistances of sensor. The results obtained for one pair of piezoresistors are shown in Fig. 2. On the assumption that given ranges[1)] of temperature and

[1)] the sensor is here proposed as two-parameter sensor - in traditional approach, when the sensor is applied as pressure sensor only, the range of the temperature changes is usually smaller pressure determines the measuring ranges, it is visible in Fig. 1, that the influence of temperature on the sensor resistance is about ten times more than the influence of pressure. It may be conclude that the sensor designed as a pressure sensor is, in fact, an excellent temperature sensor too. In this way the piezoelectric pressure sensor becomes a two-parameter pressure and temperature sensor. The obtained plots for a pair piezoresistors in investigated sensor are shown in Fig. 3.

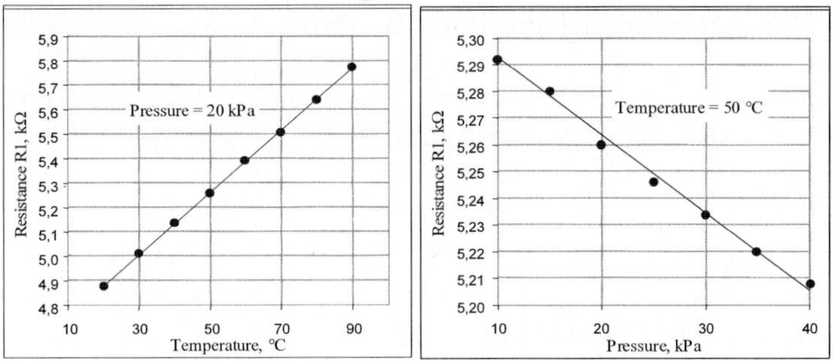

Fig. 2. The dependence of sensor resistances on temperature and pressure

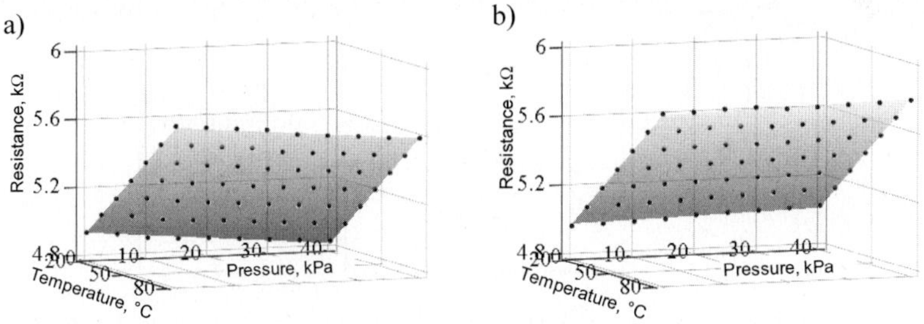

Fig. 3. Plot of resistance versus pressure and temperature for piezoresistor R1 (a) and R2 (b).

The dependence of resistance on pressure and temperature are in both cases nearly linear and could be described by the following formulas:

$$R_1 = A_1\vartheta + B_1 p + R_{01} \quad \text{and} \quad R_2 = A_2\vartheta + B_2 p + R_{02} \qquad (1)$$

where index 1 or 2 is number of piezoresistor, p is pressure, ϑ is temperature, A, B and R_0 are constant coefficients.

The values of coefficients A, B and R_0 were determined with the aid of least mean squares (LMS) method for calibration points shown in Fig. 3. Following values was obtained: $A_1 = 1.292 \times 10^{-2}$, $B_1 = -2.981 \times 10^{-3}$ and $R_{01} = 4.678$ for the resistor R_1 and $A_2 = 1.311 \times 10^{-2}$, $B_2 = 0.867 \times 10^{-3}$ and $R_{02} = 4.716$ for the resistor R_2.

4. Two-parameter pressure and temperature measuring converter based on piezoresistive sensor

The formulas (1) form a mathematical model of two-variable measuring converter of pressure and temperature to resistances R_1 and R_2 ($\{p,\vartheta\}\rightarrow\{R_1, R_2\}$). By inverting these formulas is possible to determine the values of pressure and temperature if values of resistances R_1 and R_2 are known. The inverted model has following form:

$$\begin{aligned}\vartheta &= 17.24R_1(\vartheta,p)+59.29R_2(\vartheta,p)-361.8\\ p &= -260.7R_1(\vartheta,p)+256.9R_2(\vartheta,p)+31.38\end{aligned} \tag{2}$$

In this way a two-parameter (two-dimensional, 2D) converter of sensors resistances to pressure and temperature ($\{R_1, R_2\}\rightarrow\{p,\vartheta\}$) may be realized. It is evident, that a proper conditioning circuit and computation device are necessary, as it is shown in Fig. 4.

A conditioning circuit, shown in Fig. 5, was applied.

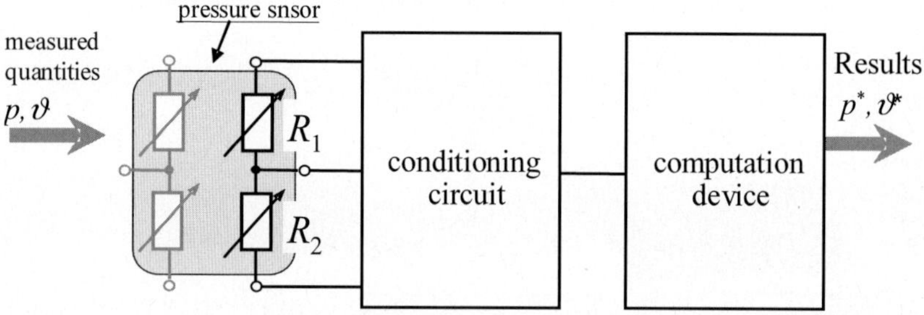

Fig. 4. Structure of measuring chain of two-variable pressure and temperature converter based on piezoresistive sensor

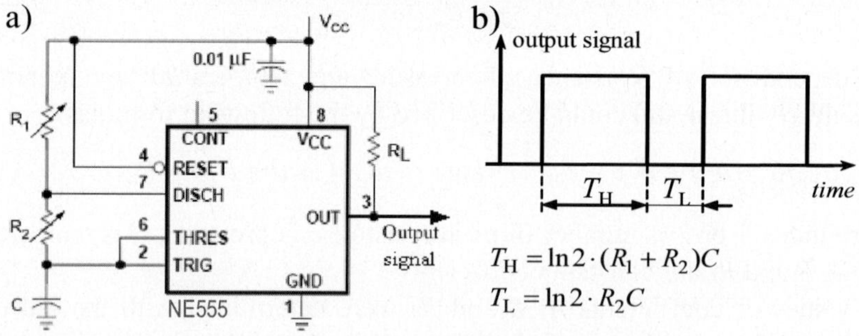

Fig. 5. Conditioning circuit based on NE555 oscillator (a), pulse output signal (b)

The output signal is a pulse signal and both of the two time intervals T_1 and T_2 depends on resistances R_1 and R_2 and in consequence on measured values p and ϑ.
As a computation device the ATmega8 microcontroller was applied to measure time intervals T_1 and T_2 and to calculate the measured values p^* and ϑ^*. Equations (2) were used.

5. Conclusions

The above described transducer was calibrated. In Fig. 6 the errors obtained for whole ranges of pressure (0...40 kPa) and temperature (20...90 °C) are plotted.

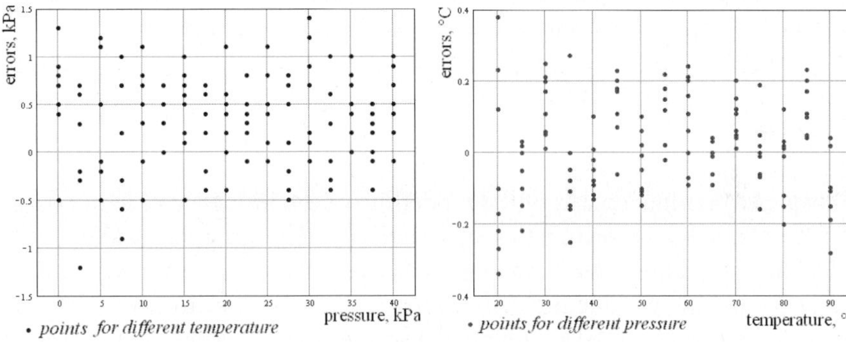

Fig. 6. Errors of investigated two-parameter transducer

As it is visible, maximum value of this errors not exceed 0.4 °C for temperature and 1.4 kPa for pressure (0.6% and 3.5% in relation to the measuring range respectively). This value was obtained with assumption of first-order (linear) model of the transducer (1). Applying second-order model allows errors decreasing, but makes calculation more complicated. Instead above proposed global models (1) and (2) other method, described for example in [2], may be considered to determine measured values.

References

[1] R. S. Figliola, D. E. Beasley, "Theory and design for mechanical measurements", John Wiley & Sons, New York, 1991.
[2] H. Urzędniczok, "The Uncertainty of an Algorythmical Reconstruction Method for the Two-parameter Measuring Converter", Proceedings of "XV Sympozjum Modelowanie i Symulacja Systemów Pomiarowych", Krynica, 2005, (in polish).

Modelling the influence of temperature on the magnetic characteristics of $Fe_{40}Ni_{38}Mo_4B_{18}$ amorphous alloy for magnetoelastic sensors

R. Szewczyk

Institute of Metrology and Measuring Systems, Warsaw University of Technology, ul. św. A. Boboli 8, 02-525 Warszawa, Poland, tel.: +48-22-234-8519, e-mail: szewczyk@mchtr.pw.edu.pl

Abstract

This paper presents results of the modelling of the influence of temperature on the magnetic characteristics of $Fe_{40}Ni_{38}Mo_4B_{18}$ amorphous alloy in as quenched state. For modelling Jiles-Athertos-Sablik model was used. Evolutionary strategies together with Hook-Jevies optimization were applied for calculation of model's parameters on the base of experimental results. To provide sufficient (from technical point of view) agreement between model and experimental data, the extension of Jiles-Athertos-Sablik model was proposed. This extension connect model's parameter k, describing magnetic wall density, with magnetic state of the material. Good agreement between experimental data and modelling confirms, that extended Jiles-Atherton-Sablik model creates the possibility of modelling of thermo-magnetic characteristics of cores of magnetoelastic sensors.

1. Introduction

Soft magnetic materials such as amorphous alloys are commonly used as a cores of the mechatronic inductive components such as cores of magnetoelastic sensors or transformers as well as inductive elements of switching mode power supplies [1]. It should be indicated, that magnetic characteristics of amorphous alloys depends significantly on temperature. This phenomenon has significant technical consequences. Functional properties of inductive component with soft magnetic material core may change during

its operation, especially if it is heated. It may cause malfunction or damage of mechatronic device.
For this reason knowledge about the influence of temperature of magnetic characteristics of amorphous alloys is very important from practical point of view. On the other hand complete model describing of temperature dependences of inductive components was still not presented.
Among four main models of magnetization process [2] only Jiles-Atherton-Sablik (J-A-S) model gives some possibility of modelling the temperature dependences of magnetic characteristics. Unfortunately, original J-A-S model do not create possibility of modelling the different magnetic hysteresis loops of the same material, with one set of model's parameters [3]. This is significant barrier in practical application of the modelling of the magnetic characteristics. Presented extension of the J-A-S model gives possibility of overcoming of this barrier.

2. Extension of the model

Total magnetization M of the soft magnetic material may be presented as the sum of reversible magnetization M_{rev} and irreversible magnetization M_{irr} [4]. In the J-A-S model the irreversible magnetization M_{irr} is given by the equation (1) [5]:

$$\frac{dM_{irr}}{dH} = \delta_M \frac{M_{an} - M_{irr}}{\delta \cdot k} \tag{1}$$

where parameter δ describes the sign of $\frac{dH}{dt}$ and k quantifies average energy required to break pining site. Parameter δ_M guarantees the avoidance of unphysical stages of the J-A-S model for minor loops, in which incremental susceptibility becomes negative [6]. Other parameters of J-A-S model, such as a, c, α, M_s, t and K_{an} are closely connected with physical properties of the material [4]. As a result J-A-S model can be used for physical analyses of the magnetization process.
It was indicated [4] that the J-A-S model parameter k changes during the magnetization process, due to change of the average energy required to break pining site [7]. On the other hand, previously presented extension of the J-A-S model [3], where the J-A-S model parameters change in the function of magnetizing field H, seems unjustified from the physical point of view. Parameter k should be connected with magnetic state of the material (described by magnetization M), not with magnetizing field H [8].

To overcome original J-A-S model limitation it should be extended by incorporation of connection between magnetic state of the material (describet by its magnetization *M*) and model's parameter *k*. Parameter *k* can be described by the vector of 3 parameters k_0, k_1 and k_2, and *k* is given as [8]:

$$k = k_0 + \frac{e^{k_2 \cdot (1-|M|/Ms)} - 1}{e^{k_2} - 1} \cdot (k_1 - k_0) \tag{2}$$

In the dependence given by equation (2), parameter k_0 determines the minimal value of *k*, parameter k_1 determines maximal value of *k* and k_2 is shape parameter.

4. Results

Experimental measurements were carried out on the ring-shaped sample made of $Fe_{40}Ni_{38}Mo_4B_{18}$ amorphous alloy in as-quenched state. Stable temperature was achieved by criostat, whereas magnetic hysteresis loops were measured by hysteresisgraph HBPL. Parameters of J-A-S model were determined during the optimization process. For optimization, the evolutionary strategies [9] together with Hook-Jevis gradient optimization were applied .

In figure 1 the experimental results (marked as dots) together with results of the modelling (marked as solid lines) are presented. One set of J-A-S model's parameters was determined for three different hysteresis loops measured in given temperature. Dependence of the J-A-S model parameters on temperature is presented in Table 1.

Results presented in figure 1 shows very good agreement between extended J-A-S model and experimental results. To confirm this agreement r^2 coefficient was calculated between model and experimental results. Due to the fact, that r^2 is higher than 0.99, it was indicated that 99% of total variation of experimental results is described by the extended J-A-S model [10].

It should be indicated, that one set of parameters enable modelling of different hysteresis loop at given temperature. If temperature changes, new set of parameters should be calculated. On the other hand, on the base of table 1, temperature dependence of J-A-S parameters can be interpolated for different temperatures from 20 °C up to 120 °C, as well as extrapolated for higher or lower temperatures.

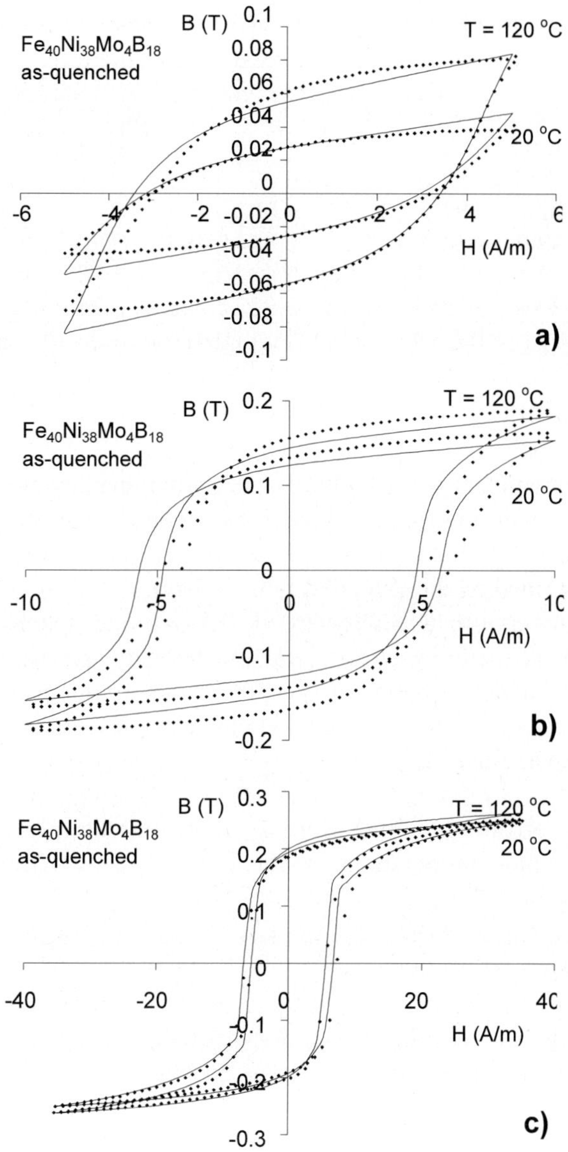

Fig. 1. Results of the modelling the influence of temperature on the shape of hysteresis loop achieved for one set of parameters of extended J-A-S model for magnetizing field Hm (♦ experimental results, — results of the modelling): a) 5 A/m, b) 10 A/m, c)35 A/m

Table 1. Results of the modelling temperature dependence of magnetic characteristics according to extended Jiles-Atherton-Sablik model.

Temp. (°C)	a A/m	k_0 A/m	k_1 A/m	k_2	c	$M_s \cdot 10^{-5}$ A/m	$\alpha \cdot 10^4$	K_{an} J/m^3	t
20	112.7	467.6	11.12	-8.26	0.466	2.97	2.39	547	0.80
40	97.3	387.5	11.46	-8.66	0.469	3.11	1.93	1 522	0.80
60	93.6	454.4	11.95	-8.97	0.494	3.20	1.83	1 769	0.80
80	92.1	830.6	11.63	-9.18	0.501	3.23	1.78	1 812	0.80
100	90.3	1238.5	11.35	-9.22	0.506	3.21	1.77	1 731	0.80
120	85.5	1905.4	9.69	-11.1	0.480	3.03	1.75	813	0.80

5. Conclusion

One set of parameters of extended J-A-S model enables modelling of the hysteresis loops of amorphous alloys for different value of maximal magnetizing field. In such a case over 99% of total variation of experimental results is described by the extended J-A-S model.
Presented temperature dependences of J-A-S model parameters enables modelling of magnetic hysteresis loops for temperatures from 20 °C up to 120 °C. Such model may be very useful for determining the temperature dependence correction factors for magnetoelastic sensors or magnetic sensors (e.g. such as fluxgates) .

Calculations for the modelling were made in Interdisciplinary Centre for Mathematical and Computational Modelling of Warsaw University, within grant G31-3.

References

[1] R. O'Handley "Modern magnetic materials" Wiley, 2000.
[2] F. Liorzou et al., IEEE Trans. Magn. 36 (2000) 418.
[3] D. Lederer et al., IEEE Trans. Magn. 35 (1999) 1211.
[4] D. C. Jiles, D. Atherton, J. Magn. Magn. Mater 61 (1986) 48.
[5] D. C. Jiles, D. Atherton, J. Appl. Phys. 55 (1984) 2115.
[6] J. Deane, IEEE Trans. Magn. 30 (1994) 2795.
[7] P. Gaunt, IEEE Trans. Magn. 19 (1983) 2030.
[8] R. Szewczyk, J. Phys D. (2007), in printing.
[9] H. P. Schwefel "Evolution and optimum seeking" Wiley 1995.
[10] M.Dobosz M. „Wspomagana komputerowo statystyczna analiza wynikow badan" EXIT, Warszawa 2001.

“Soft particles” scattering theory applied to the experiment with Kàrmàn vortex

J. Baszak, Prof. Dr. R. Jablonski

Warsaw University of Technology, Faculty of Mechatronics, 8 Boboli Str. Warsaw, 02-525, Poland

Abstract

The paper presents consideration of light scattering on Kàrmàn vortex street phenomenon [1, 2]. The analytical approximation of the scattering is provided. It is based on a method using expansion into spatial spectrum Kotelnikov-Shannon sampling functions [3]. The final result is referred to the data acquired from gaseous flow installation.
The method described herein is applied to count a number of vortices appearing behind a bluff body. The best signal to noise ratio (SNR) is searched at the angular scattering characteristic.

1. Introduction

There is no single and fruitful theory of scattering but plenty of different approaches, like Rayleigh or Mie theory, which solve partially some particular problems but do not explain physical mechanism entirely [4]. Existing methods of approximation solution like anomalous diffraction (AD, by van de Hulst), T-matrices, DDA are complex and time consuming [5]. However AD approximation is very popular since it gives very good estimation for medium size particles.
In this paper the approach recently developed and described in [6] is put into practice to specific application. It allows separating process of intrinsic scattering from diffraction process (in its immediate sense). Accordingly, both processes are also considered independently and their analytical descriptions are conducted separately. It significantly reduces the complexity of final, numerical calculations but still preserves acceptable level of accuracy.

The entire calculation is applied to coherent, linearly polarized light scattered on water droplets floating injected into air flow in a pipe. The droplets are distributed in space by Kàrmàn vortices generated by bluff body. The experimental verification of this theory proves the correctness of the applied analytical method.

2. Analytical calculation

The particle-light interaction is split in two processes. The scattering restrictively is considered as secondary radiation of the light from enlightened particle. It means the part of incident radiation hits the particle, then is absorbed and afterward re-radiated out from the particle. The remaining part of the incident radiation takes part in diffraction. In this sense the total scattering intensity (in its wider meaning) can be expressed as simple summation of scattered and diffracted light intensity. For simplicity only electric part of radiation is taken into account.

The diffracted field is expressed by Kirchhoff integral in Fraunhofer zone. The scattered field can be expressed without complicated integration as follow: $E_{scatt} = \nabla\times\nabla\times\Pi_e$, under assumption of lack of Π_m (magnetic Hertz vector) and electric and magnetic currents as well. Consequently, for monochromatic wave with linear polarization in frequency domain E_{scatt} is linear function of Π_e with a certain scaling factor adjusting them into same coordinates. And in Fraunhofer zone both vectors can be considered as parallel. Next, the Hertz vector can be derived from electric polarization p of the particle. Referencing briefly to [3] it is easier to consider polarization p in its spatial spectrum expansion based on Kotelnikov-Shannon sampling functions.

Then, there is no complicated interference among infinite amount of field vectors (coming from point vectors of elementary volume of the particle) with promptly alternating phase. Hence there is no interaction of components and simple summation can be executed.

Under assumption of dilute non-polar isotropic gases the linear and local relation between polarization p and incident electric field amplitude is simple: $p(r) = \alpha \cdot E_{in}(r)$, where α is a tensor of polarizability in general and r is spatial variable.

After additional several transformations the final equation for the scattering component of intensity from a single, spherical particle and at unity input power is following:

$$E^2_{scatt\_1}(\rho,\theta,m) = \left(\frac{m^2-1}{m^2+2}\right)^2 \cdot \frac{1}{2\cdot\pi}\rho^4 \cdot f^2(\rho,\theta,m) \qquad (1)$$

where

$$f(\rho,\theta,m)=3\cdot\left[\frac{\sin(\rho\cdot B(\rho,\theta,m))}{\rho\cdot B(\rho,\theta,m)}-\cos(\rho\cdot B(\rho,\theta,m))\right]\cdot\left[\frac{1}{\rho\cdot B(\rho,\theta,m)}\right]^2$$

$$\theta\cdot B(\rho,\theta,m)=\arcsin\left(\frac{\sin(\theta)}{B(\rho,\theta,m)}\right)$$

$$\rho\cdot B(\rho,\theta,m)=\rho\cdot 2\cdot\sqrt{\sin\left(\frac{\theta}{2}\right)^2\cdot m+\left(\frac{m-1}{2}\right)^2}$$

The parameter $\rho=4\cdot\pi\cdot a/\lambda$ where a– characteristic size of the particle and λ is incident light wavelength, θ is the angle of scattering, m denotes relative refraction index ($=n_1/n_2$) which for water drops in the air is approximately equal to 1.3. The final equation for the diffracted component of intensity from the single, spherical particle and at unity input power is as follow:

$$E^2_{diff\_1}(\rho,\theta)=\left(\frac{J_1(\rho\cdot\sin\theta)\cdot T(\rho,\theta)}{\sqrt{\pi}\cdot\sin\theta}\right)^2 \qquad (2)$$

where T filters are expressed by the equations:

$$T(\rho,\theta)=T_1(\rho)\cdot T_2(\theta)$$

$$T_1(\rho)=\left(\frac{\exp\left(\rho/3\right)^2-1}{\exp\left(\rho/3\right)^2}\right),\qquad T_2(\theta)=\exp\left(-3\cdot\theta/\pi\right)^6$$

The intensity calculation was done for population of particles with symmetrical Gaussian distribution (2.3, 1.0) assumed, where 2.3μm is the average value of a droplet diameter and 1.0μm is its variation (1·σ).

3. Experimental result and data comparison

The experimental setup is presented in Fig. 1 [1, 2]. The collimated laser beam from 650nm laser diode was forward scattered at water aerosol floating in an air discharge in a pipe.

Two series of measurements are scaled along ordinate and compared with calculated value (Fig.2). The data depicted *air1* and *air2* are related to discharge in range Re=7000÷15000. The calculated value at each point is integrated over the aperture of the collecting optics.

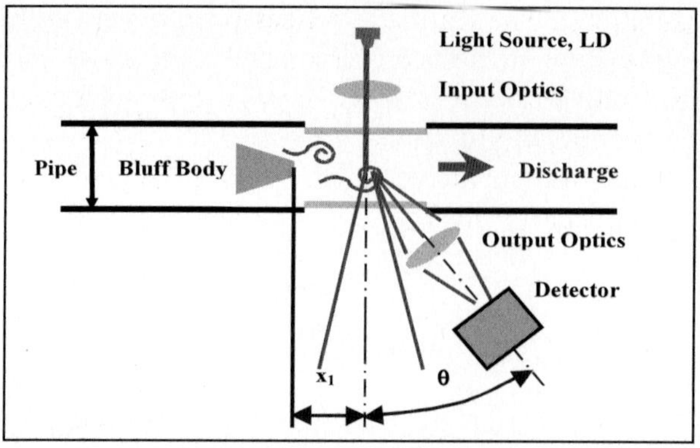

Fig. 1. Scheme of experimental setup

Each indicatrix is very asymmetric rather like in Mie theory. The widely-known multi-lobes shape of the diffraction indicatrix has changed in case of Gaussian distribution of the particles into soft, one lobe plot. This is similar to an envelope of diffraction of medium size particles, which conforms to both: assumed distribution and the aerosol parameters. The existing local and mild humps remain sharp lobes of interference at single particle diffraction plot. All plots are parallel to line with the slope of -1/5.

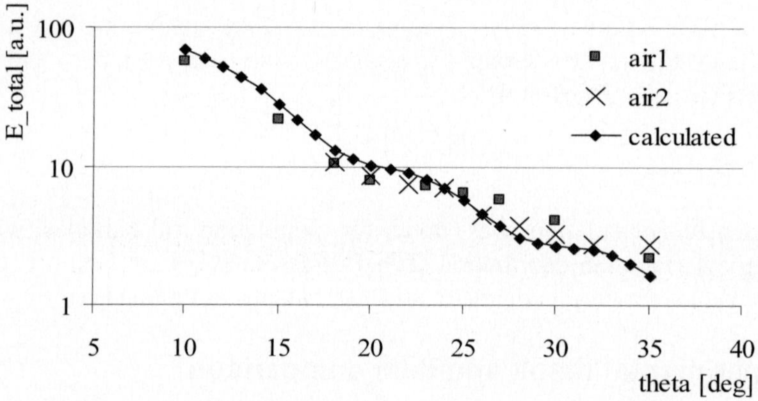

Fig. 2. Analytical and experimental indicatrices

The impact of the diffracted field on the total field is negligible. It has share of less then 7% at angle θ =7° and goes below 1% at 18°.

Fig. 3 shows the indicatrix and SNR plot for the same data. SNR is calculated from the Fourier spectrum of the signal (intensity oscillation

along Kàrmàn street). The Kàrmàn vortices were generated by bluff body immersed in the flow in countercurrent and torn away modulating the scattered field. The intensity is about 7 times lower at angles bigger then 20° compared to almost forward scattered one.
However SNR has its maximum at the angle 20° (roughly). Usually the less light the more significant share (if expressed by coefficient of

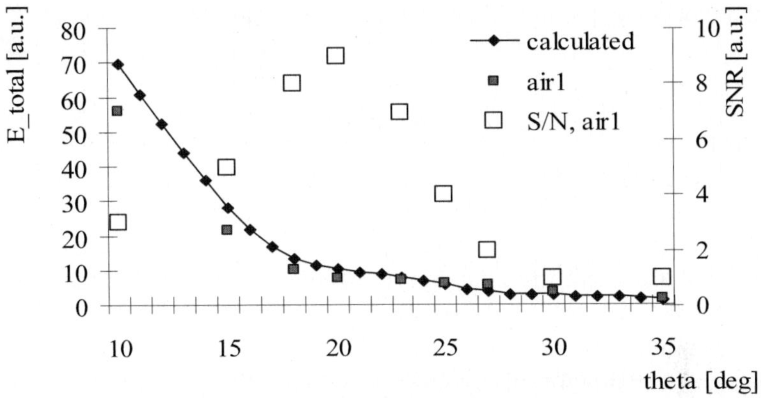

Fig. 3. Indicatrices and SNR in the air

variation) of photon shot noise. The presented phenomenon gives the idea of additional asymmetry of observed scattered field where at certain angles the intensity contains less noise.

4. Conclusion

The research indicated the strong relationship between the light noise level and the vortex geometry was expected. This idea needs to be further investigated.

5. References

[1] J. Baszak, R. Jablonski, Jour. of Phys. 13 (2005) 86
[2] J. Baszak, R. Jablonski, Optoel. Instrum. & D. Process. 40, 5 (2004) 12
[3] V. M. Rysakov, Jour. of Quant. Spectr. & Rad. Trans. 87 (2004) 261
[4] G. H. Meeten, Opt. Comm 134 (1997) 233
[5] P. Nitz et al., Solar En. Mat. & Solar Cells 54 (1998) 297
[6] V. M. Rysakov, Jour. of Quant. Spectr. & Rad. Trans. 98 (2006) 85

Measurement of cylinder diameter by laser scanning

R. Jabłoński*(a), J. Mąkowski(a)

(a) Warsaw University of Technology, Institute of Metrology and Measurement Systems, ul. św. A. Boboli 8, Warsaw, 02-525, Poland

Abstract
The existing diffraction theories do not fit to engineering applications because they do not concern the phenomena on 3D bodies. When using Fresnel theory for volumetric obstacles, the corrections depending on the shape of obstacle have to be taken into account. We proved experimentally the hypothesis that the distance between two parallel edges of curved surfaces depends on the curvatures of these edges. For instance, when measuring the diameter of cylindrical object the appropriate correction dependent on the size of object must be taken into account. The experiments were performed with cylinders 1.1mm, 2.8mm and 4.8mm - used as reference standards.

1. Introduction
In laser measuring scanners the measurement information is transformed several times and finally the detector output signal has the form of shadow of an object. Its ideal trapezoidal shape is distorted due to the dynamic changes of signal components (diffracted, reflected and geometrical beam) especially when object size is much bigger than beam waist diameter.

In previous papers [1,2,3], two the most important errors concerning laser measuring scanners were pointed out: instability of rotational velocity of deflector and output signal errors. The problem of instability of rotational velocity can be solved by direct angular measurement of deflector. This procedure was described in [1]. The output signal errors were firstly partially analyzed in [2,3]. But still there are several questions left without answer, like for instance:

- what is the minimum required area of intensity distribution pattern to be measured by photodetector (in order to obtain the measurement on satisfactotu uncertainty level).
- what are the upper and lower limits for the diameter of the cylindrical object to be measured,
- what in the influence of object surface finish.

Recently we designed a modified version of laser measuring scanner for measurement accomplished in our institute - designed to measure the cross-section of bars of various shape. When measuring the radii of oval bars, it was discovered that the measured dimension depends on the curvature. The results were confirmed by classical contact method using laser interferometer. We made a hypothesis, that the distance between two parallel edges of curved surfaces depends on the curvatures of these edges. Following that hypothesis, we assumed that the accurate measurement of a diameter of cylindrical object requires the correction dependent on the size of object.

2. Theoretical considerations

The present applications of diffraction theory in metrology are fragmentary and may be concluded that the existing solutions for diffraction of 3D bodies do not fit to engineering applications.

The Fresnel diffraction theory [4] concerns the light diffraction phenomenon on the straight edge of half-plane. The edge of cylindrical object is straight, but it can be treated as sharp only when boundary condition is fulfilled (radius of curvature = 0).

In Kirchhoff's theory there are no limits concerning the shape or thickness of obstruction. The influencing quantity is only the edge of obstruction. Following the above considerations[4] and dealing with vector potential for compound phase and amplitude structure, the arbitrary wave field can be presented as a superposition of plane waves. Since such possibility exists also for Gaussian beam, it can be assumed that diffraction on cylindrical object is described as a decomposition of a spatial distribution of light into a series of plane waves.

The distinct maxima (corresponding to diffraction fringes of Fraunhofer diffraction pattern laboriously obtained from Cornu spiral, Fig.1) are easily obtained in our experiments.

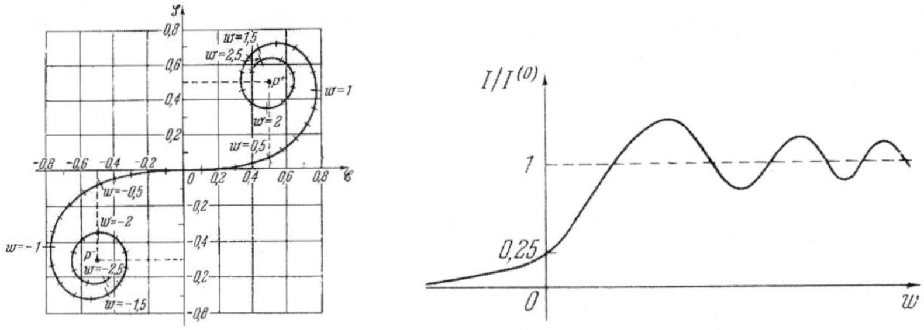

Fig.1. Cornu spiral and Fraunhofer diffraction pattern

The paper concerns the experimental method of defining the approximation errors of applying Fresnel theory to the cylindrical objects. Having the above in view, the close analysis of detector signal was carried out.

3. Experimental set-up

The experimental set-up is shown in Fig.2. An object ***O***, polished steel cylinder, is placed at the focal distance of scan lens ***SL***. He-Ne laser with beam expander and scan lens ***SL*** create the laser head ***LH***. Scan lens, of the focal length 48mm, forms the beam waist $2w_0$=47μm along the ***Z*** axis.

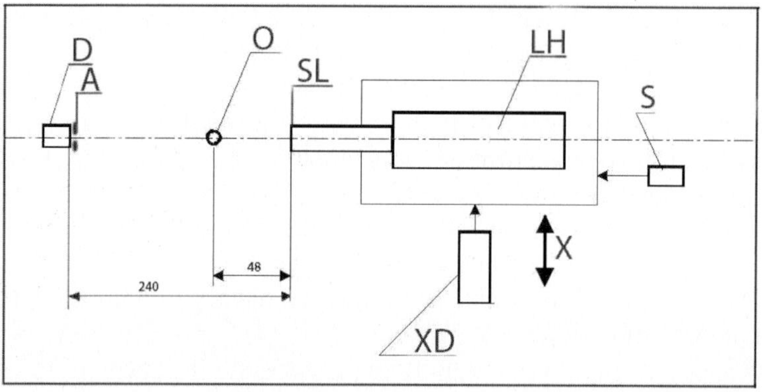

Fig.2. Measuring set-up

In order to avoid the inaccuracies caused by the instabilities of angular deflection, the entire laser head is scanned across the ***Z***-axis, by drive XD (***X*** linear scanning). The sensor ***S*** .is used to control straightness of laser head movement. The detector unit (set coaxially with laser beam axis) is composed of photodiode ***D***, aperture 0.3mm ***A*** and electronic circuit. It was calibrated with the power meter (LaserMate-Q, Coherent) and obtained signal (in V) is proportional to the measured light intensity. Computer controls scanning movement and is used to process the obtained data.
Fig. 1 does not show many auxiliary components like: system assuring parallel laser travel, vibration protection set-up, dark chamber, detector electronics, etc.

4. Measurements

The measurement begins with the laser beam axes placed far behind the object edge (in this position detector records the minimum signal). During scanning, the laser beam gradually approaches the edge of object and the diffraction wave appears and interferes with geometrical wave [3]. In accordance to Fresnel theory interference fringes appear and the first order fringe approaches the detector. Then the registered intensity slowly

decreases and reaches (theoretically) zero value (it corresponds to the central position of laser head, when the laser beam is obstructed by object). After passing the object detector registers the signal on the others side of an object (and again first order, distinct interference fringe is registered).
In this experiment, the laser head is scanned parallel to beam axes and intensity distribution pattern is collected by the stationary detector.

Laser head is scanned in 2,5μm steps. At each detector position 500 measurements are recorded with the rate 1000Hz. Only the middle part of measurements (between 250-400 is taken for further processing).
An advance data acquisition system (and computer program) was accomplished. This system realizes the following functions:

1. reading the required measurement data
2. determinates the maxima on both sides of an object
3. calculates the distance between maxima
4. takes into account the correction coefficient and calculates of the corrected dimension
5. drawing (in optional magnification) the result

5. Results

Fig.3 shows an exemplary result obtained by scanning cylinder of diameter 1.1mm.

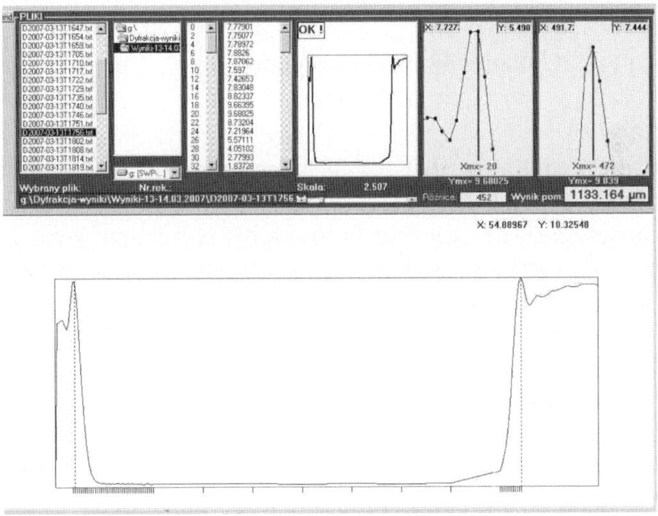

Fig.3. Exemplary cylinder diameter measurement

The upper right graphs (and data) present a close look at the left and right maximum. The difference in amplitudes gives evidence to non-coaxial setting of object and detector in reference to laser beam axes. This has no influence on result, but can be used for future adjustments.
The experiments were performed with cylinders 1.1mm, 2.8mm and 4.8mm - used as reference standards (uncertainty 1μm). In Tab.1 and Fig.4 the nominal and experimental values of diameter are presented.

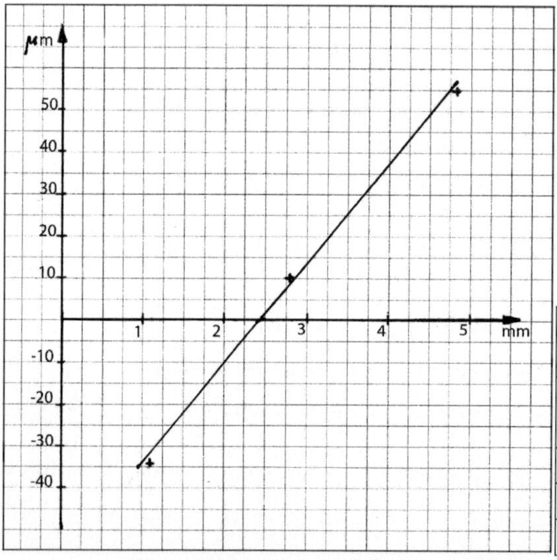

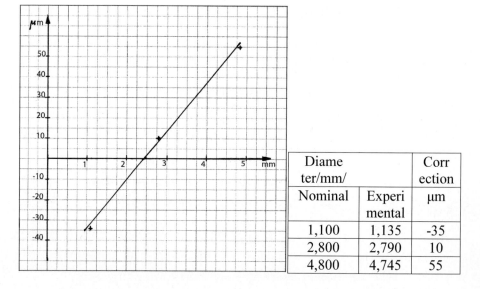

Diameter/mm/		Correction
Nominal	Experimental	μm
1,100	1,135	-35
2,800	2,790	10
4,800	4,745	55

Fig.4. Cylinder diameter corrections

6. Summary

We proved the hypothesis, that the distance between two parallel edges of curved surfaces depends on the curvatures of these edges. In case of he accurate measurement of cylindrical objects the correction, dependent on the size of the object, must be added. We assume that correction function has sinusoidal shape.

References

[1] Jablonski R.: SPIE vol. 2101, 1993, 741-749
[2] Jablonski R. Key Engineering Materials, Vo.l 295-296, pp.209-214
[3] Jabłoński R. Mąkowski J., Proc. 8th ISMTII, Sendai, Japan, 2007, p.6
[4] Born M., Wolf E. Principles of optics –Pergamon press- 1964

Tyre global characteristics of motorcycle

F. Pražák (a) *, I. Mazůrek (b)

(a, b) Faculty of Mechanical Engineering, Brno University of Technology
Technická 2, Brno, 616 69, Czech Republic

Abstract

Identity of simulation model is influencing duality of single parameters. This paper is describing on measurement and evaluation stiffness of tyre depending on speed of loading, so-called dynamic stiffness tyre. In this case, we were orient on tyre of sport road motorcycle, where we are at first watched influence camber of tyre from vertical surface on resulting characteristic. The measuring characteristic we are express as force dependent to compression and speed of compression, so-called tyre global characteristic. Mathematical description of global characteristic makes it possible simply and more accurately creation virtual model of tyre used in the multibody software.

1. Introduction

In most of mechatronics systems are subsystems and parts, which embody expressive non-linear behavior. This behavior must be defaced at modeling and design of these systems. Common non-linear effects are characterized simply functional dependence, such as saturation or non-sensitivity. Models of these non-linearity are simple and they are common to disposition in libraries different simulation systems as is Matlab/Simulink. More complicated non-linear effects are for example sliding friction and hysteresis are described simple. In several cases need not constitute enough loyal - adequate model of relevant non-linear effect.
In simple mathematic model of vehicle suspension wheel is tyre as linear spring (fig. 1a), more realistic models reason with non-linear characteristics (fig. 1b). If it is reason with inner damping of tyre, it is most often with friction damping model. Size of friction force is constant and depends only on direction of deformation velocity (fig. 1c). Endeavour by supplying of

hysteretic rubber properties with the help friction force (fig. 1d) is largely distant by real dependence of deformation force on deformation at cyclic loading of tyre (fig. 1e).

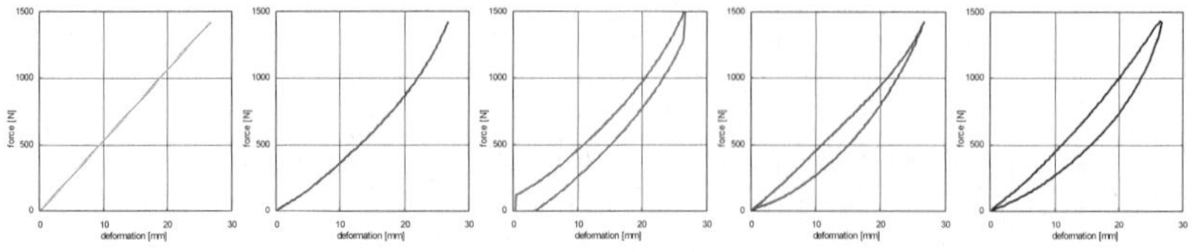

a) b) c) d) e)

Fig. 1: Different tyre characteristic in model of suspension wheel

A vehicle wheel with tyre is not as simple air tyre. To properties of tyre are shows viscoelastic properties inner tyre structure. Rubber os non-linear and hyperelastic material, whose mechanical properties are floating with temperature, but with operating time too. It is necessary know mechanical properties of tyre. Against this properties, it is possible create computation model for numeric simulation.

Fig. 2: Tested tyre in tester

Results of height mentioned, it is necessary measuring of force in dependence on deformation, but on velocity of deformation too. In laboratories Brno University of Technology, Institute of Mechanical Design, on slightly modified damper tester, we are identification viscoelastic parameters of Barum Bantam tyre with size 12x4 (Fig. 2). The bantam tyres are characteristic in that they are changing at compression size of contact surface to road. Between tyre and road are generating tangential force. This type of tyre is using for agricultural engineering, plane and small motorcycle.

In the first phases, spring and dampening properties were finding separately, similarly as with vehicle or motorcycle shock absorber strut. Stiffness characteristics at very low velocity of deformation (< 0.005m/s) are different, according to select stoke they show different hysteresis properties. In figure 3, hysteresis loop are unloaded in dependence on force-deformation for tested strokes.

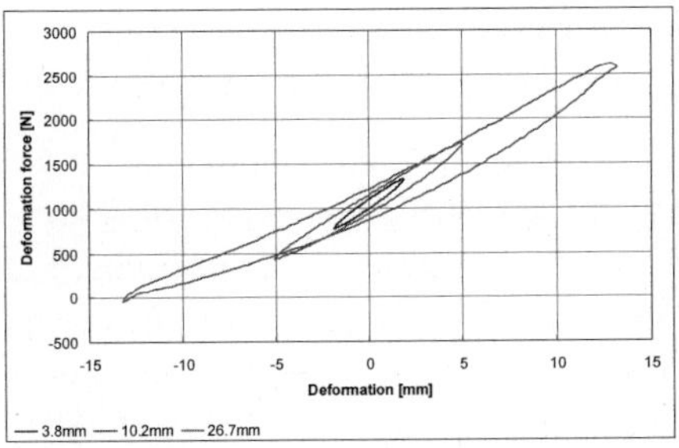

Fig. 3: Tyre characteristics for three different compressions

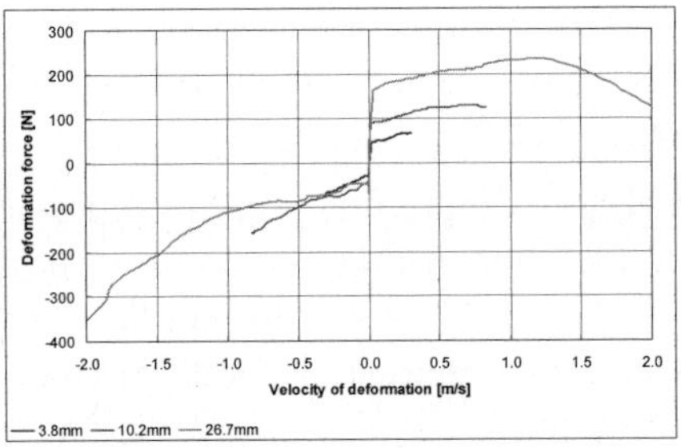

Fig,. 4: Dampening tyre characteristics

On figure 4 are F-v curves of these variants. Dampening characteristic is markedly dependent on test stroke and it is impossible use flat dampening characteristics. It is not proper tyre model as system with one F-s and F-v characteristics.

2. Globální charakteristika pneuamtiky

Therefore, we decided to solve the problem adequately to global damper characteristics [1] to measurement so-called global tyre characteristics. The test of tyre proceed by method start-finish, where in time one minute we are increase frequency of tyre compression to maximum frequency and then the frequency was decreasing until to stopping the tester. The test was made for three different value of compression (4, 10, 26mm) and four different tyre pressure (0.7, 1, 1.5 a 2bar). The assessment of test conditions was from request step over due nature frequency of suspension wheel (c. 20Hz).

In figure 4 (for pressure 1.5bar), aftermath of hysteresis tyre behavior are visible. To describe of mechanical properties, it is possible fenomenologic access, which is found on mathematic model [2]. Mathematic model is voiced by the help of deformation energy density and it describes linear relation between strain and shear deformation. Thereby calculation for each interactive step would be high overloaden superior model of wheel suspension, eventually vehicle.

A quick approach seems to be procedure parameterization of regression analysis real global characteristics with one departure from reality. Before regression analysis it is eliminated hysteresis loop by numeric. Then, for superior model is available smooth 3D function, which it is providing values of viscoelastic force F in depending on both deformation d and velocity of deformation v. It is multiple regression function of second-rate:

$$y = a + bx_1 + cx_2 + dx_1^{\ 2} + ex_2^{\ 2} + fx_1x_2 \qquad (1)$$

Calculation of response in superior models is fast and enough accurate with this simple function. In figure 5c is shown expressive wave of regression surface at high velocity of deformation. It may result in influence at simulation of over-cross hurdle high velocity. Influence of simplification will test on quarter car model [3] and next on vehicle.

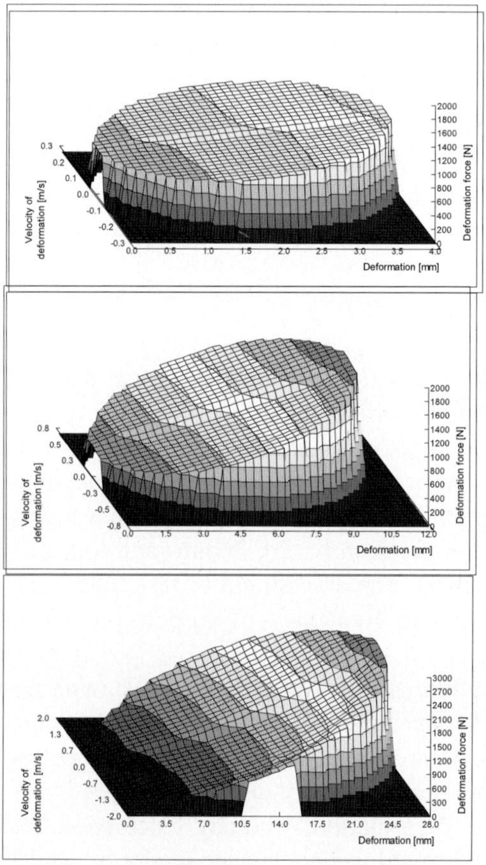

Fig. 5: Global tyre characteristics for stroke 4, 10 a 26mm

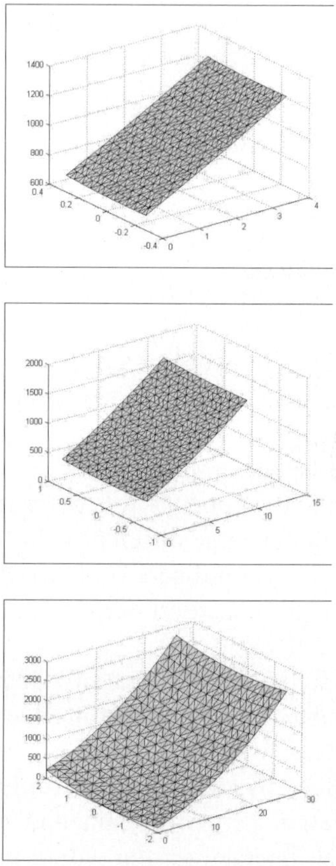

Fig. 6: Surface adventitious by regression analysis from global characteristics

3. Conclusion

In our test-room, the new methodic of measurement of viscoelastic tyre properties was created. Resulting dependence was called as global tyre characteristic. The method of analysis makes it possible separate hysteretic properties of tyre from basic deformation characteristic. Identification of tyre parameters with method proposed regression analysis makes it possible simple implementation to superior models. Before personal design and testing of simulation models, it is necessary plan experiments so they provided all information for identification of needed parameters.

This work was developed with the support of the grant project GA□R No.: 101/03/0304

References

[1] Votrubec R.: Globální charakteristika tlumi□e, dissertation work, Technical University of Liberec, Liberec, 2005
[2] Bergstršm, J., S.; Boyce, M., C.: Constitutive modeling of the large strain time – dependent behavior of elastomers, J. Mech. Phys. Solids., Vol. 46, 1998, ISSN 0022-5096
[3] MAZ□REK I, DO□KAL A., Pražák F.: Diagnostic model of a shock absorber, Engineering Mechanics, 2005, ISSN 1210-2717, str. 71-76

Magnetoelastic torque sensors with amorphous ring core

J. Salach

Institute of Metrology and Measuring Systems
Warsaw University of Technology,
sw. A. Boboli 8,
02-525 Warsaw, Poland

Abstract

This paper presents new type of magnetoelastic torque sensors. In this sensors the amorphous ring cores were applied as sensing elements. The ring cores were in as-quenched state and after thermal relaxation. For the investigation three type of tape were used, iron based ($Fe_{78}B_{13}Si_9$), nickel based ($Fe_{40}Ni_{38}Mo_4B_{18}$) and Finemet ($Fe_{73.5}Si_{13.5}Nb_3Cu_1B_9$). Sensor investigation indicate interesting properties and high sensitivity. Results prove that the thermal relaxation is necessary and select the best compose of ring core.

1. Introduction

Idea of application of amorphous magnetic materials as compressive sensors proved, that these materials exhibit significant influence of stresses on magnetic properties [1]. Moreover such alloys have very high stress strength together with rust resistance [2]. For these reasons it seems, that amorphous magnetic alloys are the best known material for magnetoelastic sensors applications.

On the other hand industrial application of the amorphous magnetic alloys for torque measurements seems to be still not possible from engineering point of view. The main barrier in development of torque sensors made of metallic glasses is connected with the lack of the method of applying an uniform, torque moment to magnetoelastic sensing element with closed magnetic circuit. The problem is additionally complicated due to the fact, that amorphous magnetic materials are produced in the form of thin ribbons with thickness less than 25 μm [3]. This paper presents ideas of overcoming these problems.

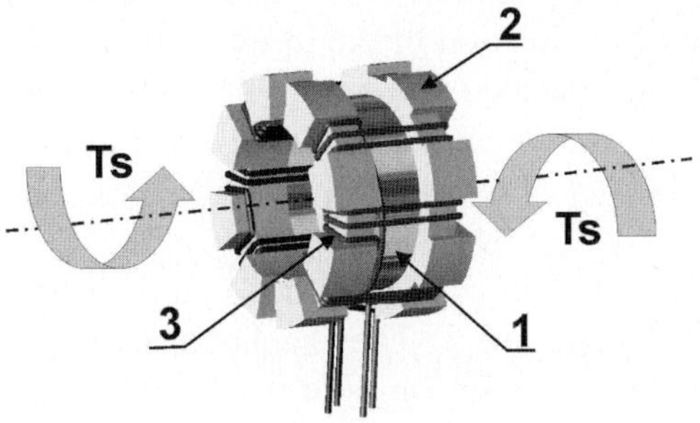

Fig. 1. Device for application of the torque to magnetoelastic ring-shaped core [4]: 1 – magnetoelastic sensing core, 2 – non-magnetic backings with grooves for winding, 3 – magnetizing and sensing windings

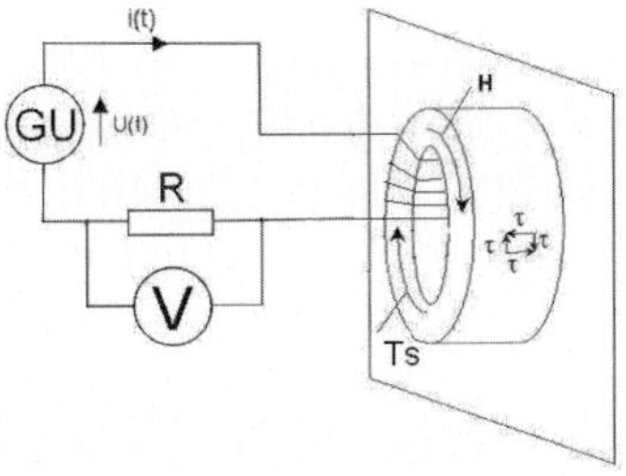

Fig. 2. Magnetoelastic sensor in single-coil configuration: GU – Voltage Generator, R – resistor, V – voltmeter

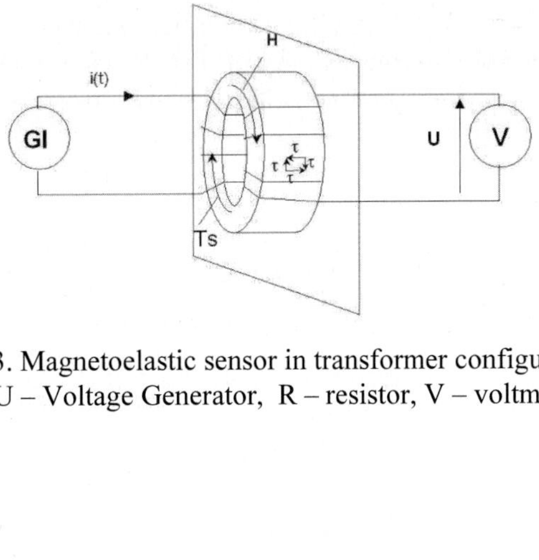

Fig. 3. Magnetoelastic sensor in transformer configuration: GU – Voltage Generator, R – resistor, V – voltmeter

2. Method of investigation

New methodology of application of the torque to the ring shaped core is based on the idea of application of the torque moment T_s in parallel direction to the direction of the axis of the ring [4]. As a result uniform distribution of the shear stresses τ can be achieved. This uniform distribution of stresses (non-uniformity is less than 5%) was confirmed during the thermovision tests [5].

Device for practical utilization of developed method of application of the torque is presented in figure 1. The magnetoelastic ring-shaped sensing core (1) is mounted to the base planes of the special nonmagnetic backings (2). Due to these grooves the core can be winded by magnetizing and sensing winding (3) and the changes of sensing element's parameters under influence of torque T_s can be measured. These backings are similar to presented previously backings for application of the uniform compressive stress to the ring core [6].

For development of magnetoelastic torque sensors two configuration of electronic transducer were proposed: single-coil configuration and transformer configuration. In single coil configuration changes of the impedance of amorphous alloy based inductive element result in changes of the value of current in the circuit. This current is measured on resistor R with 1 ohm resistance, whereas it is supplied from the sine-wave voltage generator GU with given maximal voltage. In the transformer configuration primary winding is supplied from sine-wave current source GI with given value of maximal current. Output signal is generated by the secondary winding of the transformer.

In both cases the TrueRMS value of output signal were measured by specialized multimeter. It should be indicated, that TrueRMS value of output voltage is much more robust on clutter than for example peak-to-peak amplitude. For this reason measurements of TrueRMS signal increase accuracy and repeatability of magnetoelastic torque sensors.

3. Results

In figure 4 the influence of torque Ts on the magnetic characteristics is presented. Under the torque up to 4 Nm shape of hysteresis loop of $Fe_{40}Ni_{38}Mo_4B_{18}$ amorphous alloy annealed in 380 °C for 1 hour changes significantly. As it is presented in figure 4a the value of both remanence B_r and maximal flux density decreases whereas coercive force H_c is nearly constant. The influence of torque T_s on the value of the maximal flux density B (achieved for magnetizing H_m) is presented in the figure 4b. It should be indicate, that for magnetizing field H_m equal 5 A/m, value of flux density B decreases up to 75%.

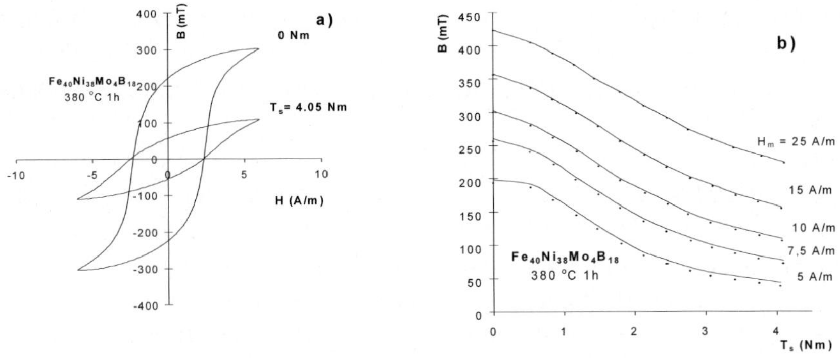

Fig. 4. The magnetoelastic characteristics of ring-shaped cores made of $Fe_{40}Ni_{38}Mo_4B_{18}$ amorphous alloy annealed in 380 °C for 1 hour
a) torque dependence of hysteresis loops, b) influence of torque on maximal value of flux density in the core, $B(T_s)_H$ characteristics

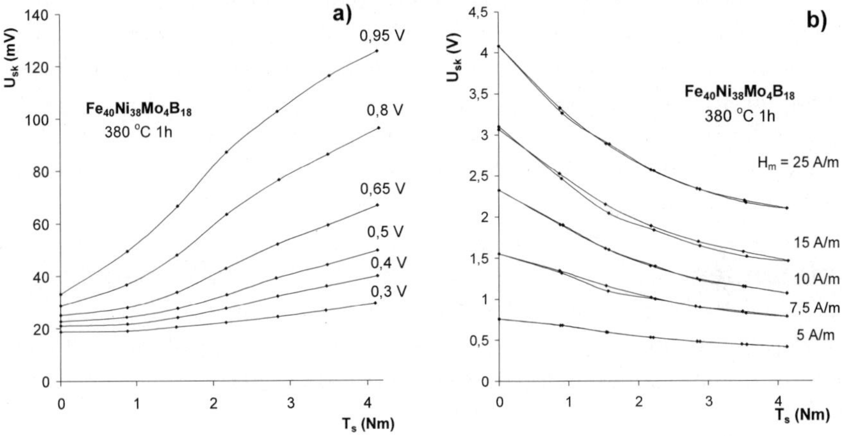

Fig. 5. The characteristics magnetoelastic torque sensors with core made of $Fe_{40}Ni_{38}Mo_4B_{18}$ amorphous alloy annealed in 380 °C for 1 hour
a) in single-coil configuration, b) in transformer configuration

In figure 5a the characteristics of the torque sensor operating in single-coil configuration are presented. It should be indicated, that both high sensitivity as well as negligible value of hysteresis were observed. In figure 5b the characteristics of sensor in transformer configuration are presented. Maximal value of the current in the primary coil of transformer is presented as value of magnetizing field in the core generated by this current. This gives possibility of connection between sensor characteristics and characteristics of the material presented in figure 4. Also in this case high sensitivity of the sensor was observed. On the other hand in transformer configuration significant hysteresis on the sensor characteristic was observed. For this reason single-coil configuration is much more useful for practical applications.

4. Conclusion

Presented method of application of the torque to the ring shaped sensing element may be successfully utilized in development of torque sensors. In such sensors the uniform distribution of shearing stress can be achieved as well as core can be winded for both single-coil as well as transformer configuration operation
Experimental results indicate high magnetoelastic sensitivity of cores made of $Fe_{40}Ni_{38}Mo_4B_{18}$ amorphous alloy annealed in 380 °C for 1 hour. In the core subjected to torque Ts up to 4 Nm and magnetized by the magnetizing field H_m equal 5 A/m, value of flux density B decreases up to 75%.
Magnetoelastic torque sensor operating in the single-core configuration exhibit both high sensitivity as well as negligible magnetoelastic hysteresis. For this reason single-coil configuration seems to be the optimal configuration for such sensors.

This work was supported by Polish Ministry of Education and Science under the Grant financed in the years 2005-2007.

References

[1] A. Bienkowski, R. Szewczyk, Sensors & Actuators 113 (2004) 270.
[2] R. Hasegawa J. Magn. Magn. Mater. 41 (1984) 79.
[3] T. Meydan J. Magn. Magn. Mater. 133 (1994) 525.
[4] J. Salach, A. Bienkowski, R. Szewczyk, Patent Pending, P-370124, 2004.
[5] J. Salach, A. Bienkowski, R. Szewczyk, Proceedings of IEEE, Sensors (2004) 505.
[6] R. Szewczyk, A. Bieńkowski, R. Kolano, Cryst. Res. Technol. 38 (2003) 320.

Subjective video quality evaluation: an influence of a number of subjects on the measurement stability

R. Kłoda, A. Ostaszewska

Warsaw University of Technology, Institute of Metrology and Measurement Systems, Sw. Andrzeja Boboli 8, Warsaw, 02-525, Poland

Abstract

Authors present results of examination of bitstream influence on compressed video quality. Discussed experiments were carried out by DCR method on a work station designed in Institute of Metrology and Measurement Systems. For data analysis the Wilcoxon matched pairs test was used.

1. Introduction

Commonly accepted way of compressed video quality evaluation is test with observers' participation. Results of such examination enable for finding bitrate appropriate for given material. Studies on relationship between coding parameters and video quality are conducted by numerous research centers all over the world. Published test results are limited only to mean observers' score MOS [1, 2]. In the paper authors make an attempt of statistical interpretation of obtained results, inter alia to answer the question: what the observers' number should be.

2. An example of statistical analysis of test results

The example presented here concerns quality evaluation of sequences coded in MPEG-2 format, which is used in DVD-Video standard. Researches were conducted with the use of working station constructed by authors. The DCR (Degradation Category Rating) method was used, in accordance with ITU-T Recommendation P.911 [4]. In case of DCR method a pair of sequences is presented. The first is always the source reference, the second (displayed after 2 seconds) – is the coded sequence as it is presented in figure 1.

Fig. 1. Scheme of DCR method

The subjects are asked to rate the impairment of the second stimulus in relation to the reference. The five-level scale for rating the impairment is used: 5 - imperceptible, 4 - perceptible but not annoying, 3 - slightly annoying, 2 - annoying, 1 - very annoying. The example presents results of three 10-seconds long sequences evaluation – each of them of different amount of temporal and spatial information. The bitrate (Vb) was rated as follows: 1, 2, 3, 4, 5, 7, 9 Mbps, which spans all range of Vb in DVD-Video standard. The observers were mostly students of Mechatronics Department. The mean opinion score given by 52 observers is presented in figure 2.

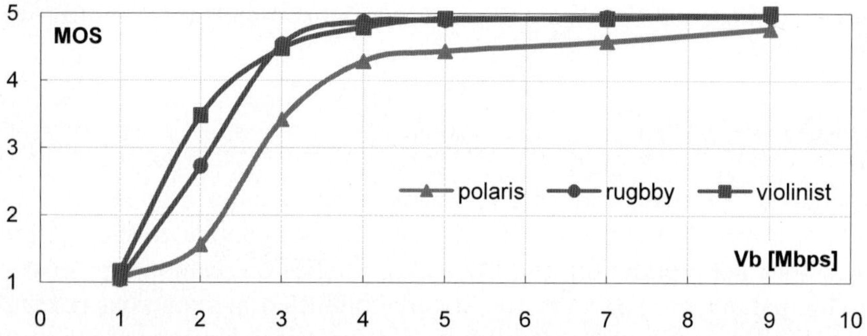

Fig. 2. Mean opinion score from 52 observers for three sequences

All curves are monotone but their shape is different. The shape of the curves is dependant on the video content, its dynamic and the level of complexity.

In the 'violinist' sequence there is a musician playing the violin. The picture tends to be static with a great amount of details. The 'rugby' sequence is a GSM operator's commercial. It is made up of several dynamic shots and contains lots of details. The 'polaris' sequence is a computer animation- dynamic, colorful with a strong noise.

The plot can be divided into two parts. The first ($1 \leq Vb < 4$ Mbps) is a rapid monotone of quality. The second ($Vb \geq 4$ Mbps) is the range of bitrate where subjects' scores become stable.
The increase of bitstream results in augment of file size. The issue is to find the value of bitrate for which the growth of quality stops being statistically significant. In this case it is necessary to conduct tests. To choose a test type, the distribution of scores must be recognized, with set of statistical parameters. For 'polaris sequence' (Tab. 1) the standard deviation depends strongly on Vb (the highest value for Vb range of 3 Mbps). The skeweness also varies (from –1.89 to 3.91), which gives the evidence that the distribution is not normal. It is also apparent in the histograms presented in figure 3.

Tab.1. Statistical parameters for 'polaris'

Vb	1	2	3	4	5	7	9
Mean	1.06	1.56	3.42	4.29	4.44	4.58	4.77
St.dev	0.24	0.76	0.75	0.72	0.61	0.50	0.47
Skewness	3.91	1.23	-0.60	-0.83	-0.59	-0.32	-1.89

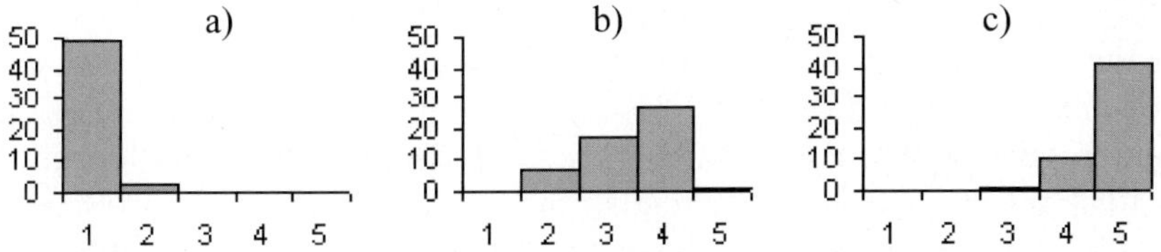

Fig. 3. Histograms for scores given to 'polaris':
a) Vb = 1 Mbps, b) Vb = 3 Mbps, c) Vb = 9 Mbps

Because the scores doesn't have the normal distribution, no parametric test may be used. The Wilcoxon matched pairs test is appropriate for this purpose. With this test it was checked if the increase of quality is statistically significant in case of two adjacent Vb values.
The null hypothesis is H_0: $median_1 = median_2$, the alternative hypothesis is H_1: $median_1 < median_2$. The significance level was $\alpha = 0{,}05$. The test results are presented in table 2.

Tab. 2. The results of Wilcoxon test and the interpretation

i	Pairs of analyzed sequences	p-Value	interpretation
1	Vb = 1 & Vb = 2	2.70E-05	D
2	Vb = 2 & Vb = 3	1.12E-09	D
3	Vb = 3 & Vb = 4	1.69E-07	D
4	Vb = 4 & Vb = 5	0.161796	ND
5	Vb = 5 & Vb = 7	0.191343	ND
6	Vb = 7 & Vb = 9	0.050008	ND
7	Vb = 4 & Vb = 7	0.014894	D
8	Vb = 4 & Vb = 9	0.000283	D
9	Vb = 5 & Vb = 9	0.002962	D
D – significant difference, ND – no significant difference			

The results indicate that the quality improvement caused by bitrate increase in a range of 1 to 4 Mbps is statistically significant ($p \le 0.05$). For higher Vb values there is no sufficient evidence for quality improvement.

3. The subjects number influence on the experiment results

The number of subjects, who participate in research, according to Recommendations [4] should be 6 – 40. Usually the number of observers in test is a compromise between costs and precision of assessment i.e. confidence of the final result. For the normal distribution the confidence level goes down by the square root of the number of observers. Because in this case data are not based on the normal distribution, the influence of the number of subjects on the results of statistical deduction can be examined in an experimental way only.

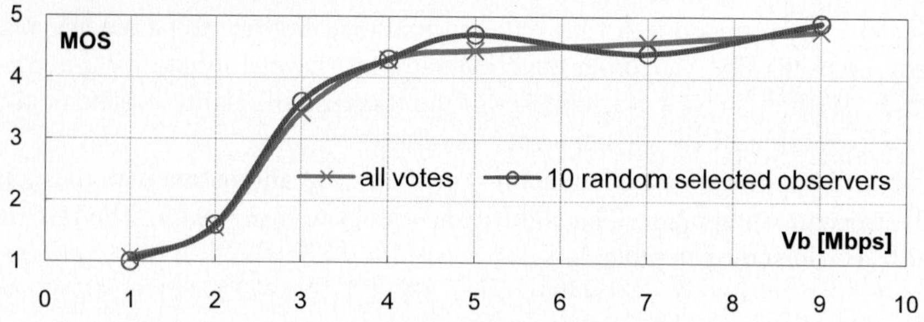

Fig. 4. Scores for 'polaris' sequence

The conducted experiment consisted in sampling of 10 observation subsets from the set of 52 observations. The plots were of similar character to the plot of all observations, but some curves are not monotone. Such a case is shown in figure 4.

Tab. 2. The results of Wilcoxon test and the interpretation for 10 subject

i	Pairs of analyzed sequences	p-Value	interpretation
1	Vb = 1 & Vb = 2	0.027715	D
2	Vb = 2 & Vb = 3	0.005065	D
3	Vb = 3 & Vb = 4	0.027715	D
4	Vb = 4 & Vb = 5	0.067898	ND
5	Vb = 5 & Vb = 7	0.224925	ND
6	Vb = 7 & Vb = 9	0.043123	D
7	Vb = 4 & Vb = 7	0.685833	ND
8	Vb = 4 & Vb = 9	0.043123	D
9	Vb = 5 & Vb = 9	0.361317	ND
D – significant difference, ND – no significant difference			

The result of Wilcoxon test is that there is no quality increase for the pair i = 5 (the average score even decreased), whereas there is an unexpected growth of quality for the pair i = 6. The Wilcoxon matched pairs test for wider ranges i = 7 .. 9 in the second part of the plot (Vb > 4 Mbps) also revealed the lack of stability of scores as it presented in table 3.
The analysis carried out with this method on groups of 10 randomly selected subjects corroborated that for 1 Mbps $\leq$ Vb < 4 Mbps the decrease of the subjects number has no influence on the stability of scores.
For Vb $\geq$ 4 Mbps it is not possible to claim what the influence of a subjects number on stability is, because there is no significant increase of quality.

References

[1] Q. Huynh-Thu and M. Ghanbari (UK): A Comparison of Subjective Video Quality Assessment Methods for Low-Bit Rate and Low-Resolution Video Proceeding (479) Signal and Image Processing – 2005.

[2] N. Suresh and N. Jayant (USA): Subjective Video Quality Metrics based on Failure Statistics Proceeding (493) Circuits, Signals, and Systems – 2005.

[3] Y. Kato and K. Hakozaki: A Video Classification Method using User Perceptive Video Quality, Proceeding (516) Internet and Multimedia Systems and Applications – 2006.

[4] ITU-T Recommendation P.911 (1996), Subjective audiovisual quality assessment methods for multimedia applications.

The grating interferometry and the strain gauge sensors in the magnetostriction strain measurements

L. Sałbut (a) *, K. Kuczyński (b), A. Bieńkowski (b), G. Dymny (a)

(a) Institute of Micromechanics and Photonics
Warsaw University of Technology
8 Św. A. Boboli St.,
02-525 Warsaw, Poland

(b) Institute of Metrology and Measurement Systems
Warsaw University of Technology
8 Św. A. Boboli St.,
02-525 Warsaw, Poland

Abstract

The paper presents newly developed measuring systems and theirs applications for testing of the strain distribution in the magnetostrictive materials. The system uses three measurement techniques: strain-gauge sensor for local strain measurement, grating interferometry for determination of in-plane displacement/strain distribution and classical two-beam interferometry for outt-of-plane displacement and total elongation measurement.

1. Introduction

Magnetostrictive effect is connected with dimension changes of soft magnetic materials during the process of theirs magnetization. Such materials are used in civil and military mechatronic devices as actuators and sensors. The paper presents newly developed measuring systems for testing of the magnetostrictive properties of the soft magnetic materials. The system uses simultaneously two measurement techniques: strain-gauge sensors for local strain measurement and optical: grating interferometry for in-plane dis-

placement/strain distribution determination and two-beams interferometry for out of plane and total elongation of the specimen under test.
Developed measuring installation creates new possibility of testing of the magnetostrictive properties of the soft magnetic materials for inductive components. Due to the simultaneous measurement of λ as a function of the magnetizing field H and monitoring the strain distribution in the measurement area, the new possibilities of the experimental verification of the theoretical models of the magnetomechanical effects can be obtained.

2. Strain gauge technique

The schematic block diagram of the installation utilizes semiconductor strain-gauge measuring techniques is given in Fig. 1. This solution creates possibility of simultaneous measurement of the magnetistriction λ and the flux density B in the function of magnetizing field H. Moreover developed measuring installation gives possibility of testing of the initial and the reversal hysteresis λ(H) loops. As a result it gives more complete information about magnetostrictive properties of tested materials, than known methods of measurement saturation magnetostriction λ_s, In addition simultaneous measurements of B, λ and H gives more complete information on process of the magnetization of the soft magnetic materials.

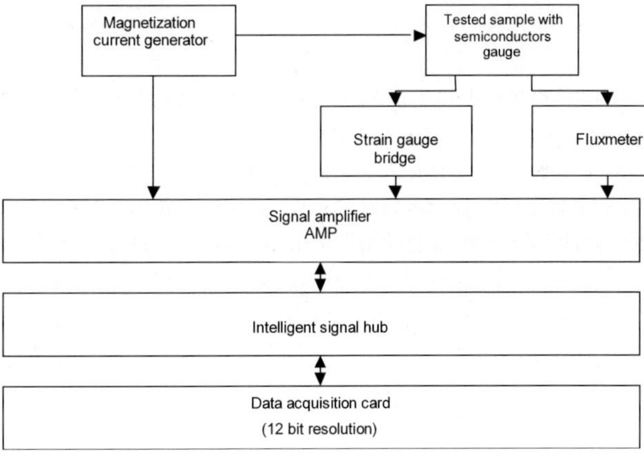

Fig. 1. Schematic of measuring system

Moreover results of this tests of the properties of the soft magnetic alloys is important for the constructors of the mechatronic inductive components with the soft magnetic materials cores [3].
One of the most important element, of utilized measuring methodology, is a digital method of the compensation of the influence of the heat generated

by current in magnetizing winding. This heat is the reason of increasing of the offset error of the strain-gauges. This problem is especially important during the measurements of the properties of nearly zero magnetostrictive materials. In this case the compensation of the influence of a heat generation is absolutely necessary to achieve acceptable resolution of the measuring system.

3. Interferometry

Grating interferometry (GI) is an optical method for in-plane displacement measurement with submicron sensitivity in the full field of view [1]. On a specimen subjected to analysis, a high frequency grating of equidistant lines is deposited. When the specimen is subjected to stresses, deformation of the specimen, and consequently of the grating applied to it, occurs. The deformed grating is then symmetrically illuminated by two mutually coherent beams with plane wave fronts. The incident angles of these beams are equal to the first diffraction order angle of the grating. In such configuration +1 and –1 diffraction orders beams propagate co-axially along the grating normal. The wave fronts of these beams are now not plane due to specimen grating deformation and the intensity distribution of the interferogram can be described as follow:

$$I(x,y) = a(x,y) + b(x,y)\cos\left[\frac{4\pi}{p}u(x,y)\right] \tag{1}$$

where a(x,y) and b(x,y) are the local values of background and contrast in an interferogram, u(x,y) represents in-plane displacements vector in direction perpendicular to the grating lines and p is the grating period. Note that the interference fringe represents a line constant displacement u(x,y). Along a fringe we have u = constant, and the difference in the u value between two consecutive fringes is $\Delta u = p/2$. For example, when using the specimen grating of spatial frequency 1200 lines/mm the basic sensitivity is 0.47 μm per fringe order.

GI is insensitive to out-of plane displacement, so if the information about it is required other type of interferometrers should be used. The most popular are classical interferometers working in Fizeau or Twyman-Green configurations [1]. In this case the intensity distribution can be expressed as follow:

$$I(x,y) = a(x,y) + b(x,y)\cos\left[\frac{2\pi}{\lambda}w(x,y)\right] \tag{2}$$

where λ is the wavelength of illuminating beam and w(x,y) represents out-of-plane displacement. These interferometers are insensitive to in-plane displacement, so both grating and classical interferometry, can be used simultaneously. Fig. 2 shows a schematic representation of the basic configuration of measurement system, combined GI andTwyman-Green interferometer (TGI), which may be used for sequential u(x,y), v(x,y) and w(x,y) displacement measurement.

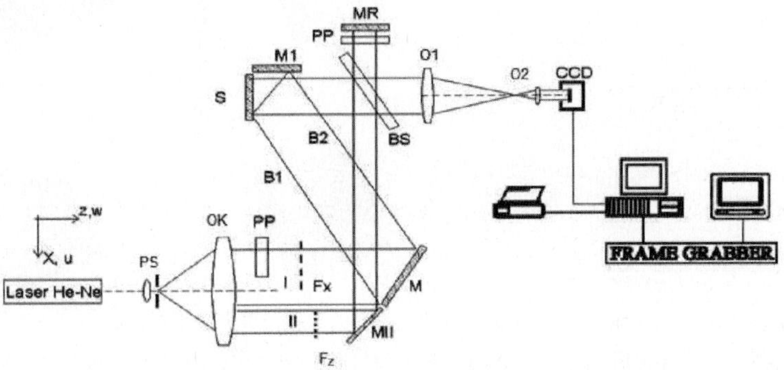

Fig. 2. Optomechanical configuration of the three-mirror four-beam grating interferometer modified for w(x,y) measurements

4. Experimental results

The specimen under test is shown in Fig.3. It is made from the (Fe2O3)50 (NiO)17,5 (ZnO)32 (CoO)0,5 ferrite and has dimensions 70 mm x 23 mm x 15 mm. The position of the specimen with diffraction grating and GI field of view (15 mm x 15 mm) is shown in Fig. 3a and placement of the strain gauges is presented In Fig. 3b.

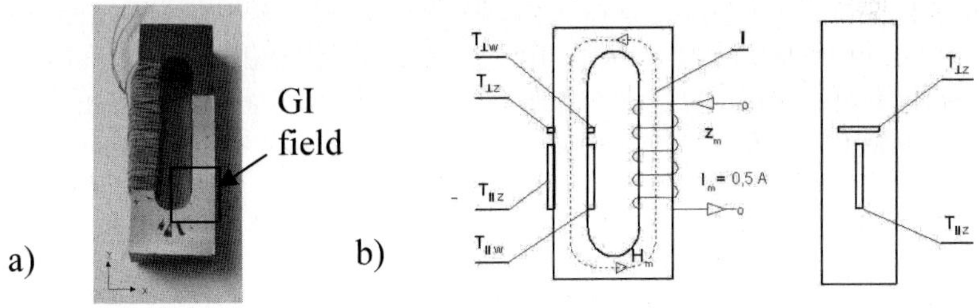

Fig. 3. Specimen under test

Exemplary results of testing of the initial and the reversal hysteresis λ(H) loops and in-plane displacement vector component and strains in y direction are shown in Fig. 4.

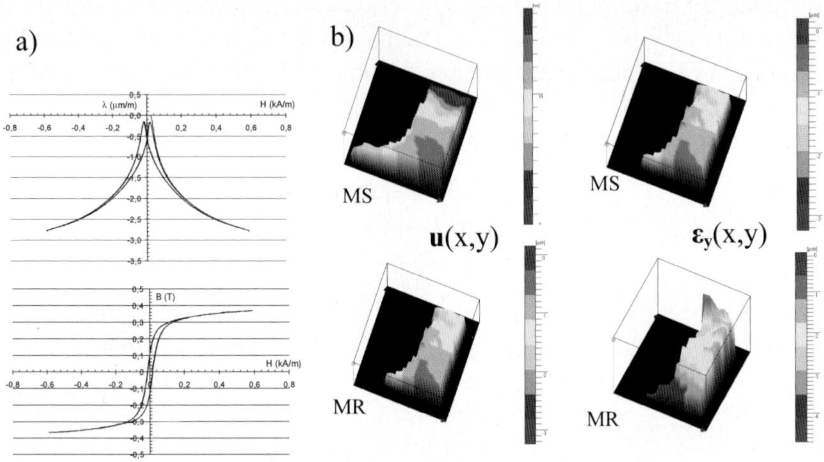

Fig. 4. Example of the $(Fe_2O_3)_{50}$ $(NiO)_{17,5}$ $(ZnO)_{32}$ $(CoO)_{0,5}$ ferrite specimen measurements: a) the initial curve and the magnetostrictive hysteresis loop obtained by semiconductor strain-gauge b) in plane displacements **u** and strains ε_y maps in magnetic saturation (MS) and remanence (MR) obtained by GI

5. Conclusions

Application of interferometric methods combined with strain gauge sensors for testing of magnetostrictive efect is presented. Grating interferomtry (GI) modified by adding Twyman-Green interferometer enables measurement of **u**, **v** and **w** displacements distribution. Measuring installation utilizes semiconductor strain-gauge measuring techniques. This solution creates possibility of simultaneous measurement of the magnetistriction λ in the function of magnetizing field H. Moreover developed measuring installation gives possibility of testing of the initial and the reversal hysteresis λ(H) loops. Results of this tests are important for the constructors of the mechatronic inductive components with the soft magnetic materials cores.

References

[1] K. Patorski, M.Kujawińska, L. Salbut „Interferometria Laserowa z Automatyczną Analizą Obrazu" OWPW, Warszawa, 2005.

[2] K. Kuczyński, A. Bieńkowski, R. Szewczyk "New measurening system for testing of the magnetostrictive properties of the soft magnetic materials"; 14th IMEKO TC4 SYMPOSIUM (2005) 434.

Micro-features measurement using meso-volume CMM

A. Wozniak (a) *, J.R.R. Mayer (b)

(a) Institute of Metrology and Measuring Systems, Warsaw University of Technology, Św. A. Boboli 8 Street, 02-525 Warsaw, Poland

(b) Département de génie mécanique, École Polytechnique de Montréal, C.P. 6079, succ. Centre-ville, Montréal, Canada

Abstract

Paper will discuss an understanding for the compensation of the probe ball radius in a scanning process of micro-features. As will be shown, the indigenous CMM software does not always adequately compensate the stylus tip radius. As a result, the information about the real shape of the measured features can be distorted. In order to accurately measure precise geometric features, and in particular micro–features, a new algorithm for the compensation of the stylus tip radius in a CMM scanning process will be proposed. To demonstrate the performance of the indigenous CMM software as well as feasibilities of our new algorithm, we will show the results of measurements of the profiles of precise micro-feature such as silicon micro-grove. Tests will be carrying out on a fixed bridge, moving-table Mitutoyo LEGEX 910 CMM equipped with a MPP-300 scanning probe and also on a Zeiss ACCURA CMM equipped with a VAST GOLD scanning probe.

1. Introduction

The new generation high performance coordinate measuring machines (CMM) equipped with scanning probes offers new and effective possibilities for shape measurement. The accuracy of these high-end CMM scanning probes is in the sub-micrometer range and thus could be useful for the measurement of precise geometric features like micro–features. However, CMM data processing software is often designed and tested on rather large features relative to the machine meso-volume (usually up to one meter sides) and to the stylus tip diameter (up to few millimetres). As a result, maintaining this accuracy for freeform surface measurement of small size features is hard to obtain because of the CMM built in algorithms for probe radius compensation.

The specified accuracy of these CMMs with scanning modes is in the sub-micrometer range. However, the probing system accuracy depends on the design parameters of the transducer, probe configuration, measuring strategy and the algorithms of probe radius compensation. The usual built-in real-time CMM software for processing scanned data points results in some distortion of the real shape of the surface. Most of these are based on NURBS [1] or other surface or 2D contour models to estimate the direction in which to apply the effective tip radius correction. The direction is etimated as the normal vector to the fitted indicated measured point surface. As a result, the information about the real shape of the measured surface can be distorted in certain cases.

In order to accurately measure on coordinate measuring machine we propose a new method for corrected measured point determination in a CMM measuring process.

2. Specified accuracy of CMMs and disadvantages of normal vector based methods of probe tip radius correction especially during micro-feature measurements

Nowadays, the methods that have been used to calibrate the CMM have relied on checking the standard calibration artifacts: like plate masters with reference balls or rings or gauge blocks (also according International Standards ISO 10360-2:2001[2]). This kind of calibration limits the information about the probe inaccuracy only to the selected shapes and gives too little information required for form measurements of any surface.

The high accuracy of CMMs obtained using tests as per ISO standards [2] is in the sub-micrometer range and thus could be useful for the measurement of precise geometric features. However, CMM data processing software is often designed and tested on rather large features (like a reference ball) relative to the stylus tip. As a result, maintaining this accuracy for freeform surface measurement is challenging for small features because of the CMM built in algorithms for probe radius compensation.

The stylus tip radius correction is an offset vector of norm equal to the effective stylus tip radius which is added to the indicated measured point (the measured stylus tip centre point) to estimate the actual contact point (the stylus tip contact point on the real surface), i.e. the corrected measured point on the surface. The contact between a sphere and a surface occurs where both surfaces have a common tangent. It follows that the offset vector is normal to the surface at the point of contact so the primary task for correction is to estimate this vector at each data points. However, in case of freeform surfaces the normal vectors have to be calculated taking into consideration the set of stylus tip centre points. Thus, because of inherent measuring machine inaccuracy, small deviations of centre point coordinates can cause big deviation in the direction of the normal vector calculated by the CMM software. As a result, we observe incoherently connected measured point patterns

3. Principle of the new method of corrected measured point determination in coordinate metrology

In order to accurately measure, amongst other things, small features, we propose a new algorithm for the compensation of the stylus tip radius in a CMM scanning process (as shown in Fig. 1). The proposed algorithm is dedicated to high definition measurement. Advantages of the algorithm are that we do not calculate the normal vector and we do not use a NURBS for smoothing (filtrating) of the measured shape.

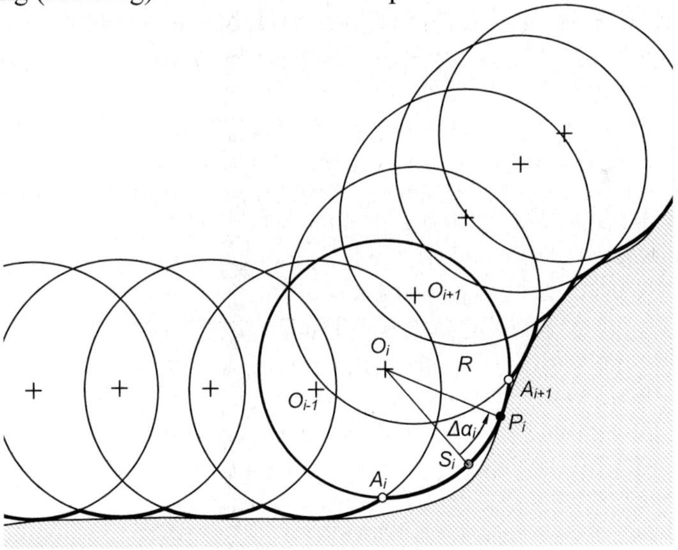

Fig. 1. Analysis of geometry of scanning path for corrected measured point determination in a CMM scanning process

The proposed compensation method involves the following steps:

A. Performing a series of high density measurements on the geometric characteristic to be measured with a spherical stylus tip;
B. The outline of the probe ball defines an arc for each measurement point;
C. Calculating for each arc the points of intersection A_i and A_{i+1} with the preceding and the following arcs;
D. For each arc, estimating the point of contact S_i with the characteristic as being the mid-point of the arc;
E. Determining an angular adjustment $\Delta\alpha i$ using fuzzy knowledge base and applying respectively a further compensation based on the corresponding angular adjustment.

The calculation of $\Delta\alpha i$ may be performed using a variety of known rule-based or other AI techniques such as neural networks. In the experimental implementation of our method we have opted for a calculation of $\Delta\alpha i$ using a fuzzy logic algorithm.

4. CMM machine set-ups and measuring results

Tests were carried out on two CMMs: high-end accuracy, fixed-bridge moving-table Mitutoyo LEGEX 910 equipped with a MPP-300 passive scanning probe and good accuracy, moving-bridge, made by Zeiss ACCURA equipped with a VAST GOLD active scanning probe. The LEGEX machine maximum permissible error of indication for size measurement MPE_E = 0.48+L/1000 μm (with L as size length in meters) whereas for ACCURA MPE_E = 1.7+L/333 μm as per standard ISO-10360 [2].
However, in case of measurement of small features in scanning mode the maximum permissible scanning probing error, MPE_{Tij} [2] equal 0.32μm and 2.7 μm for LEGEX and ACCURA respectively, is a more appropriate parameter to describe the expected precision.
A silicon micro-V-groove artefact has been used to demonstrate the feasibility of micro-features measurement using abovementioned meso-volume CMMs. Fig. 2 shows the results obtained with the Mitutoyo LEGEX. A small 0.3 mm probe stylus tip diameter was used. The measuring velocity and the sampling step were set to 0.2 mm/s and 11 μm respectively. Both the raw (uncompensated) data and corrected measured points obtained by the CMM's own software have been stored in computer files by the CMM. The dashed line and square points represent the corrected measured points using the CMM's own software. As can be seen, the corrected measured points obtained by the CMM's own software suggest that some distortion

occurs during the processing of the scanned indicated measured points resulting in an incorrect representation of the surface with points that do not exist. The errors are of the order of several tenths of millimetres. However, points processed with the proposed new method of compensation, represented by black dots, are all on the expected profile when using the same graphic scale and a 0,2 mm offset to graphically distinguish the two sets of data. In this case, the distortion of the measurement results is not visible. The new algorithm of compensation gives a better estimate of the measuring features because it does not show the unreal, non existing measurement data.

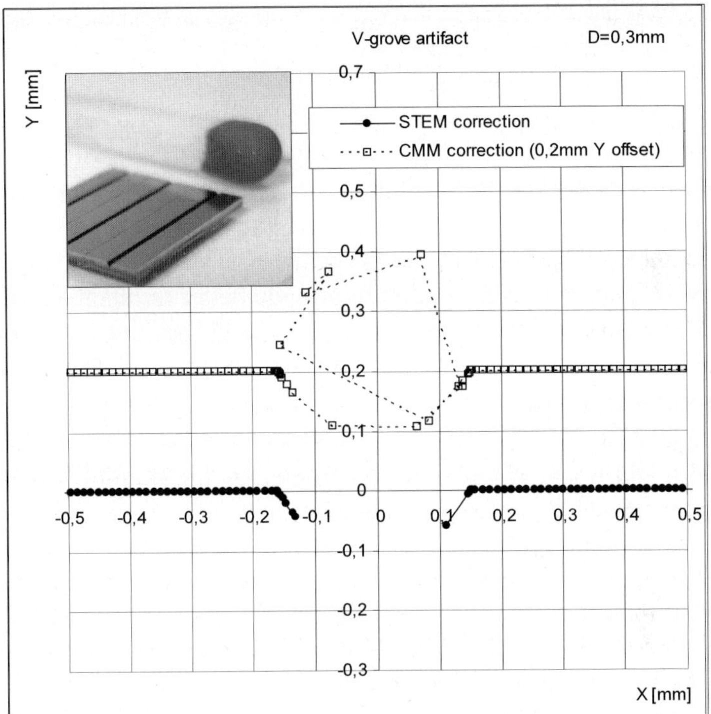

Fig. 2. Graph of the measurement result of a plane section of the micro-V-groove artefact using the LEGEX CMM built-in method and proposed correction method

Fig. 3 shows the results obtained with the Zeiss ACCURA CMM. Built-in compensation algorithm use splines to obtain a best fit curve (curled up continuous line) and its normal vectors (arrows) in order to compensate as a post-measurement operation. As can be seen, the corrected measured points obtained by the CMM built in algorithms for probe radius correction suggest that some points are performed in a non-existing continuous characteristic. Because of larger inherent measuring machine inaccuracy, deviations of centre point coordinates cause bigger deviation in the direction of the normal vector calculated by the CMM software. As a result, we observe incoherently connected corrected measured point patterns.

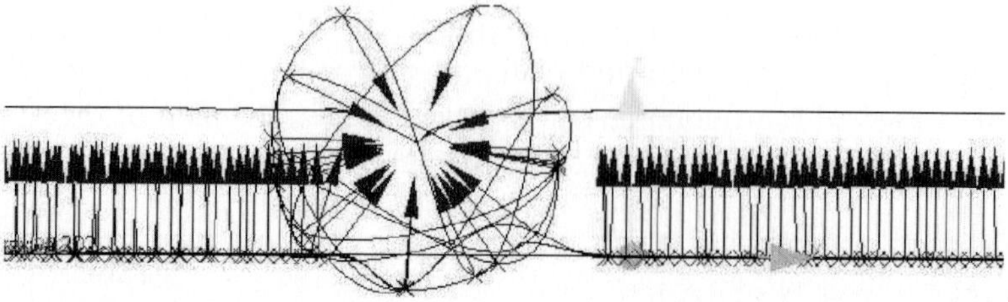

Fig. 3. Print screen of Zeiss ACCURA software with graph of the measurement result of section of the micro-V-groove artefact using the CMM built-in method

Currently used methods for tip radius correction may result in serious disordering of the corrected points. We proved that this is especially true for small feature measurements. The proposed new method of compensation could be an important contribution to provide better results then using current CMM built-in method.

Acknowledgment

This work and Adam Woźniak was supported by the Homing and Supporting Grant of Foundation for Polish Science.
René Mayer thanks Canada's NSERC for a Discovery Grant RGPIN-155677-02 and the Canadian Foundation for Innovation for equipment grant FCI-3000.
The authors would like to thank Mélissa Côté, research associate at École Polytechnique de Montréal for valuable discussion and assistance in some of the tests.

References

[1] Y. Zhongwei, Z. Yuping, J. Shouwei, "Methodology of NURBS surface fitting based on off-line software compensation of errors of a CMM", Precision Engineering 27 (2003), 299-303.
[2] ISO 10360-4 2000 Geometrical Product Specifications (GPS) – Acceptance and reverification tests for coordinate measuring machines (CMM) – Part 4: CMMs used in scanning measuring mode, Switzerland, 2000.

Distance measuring interferometer with zerodur based light frequency stabilization

M. Dobosz

Institute of Metrology and Measuring Systems,
Faculty of Mechatronics
Warsaw University of Technology
Sw. A. Boboli 8 St.
02-835 Warsaw, Poland

Abstract

A prototype of an interferometrically stabilized diode laser interferometer for linear displacement measurements is presented. We have proposed stabilization of laser diode output wavelength by using external interferometer that measures wavelength changes and controls laser diode supplying current in a feedback loop. Control interferometer is formed as a wedge plate made from zerodur that is characterized by a nearly zero temperature expansion coefficient. To obtain constant optical path difference Zerodur plate is thermally stabilized. Results of tests of metrological feasibilities of the interferometer are presented.

1. Introduction

Typical single mode emitting visible light laser diode (LD) coherence distance reaches about of 2m [1]. Potentially this source of light could be used as a cheap, long lifetime and a very small source of light in an laser interferometer for distance measurement. However LD is characterized by a strong dependence of the frequency of emitted light on junction temperature and current.

Since the laser diode light frequency can be easy tuned by controlling junction current we propose laser diode wavelength stabilization by means of an external interferometer, that is used to measure wavelength changes and control laser diode current in an feedback loop [2].

2. Control interferometer for LD frequency stabilization

Optical path difference in this external control interferometer is kept as stable as possible by using a temperature stabilized zerodur. Zerodur is a kind of glass having a very small temperature expansion coefficient (in the range from 0°C to 60°C class II zerodur expansion coefficient equals $=0\pm0.1\cdot10^{-6}K^{-1}\pm0.005\cdot10^{-6}K^{-1}$, with inaccuracy of its evaluation equal to $=0.025\cdot10^{-6}$), so that the geometrical dimension of the control interferometer remain constant when the ambient temperature changes. However the refraction index of zerodur is significantly influenced by its temperature. The coefficient of temperature change of the zerodur refraction index equals $14.4\cdot10^{-6}K^{-1}$. Thus to keep unchangeable optical path difference in the zerodur control interferometer the refraction coefficient of the zerodur plate is precisely temperature stabilized by the accurate temperature controller.

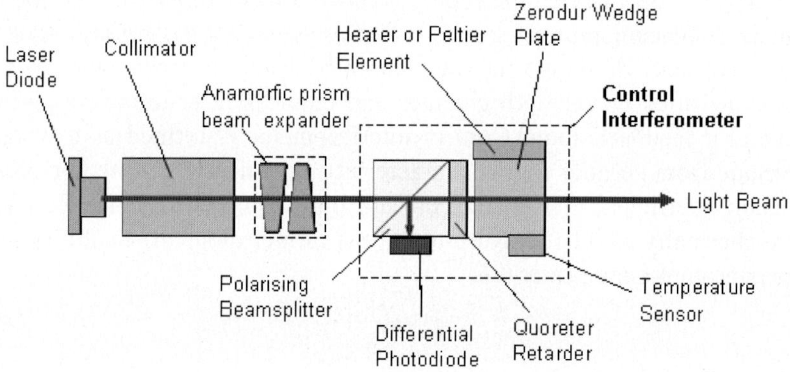

Fig. 1. Scheme of the control interferometer.

Scheme of the proposed control interferometer is shown on figure 1. Light beam emitted be the laser diode is collimated by collimating lens CL and circulated by means of an anamorfic prism beam expander. Next the light incidents on polarizing beam splitter BS. The incident LD beam is in high degree linearly polarized. The polarization is directed 45 degree relative to the base of the beamsplitter. As the result a majority of the light power is transmitted by the beamsplitter BS. Next the beam is going through the quarter retarder QR that changes its polarization into the circular. Next the

beam is transmitted by the wedge plate made from zerodur (ZWP). The light reflected back by facet plane surfaces of the wedge plate ZWP passes through the QR retarder, changing its polarization back to the linear. But the direction of this polarization is perpendicular to the polarization of the input beam. Thus the reflected back light is directed by the beamsplitter into a photodetector in a form of differential fotodiode DP. In the plane of the photodetector interference fringes are observed. Angle between zerodur wedge plate outer surfaces is adjusted in manufacturing process in such a way that the constant of the observed fringes is equal to twice of the distance between photodetector photodiodes. Differential photodiode DP analyses continuously a single interference fringe position that is depending on the light wavelength. Photodetector DP is connected to the LD driver by means of an electronic circuit that adds the required DC correction component to the laser diode supplying current. During normal operation the photodetector output signal controls in the feedback loop the laser diode driver to keep the stable position of the fringe relative to the photodetector axis.
Described above optical set of the control interferometer is mounted in a copper housing and thermally stabilized, using a heater or Peltier element and an accurate temperature sensor. View of the control interferometer is shown in figure 2.

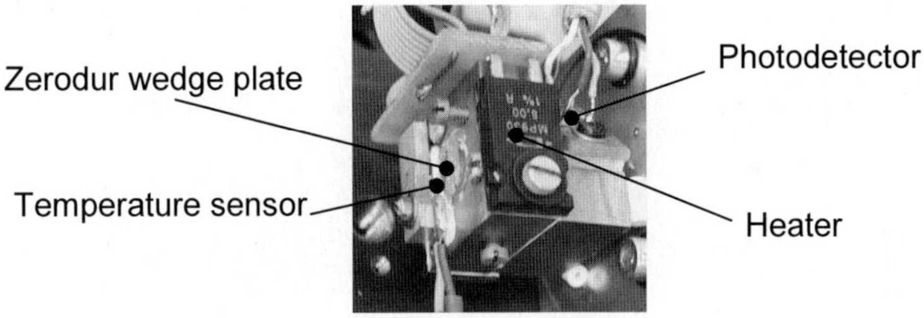

Fig. 2. View of the control interferometer

Finally, obtained level of the laser diode wavelength stabilization was equal to about 10^{-7}.

2. Distance measuring interferometer metrological feasibilities

View of the developed distance measuring interferometer with described above zerodur based light frequency stabilization is show In Fig.3. Interferometer electronic circuit allows distance measurements with resolution of about 1 nm. LD interferometer repeatability has been evaluated by means of a reference HeNe laser interferometer having resolution 5nm (system ZLM500 manufactured by Zeiss). During the performed tests both interferometers measured displacements of the same vibrating element.

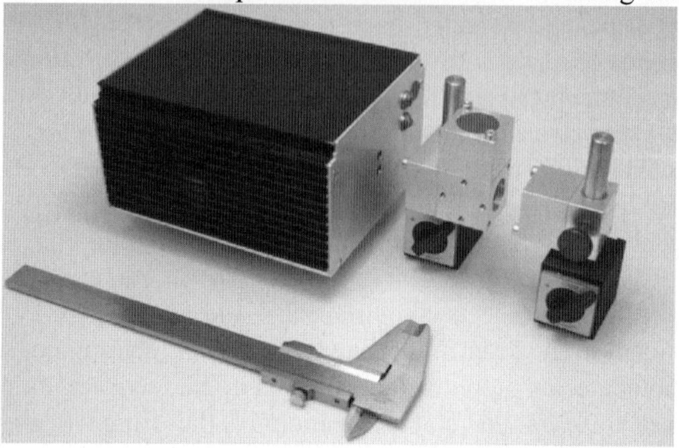

Fig. 3. View of the developed interferometer

Exemplary results of both interferometer's measurements comparison is show in fig. 4.

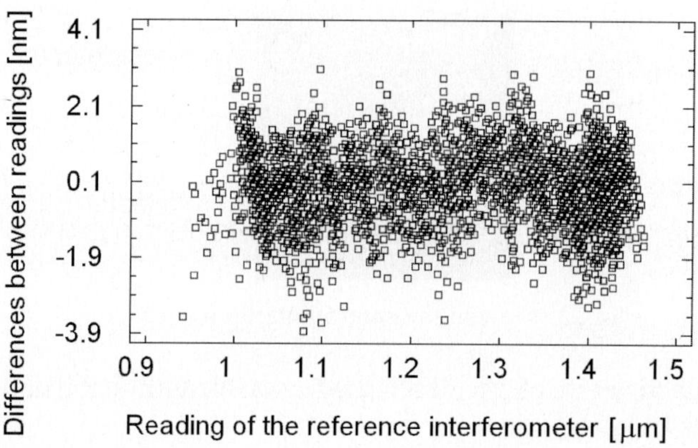

Fig. 4. Differences between interferometer's readings

Observed difference in interferometers reading were caused mainly by this discretization errors of the tested and reference interferometers.
Position evaluation repeatability of the developed interferometer is illustrated in Fig. 5. Presented dependence shows that the position evaluation repeatability decreases proportionally to the distance between reference

and measuring interferometer reflector. This dependence is caused by the noise inherent to the LD injection current. 2 meter measuring range of the system is visible.

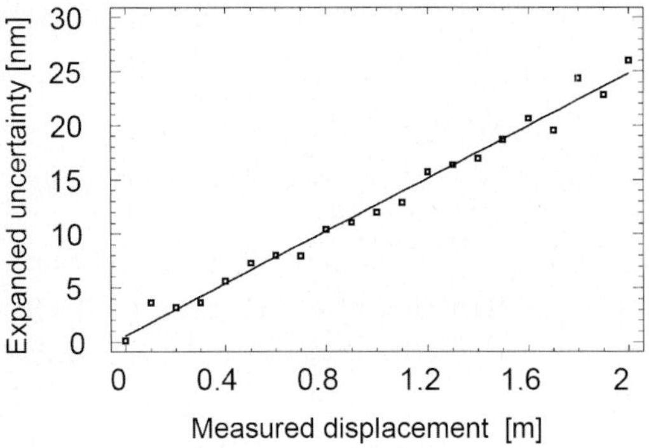

Fig. 5. Expanded uncertainty of the interferometer reflector position evaluation calculated using coverage factor k=2 versus measured displacement of the interferometer.

3. Conclusions

A model of interferometrically stabilized diode laser interferometer for linear displacement measurements has been presented. Laser diode has been stabilized by means of the zerodur wedge plate based external interferometer. Obtained relative level of wavelength stabilization is equal to about 10^{-7}. Resolution of the developed interferometer is equal 1nm. Expanded uncertainty of the interferometer reflector position evaluation changes from about ±3 nm, ±30 nm at the beginning and at the end of the measuring range respectively. Two meter measuring range of the system has been shown.

Acknowledgments

This work was sponsored by the Polish Science Research Committee, Grant No. T07D 021 23

References

[1] CH Henry. Theory of the linewidth of semiconductor lasers. IEEE J Quantum Electron 1982; QE-I8: 259-264.

[2] Dobosz M, Układ interferometru zwłaszcza do pomiaru przemieszczeń liniowych i kątowych. Patent application No. P 354270.

Application of CFD for the purposes of dust and mist measurements

M. Turkowski (a)

(a) Warsaw University of Technology, Institute of Metrology and Measurement Systems, ul. Boboli 8, 02 525 Warszawa, Poland

Abstract

The dust and mist load in the natural gas becomes increasingly important parameter of the gas quality. The contemporary methods of the dust and mist load measurement consist on the analyse of the effects of light scattering in the dust & mist particles. These methods need however the precise shaping of the stream profile and path lines in the region where the measurements take place. The paper presents the results of designing the nozzle for these purposes with the use of the CFD modelling.

1. Introduction

As the dust and mist load in the natural gas becomes increasingly important parameter of the gas quality the new methods enabling the measurements of this parameter on-line are developing now very fast. One of the contemporary methods consist on the measurements of the effects of light scattering in the dust & mist particles. These methods need however the precise shaping of the stream path lines and profile in the research and calibration facility. The Computational Fluid Dynamics (CFD) has proved to be very useful tool for the design of the nozzle assuring the appropriate profile of the stream.

2. Desirable flow patterns in the region of measurement

It has been established that the measurement area should have about 2 mm diameter. Through the nozzle of about 2 mm diameter will flow the air loaded with particles. This is the reference stream, so the quantity and dimensions of particles are precisely known. The velocity of the stream should be close to the velocity of the surrounding clean air to minimize the turbulent mixing of both central and surrounding stream. Moreover, the velocity profile of the central stream loaded with particles should be as far as possible flat. The zone of measurement should be at a distance about 10 mm from the stream outlet.

In order to obtain such flow characteristics the converging nozzle was used. Its profile was close to the profile of the Witoszyński nozzle [1]. This type of nozzles is widely used in flow metrology to obtain flat velocity profile in the primary standards [2] and, for the same reasons, in automotive wind tunnels [3].

To resolve the problem the computational fluid dynamics (CFD) methods were used. The proposed geometry of the nozzle was meshed, using both 2D and 3D meshes – see fig. 1 and 2. Finally, after some first modeling experiments the 2D model has been chosen as for the axisymmetrical flow it is sufficient accurate and need less computational power.

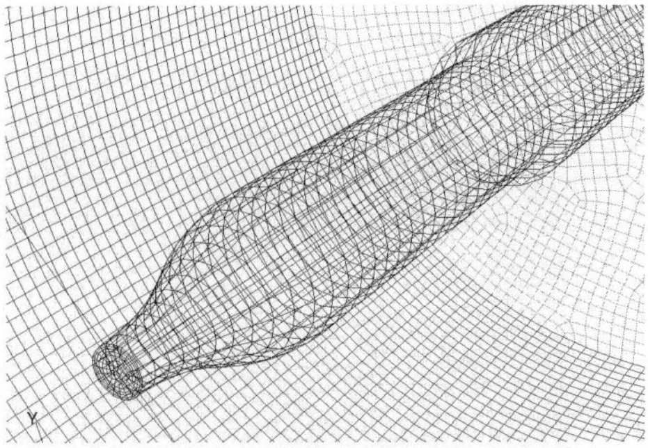

Fig. 1 The 3D mesh used for CFD simulation

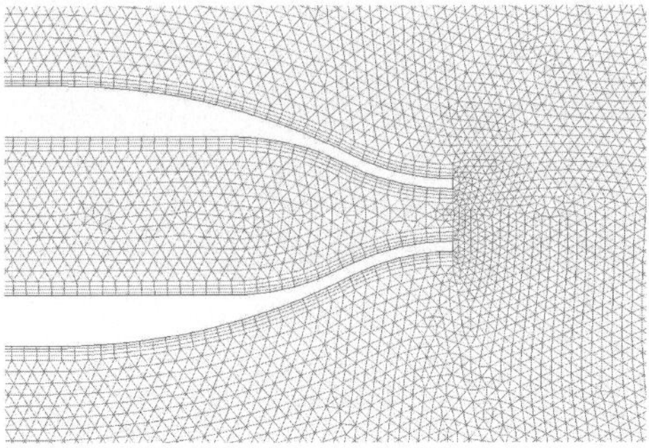

Fig. 2 The 2D mesh used for CFD simulation

3. Results of the modeling

The calculated pressure and velocity fields have been shown in fig. 3 and 4. The gauge pressure at the inlet of the nozzle was 2 Pa. The resulting outlet velocity was about 2.2 m/s. There is almost exact relation between dynamic pressure p and velocity v according to the Bernoulli law ($p = \rho v^2/2$), so this relation can be used as a thumb rule to operate the stand.
The resulting velocity profile at nozzle exit is however rather sharp, almost parabolic as for laminar flow (fig. 5). It is no surprise – the Reynolds number low (Re = 200), in the laminar region. At a distance 10 mm from the nozzle the velocity profile is more flat. The difference of velocities between the central and surrounding stream is at this location not very high (about 30 %). The flow paths (fig. 6) downstream the nozzle exit show, that at a distance of 10 mm the mixing with surrounding clean air will be still not intensive.

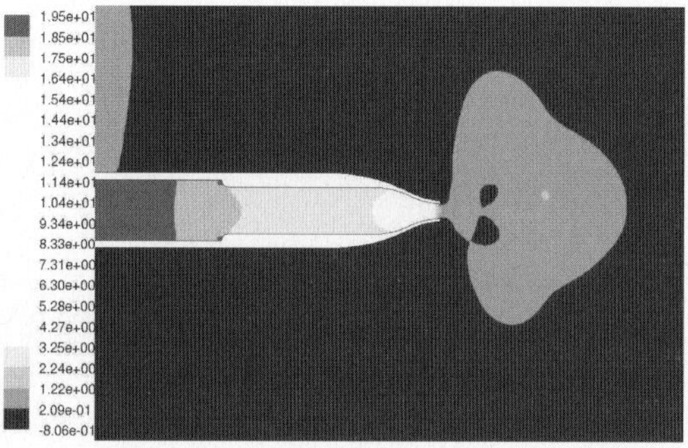

Fig. 3 Pressure field in the nozzle region. Inlet pressure 2 Pa.

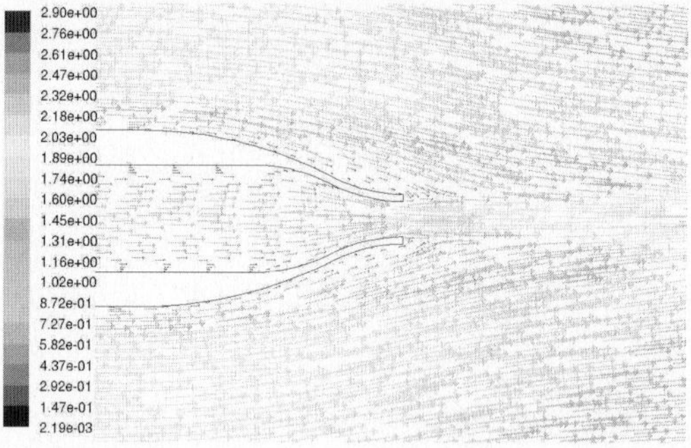

Fig. 4 Velocity field in the nozzle region

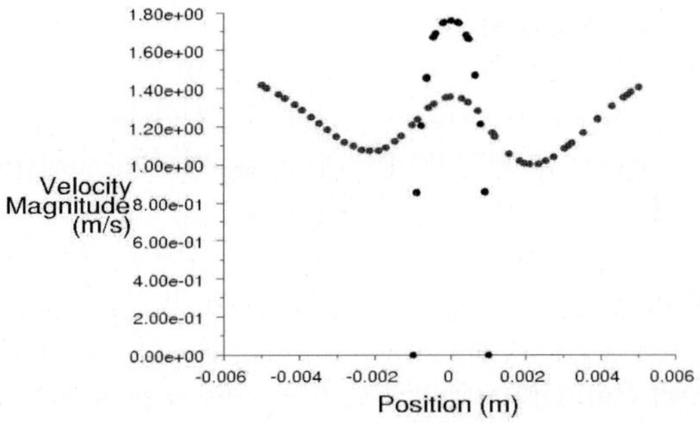

Fig. 5 Velocity profiles at nozzle exit (sharp profile) and 10 mm form the exit (wavy profile) for the case presented in fig. 4

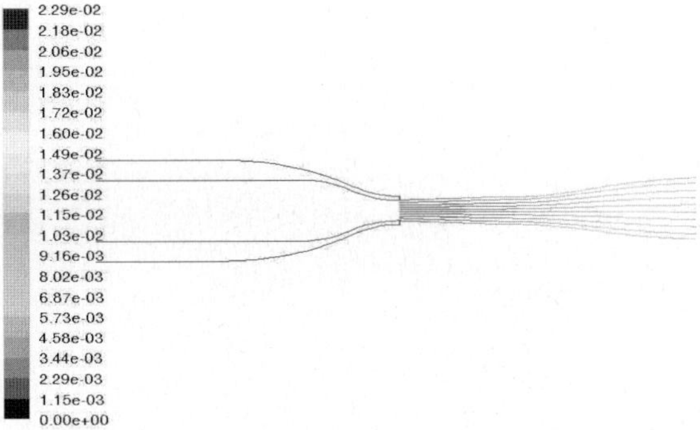

Fig. 6 Flow paths lines downstream the nozzle exit

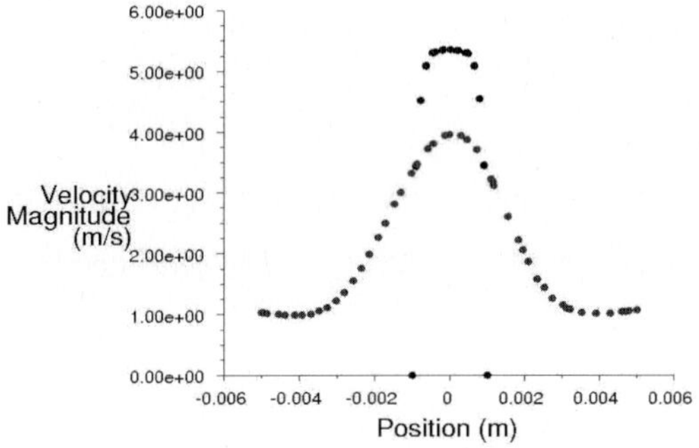

Fig. 7 Velocity profiles at nozzle exit (black) and 10 mm form the exit (bell shape) for the inlet pressure 20 Pa

The fig. 7 shows, that at inlet pressure 10 times higher (20 Pa) the velocity rises about 3 times, up to almost 6 m/s (that is consistent with Bernoulli law). Much more flat, better velocity profile can be observed, although the Reynolds number is still low ($Re = 600$), so still in the laminar flow region. The higher velocities can however bring more problems in point of view of particle measurement so some compromising solution has to be accepted after first experiments.

4. Conclusions

The Computational Fluid Dynamics (CFD) has proved to be very useful tool for the design of the nozzle assuring the appropriate profile of the stream. The further research will be conducted with the use of the designed nozzle presented at fig. 8.

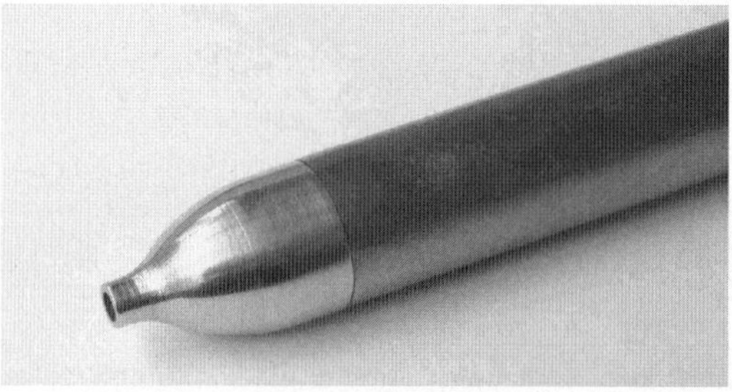

Fig. 8 The view of the designed flow nozzle

References

[1] Witoszyński Cz.: Prace wybrane. PWN, Warszawa, 1957

[2] Dopheide D., Taux G., Krey E.A: Development of a New Fundamental Measuring Technique for the Accurate Measurement of Gas Flowrates by Means of Laser Doppler Anemometry, Metrologia, 27, 1990, pp. 65-73

[3] Wolf T.: Design of a variable contraction for a full scale automotive wind tunnel. Journal of Wind Engineering and Industrial Aerodynamics, vol. 56, Issue 1, April 1995, pp. 1-26

Optomechatronics cameras for full-field, remote monitoring and measurements of mechanical parts

L.Salbut *, M.Kujawińska, A.Michałkiewicz, G.Dymny

Institute of Micromechanics and Photonics,
8 Sw. A. Boboli St., 02-525 Warsaw, Poland

Abstract

Optomechatronics cameras based on digital holography and grating interferometry are presented. Their compact design, "black box" operational approach and interferometric data processing are described. The representative examples of applications in experimental mechanics, microelements testing and outdoor engineering measurements are presented.

1. Introduction

Fast growing technology and requirements for testing of different types of materials and devices require new methods and systems for investigation of their parameters in laboratory, workshop and outdoor conditions. Main interesting quantities are: shape and shape deformation, local materials constants and displacement/strain distributions. Several optical techniques have been already developed for these measurements [1]. One of them is holographic interferometry and in particular its digital version (DHI), which provides a simple way to record and restore amplitude and phase of an investigated object [2]. It allows, using amplitude information, distant monitoring of object and remote optoelectronic reconstruction. Proper manipulation of phase could provide information about shape of an object and its displacement during loading. Other technique, developed for in-plane displacements/strain measurement and monitoring, is grating interferometry (GI) [3]. In the paper a new optomechatronics cameras based on both, DHI and GI, techniques are introduced. Those cameras systems with compact design and "black box" measurement approach allow fast and accurate measurement performed directly at the object under test and often in outdoor environment. Design details and errors analysis of cameras as well as results of their initial applications are presented and discussed.

2. Digital holographic cameras

Digital holography provides a way to record and restore amplitude and phase of investigated object [2]. It allows, using amplitude information, distant monitoring of object and remote optoelectronic reconstruction. Proper manipulation of phase could provide information about shape of an object and its out-of-plane and in-plane displacements during internal or external loading. The architecture of holographic system (Fig.1) which consists of two types of measurement heads, fibre optics link and control/illumination module is presented. One of the measurement head is configured to perform out-of-plane displacement and shape measurement. The second one (with four illumination beams) allows to measure (u,v,w) displacement fields. The data from both cameras can be transferred remotely to optoelectronic reconstruction station based on LCOS spatial light modulator. Due to fibre optics light delivery system the DHI head is able to work in a distance from its electronic/processing part and it allows direct access to all mechanical parts of machinery. The DHI head can be hand-held or can be mounted directly at a machine.

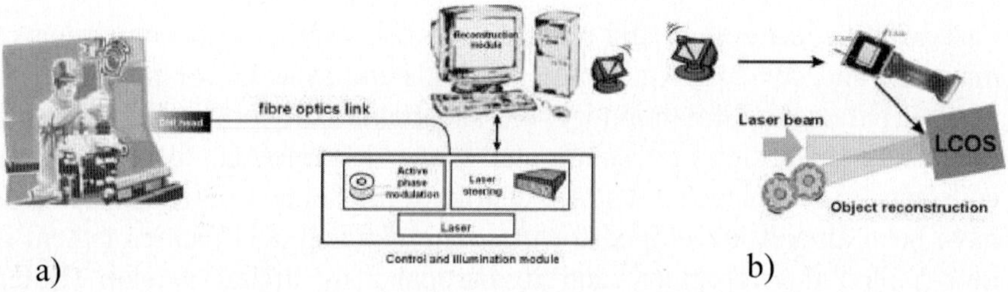

Fig. 1. Scheme of DHI system for mechanical parts monitoring a) measurement module, b) optoelectronic reconstruction module.

In practice two main configurations are applied depending on the choice of measurand, namely: the out-of-plane displacement/shape measurement setup and full displacement vector (u,v,w) setup. The hardware and software solutions are different for both systems and therefore in order to provide user friendly cameras for industrial inspection two separate devices have been designed and built: DH_SHAPE and DH_UVW. All the optical elements and detector are closed in a tube with the diameter 50 mm. Electronic module with laser and PZT controllers and synchronization is separated from holographic head. Also for specific applications the cameras are equipped with the remote data transfer and used for both numerical or optoelectronic distance reconstruction systems. The holocamera DH_SHAPE, designed for shape determination and out-of-plane displacement measurement, is shown in Fig. 2a.

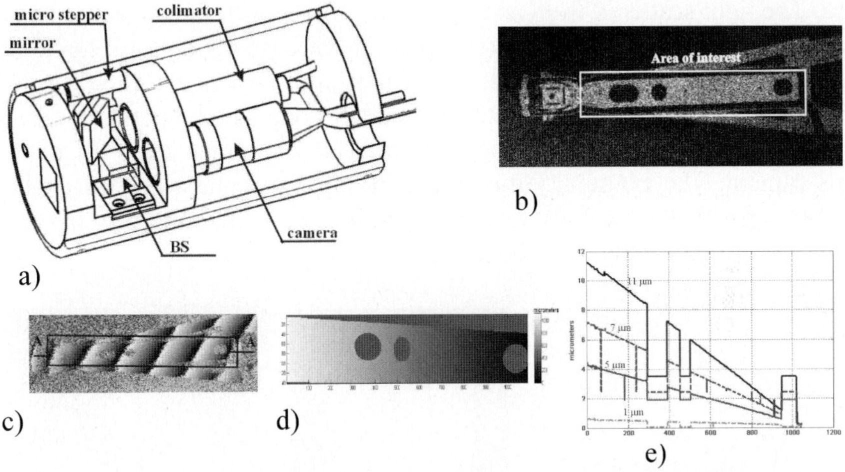

Fig. 2. The DH_SHAPE camera for out-of-plane displacement measurement and shape determination. a) the scheme of camera and the exemplary measurements of shape and out-of-plane displacement of the ink-jet print tip: b) the photo of the object, c) the phase mod 2π map representing the shape, d) the shape map and e) the central crossections of displacements for two different loads.

The source is pigtailed laser with output power 7mW and operating wavelength 532 nm. The beam is delivered by a single mode fiber SM450 and is splitted by single mode fiber optic coupler in 90/10 ratio to form the object and reference beams respectively. The tip of object illumination fibre can be subjected to the linear shift introduced by micromotor. Reference fiber tip is placed in the focal point of collimator lens and the plane reference beam directed through mirror and beam splitter is impinging at CCD matrix. The light scattered from object recombine with the reference beam at the beamsplitter cube and the resultant interference field is captured by CCD matrix. The CCD is the standard microhead B/W camera JAI M536 CCIR with pixel size 8.6 µm and resolution 752x582 pixels. The dimensions of measurement head are: diameter 50mm and length 100mm. The

exemplary measurements of the out-of-plane displacement of the ink-jet print subjected to bending force are shown in Fig.2b-d
The second holocamera DH_UVW designed for an arbitrary object displacement vector measurement is shown in Fig. 3a. The source is the pigtailed laser with output power 7mW and operating wavelength 1064 nm The beam is splitted into object and reference beams by single mode fiber optic coupler in 90/10 ratio. The object beam goes to optical switch 1x4 and produces 4 object illumination beams, which one-by-one illuminates an object. The reference beam is directed similarly as in PH_SHAPE camera. The light scattered from an object recombines with the reference beam at beamsplitter cube and the resultant interference field is captured by CCD matrix. The CCD applied here is the same camera as in the case of DH_SHAPE camera as its spectral sensitivity covers both operating wavelengths. The dimensions of measurement head are similar as in the previous camera. The exemplary measurements of u(x,y) and w(x,y) displacement maps of the composite specimen subjected to fracture test are presented in Fig. 3 c and d respectively.

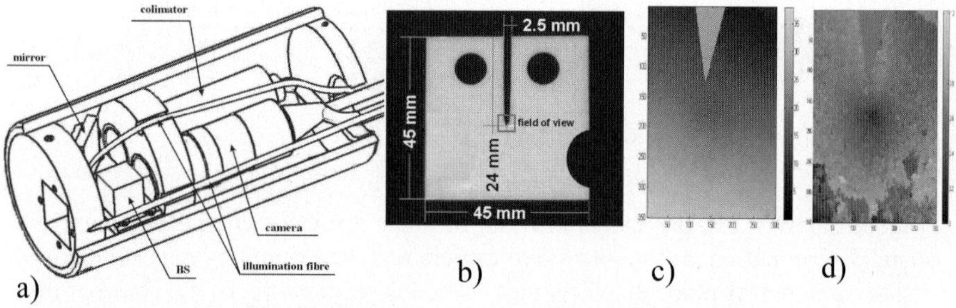

Fig. 3. The holographic camera DH_UVW for arbitrary vector displacement determination a) and exemplary measurements: the specimen made of short fibre reinforced composite prepared for the fracture test b) and the in-plane displacement (u) map c) and out-of-plane (w) displacement map d) obtained for tensile load 50-10 N.

3. Camera with waveguide grating interferometric head

Grating interferomtry (GI) is one of the most useful optical methods for in-plane displacement/strain measurement and monitoring [1]. Below we present the new camera based on low cost waveguide grating interferometric head. Due to insensitivity to vibration, small dimensions and weight, wireless connections and automated fringe pattern analysis it can be used for constant or periodic strain monitoring in crucial points of engineering structures.

Design of the grating interferometric head (WIH) is shown in Fig.4a. It bases on the glass or plastic (e.g. PMMA) plate which guides beams diffracted on grating CG [4]. Next, these beams symmetrically illuminate the specimen grating SG deposited on the surface of object under test. The incident angles of beams are tuned to the plus first and minus first diffraction order angle of the gratings. The diffracted beams, propagate co-axially along the grating normal, interfere and generate the fringe pattern. The interference fringes represent lines of constant in-plane displacement with basic sensitivity of $p/2$ (p – the grating period).

The GI camera for in-plane displacement/strain monitoring (Fig.4b) consists of: WIH, illuminating module IM with micro-laser (λ = 532 nm), detection module DM with wire-less CCD camera and phase shifter module PSM. The phase shift required for automatic fringe pattern analysis by temporal PSM method [5] is realised by tilting of the illuminating module IM towards to WIH.

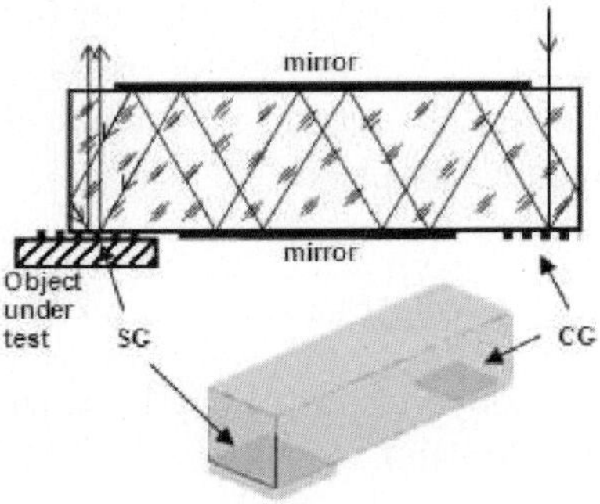

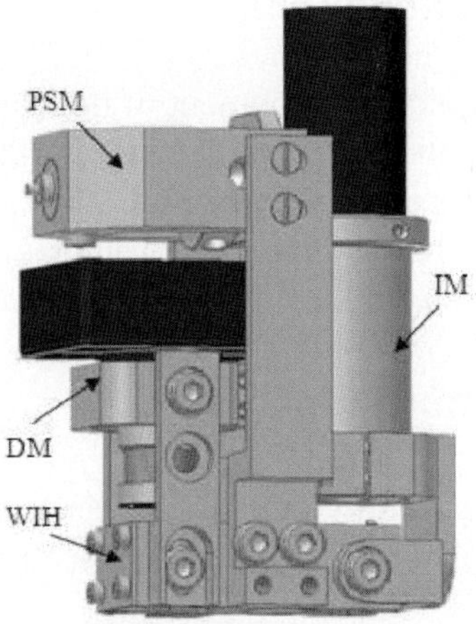

Fig. 3. The scheme of waveguide interferometric head (a) and design (b) of the GI camera for in-plane displacement/strain measurement

Technical parameters of the GI camera are as follow: the field of view - 2 mm x 2 mm, displacement range - 20 μm, specimen grating - 1200 lines/mm, basic sensitivity - 417 nm/fringe and accuracy - ±20 nm. It enables monitoring in continuous (network of sensors constantly mounted on the structure) or periodic one camera periodically mounted in the same position) modes. The exemplary result of displacement and strain distribution measurements at the cracked steel specimen under tensile load is presented in Fig. 5. Distribution of horizontal component of the in-plane displacement vector around the tip of the crack is shown in Fig. 5c and strain distribution calculated by numerical differentiation of displacement field is presented in Fig. 5d.

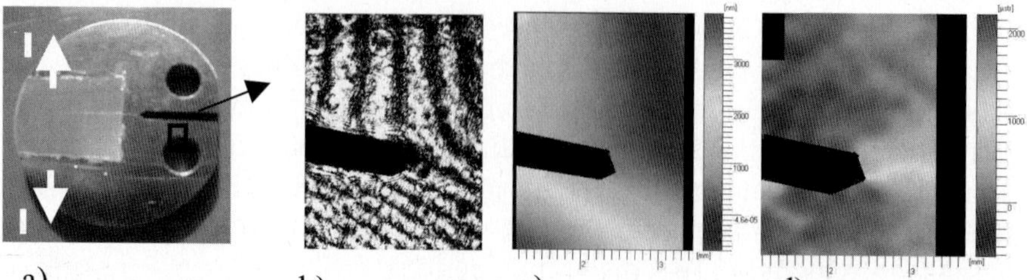

a) b) c) d)

Fig. 5. Exemplary results of measurement. a) specimen under test (black rectangular denotes the measurement area), b) interferogram, c) in-plane displacement map, d) strain distribution

4. Conclusions

The quick development of novel photonics, mechanical and optoelectronic elements allows to design and build compact, low-cost, multifunctional devices providing the high sensitivity measurements in outdoor conditions. In the paper three novel cameras which take full advantage of optomechatronics approach to design and data processing are presented.

Acknowledgements

We gratefully acknowledge the financial support of Ministry of Science and Higher Education within the Research Project PBZ-MiN-009/T11/2003

References

[1] L. Salbut, M. Kujawinska, Opt. Lasers Eng. 36 (2001) 225.
[2] T. Kreis „Holographic Interferometry“ Akademie Verlag, Berlin, 1996.
[3] D. Post, B. Han, P. Ifju "High Sensitivity Moiré" Springer-Verlag, 1994.
[4] L. Salbut, Optical Engineering 41(2002) 626.
[5] K. Patorski, M. Kujawińska, L. Sałbut "Handbook of moiré technique" Elsevier, Holland, 1993

The studies of the illumination/detection module in Integrated Microinterferometric Extensometer

J. Krężel*, M. Kujawińska, L. Sałbut (a), K. Keränen (b)

(a) Institute of Micromechanics and Photonics, Warsaw University of Technology, 8 Sw. A.Boboli St., 02-525 Warsaw, Poland

(b) VTT Technical Research Centre of Finland, Kaitoväylä 1, 90570 Oulu, Finland

Abstract

In this paper the novel, miniaturized full-field optical extensometer for full-field in-plane displacement measurement is presented. It consists of two modules: illumination/detection module (IDM) and Measurement Head (MH) based on grating interferometry. In the paper we study the performance of IDM, which provides the object illumination beam and captures the output interferogram. The characterization of VCSEL comprises determination of polarization state, intensity distribution, coherence path length and shape of the wavefront after being collimated by beam forming optics. The detector matrix has been tested mainly in respect of the signal to noise ratio at various exposure times and its spatial uniformity of sensitivity.

1. Introduction

Growing microtechnology raises a demand for development of novel measurement instrumentation allowing for determination of displacement field in microspecimens or microsystems under test. New strict measurement requirements defined by a size of microobjects, their fragility and need for high measurement resolution and accuracy have been determined. On the other hand microtechnology offers new means for development of instrumentation with innovative architecture which lead to a design of low-cost extensometer presented on figure 1a.

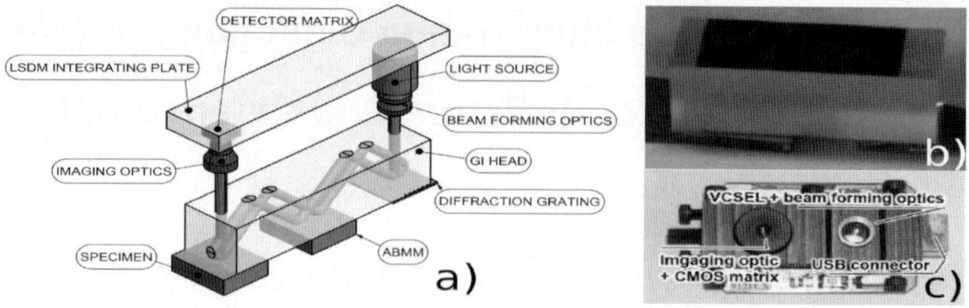

Fig. 1. Integrated Microinterferometric Extensometer (IME): a) concept, b) Measurement Head (MH), c) Illumination/Detection Module (IDM)

The Measurement Head (fig. 1b) is made of the homogeneous beam guiding block (glass or PMMA) integrated with diffraction grating which forms the object illuminating beams. The extensometer provides in-plane displacement/strains maps of microobjects in 1.4 mm x 1.4 mm field of view. The description of MH design is given in [1], here we focus on characterization of IDM (fig. 1c).

2. IDM characterization

The IDM integrates VCSEL emitting 665 nm light with one or many longitudinal and transverse modes, beam forming aspheric optics, imaging optics, detector matrix (CMOS B&W 1280x1024 pixels camera) and evaluation board.

2.1. VCSEL characterization

Coherence path length (CPL):
Preliminary numerical simulations [1] revealed that the total optical path difference (OPD) between the beams guided in MH might be up to 0.6 mm. The coherence path length (CPL) was tested by monitoring fringe pattern contrast in the function of OPD in T-G interferometer. In the case of monomode VCSELs, the measured CPL equals a few centimeters (fig. 2 a). In the case of multimode VCSELs the contrast varies (fig. 2b), however the high contrast region around 0 OPD was larger than 0.6 mm (fig. 2c).
All characterized light sources meet requirements of IMS in respect of the time and spatial coherence

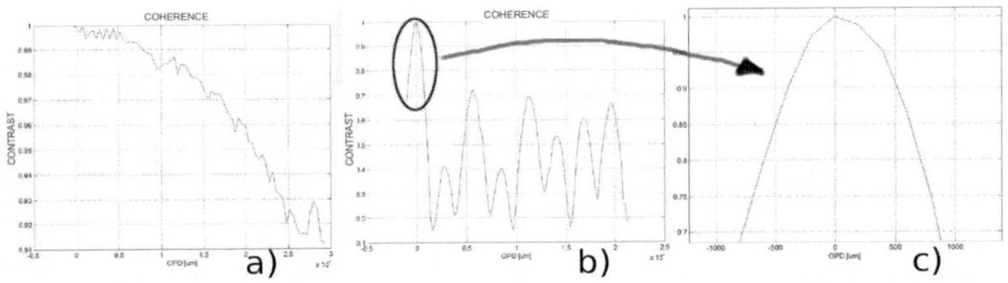

Fig. 2. The measurements of coherence path length: a) monomode VCSEL, b) multmode VCSEL, c) multimode VCSEL 0 OPD peak width

Polarization state measurement:

The polarization state of the beam emitted by VCSEL was determined in the system shown in fig. 3a.

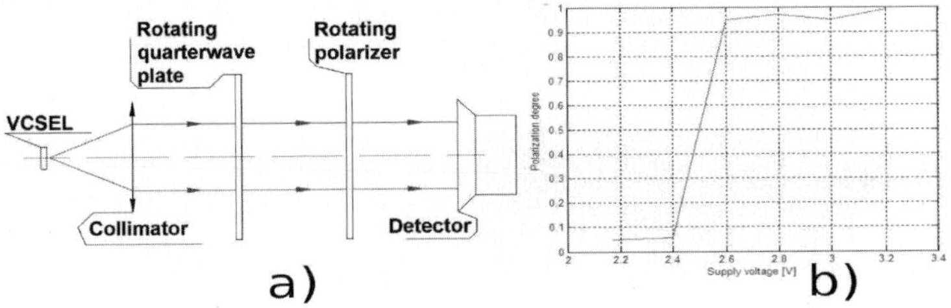

Fig. 3. The measurement of beam polarization: a) the setup, b) linear polarization degree vs supply voltage

Polarization state studies revealed that every measured VCSEL emits light with partially linear polarization regardless of supply voltage value, however the degree of linear polarization is very high (fig. 3b)

IMS requirements concerning polarization are met. The linearly polarized light with proper orientation of polarization plane makes it able to avoid beam deploarization or polarization changes while being guided in MH medium, and consequently leads to high interference pattern contrast.

Collimated beam wavefront investigation:

The collimating lens forms the plane beam with diameter of 6.5 mm. To minimize the measurement errors and unwanted parasitic light and to provide uniform measurement sensitivity in the area of interest the collimated beam is required. The investigation of wavefront quality was performed in grating shearing interferometer (fig. 4), in which the first derivative of wavefront is given by the phase resulting from interference of two slightly displaced beams (fig. 5.).

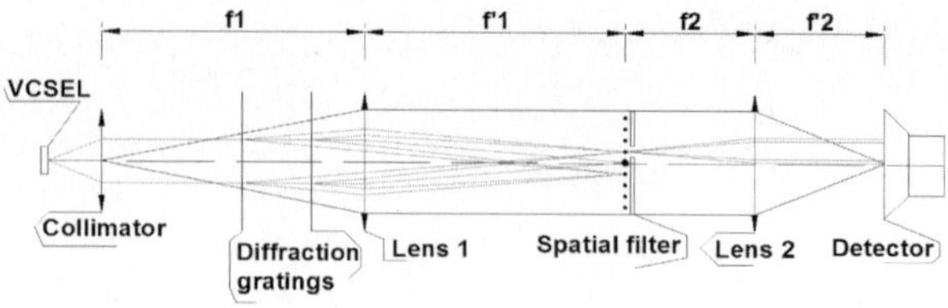

Fig. 4. Wavefront measurement setup based on the shearing interferometer

The phase encoded in the interferogram was determined by temporal phase shifting method of fringe pattern analysis [2]. The transverse aberration was estimated for less than 0.06λ. It proves that the quality of the investigated wavefront is fully sufficient for grating interferometry based measurement.

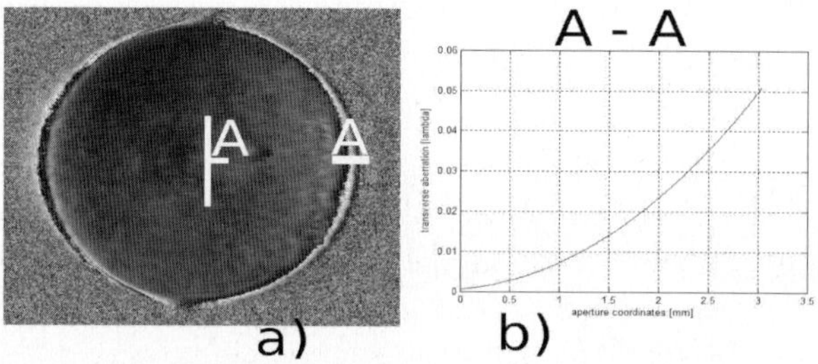

Fig. 5. Wavefront investigation: a) measured phase, b) transverse aberration in x direction

3. Detector characterization

CMOS matrix applied in IDM is a low-cost B&W camera of resolution 1280 x 1024 pix. The active data acquisition area is 6.66 mm x 5.32 mm with 5.2 um x 5.2 um pixel size. According to the hardware data-sheet the signal to noise ratio is 54 dB. The camera is intended to be used for interferometric measurement with nanometric sensitivity, therefore the comparison of its performance with high quality analog CCD camera was performed in T-G interferometer [3] with flat mirror as a test object. Temporal phase shifting method and FFT method were used for interferogram analysis. No difference in measurement accuracy was observed, when we compare the results obtained from data captured with both cameras (fig. 6.). The noise generated in the detection part of IDM is 3 bit for 8 bit depth. The increment of exposure time resulted with increment of number of noisy pixels which concentrate mainly near the border of active detection area.

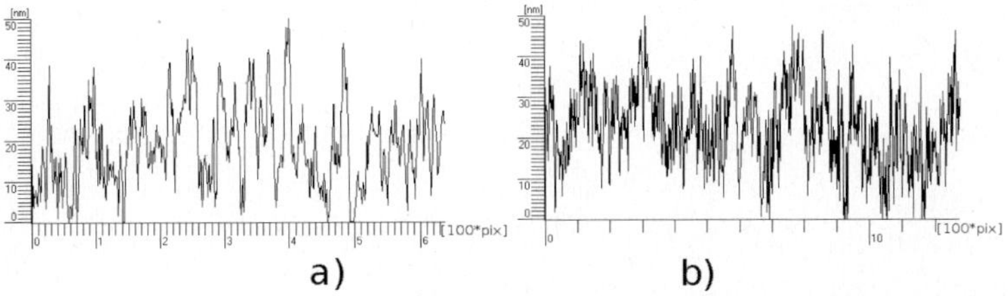

Fig. 6. Measurement result comparison: a) high quality CCD matrix, b) IDM's CMOS matrix

4. Conclusions

The source and detector applied in IDM meet the requirements of the optical extensometer set by MH properties and grating interferometry principle. The presumable error arising form IDM's output beam may by subtracted form the measurement result as IME systematic error, although the IME with glass MH revealed neglectible systematic error.

Acknowledgments

We gratefully acknowledge the financial support of EU within NoE for Micro-Optics NEMO and the Dean of the Faculty of Mechatronics, Warsaw University of Technology.

References

[1] J. Krężel, L. Sałbut, M. Kujawińska, "M(O)EMS-based integrated micro-interferometric system", Proceedings of SPIE, Vol. 6348-22

[2] K. Patorski, M. Kujawińska, L. Salbut, "Interferometria Laserowa z Automatyczną Analizą Obrazu", Oficyna Wydawnicza PW, Warszawa (2005)

[3] J. Kacperski, M. Kujawińska, "Multifunctional interferometric platform for static and dynamic MEMS measurement", Proc. SPIE 5878, s. 64-73, 2005

[4] L. Salbut, „Waveguide grating (moire) interferometer for in-plane displacement/strain fields investigation" Opt. Eng. 2002 41, 626-631

Analysis and design of a stationary Fourier-transform spectrometer using Wollaston prism array

L. Wawrzyniuk, M. Dwórska
Institute of Micromechanics and Photonics
Warsaw University of Technology
8 Sw. A.Boboli St., 02-525 Warsaw, Poland

Abstract

In the paper the design of a stationary Fourier-transform spectrometer using Wollaston prism array is proposed. Interferogram with maximum path difference is divided into components composed from each sub-prism provides high resolution in recovered spectrum. Those components are located on rows of the CCD camera producing individual interferograms, respectively. Such a configuration allows us to merge individual interferograms into one using computer procedures. The analysis of errors and their influence on the recovered spectrum is performed. Due to significant influence on the spectrum errors in design of the prism array were chosen and the analysis of the error of the merging point selection was also discussed.

1. Introduction

The Fourier-transform spectrometer (FTS) is a powerful spectroscopic tool that provides high spectral resolution measurements, especially in infrared region of radiation. The conventional FTS is based on the Michelson interferometer [1,2]. During a measurement one mirror is continuously scanned which induces a phase delay between two beams of the interferometer. The spectrum can be found by applying a Fourier transform on the measured intensity pattern. The main advantages of FTS's are high throughput and high signal-to-noise ratio. However, the technical requirements to the scanning mechanism are very high, especially for the stable angular position of the scanned mirror. One of the known possibili-

ties to avoid such troubles is the design of a stationary Fourier-transform spectrometers (SFTS) without any moving elements. A disadvantages of such a solution is smaller spectral resolution. Nevertheless, for many applications, namely, for the study of air pollutions, mentioned low resolution can be quit sufficient.
The most widespread designs of SFTS's are based on the polarization interferometer with a Wollaston prism. Lack of mirrors and beam splitters, no influence of parasite light on the source characteristics belong to advantages of such a solution. Moreover, common path interferometer is insensitive to mechanical vibration. However, to achieve large optical path difference for higher spectral resolution large sizes of the Wollaston prism is indispensable. Therefore the spectrometer becomes larger and more expensive. To enlarge spectrometer possibilities a Wollaston prism array is proposed in the form of some lines. To achieve the continuous spectrum the proper merging of individual spectral lines has to be done. The paper subject is the analyzes of the influence of merging errors on the measurement accuracy.

2. Design principles

The optical scheme of the spectrometer with the Wollaston prism array is shown in Fig.1 [3]. The Wollaston prism array is composed of some Wollaston sub-prisms. Each sub-prism generates the interferogram for different optical path. The maximum path difference of any chosen sub-prism corresponds to the minimum one of the adjacent sub-prism. In such a way the continuous optical path difference can be received, and in consequence, the continuous interferogram, which is registered by 2D detector. The Wollaston sub-prism set can be treated as one long Wollaston prism. All sub-prisms have, besides their thickness, the same angular shape, what facilitates the joining of particular interferograms generated by all sub-prisms into one virtual interferogram.

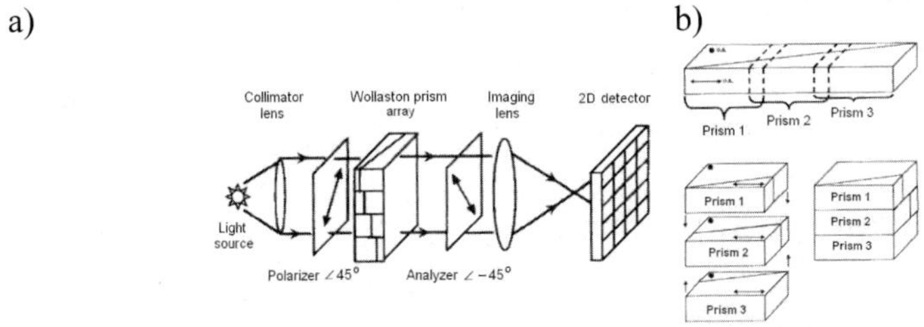

Fig.1. Optical scheme of the spectrometer: a) Wollaston prism array based interferometer, b) ideal design of the equivalent long Wollaston prism

In a proposed lay-out of the SFTS precision of joining separate pieces of the interferogram is strictly required. To achieve correct interferogram for enlarged optical path difference the merging point must be selected precisely. Moreover, to provide high spectral accuracy of measured spectrum each sub-prism must be precisely made and stacked into the Wollaston prism array. The most important physical requirement is that the wedge angles for all sub-prisms must be equal. If this is not a case, the spatial frequencies of the several partial interferograms will not be the same. That causes difficulties in assigning the merging point for corresponding interferograms correctly. As is understood in the art, achieving the continuous interferogram, and in consequence, the correct analysis of the spectrum will not be possible. Furthermore, different value of the wedge angles in several sub-prisms will change a direction and a localization of the interference fringes. To overcome this problem, the fringe direction and the spatial frequencies must be compensated.

In this paper the analysis of the influence of the errors on the measured spectrum is discussed. The sampling error and sub- prisms design error are selected.

3. Experimental results

To demonstrate and analyze the performance of the Wollaston prism array based spectrometer the numerical simulation was prepared. We assumed the set of two sub-prisms with lengths and wedge angles equal $L = 40$ mm and $\theta = 3^\circ$, respectively. Quartz was selected as the bierefringent material. We assumed a monochromatic light source with a wavelength $\lambda = 589$ nm. Intensity of the ideal interferogram is given by

$$I = 2\left[1 + \cos(\frac{2\pi}{\lambda} L\varepsilon)\right],$$

where: $\varepsilon = 2(n_e - n_o)\tan\theta$ is the Wollaston prism split angle, n_e and n_o are the extraordinary and the ordinary refractive indexes of the bierefringent material, respectively.

3.1. The merging point effect

In the proposed lay-out of the SFTS there is a requirement of assigning the merging point precisely. In the merging point the maximum path difference of chosen sub-prism must correspond to the minimum one of the adjacent sub-prism. If it is not a case, the merging algorithm will not be succeed correctly. This problem is caused by errors in sampling algorithm of the interferograms generated by each sub-prism. The simulation of described effect was carried out for two Wollaston prisms in the array, by applying a different sampling step Δz to one of the merging interferograms. Fig.2 shows hypothetical results obtained for a single wavelength.

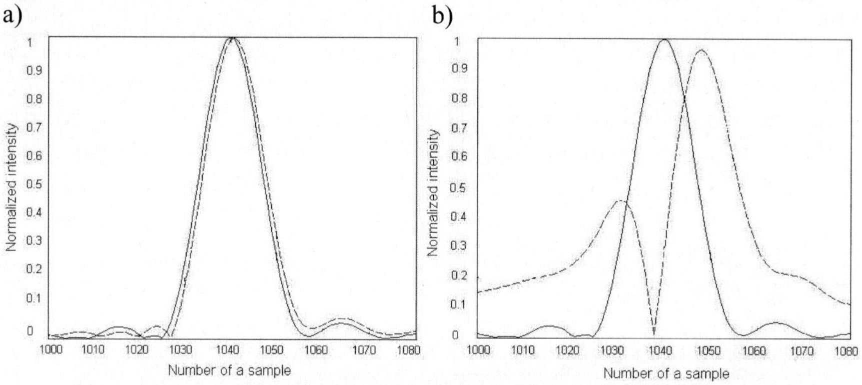

Fig.2 .Spectrum for two-Wollaston prisms array based spectrometer where: (continuous line) correct spectrum, (discontinuous line) spectrum for wrong assigned merging point; a) $10\%\Delta z$, b) $100\%\Delta z$

As shown in Fig.2, the more difference in the value of the sampling step Δz, the larger distortion in the spectrum. Furthermore, a decrease of the normalized intensity and a displacement of instrument line shape are also observed. For a large value of the sampling error the second wavelength is seemingly detected. The observed effect is difficult to avoid in a real device, hence defining the sampling error that can be accepted is required. To provide accuracy of the merging algorithm procedures described error shouldn't exceed 10% of the sample step value.

3.1. The wedge angle effect

The value of the wedge angle in each sub-prism has the significant influence on the spectrometer performance. Differences in several wedge angles value change the spatial frequency of the interference fringes made by appropriate sub-prisms. That causes problems during the merging algorithm. When two Wollaston prisms are used in the array, differences in wedge angles value $\Delta\theta$ lead to the decrease of the normalized intensity in the spectrum. The distortion in the instrumental line shape is also observed. Alike in the sampling error effect the seemingly detection of the another wavelength appears, as shown in Fig.3. The larger difference in value of wedge angles in several sub-prisms, the more significant participation of the false wavelength in the spectrum. It is obvious that the distortion of the measured spectrum will be larger according to raising amount of Wollaston prisms used in the array and stronger geometrical inaccuracy of each component.

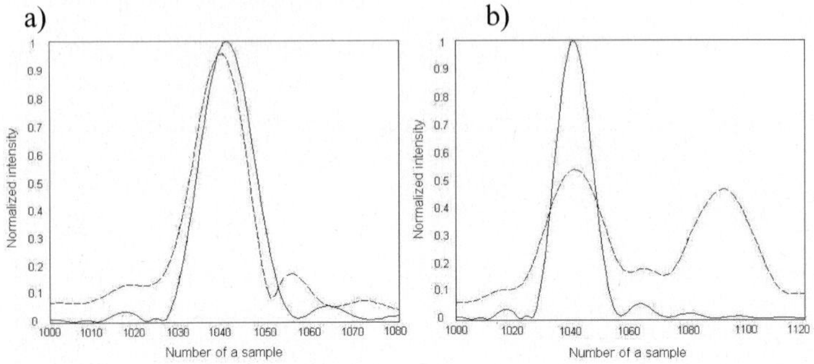

Fig.3. Correct spectrum (continuous line) and spectrum with the wedge angle error (discontinuous line); a) $\Delta\theta = \theta / 200$ b) $\Delta\theta = \theta / 10$

4. Conclusions

The Wollaston prism array based nonscanning spectrometer has been presented. It has been shown that errors in procedures of the merging algorithm and the inaccuracy of the prism array geometry lead to distortion in the shape of the spectrum and depreciate it's spectral resolution. The seemingly detection of another wavelengths can make it impossible to analyze the measured spectrum correctly. To alleviate this problem and provide required spectral resolution the numerical procedures that enable correction of the interferogram and the spectrum, respectively, are commonly used.

References

[1] R.J.Bell Introductory Fourier Transform Spectroscopy. Academic Press. New York 1972

[2] R.Jóźwicki, M.Rataj Fourier Spectrometry and its applications. Opto-Electr Rev. Vol.6.No.4. (1998)

[3] D. Komisarek, K. Reichard, D. Merdes, D. Lysak, P. Lam, S.Wu, S.Yin High - performance nonscanning Fourier - transform spectrometer that uses Wollaston prism array. Appl.Opt. Vol.43, No.20. (2004).

Modeling and design of Microinterferometric Tomography

Tomasz Kozacki (a), Youri Meuret (b), Rafał Krajewski (a), Małgorzata Kujawińska (a)

(a) Warsaw University of Technology, Institute of Micromechanics and Photonics,
8 Sw. A.Boboli St.,02-525 Warsaw, Poland

(b) Vrije Universiteit Brussel, Dept. of Applied Physics and Photonics (IR-TONA),
Pleinlaan 2, 1050 Brussels, Belgium

Abstract

In this paper we discuss modeling and design issues of microinterferometric tomography system (MiTS) for out-door non-destructive optical fiber inhomogeneity inspection. Presented system is a monolithic microdevice composed from a number of optical microelements: light source, gratings, system body, inspected fiber, imaging optics and detection device. In order to select optimal design an accurate and efficient simulation method of complete system is necessary. There are many optical modeling methods basing on Maxwell equation, wave optics or ray optics. Unfortunately none of them could be used to simulate complete system. These methods either require to much computer power or simply would produce inaccurate results. Therefore to simulate complete system in a reliable way a number of methods is used in different system regions. The received final results are used to study the reconstruction and imaging problems of the MiTS system.

1. Introduction

Optical Diffraction Tomography (ODT) [1] used for measurement of internal 3D complex refractive-index distribution is now well established technique. Although the ODT can be successfully used in tasks such as fiber, waveguide refractive index characterization, it is still a laboratory tool with limited utility for the outdoor applications. Therefore in this paper we explore design of integrated microinterferometric tomography system. The system is built as a monolithic block with reduced vibration and it suits for tasks of outdoor measurements [2, 3].

In the paper we compare two different system designs, with on and off axis illuminations. We compare both configurations by means of numerical simulations. Therefore accurate and efficient modeling method of the system is necessary. Unfortunetly no single modeling tool can accomplish such a task. We use combination of three methods: free space propagation (FSP) [4], finite difference time domain method (FDTD) [5] and wave propagation method (WPM) [6]. With the FDTD method the grating diffraction orders magnitudes are characterized. The WPM is numerically much more efficient tool and it is applied at the object structure region. At all the other system regions the FSP method is applied.

2. Off axis microinterferometric tomography system

In the Fig. 1 below the setup configuration of microinterferometric tomography system with off axis illumination is presented. The system is built from monolitic integrated Mach-Zehnder interferometer using the grating beamsplitter and recombiner (interferometer body), illumination and detection units. The interferometer body is made from PMMA (index of refraction 1.48352). The measured element is placed under off axis illumination, when comparing to the optical axis of the detection unit. Therefore there is a difficulty in obtaining undistorted integrated image of measured structure.

In tomography each integrated image must be obtained using imaging system with imaging plane parallel to the optical axis and in central region of object structure [7]. In the system the object is illuminated with off axis illumination beam and then the beam propagation direction is changed at the recombining grating. Therefore the obtained integrated object images are highly distorted. Such a distortion must be corrected using numerical algorithms. The main purpose of such an imaging correction algorithm is introduction of the object beam deflection at recombining grating.

The image correction algorithm consists of three steps. In the first step distorted experimentally obtained complex image is propagated to the recombining grating using free space propagation algorithms. In the second step beam deflection is introduced, what simulates the grating in the imaging system. Then the beam is propagated to the object centre. It gives object integrated image with inclined imaging plane. At the final procedure step this inclination is removed using free space propagation algorithm to a tilted plane [8].

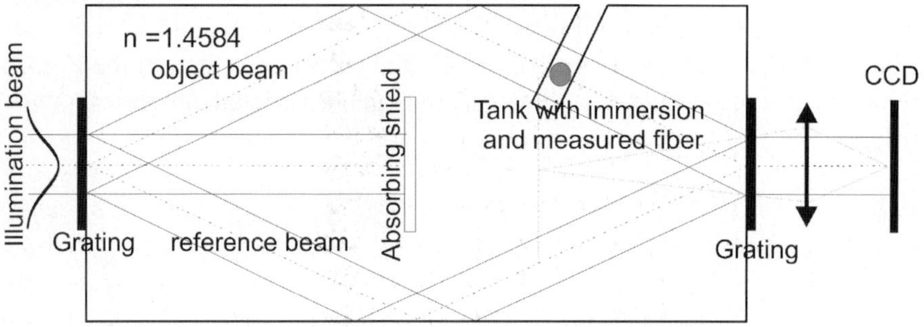

Fig. 1. Off axis microinterferometric tomography system

The above system was simulated using the combination of two methods: free space propagation and wave propagation methods. For this system the grating can be treated approximately with thin element grating equation coefficients. The simulated object is a fiber of 50 mm^{-3} diameter with the refractive index of cladding n_{cl}=1.48352, and core n_{co}= 1.49352.
In Fig. 2 below plots of projection images received from standard optical system imaging (a) and with application of mentioned above imaging correction algorithm (b) are presented using solid line. The projection plots are compared with the theoretical object projections (doted lines). The results show the necessity of application of the imaging correction algorithm.

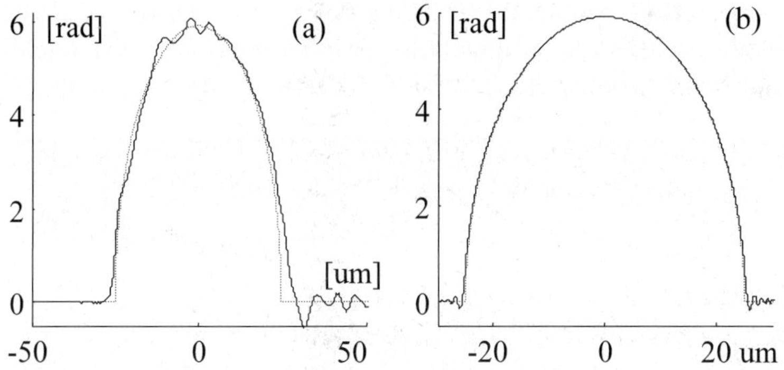

Fig. 2. The projection images obtained in the system Fig. 1 using optical imaging only (a), using imaging correction method (b)

3. On axis microinterferometric tomography system

In the section above the imaging problems with off axis object illumination were presented. These problems can be eliminated by application of a numerical imaging correction algorithm if two conditions are fulfilled: the system geometry is well known, entire object diffracted field is captured. The imaging problems can be solved in the system presented in the Fig. 3, with on axis object illumination without any additional numerical algorithms.

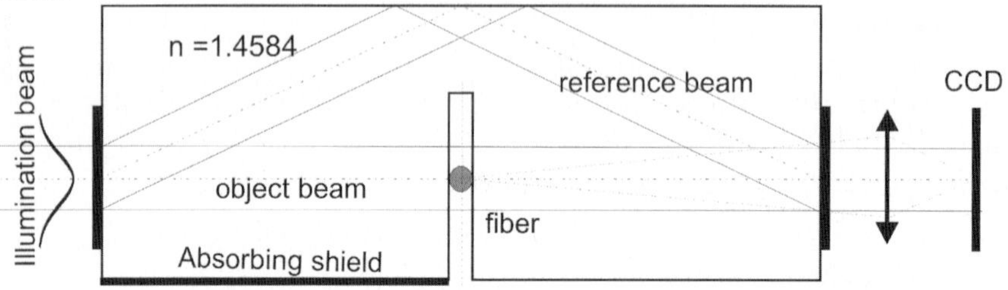

Fig. 3. On axis microinterferometric tomography system

In the system the grating with profile depth 0.25 mm^{-3} etched in the resist with the refr. index 1.62 producing equal orders -1, 0 and 1 was applied at both system sides. Unfortunetly the asymmetric grating orders are used in the system. Therefore the major problem with the system is maintaining the high interference fringes contrast at system output. To examine the contrast we have analyzed the grating orders using finite-difference time domain method at both microinterferometer sides. In Fig. 4 exemplary amplitude of the optical filed computed using FDTD method is presented. Combining all of the simulation results small drop of the fringe contrast was received i.e., 0.999. Unfortunetly to preserve fringes of such a quality the gratings in the system would have to be produced very accurately.

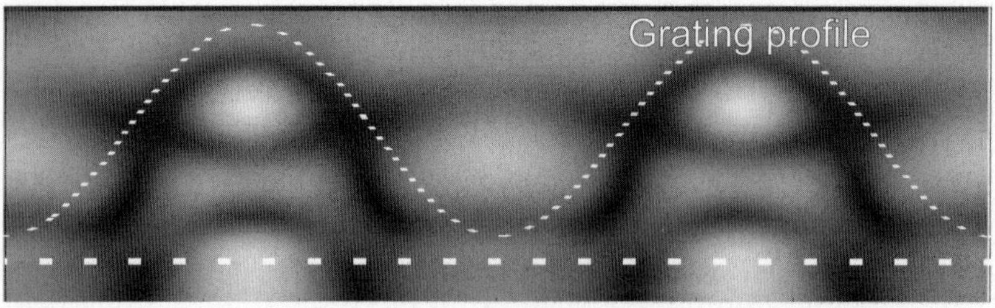

Fig. 4. Amplitude of the optical field received at the left system grating

4. Conclusion

In the paper we have examined two possible configuration of the MiTS system: with off and on axis illumination of measured structure. We have shown that the system with off axis illumination requires an introduction of additional imaging correction algorithms. For the system with on axis illumination the requirements imposed on grating are high.

5. Acknowledgements

We would like to acknowledge for support the European Network of Excellence in Micro-Optics NEMO and the Ministry of Science and Information Technologies within project N505 008 31/1374.

References

[1] M. Born, E. Wolf, "Principles of Optics" 7th (expanded) edition, Cambridge University Press, 1999.
[2] R. Krajewski, M. Kujawińska, B. Volckaerts, H. Thienpont, "Low-cost microinterferomtric tomograpy system for 3D refraction index distribution Measurements in optical fiber splices, proc. SPIE 5855 (2005) 347.
[3] R. Krajewski, B. Volckaerts, Y. Meuret, M. Kujawińska, H. Thienpont „Design, modeling and prototyping of a Microinterferometric Tomography System for optical fiber inspection" proc. SPIE 6188 (2006) 138.
[4] J. W. Goodman, Introduction to Fourier Optics, New York, 1996.
[5] A. Taflove, S. C. Hagness, Computational Electrodynamics: The Finite-Difference Time-Domain Method, New York, 2005.
[6] K.H Brenner, W. Singer, "Light propagation through microlenses: a new simulation method", *Appl. Opt.*, 32 (1993) 4984.
[7] T. Kozacki, M. Kujawinska, P. Kniazewski, "Investigating the limitation of optical scalar field tomography ", Optoelectronics Review, 15 (2007) 102.
[8] N. Delen, B. Hooker, "Free-space beam propagation between arbitrarily oriented planes based on full diffraction theory: a fast Fourier transform approach," J. Opt. Soc. Am. 15 (1998) 857.

Technology chain for production of low-cost high aspect ratio optical structures

R. Krajewski (a), J. Krezel (a), M. Kujawinska (a), O. Parriaux (b), S. Tonchev (b*), M. Wissmann (c), M. Hartmann (c), J. Mohr (c)

(a) Warsaw University of Technology Institute of Micromechanics and Photonics 8 Sw. A.Boboli St.,02-525 Warsaw, Poland

(b) Saint-Etienne University 10, Rue Barrouin, Bâtiment F F-4200 Saint-Etienne France *) on leave from the ISSP, Sofia, Bulgaria

(c) Forschungszentrum Karlsruhe Institut für Mikrostrukturtechnik Hermann-von-Helmholtz-Platz 1 76344 Eggenstein-Leopoldshafen, Germany

Abstract

The paper covers the studies on the replication process of the high aspect ratio optical structures performed by Hot Embossing technology. Considered structures are the microinterferometric heads shaped in the form of PMMA blocks with a high frequency diffraction grating placed on the side walls. We propose the full technology chain including replication of the glass machined prototypes. Additionally the preliminary characterization tests of exemplary replicated optical structures are presented.

1. Introduction

Recently there is a need of low cost optoelectronic sensors providing full-field measurements of such quantities as shape, displacement or refractive index distribution. It can be provided by microinterferometers integrating all optical components in single massive block of optical material with the source & detection module. Such configuration assures high vibration and external factors insensitivity and requires much less packaging efforts comparing to the traditional bulk optic-based systems.

There is a variety of possible microinterferometric configuration. In this paper two such approaches (fig.1) are considered in detail: the compact microinterferometric tomography system for 3D refractive index distribution measurement in optical fibers [1] and the grating microinterferometer for in-plane displacements/strain studies in microelements and microregions of large engineering structures [2].

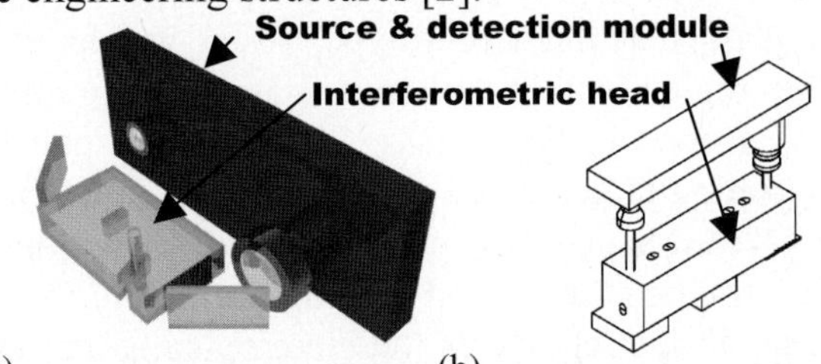

(a) (b)

Fig. 1 The concepts of Microinterferometric Tomography System (a) and Integrated Microinterferometric Optical Extensometer (b)

In this paper we focus on the fabrication and replication issues of the interferometric heads. Both designs are few millimeters size cuboids with high frequency gratings placed on sidewalls. Due to their high aspect ratio and requirement on the sidewall surface optical quality the technology chain of such structures is not established yet.

Therefore at first we propose this chain with special attention paid for usage of low cost replication methods. To verify the applicability of the proposed replication technology we present the characterization results of the most crucial structure parameters in the exemplary replicated optical structures.

2. Technology chain for microinterferometric heads

Preliminary calculations [3] allowed to determine some of the requirements to be fulfilled by the fabrication technology. These are high dimensions accuracy and optical quality of the sidewall (flatness less than $0.12\mu m$ over the $4mm^2$).

The requirement of low cost of the final components set a 2 steps technology chain consisting of prototyping and further replication with one of the low cost replication technologies. Because of relatively small material stress during the replication process we decided to use hot embossing (HE) [4, 5]. This replication technology requires a previously prototyped structure.

Although there are several technologies of low-cost microstructure prototyping technologies, but just a few of them are capable to produce structures with the features demanded by the presented interferometric head i.e. large volume and high aspect ratio, side wall featuring optical flatness ($R_q \leq 0.05\mu m$). Additionally the demand of a high frequency grating inte-

grated on the sidewall skips the most popular rapid prototyping technologies such as Deep Proton Writing or LIGA technology.
Therefore it was decided to propose a new technology chain (fig. 2) based on well known glass machined prototype, resist grating deposition followed by further replication in polymer material. Microinterferometric head prototype is shaped in BK7 glass. Sidewalls are polished and high frequency resist grating is deposited on a side wall. Then the prototype is electroplated to get the shape of the mask. When the electroplating process is finalized the prototype is dissolved making a molding insert ready for a replication process. Proper choice of the electroplating and the replication parameters should provide the polymer structure with well replicated sidewalls and the grating profile. Preliminary tests provided the proof-of-principle for the proposed replication chain however significant process tuning is necessary.

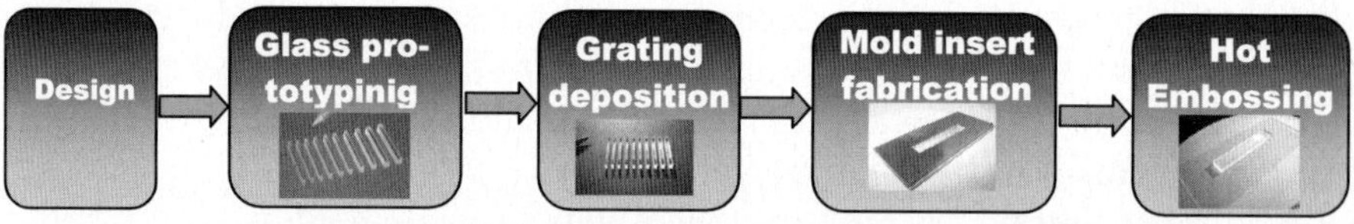

Fig. 2. The technology chain of the microinterferometric head

Beside the parameters mentioned above the replicated structure should assure an optical homogeneity of the entire volume. To make the inteferometric measurements possible the deformation of a plane wave front transmitted through the replica should be not larger then $\lambda/2$ (0.32µm) over the total propagation length 13mm. Therefore simultaneously to the technology chain optimization the homogeneity tests of the already replicated structure were performed.

3. Optical characterization of hot embossed microstructures

To test the applicability of HE for the replication of the high aspect ratio optical components, an assessment of the most important optical features was based on the already replicated structures. These tests include surface quality inspection, surface topography, an integrated refractive index measurement and quantitative test of stresses generating birefringence in the replicated structures.

3.1 Surface quality inspection and topography

The surface quality test was performed with the non-contact optical profiler WYKO NT-2000. We performed the measurement of the same region of the sidewall for both prototype, and the replicated structure (fig.3). The observed R_q roughness increases in the replica and visible longitudinal scratches (represented as R_t) along the sidewall are observed. These values in replica have to be significantly improved by the technology optimization.

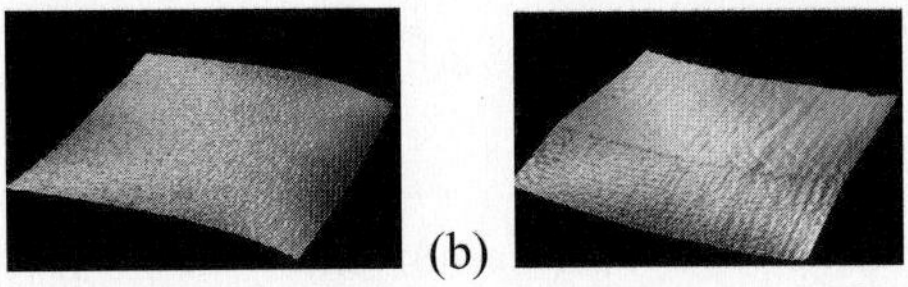

(a) (b)

Fig.3 Roughness measurement results of the demonstrator's sidewall (area: $62.5*10^{-3}$ µm): (a) prototype R_q74nm R_t=470nm (b) replica R_q=180nm R_t=1.1µm

To test the deformation of the replicated structure during the hot embossing process, the topography test of the sidewall neighborhood was performed with Twyman–Green interferometry using the methodology presented in [6]. The results show material deformation in the sidewall region (fig. 4). Deformation values vary up to ±4 µm depending on the structure geometry.

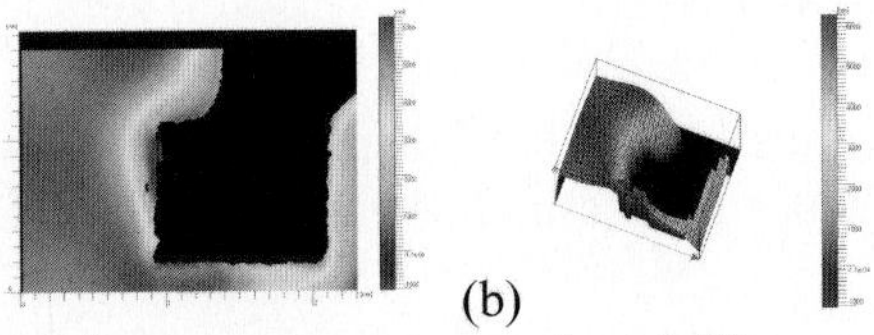

(a) (b)

Fig. 4 The local HE microprism topography (a) 2D map (b) 3D plot

3.2 Integrated 2D refractive index and birefringence distribution measurement

The microinterferometer architecture requires small variations of the refractive index. Therefore the tests of optical homogeneity of the interferometric block were performed by measurements of the integrated refractive index distribution with Mach-Zehnder interferometer. The exemplary observed local inhomogeneity Δn=0.06µm over 2mm has relatively small impact on the light beam propagating in the replicated structure. The phase error can be corrected in the systematic error removal process. To make an optical characterization complete, a qualitative studies of birefringence distribution was performed (fig.5). Hot Embossing process should be accurately monitored in order to avoid the eventual increase of refractive index inhomogeneity and birefringence distribution.

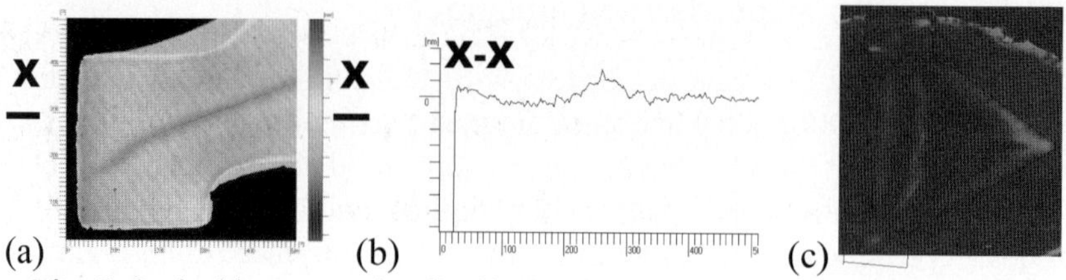

Fig. 5. Optical homogeneity distribution in the replicated structure (a) 2D map (b) cross-section n(x) (c) quantitive 2D birefringence distribution result

4. Summary

We have demonstrated the fabrication chain of large volume high aspect ratio microstructures in PMMA. Presented novel methodology combines the accuracy and high quality of the glass machined prototypes and low cost replication by hot embossing technology. The presented results show the potential applicability of HE for the fabrication of high aspect ratio optical structures, however the technology process has to be optimized. Further work will include the studies on improvement both the sidewall quality and the optical homogeneity of replicated interferometric heads.

Acknowledgements

The authors are grateful for the financial support within the European Network of Excellence in Micro-Optics NEMO (EC FP6 program).

References

[1] R. Krajewski, et al. „Design, modeling and prototyping of a Microinterferometric Tomography System for optical fiber inspection", Proc.SPIE vol.6188, pp.138-148, 2006

[2] J. Krężel, et al. "M(O)EMS-based integrated micro-interferometric system", Proc. SPIE, vol. 6348-22, 2006

[3] R. Krajewski et al. "Low-cost microinterferomtric tomograpy system for 3D refraction index distribution Measurements in optical fiber splices", Proc. SPIE vol. 5855, pp.347-351, 2005

[4] M. Heckele, "Hot embossing - a flexible and successful replication technology for polymer MEMS", Proc. SPIE Vol. 5345, pp. 108 – 117, 2004

[5] M. Heckele, et al. "Review on micro molding of thermoplastic polymers", J. Micromechanics and Microengineering, 14 R1-R14, PII:S0960-1317(04)68657-8, 2004

[6] J. Kacperski, et al. "Multifunctional interferometric platform for static and dynamic MEMS measurement", Proc. SPIE vol. 5878, pp. 64-73, 2005

Automatic color calibration method for high fidelity color reproduction digital camera by spectral measurement of picture area with integrated fiber optic spectrometer

M. Kretkowski(a) *, H. Suzuki (b), Y. Shimodaira (c), R. Jabłoński (d)

(a), (b), (c) Shizuoka University, Department of Electrical and Electronic Engineering, 3-5-1 Johoku, Hamamatsu, 432-8561, Japan

(a), (d) Warsaw University of Technology, Department of Mechatronics, Boboli 8, Warsaw, 02-525, Poland

Abstract

Keywords: human vision gamut, conversion matrix, high fidelity color reproduction

High fidelity color reproduction requires modern approach to color representation. Our research is aimed to develop a digital camera system capable of reproducing high fidelity colors from whole range of human vision gamut. The camera is equipped with three originally designed RGB filters, which spectral characteristics match human eye spectral response covering nearly all colors from human vision color gamut. This paper describes automatic method of color calibration with use of an integrated spectrometer inside the camera with a fiber optic measuring probe. The measurements are carried out in the acquiring conditions due to implementation of a semi-transparent mirror, splitting the image into two perpendicular directions to separate spectra measuring area from image acquisition area projected onto CCD. As a reference for color, a color checker is used for numerical color description. The fiber optic probe is positioned over reflected picture area for measurement of the color checker pads spectra. For accurate color reproduction a conversion matrix is used, which is estimated by least square sum analysis involving spectral data and values obtained from camera CCD sensor for each color reference.

Introduction

For accurate color reproduction there is a need of developing new tools for calibrating a digital camera system capable of reproducing high fidelity colors for telemedicine, internet shopping, industrial design and other high accuracy color reproduction demanding applications. The camera is equipped with three filters S1, S2 and S3, which correspond to red, green and blue band. These filters have relative spectral sensitivity similar to primary color sensitive cones in human eye (Fig. 1.).

Macbeth ColorChecker® Chart is used to determine numerical color description for reference. As an illuminant there is a CIE Daylight 65 (D65) standard used. Under this certain conditions, automatic method of calibration with use of an integrated spectrometer inside the camera based on a fiber optic measuring probe has been developed.

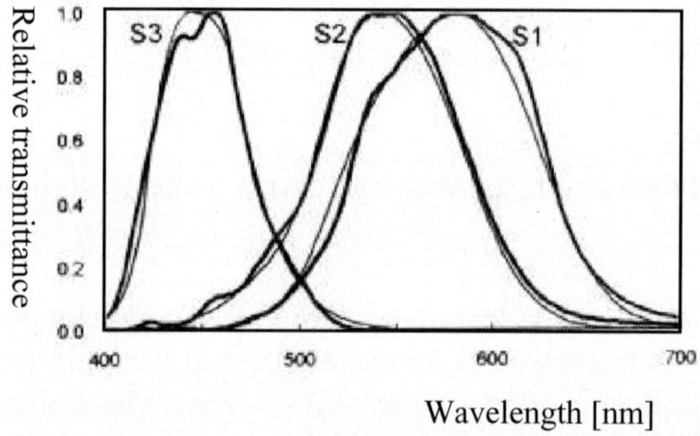

Fig.1. Filter spectral characteristics.

Measured color is represented by the tristimulous XYZ values [1]. Color captured by the camera is represented by S1, S2, S3 values, corresponding to pixel value under each filter acquisition. To provide high fidelity color reproduction, a conversion matrix between XYZ and S1, S2, S3 values must be obtained by means of least square sum analysis [2].

XYZ camera principle

For total description of color there can be used so called tristimulous values XYZ [1]. A digital still camera with output containing these values, supported by proper filters and equipped with internal spectrometer is able to reproduce high fidelity colors in the same way as human eye would perceive it. Figure 2 shows principle of the camera.

Light passing through the lens is split by a semi-transparent mirror. Part of the light reflected by the mirror is used for spectral measurement, while light passing through is projected onto CCD imaging device with S1, S2, S3 filters put on its way.

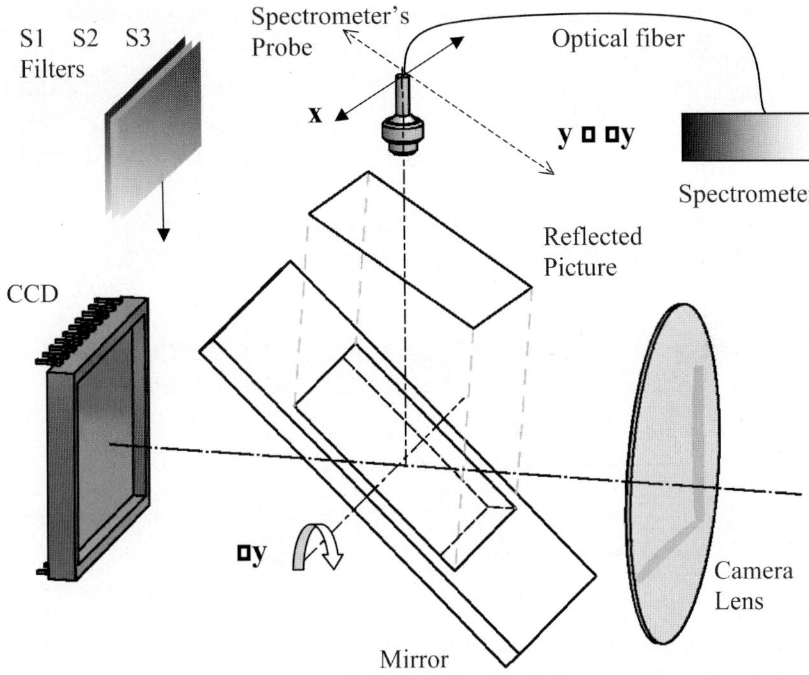

Fig. 2. XYZ camera principle.

Basic principle is to take three pictures of photographed scene under each filter acquisition. This gives three layer image called: "S1, S2 and S3 image".

The spectrometer is used for calibration of the camera. With an algorithm described further, by mathematical conversion S1, S2, S3 image is converted to XYZ image which describes colors contained in the scene in XYZ color space [1] which is a base for any other color representation.

Proposed calibration algorithm

Spectral data is collected automatically from Macbeth Chart's 24 colors by fiber optic probe positioned over picture area along the x and y

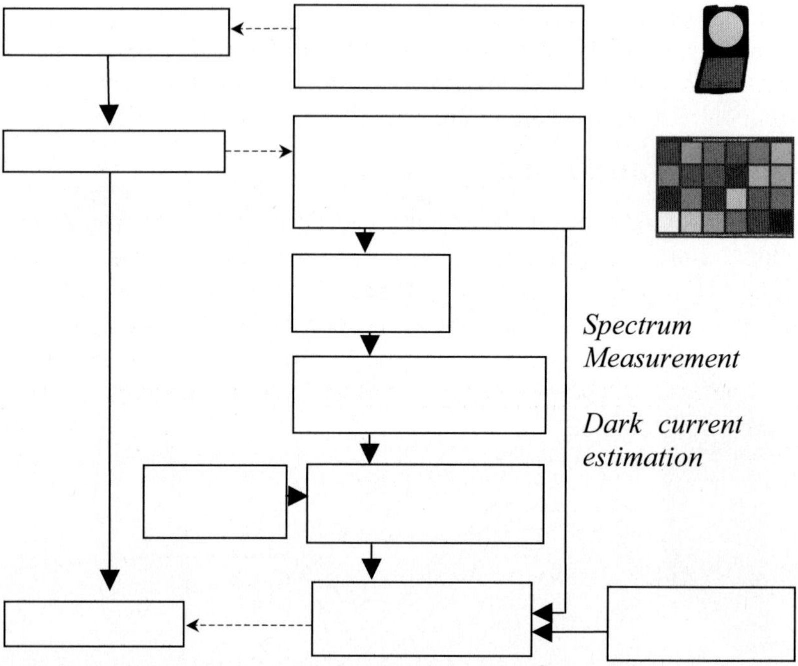

axes (Fig. 3.). The positioning mechanism is built inside the camera and its coordinate system has been calibrated with picture area. This gives a possibility of acquiring spectral data from any point of the photographed scene. Described algorithm uses image processing for recognition of each color on Macbeth Chart and to find coordinates for spectrometer's probe to be positioned and perform measurement of each color automatically.

Fig. 3. Block scheme of calibration algorithm.

Calibration starts with acquiring illuminant data (Fig. 3.) by perfect reflecting plate. The Macbeth Chart then is placed into the scene and color pads are recognized and measured. XYZ and S1, S2, S3 data is tabulated and stored into memory. Basic relation between the values is:

$$\begin{bmatrix} X \\ Y \\ Z \end{bmatrix} = \begin{bmatrix} a_{11} & a_{21} & a_{31} \\ a_{21} & a_{22} & a_{23} \\ a_{31} & a_{32} & a_{33} \end{bmatrix} \begin{bmatrix} S_1 \\ S_2 \\ S_3 \end{bmatrix} \qquad (1)$$

The color difference between reference and values obtained from calibrated camera, is well described by CIE Lab color space derived from XYZ [1]. The color difference □E principle which describes accuracy of color reproduction can be expressed with equation (2) (CIE 1932). Where

$$\Delta E = \sqrt{(L_1 - L_2)^2 + (a_1 - a_2)^2 + (b_1 - b_2)^2} \quad (2)$$

Li is luminance of compared colors and ai, bi are chroma coefficients. However at present CIE 2000 is used, to describe □E in more perceptually uniform color difference representation.

Results and discussion

The calibration of the spectrometer is most important in order to develop a conversion matrix giving the smallest color difference between measured XYZ color and X'Y'Z' obtained from S1, S2 and S3 [2]. Other major influencing factor is shading of the camera lens (winietting).

Present results show strong dependency between F number and color difference (Tab. 1.). As we observed color difference decreases as the F number increases.

Table 1. Color difference and F number.

F of the camera lens	2.8	5.6	8
Color difference (average) CIE 2000	2.04	1.35	1.13

This is caused by shading characteristics of whole optics including camera lens, mirror, fiber optics and spectrometer spectral response. To improve the accuracy of reproduction and reduce □E, amount of light reflected from the semi-transparent mirror must be increased to provide better signal to noise ratio of the spectrometer.

Above facts lead to conclusion that accurate color reproduction for still digital camera requires careful calibration. Proposed algorithm uses color reference and includes various factors for maximum performance. However in further development it will be investigated for possibility of calibration using colors of the photographed scene as a reference.

References

[1] R.W.G Hunt "The Reproduction of Colour" ISBN (1995)
[2] T. Ejaz, T. Horiuchi, G. Ohashi, Y. Shimodaira IEICE Trans. Electron., E89-C (2006)

Coherent noise reduction in optical diffraction tomography

A. Pakuła*, T. Kozacki

Warsaw University of Technology, Institute of Micromechanics and Photonics, 8 Sw. A.Boboli St., 02-525 Warsaw, Poland

Abstract

Optical Diffraction Tomography (ODT) is the method for characterization of 3D distribution of refractive index in micro optical elements. The 3D distribution is obtained from several measurements taken for different object angular orientation taken by means of laser interferometry. Unfortunately the interferometry as a coherent technique suffers from undesirable coherent noise in measurements results. Additionally the rotation of a measured object increases the noise influence on the quality of refractive-index reconstruction due to its amplification in the object area. In this paper we propose the coherent noise suppression technique. The technique involves modification of ODT setup configuration by introducing off center object rotation and applying a modified numerical algorithm of tomographic reconstruction. The algorithm modification involves introduction of numerical imaging and detection of element position from its diffraction spectrum. The received results of noise reduction are shown through numerical simulations and experiments involving optical single mode fiber measurements.

1. Introduction

The recent rapid growth of microelements with three-dimensional phase distribution, photonics structures and materials (e.g. photonics crystal fibers, GRIN lenses etc.) which are vastly used in photonics devices results in need for fast, nondestructive method of 3D refractive index distribution. The ODT is the method suitable for such measurement tasks [1-6].

Using classical interferometry 2D refractive index distribution, integrated along optical axis, is obtained for variable angular object orientation. For every single object angular orientation (range within 1°÷180°) interferogram is recorded and analyzed. Then the 3D refractive index distribution is obtained using tomographic reconstruction algorithm.
The results obtained by ODT have been recently improved by introduction of numerical procedure of refocusing to the best focus plane and sample radial run-out correction algorithm [6], although there is still unsolved problem of coherent noise, especially if measured sample introduces slight changes (order of 10^{-3} or lower) in refractive index.

2. Coherent noise reduction technique

In principle the ODT involves sample rotation along the axis in the object centre. This causes the single speckle magnification during the tomographic algorithm what results in semicircular shape in reconstructed refractive index map (Fig. 1).

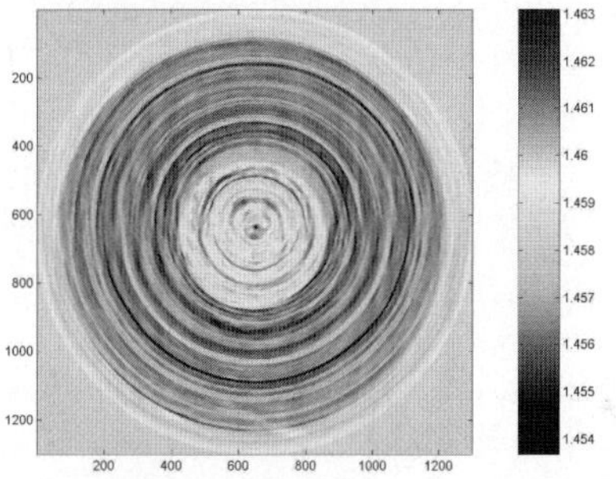

Fig. 1: Coherent noise influence on refractive index reconstruction by tomographic algorithm – multimode optical fiber [5].

Introduction of substantial radial run-out into sample rotation process along with sinogram correction correlation technique allows the coherent noise influence to be reduced. According to simulation results (simulated object: refractive index distribution: step $\Delta n=0.01$, diameter $\phi=100\lambda$) the optimal radial run-out range is 0.75ϕ - 1.25ϕ, Fig. 2. In this range the RMS factor decreases and the S/N ratio rises – Fig. 3.
Although introducing run-out into measurement, according to simulations, is a significant benefit, it gives some experimental difficulties. Such a

modified setup requires larger measurement field of view, which results in lower magnification. Additionally the defocusing of the sample is inserted by the sample rotation. For decreasing the influence of factors mentioned above the numerical sinogram correction and refocusing to the best focus plane algorithms need to be applied.

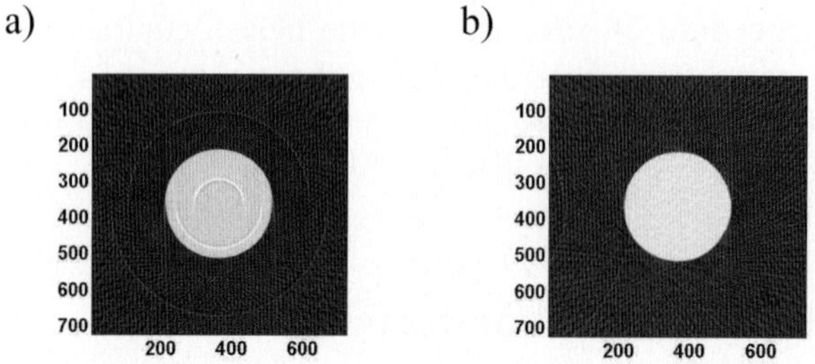

Fig. 2: Simulations results of tomographic numerical reconstruction: (a) without radial run-out (b) radial run-out - 2ϕ.

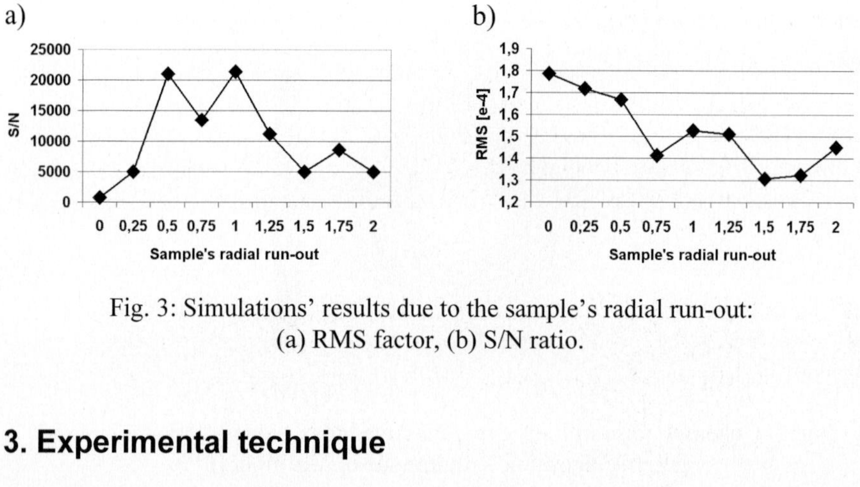

Fig. 3: Simulations' results due to the sample's radial run-out: (a) RMS factor, (b) S/N ratio.

3. Experimental technique

Experimental tomographic setup is based on classical Mach – Zehnder interferometer setup (Fig. 4.). He – Ne laser beam formed by microscope objective (OB1) is spitted into reference and object beams by a coupler (FC). The measured object (O) submerged in immersion liquid (n_{633}=1.4584) is illuminated by a plane wave and rotated during the measurement. The imagining system is focused in the sample's centre area. The measurement is performed in the following steps. For every sample angular orientation interferogram are grabbed and analyzed forming phase pro-

jection images data set. Finally from this data set 3D refractive index map is reconstructed by means of thomographic reconstruction algorithms.
As a proof of principle of our method the experiment involving characterization of refractive index distribution of single mode telecommunication fiber (SMF 28) was performed (core diameter 8.2 μm, cladding diameter 125 μm, Δn=0.0053) [7]. The experimental run-out was 129.2μm.

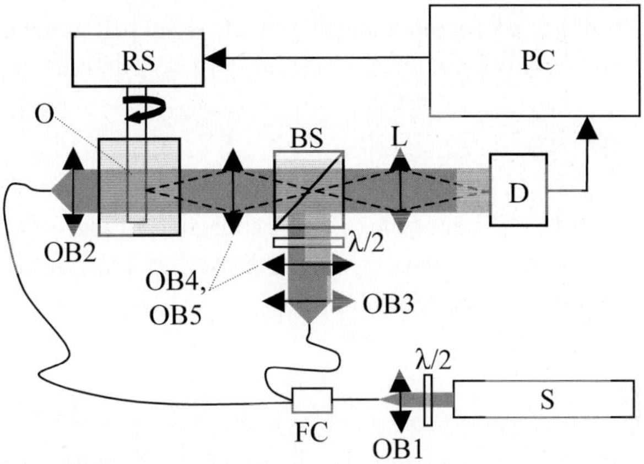

Fig. 4: Experimental tomographic setup: S – He-Ne laser, λ/2 – halfwave plate, OB1, OB2, OB3 – fiber coupling objectives, FC – fiber coupler, OB4, OB5 – microscopic imagining objectives, BS – beam splitter, L – camera objective, O – measured object, RS – rotation stage, D – detector, PC – central unit.

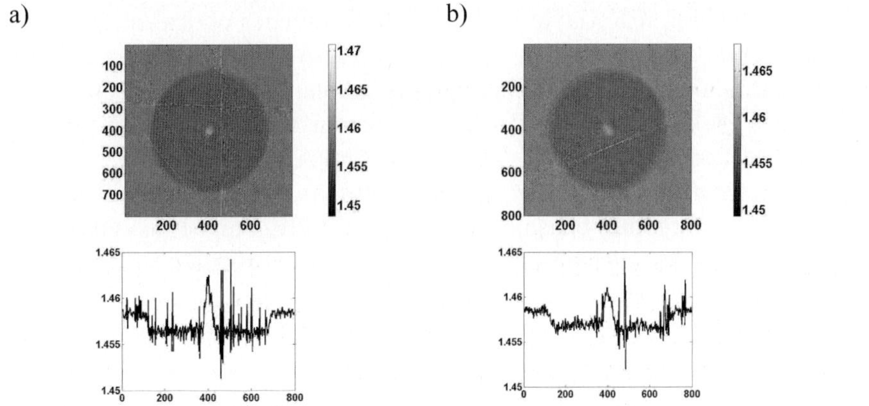

Fig. 5: Results of measurement of single mode fiber (1 pix = 0.2315 μm): (a) reconstruction and cross section without radial run-out, (b) reconstruction and cross section with 129 μm radial run-out.

The core diameter and difference of average refractive index between the core and the cladding of the reconstructed fiber from measurements without and with radial run-out are equal adequately: ϕ = 9.03 μm (39 pix), Δn=0.006 and ϕ=9.26 μm (40 pix) (elliptical deformation occurs), Δn=0.005. However, as it is shown in Fig. 5b the semicircular shape in tomographic reconstruction which has its origins in the coherence noise is removed by introduction of radial run-out the proposed technique introduces deformation of the measured object. The full uncertainty analyses for the case of the modified procedure have to be performed.

5. Conclusions

The novel method of reduction of the coherence noise influence on tomographic reconstruction reduction was proposed and verified by the numerical simulations and the experiment.

References

[1] M. Kujawinska, P. Kniazewski, T. Kozacki "Enhanced interferometric and photoelastic tomography for 3D studies of phase photonics elements", Proc. of the Symposium on Photonics Technologies for 7th Framework Program, 467- 471, Wroclaw, 2006

[2] W. Gorski "Tomographic microinterferometry of optical fibers", Opt. Eng. 45 (12), 2006

[3] B.L. Bachim, T.K.Gaylord „Microinterferometric optical phase tomography for measuring small, asymmetric refractive-index differences in the profiles of optical fibres and fiber devices", App. Opt., Vol. 44, 2005

[4] P.Guo, A.J.Devaney „Comparison of reconstruction algorithms for optical diffraction tomography", J. Opt. Soc. Am. A., Vol. 22, 2005

[5] P.Kniazewski, W. Gorski, M. Kujawinska „Microinterferometric tomography of photonics phase elements" Proc. SPIE Vol. 5145, 2003

[6] T.Kozacki, M.Kujawińska, P.Kniażewski „Investigation of limitations of optical diffraction tomography", Opto-Electron. Rev., 15, 2007

[7] Cornig Inc. "Cornig SMF 28 Optical Fiber Product Information", 2002

On micro hole geometry measurement applying polar co-ordinate laser scanning method

R. Jabłoński, P. Orzechowski

Warsaw University of Technology, Institute of Metrology and Measurement Systems, Sw. A. Boboli Street 8, Warsaw, 02-525, Poland

Abstract

The measurement of long micro hole is often problem in contemporary technology. Particularly difficult is the measurement of micro holes of the length to diameter (l/d) ratio higher than 10. The paper presents the new measurement method based on polar co-ordinate scanning with photon counter as a detector. When cylindrical micro hole is scanned with elliptical laser beam, and the hole axis and measurement device axis is not coaxial, the measurement results can be ambiguous and dependent on a distance between axes. This problem can be solved by expansion of results series into a Fourier series. The ratio of zeroth and first order coefficient of Fourier expansion, strongly depends on the distance between axes.

1. Introduction

Contemporary technologies make possible the production of small and long holes. The commonly used parameter in micro hole technology is length to diameter ratio (l/d). Holes made by photolithographic technologies (LIGA, DRIE), could have smallest dimensions of single micrometers, but its length is limited to 0,5 mm (DRIE), or 3 mm (LIGA). Another technologies like laser drilling, or EDM, ECM make possible the production of micro holes, but l/d ratio is limited for diameter smaller than approx. 30 μm [1]. Measurement techniques, applied for such small objects are mainly microscope methods [2], but also special contact methods [3], volumetric, and diffraction methods. Measurement becomes very difficult when the hole is small and long (l/d>10). The new method presented in this paper enables measurement of holes of 30 – 100μm (diameter), and l/d>10, with accuracy 5% of measured value

2. Theoretical Considerations

The proposed method consists on illumination of hole by specially shaped laser beam and measurement the radiation energy passing the hole. It is obvious that hole light transmission efficiency, understood as ratio of light radiation energy illuminating the hole, and passing through it, depends mainly on the hole geometry. The energy passing through the plane (2D) hole is described by following expression:

$$E = \iint_{S_h(x,y,r_1,\ldots,r_n)} I(\xi,\eta,p_1,\ldots p_n)\,d\xi\,d\eta \qquad (1)$$

Where:
E – energy passing measured hole (expressed in pcs. of photons)
$I(\xi,\eta,p_1,\ldots, p_n)$ – distribution of illuminating light radiation energy (Fig. 1)
$S_h(x,y,r_1,\ldots, r_n)$ – function expressing the hole edge (Fig. 1)
ξ,η - co-ordinates in illuminating light beam system (input beam system)
x,y - co-ordinates in measured hole system (output system)
$p_1,\ldots, p_n$ – input parameters (diameter, position, orientation of input beam)
$r_1,\ldots, r_n$ – parameters of function expressing measured hole edge
However the above expression is valid for the plane objects it was proved [4] that, under some conditions, this dependency can be applied for three-dimensional objects.

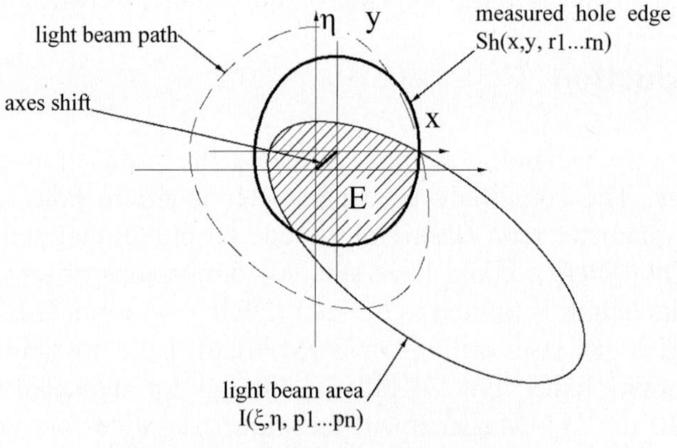

Fig.1 Measurement idea scheme.

The most essential condition is small divergence (less than 0,1 mrad) of laser beam.

$S_h(x,y,r_1,\ldots, r_n)$ is an unknown function and it can be represented by, for example as a polynomial. It results from (1) that in order to get more than one of hole edge function parameters, at least one of input parameters $(p_1,\ldots., p_n)$ has to be variable. The simplest way to change one of the parameters p_i is to move the beam. The principle of new method is to rotate the beam around particular axis so that, the beam can be considered as scanning in polar co-ordinates. It fits very well to the objects of radial symmetry (the beam path has also the radial symmetry). As the result the function radiation efficiency vs. angle of rotation ($Q_{(\alpha)}$) is obtained, after that function S_h and its parameters (r_i) can be calculated using numerical methods.

The problem discussed in this paper concerns ambiguity caused by uncertainty of the position of hole and beam rotation axis. It can be solved by expansion the $Q_{(\alpha)}$ function to Fourier series and than examination the first to zeroth harmonic ratio. It is obvious that displacement between axes causes the increase of first harmonic, unfortunately it results in the decrease of zeroth harmonic, so that only the ratio between them determines the axes distance.

3. Measurement stand

Measurement stand used in this experiment is shown on Fig. 2. The stand consists of laser 1 with beam expander 2, cylindrical lens motor 3, polarizers used as density filter 4, beam splitter 5, measured hole 6, photon

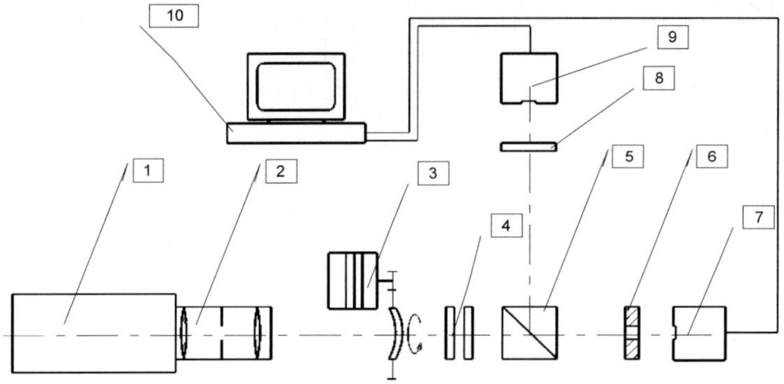

Fig. 2 Measurement stand for polar co-ordinate scanning method

counter 7, additional density filter 8, reference photon counter 9, PC computer 10. The laser beam is shaped by beam expander. The cylindrical lens is to transform circular beam to elliptical and its rotation makes the laser beam rotation. Two crossed polarizers are used to adjust the beam intensity. Beam splitter splits beam into two parts, one of them passes the hole, and its energy is measured by photon counter 7. The energy of the second beam is measured by photon counter 9; the density filter 8 is used to decrease the level of light energy to the required level of photon counter. The measurement data are collected by PC computer. The measurement results (light efficiency) are given as ratio of data from two photon counters.
As it was mentioned above, theoretically the incident beam should rotate around its optical axis and around the hole axis. However unavoidable uncertainties of rotary movement and cylindrical lens shape, makes the beam path more complex. The path has been determined by CCD camera, and applied to calculation of hole real dimensions (Sh). Beam axis moves with very small apex angle, (less than 1 mrad) and so this error component was not taken under consideration. The beam dimensions was determined by the moving edge method as 100 x 160 □m.

4. Results and Discussion

Fig. 3a shows the measurement results of hole shown on Fig. 3b. The hole nominal diameter is 80 μm and the length (material thickness) is 3,5 mm. The hole position is located at the minimum first to zeroth Fourier ratio (±1μm point "0,0"on the graph in Fig. 5). Fig. 4 shows the light efficiency ($Q_{(\alpha)}$) measurement results after shifting the hole of 20 and 40 μm, from initial position ("0,0" point), comparing to the results for initial position. One can see that the line representing measurement results after 20 and 40 μm shift differs comparing to line obtained in the point "0,0" (with minimum first to zeroth harmonic ratio). This difference will create the measurement errors, if the measurement device and measured hole are not coaxial.
Fig. 5 shows the 3D graph first to zeroth harmonic ratio vs. hole position. There is local minimum in the point mentioned above as "0,0" and the line representing measurement results in this point (Fig. 3a) the best fits the hole shape shown on microscopic image (Fig. 3b). However small (-5 μm) systematic error has been observed.

a) b)

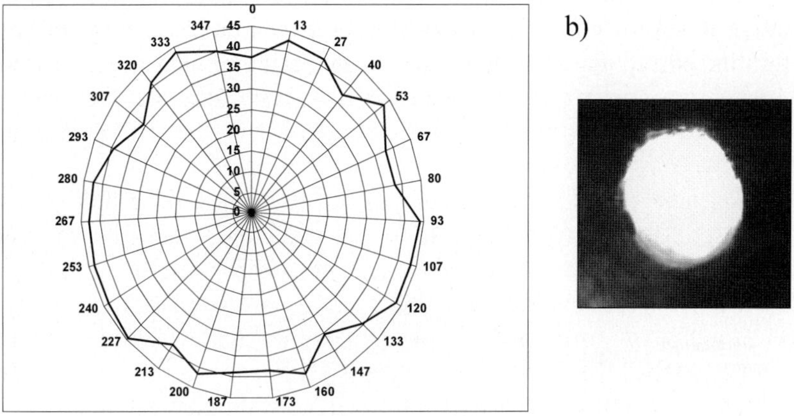

Fig. 3 Results of hole measurement after recalculations taking real path of beam move: a) dimensions [μm] vs. beam position angle [degrees] b) microscopic image of measured hole in transmitting light

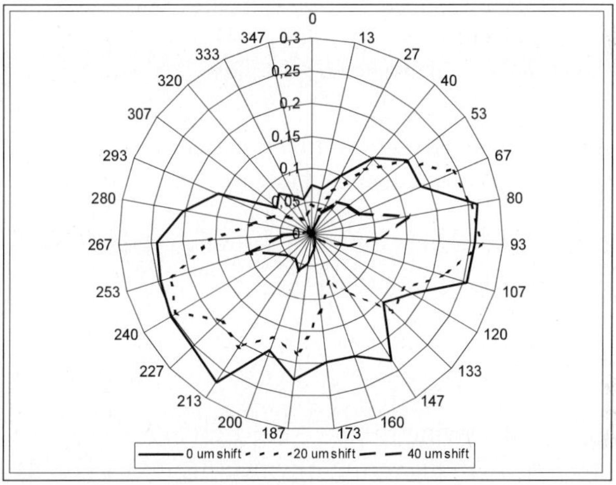

Fig. 4 Results of hole measurement, for increasing shifts between axes of hole and measurement stand. (light transmission efficiency $Q_{(\alpha)}$) [-] vs. beam position angle [degrees])

5. Conclusions

The above investigations indicated that, when applying the method with scanning in polar co-ordinate, the measurement results can be ambiguous and depends on the hole and the measurement device mutual position. However the Fourier analysis lets to determine the distance between axes of hole and measuring device. The results of Fourier analysis can be used either to set the correct position of hole during the measurement, or to determine corrections used for further calculation of measurement results.

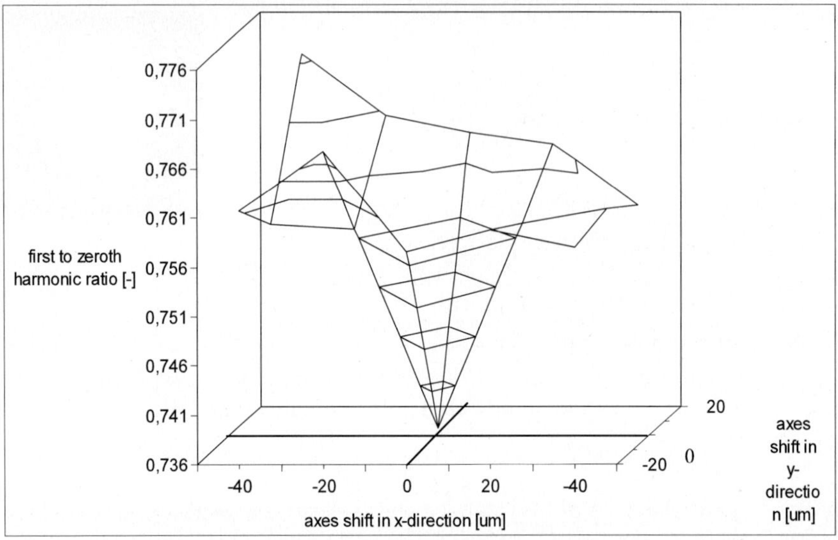

Fig. 5 The first to zeroth harmonic ratio for various distances between nominal hole axis and nominal measurement stand axis.

The above mentioned systematic error is not explained yet, and it will be the subject of further investigations.

References:

[1] B. Odom, Manufacturing Engineering 126/2 (2001) 88-102
[2] P. Waurzyniak, Manufacturing Engineering 133/1 (2004) 107-114
[3] M. Yamamoto, I. Kanno, S. Aoki, Proceedings of 30-th Conference on MEMS (2000) 217-222.
[4] R. Jabłoński, P. Orzechowski, Precision Engineering 30 (2006) 180-184.

Silicon quantum detectors with large photosensitive surface

A. Baranouski (a), A. Zenevich (b) , E. Novikov (b)

(a) Institute of Applied Physical Problems, Kurchatov str., 7,
Minsk, 220064, Republic of Belarus

(b) Higher state college of communication, Skorina str. 8/2,
Minsk, 220114, Republic of Belarus

Abstract

Semiconductor light detectors are the part of vision and image processing systems. Pulse amplitude distribution of silicon avalanche photodiodes with photosensitive surface 7 mm^2 is investigated using measurement computer system. Amplitude characteristics in the photon-counting mode are studied depending on the supplied overvoltage, laser intensity and photosensitive surface area. It is shown that changing the supplied overvoltage and photosensitive surface area causes increase / decrease of the peaks number on pulse-amplitude distribution curve due to several microplasmas in the region of space charge.

1. Introduction

Computer vision and pattern recognition systems require application of various combinations of optical sensors, laser rangers, microwave sensors. In this case very simple and inexpensive solution is utilization of charge-coupled devices, manufactured as multiple-unit matrices, and photodetectors with large area of the photosensitive surface. Progress of microelectronics in the area of such image registration devices development ensures combination of the high resolution and high rate of imaging with the possibility for registration of separate photons. A single-quantum registration or the photon counting method is the most frequently used for registration of the optical radiation with the extremely week intensity by application of the solid-state photodetectors with the internal amplification [1].

An ordinary silicon photodetectors with large photosensitive surface areas possess sufficiently high thermoelectric noise, thereby preventing application of the photon counting mode at room temperatures. Hence, the purpose of this work is to show the possibility of realizing the photon counting mode using photodetectors with photosensitive surface area up to several square millimeters with a view to register very low intensity light.

2. Experiment and discussion

The avalanche photodetectors with a 7 mm^2 photosensitive area were used. They featured a metal–resistive layer–semiconductor structure [2] based on single-crystal silicon substrate with a 1 $\Omega \cdot$cm resistivity. Thin undoped zinc–oxide film of *n*-type conductivity (d=30 nm, and $\rho_1 = 10^7 \Omega$ cm) was locally formed, ensuring the formation of an iZnO–Si heterojunction and acting as a resistive layer, and ZnO : Al film (d□ 0.5 μm, and $\rho_2 = 10^{-3}$ Ω cm) as a transparent conducting electrode.

The photon–counting mode was realized with a passive avalanche quenching circuit [2]. The photodetector acts similarly to a Geiger-Muller quantum counter. Avalanche breakdown voltage U_{av} of photodetectors equals 76.8 V. So called overvoltage $\Delta U = U_s - U_{av}$ (U_s – supply voltage) was used to analyze amplitude characteristics under variation of experiment conditions. Semiconductor laser with $\lambda = 0.68$ μm and focusing system were utilized to light the photosensitive area completely and in part.

Hardware and software package comprising 100 MHz analog-digital converter was used for registration of pulse amplitude characteristics in the real-time mode.

Pulses with duration 1.0-1.1 μs and rise time less than 100 ns were observed. Their amplitude *A* depended on supplied overvoltage. The pulse is called dark when avalanche breakdown initiated by electron as a result of thermal excitation. And the pulse is called signal when electron generated via photon absorption.

Amplitude distribution of dark pulses was measured as a function of supply overvoltage (Fig. 1). The number of peaks on amplitude distribution increased as supply voltage rose. The similar picture was observed for the total process of dark and signal pulses.

The number of peaks depends on homogeneity of photosensitive area and charge carriers multiplication region. Heterogeneities have different gain and account for avalanche pulses with different amplitude. Such heterogeneities in space charge region are called microplasmas.

Mean M and variance D were calculated versus supplied overvoltage for dark and signal pulses to characterize statistics of pulse amplitude (Fig. 2).

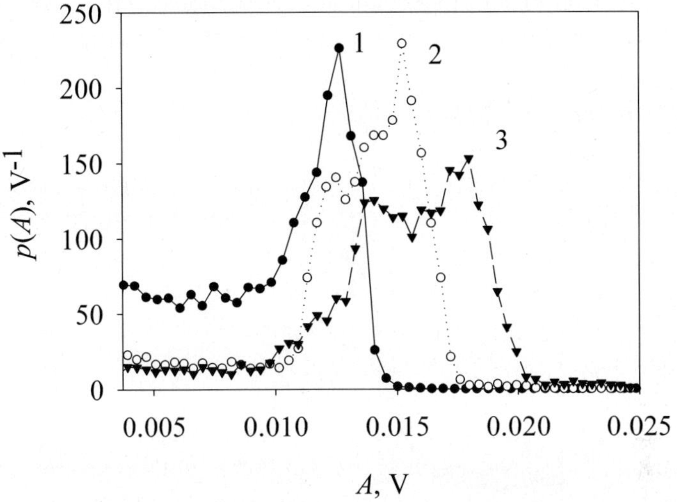

Fig. 1. Amplitude distribution of dark pulses for three supply overvoltages (1 - $\Delta U = -0.3$ V; 2 – $\Delta U = -0.1$ V; 3 – $\Delta U = 0.1$ V)

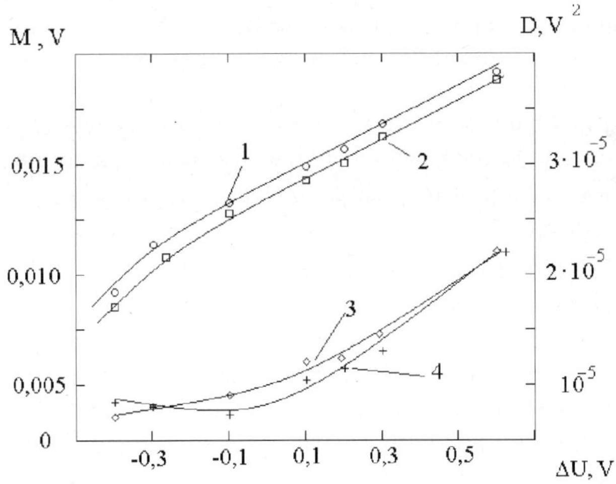

Fig. 2. Mean and variance of avalanche pulse amplitude versus supply overvoltage (1, 3 – dark pulses, 2, 4 – signal and dark pulses)

The behavior of statistical parameters is similar in presence and absence of laser irradiation. The variations of mean and variance do not exceed more then 2-3 times. In this case operating mode is chosen according to the number of peaks and their magnitude. Demonstrate this fact.
Amplitude distributions of avalanche pulses demonstrated shape changing as we varied laser stimulation area on photosensitive surface (Fig. 3).

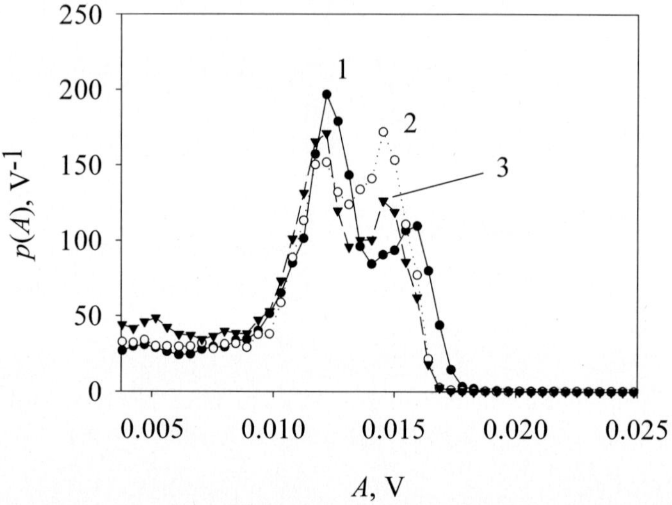

Fig. 3. Amplitude distribution of avalanche pulses for $\Delta U = 0$ V (1 – dark pulses, 2, 3 – laser stimulation in different regions of photosensitive surface area)

Each peak magnitude of the amplitude distribution depended on the laser beam position on the photosensitive area. By measuring number, position and magnitude of peaks it is possible to evaluate light intensity on the photosensitive surface.

3. Conclusion

Amplitude characteristics of avalanche pulses in silicon quantum detector with large photosensitive surface have been investigated for the purpose of using the devices in automatic vision systems operating in the photon-counting mode. It was shown that imperfections in light detection and amplification regions resulted in variation of avalanche pulse amplitude. Therefore, such photodetectors containing several multiplication regions may be used in the pattern recognition system under very low light intensity.

References

[1] J. Fraden "Handbook of modern sensors: physics, designs, and applications" Springer-Verlag, New York, 2004.
[2] I. R. Gulakov, V. B. Zalesskii, A. O. Zenevich, T. R. Leonova, Instruments and experimental techniques. 50, 2 (2007) 249.

Fizeau interferometry with automated fringe pattern analysis using temporal and spatial phase shifting

Adam Styk, Krzysztof Patorski

Institute of Micromechanics and Photonics, 8 Sw. A. Boboli St.
Warsaw 02-525, Poland

Abstract

The paper presents a novel approach to measure the parameters of quasi-parallel plates in a Fizeau interferometer. The beams reflected from the front and rear surfaces lead to a complicated interferogram intensity distribution. The phase shifting techniques (temporal and spatial) are proposed to process the interferograms and obtain a two-beam-like fringe pattern encoding the plate thickness variations. Further pattern processing is conducted using the Vortex transform.

1. Introduction

The surface flatness of transparent plates is frequently tested in a conventional Fizeau interferometer. In case of quasi-parallel plates, however, a common problem is the interference of more than two beams. They are reflected from the plate front and rear surfaces and the reference flat. Parasitic intensity distribution modulates the two-beam interferogram of the plate front surface [1,2] and makes the application of phase methods for automatic fringe pattern analysis [3-6] inefficient. On the other hand parasitic fringes contain the information on the light double passage through the plate. Several methods to suppress unwanted fringe modulations are available, for example: index matching treatment on the rear surface of the plate, short-coherence interferometry, grating interferometry, grazing incidence interferometry and wavelength-scanning interferometry [1, 2].
In this paper we present preliminary investigations of a novel proposal of processing the interferograms of quasi-parallel optical plates. It is based on

the observation that the two-beam interference pattern formed by the beams reflected from the front and rear plate surfaces only can be readily derived from the three-beam interference using temporal phase stepping (TPS) or spatial carrier phase stepping (SCPS) methods. The resulting single frame pattern can be processed using the Vortex transform approach [7,8]. The phase distribution obtained maps the plate thickness variations.

2. Principle and theory of the method

The intensity distribution formed by the three interfering beams in the Fizeau cavity can be rewritten as:

$$I = A_r^2 + A_f^2 + A_b^2 + 2A_f A_b \cos(\theta_f - \theta_b) + 2A_f A_r \cos(\theta_f - \theta_r) + 2A_r A_b \cos(\theta_r - \theta_b). \tag{1}$$

A_r, A_f, A_b, θ_r, θ_f and θ_b are the amplitudes and phases of the three beams, respectively. For notation brevity the (x,y) dependence of all terms has been omitted.

The goal is to determine the parameters of a quasi-parallel plate such us the front surface phase θ_f and back surface phase θ_r using typical fringe pattern analysis methods. Using the phase shifting method with a mechanical (PZT) phase shift one obtains the interferograms in the form:

$$I = D + 2A_f A_r \cos(\theta_f - \theta_r - n\delta) + 2A_r A_b \cos(\theta_r - \theta_b - n\delta), \tag{2}$$

where $n = 0,1,2,3,..$ and δ is the phase shift between frames. When the amplitudes of the beams reflected from the front and back surfaces are nearly equal ($A_f \cong A_b$) the terms before cosine terms in Eq. 2 are equal as well. Using trigonometric identities the intensity distribution becomes:

$$I = D + 2A_f A_r \cos\left(\frac{\theta_f - \theta_b}{2}\right)\cos\left(\frac{\theta_f + \theta_b}{2} - \theta_r - n\delta\right). \tag{3}$$

A new fringe pattern based on two fringe families multiplied by each other is obtained. It can be treated as a two beam interferogram with the bias described by the term D (constant) and the modulation distribution described by the first cosine term. The modulation distribution carries the information on the plate optical thickness variations and the main cosine term gives the information on the sum of two surfaces. Using conventional techniques for fringe pattern analysis (TPS, SCPS), it is possible to evaluate the information on the optical thickness variations separately.

Five mutually phase shifted interferograms, Eq. 3, acquired with the phase shift $\delta = \pi/2$ and put into the standard TPS five frame algorithm [9,10] for modulation calculation give the modulation distribution *Md* in the form:

$$Md = \left| 2 A_f A_r \cos\left(\frac{\theta_f - \theta_b}{2} \right) \right| . \tag{4}$$

The calculated distribution may be treated as a fringe pattern without bias. However, in the form presented by Eq. 4 it cannot be analyzed due to its highly nonsinusoidal profile as it is a modulus function. To overcome this difficulty one can square the *Md* distribution and obtain:

$$Md^2 = \frac{\left(2 A_f A_r\right)^2}{2} \left[1 + \cos\left(\theta_f - \theta_b\right)\right] . \tag{5}$$

The optical thickness variations of the quasi - parallel plate can be evaluated from Eq. 5. As the presented fringe pattern cannot be intentionally modified, only the single frame analysis methods can be applied [7,11].

3. Fringe pattern analysis method

In this Section the processing path of the three – beam interference pattern described by Eq. 3 is introduced, see Fig. 1.

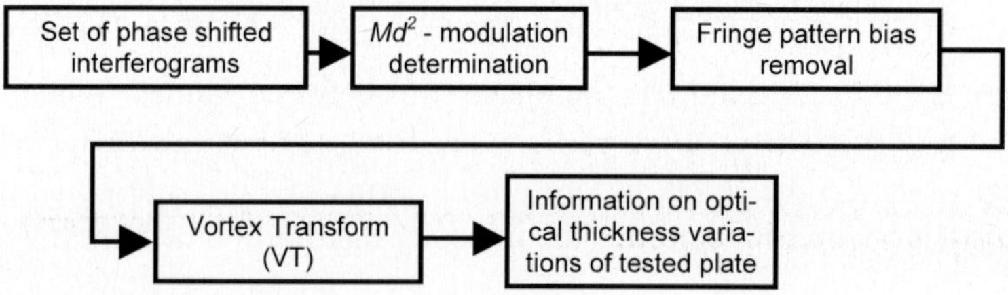

Fig. 1. Three beam interferogram processing path.

In the first step the set of five to seven mutually phase shifted interferograms (with $\delta = \pi/2$) is recorded. If the TPS technique cannot be implemented, the SCPS technique might be used. In this case only one interferogram, with intentionally introduced spatial carrier fringes, is sufficient to evaluate the desired parameters. Unfortunately the interferogram processing using the SCPS method provides lower accuracy than the TPS method.

The next step in the three beam interferogram processing path is to calculate the squared interferogram modulation distribution Md^2. This can be performed with a specially derived TPS algorithm with high resistance to the phase step error [12]. Detailed studies of systematic errors of the most common TPS algorithms applied to modulation calculations can be found in [13]. The modulation fringe pattern needs to be processed using single frame analysis methods. The method presented by Larkin et al [7,8] was chosen for calculations. This method, called the Vortex Transform (VT), is based on the two-dimensional Hilbert transform.

4. Experimental results

Experimental work has been conducted using the Fizeau interferometer with the reference element axially displaced by three PZTs placed along the optical element circumference (diameter of 50 mm). Figure 2 presents a three beam interferogram (one from the set of five phase shifted interferograms) of a microscope cover glass and the calculated squared modulation distribution Md^2. Figure 3 shows the quadrature signal of the fringe pattern (Fig. 2b) calculated using VT and the wrapped phase distribution with information about plate optical thickness variations.

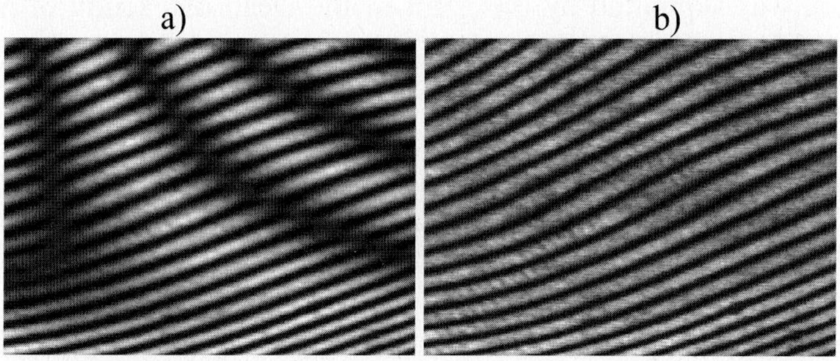

Fig. 2. Experimental three - beam interferogram (a) and the calculated squared modulation distribution Md^2 (b).

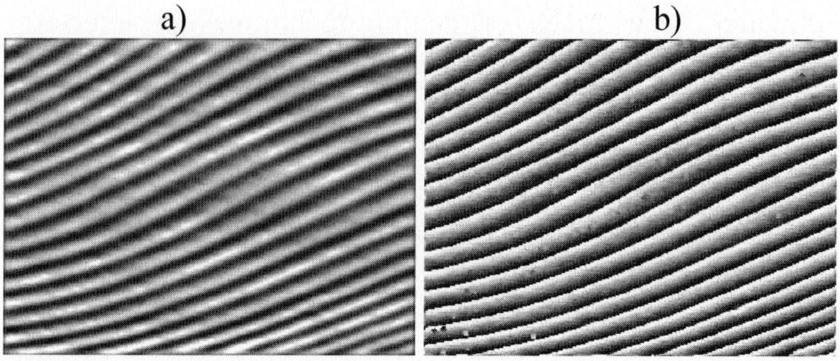

Fig. 3. Quadrature signal of the fringe pattern presented in Fig 2b (a) and the calculated wrapped phase distribution (b).

5. Conclusions

Preliminary investigations of a novel processing path of interferograms of quasi-parallel optical plates tested in a Fizeau interferometer were presented. The processing path is based on the observation that the two-beam interference pattern formed by the beams reflected from the front and rear plate surfaces only can be derived from the three-beam interference using either temporal (TPS) or spatial carrier phase stepping (SCPS) methods. The evaluated single frame pattern can be subsequently processed using the vortex transform (VT) approach. The phase distribution obtained corresponds to plate optical thickness variations. Experimental investigations corroborate the theoretical and numerical findings.

Acknowledgments

The authors want to thank Dr. Piotr Szwaykowski for performing the part of measurements and fruitful discussions.
This work was supported by the grant of the Dean of Faculty of Mechatronics and the statutory founds.

References

[1] P. de Groot, “Measurement of transparent plates with wavelength-tuned phase-shifting interferometry”, *Appl. Opt.* 39(16), 2658-2663 (2000).
[2] K. Hibino, B.F. Oreb, P.S. Fairman, and J. Burke, “Simultaneous measurement of surface shape and variation in optical thickness of a transparent parallel plate in wavelength-scanning Fizeau interferometer”, *Appl. Opt.* 43(6), 1241-1249 (2004).
[3] J. Schwider, “Advanced evaluation techniques in interferometry,” Chap. 4 in *Progress in Optics*, E. Wolf ed., 28, 271-359, North Holland, Amsterdam, Oxford, New York, Tokyo, 1990.
[4] J.E. Greivenkamp, and J.H. Brunning, “Phase shifting interferometry,” Chap 14 in *Optical Shop Testing*, D. Malacara ed., 501-598, John Wiley & Sons, Inc., New York, Chichester, Brisbane, Toronto, Singapore, 1992.
[5] K. Creath, “Temporal phase measurement methods,” Chap. 4 in *Interferogram Analysis: Digital Fringe Pattern Measurement*, D.W. Robinson and G. Reid, eds., 94-140, Institute of Physics Publishing, Bristol, Philadelphia, 1993.

[6] D. Malacara, M. Servin, and Z. Malacara, *Interferogram Analysis for Optical Shop Testing,* Marcel Dekker, Inc., New York, Basel, Hong Kong, 1998.

[7] K.G. Larkin, D.J. Bone, and M.A. Oldfield, "Natural demodulation of two-dimensional fringe patterns. I. General background of the spiral quadrature transform", *J. Opt. Soc. Am. A*, 18(8), 1862-1870 (2001).

[8] K.G. Larkin, "Natural demodulation of two-dimensional fringe patterns. II. Stationary phase analysis of the spiral phase quadrature transform", *J. Opt. Soc. Am. A*, 18(8), 1871-1881 (2001).

[9] J. Schwider, R. Burrow, K.E. Elssner, J. Grzanna, R. Spolaczyk, K. Merkel, "Digital wave-front measuring interferometry: some systematic error sources", *Appl. Opt.* 22(21), 3421-3432 (1983).

[10] P. Hariharan, B. Oreb, T. Eiju, "Digital phase-shifting interferometry: a simple error compensating phase calculation algorithm", *Appl. Opt.* 26(13), 2504-2505 (1987).

[11] M. Servin, J.A. Quiroga, J.L. Marroquin, "General *n*-dimensional quadrature transform and its application to interferogram demodulation", *J. Opt. Soc. Am. A*, 20(5), 925-934, 2003.

[12] K.G. Larkin, "Efficient nonlinear algorithm for envelope detection in white light interferometry", *J. Opt. Soc. Am. A* 13(4), 832-843 (1996).

[13] K. Patorski, A. Styk, "Interferogram intensity modulation calculations using temporal phase shifting: error analysis", *Opt. Eng.* 45(8), 085602, (2006).

Index